T0348848

Statistical Orbit Determination

Byron D. Tapley
Center for Space Research
The University of Texas at Austin

Bob E. Schutz
Center for Space Research
The University of Texas at Austin

George H. Born
Colorado Center for Astrodynamics Research
University of Colorado, Boulder

ELSEVIER
ACADEMIC
PRESS

Amsterdam • Boston • Heidelberg • London • New York • Oxford
Paris • San Diego • San Francisco • Singapore • Sydney • Tokyo

Senior Publishing Editor	Frank Cynar
Project Manager	Simon Crump
Editorial Coordinator	Jennifer Helé
Marketing Manager	Linda Beattie
Composition	Author

Elsevier Academic Press
200 Wheeler Road, Burlington, MA 01803, USA
525 B Street, Suite 1900, San Diego, California 92101-4495, USA
84 Theobald's Road, London WC1X 8RR, UK

This book is printed on acid-free paper. ∞

Library of Congress Cataloging-in-Publication Data
Application submitted.

British Library Cataloguing in Publication Data
A catalogue record for this book is available from the British Library.

ISBN: 0-12-683630-2

For all information on all Academic Press publications,
visit our Web site at www.academicpressbooks.com.

Transferred to Digital Printing in 2010

Contents

Preface

The modern field of orbit determination (OD) originated with Kepler's interpretations of the observations made by Tycho Brahe of the planetary motions. Based on the work of Kepler, Newton was able to establish the mathematical foundation of celestial mechanics. During the ensuing centuries, the efforts to improve the understanding of the motion of celestial bodies and artificial satellites in the modern era have been a major stimulus in areas of mathematics, astronomy, computational methodology and physics. Based on Newton's foundations, early efforts to determine the orbit were focused on a deterministic approach in which a few observations, distributed across the sky during a single arc, were used to find the position and velocity vector of a celestial body at some epoch. This uniquely categorized the orbit. Such problems are deterministic in the sense that they use the same number of independent observations as there are unknowns.

With the advent of daily observing programs and the realization that the orbits evolve continuously, the foundation of modern precision orbit determination evolved from the attempts to use a large number of observations to determine the orbit. Such problems are over-determined in that they utilize far more observations than the number required by the deterministic approach. The development of the digital computer in the decade of the 1960s allowed numerical approaches to supplement the essentially analytical basis for describing the satellite motion and allowed a far more rigorous representation of the force models that affect the motion.

This book is based on four decades of classroom instruction and graduate-level research. The material has been derived from and shared with numerous colleagues and students. In addition to the contributions of these individuals, the material has been influenced by the approaches used at the NASA Jet Propulsion Laboratory (JPL) to fulfill Earth and planetary navigation requirements and at the NASA Goddard Space Flight Center (GSFC) and the Department of Defense Naval Surface Weapons Center in applications of satellite tracking to problems in geodesy, geodynamics, and oceanography. Various implementations of the techniques described have been accomplished over the years to meet the requirements to determine the orbit of specific spacecraft.

This book is focused on developing data processing techniques for estimating the state of a non-linear dynamic system. The book contains six chapters and eight appendices, each covering a unique aspect of the OD problem. Although the concepts will be discussed primarily in the context of Earth-orbiting satellites or planetary transfers, the methods are applicable to any non-linear system and have been applied to such widely diverse problems as describing the state of economic systems, chemical processes, spacecraft attitude, and aircraft and missile guidance and navigation.

In Chapter 1, the inherent characteristics of the OD problem are introduced in the context of a very simple dynamic problem. The concepts of the dynamic system and its associated state are introduced. The fact that essentially all observations are non-linear functions of the state variables and the complexities associated with this are discussed. The classical or well-determined problem and the over-determined problem are described as examples of the approaches used in the following chapters.

Chapter 2 lays the background for the OD problem by introducing the relevant elements of orbital mechanics. The concept of perturbed two body motion and its role in the description of orbital motion is discussed. The fundamental forces acting on the satellite are described, especially within the context of an accurate description of the orbit. Finally, the definitions of the coordinate systems used to describe the motion and the time conventions used to index the observations and describe the satellite position are given.

Chapter 3 describes the various types of observations that are used to measure the satellites motion. This includes ground-based systems such as laser and radiometric, as well as space-based measurements from the Global Positioning System. Error sources and media corrections needed for these observation types also are described. In the problem of orbit determination, the observations described in this chapter will be the input to the estimation methodologies in the following chapters.

In Chapter 4, the orbit problem is formulated in state space notation and the non-linear orbit determination problem is reduced to an approximating linear state estimation problem by expanding the system dynamics about an *a priori* reference orbit. This concept allows application of an extensive array of mathematical tools from linear system theory that clarify the overall problem and aid in transferring the final algorithms to the computational environment. Chapter 4 also introduces the important concept of sequential processing of the observations, and the relation to the Kalman Filter, and relates the algorithm to the observation and control of real time processes.

Chapter 5 addresses issues related to the numerical accuracy and stability of the computational process, and formulates the square root computational algorithms. These algorithms are based on using a series of orthogonal transformations to operate on the information arrays in such a way that the requirement for formation of the normal equations is eliminated. The approach is used to develop both batch and sequential data processing algorithms.

Chapter 6 explores the very important problem of assessing the effect of error in unestimated measurement and force model parameters. The effects of incorrect statistical information on the *a priori* state and observation noise are examined and the overall process is known as consider covariance analysis.

The eight appendices provide background and supplemental information. This includes topics such as probability and statistics, matrix algebra, and coordinate transformations.

The material can be used in several ways to support graduate or senior level courses. Chapters 1, 3, and 4, along with Appendices A, B, F, and G provide the basis for a one-semester OD class. Chapters 5 and 6, along with Appendix A, provide the basis for a one-semester advanced OD class. Chapter 2 could serve as supplemental information for the first semester class as well as an introductory orbit mechanics class. Each chapter has a set of problems selected to emphasize the concepts described in the chapter.

Acknowledgements

The authors gratefully acknowledge the assistance of Karen Burns and Melody Lambert for performing the word processing for the manuscript. We are especially indebted to Melody for her patience and expertise in dealing with countless additions and corrections. We thank the hundreds of students and colleagues who, over the past decade, contributed by providing suggestions and reviews of the manuscript. We are deeply grateful for your help and encouragement. We apologize for not mentioning you by name but our attempts to do so would surely result in our omission of some who made significant contributions.

Byron Tapley, Bob Schutz, and George Born
April, 2004

Web Sites

Relevant information can be found on the World Wide Web (www). However, the www does not provide a stable source of information comparable to a published document archived in a library. The following list of www sites was valid just prior to the publication date of this book, but some may no longer be functional. Use of readily available search engines on the www may be useful in locating replacement or new sites.

- Ancillary Information for *Statistical Orbit Determination*

    Academic Press
    http://www.academicpress.com/companions/0126836302

    Center for Space Research (CSR, University of Texas at Austin)
    http://www.csr.utexas.edu/publications/statod

    Colorado Center for Astrodynamics Research
    (CCAR, University of Colorado at Boulder)
    http://www-ccar.colorado.edu/statod

- Online Dissertation Services

    UMI (also known as University Microfilms)
    http://www.umi.com/hp/Products/Dissertations.html

- Satellite Two Line Elements (TLE)

    NASA Goddard Space Flight Center
    http://oig1.gsfc.nasa.gov (registration required)

    Dr. T. S. Kelso
    http://celestrak.com

    Jonathan McDowell's Home Page
    http://www.planet4589.org/space/

- Earth Orientation, Reference Frames, Time

    International Earth Rotation and Reference Frames Service (IERS)
    http://www.iers.org

    Observatoire de Paris Earth Orientation Center
    http://hpiers.obspm.fr/eop-pc/

    IERS Conventions 2003
    http://maia.usno.navy.mil/conv2003.html

    National Earth Orientation Service (NEOS)
    http://maia.usno.navy.mil

    International Celestial Reference Frame (ICRF)
    http://hpiers.obspm.fr/webiers/results/icrf/README.html

    International Terrestrial Reference Frame (ITRF)
    http://lareg.ensg.ign.fr/ITRF/solutions.html

    U. S. Naval Observatory Time Service
    http://tycho.usno.navy.mil/

- Space Geodesy Tracking Systems

  International GPS Service (IGS)
  http://igscb.jpl.nasa.gov

  International Laser Ranging Service (ILRS)
  http://ilrs.gsfc.nasa.gov

  International VLBI Service for Geodesy and Astrometry (IVS)
  http://ivscc.gsfc.nasa.gov

  International DORIS Service (IDS)
  http://ids.cls.fr/

- Global Navigation Satellite Systems

  GPS
  http://gps.losangeles.af.mil/

  GLONASS
  http://www.glonass-center.ru/

  GALILEO
  http://www.europa.eu.int/comm/dgs/energy_transport/galileo/

Web pages for particular satellite missions are maintained by the national space agencies.

Chapter 1

Orbit Determination Concepts

1.1 INTRODUCTION

The treatment presented here will cover the fundamentals of satellite orbit determination and its evolution over the past four decades. By satellite orbit determination we mean the process by which we obtain knowledge of the satellite's motion relative to the center of mass of the Earth in a specified coordinate system. Orbit determination for celestial bodies has been a general concern of astronomers and mathematicians since the beginning of civilization and indeed has attracted some of the best analytical minds to develop the basis for much of the fundamental mathematics in use today.

The classical orbit determination problem is characterized by the assumption that the bodies move under the influence of a central (or point mass) force. In this treatise, we focus on the problem of determining the orbits of noncelestial satellites. That is, we focus on the orbits of objects placed into orbit by humans. These objects differ from most natural objects in that, due to their size, mass, and orbit charateristics, the nongravitational forces are of significant importance. Further, most satellites orbit near to the surface and for objects close to a central body, the gravitational forces depart from a central force in a significant way.

By the state of a dynamical system, we mean the set of parameters required to predict the future motion of the system. For the satellite orbit determination problem the minimal set of parameters will be the position and velocity vectors at some given epoch. In subsequent discussions, this minimal set will be expanded to include dynamic and measurement model parameters, which may be needed to improve the prediction accuracy. This general state vector at a time, t, will be denoted as $\mathbf{X}(t)$. The general orbit determination problem can then be posed as follows.

If at some initial time, t_0, the state $\mathbf{X}_0$ of a vehicle following a ballistic trajectory is given and if the differential equations that govern the motion of the vehicle are known, then the equations of motion can be integrated to determine the state of the vehicle at any time. However, during an actual flight, the initial state is never known exactly. Moreover, certain physical constants as well as the mathematical specification of the forces required to define the differential equations of motion are known only approximately. Such errors will cause the actual motion to deviate from the predicted motion. Consequently, in order to determine the position of the spacecraft at some time $t > t_0$, it must be tracked or observed from tracking stations whose positions are known precisely. With the observations of the spacecraft motion, a better estimate of the trajectory can be determined. The term "better estimate" is used since the observations will be subject to both random and systematic errors and, consequently, the estimate of the trajectory will not be exact.

The observational data will usually consist of such measurements as range, range-rate, azimuth, elevation, or other observable quantities. That is, the state variables (position, velocity, unknown model parameters, etc.) will not be observed, but rather the observable will usually be some nonlinear function of the state variables.

The problem of determining the best estimate of the state of a spacecraft, whose initial state is unknown, from observations influenced by random and systematic errors, using a mathematical model that is not exact, is referred to as the problem of state estimation. In this presentation, such a procedure will be referred to as the process of *orbit determination*. The word "best" is used to imply that the estimate is optimal in some as yet undefined statistical sense.

When an estimate of the trajectory has been made, the subsequent motion and values for the observations can be predicted. In the orbit determination procedure, the process of predicting the state of a vehicle is referred to as "generating an ephemeris." An ephemeris for a space vehicle is a table of its position and velocity components as a function of time. The predicted values will differ from the true values due to the following effects:

1. Inaccuracies in the estimated state vector (i.e., position and velocity vector) caused by errors in the orbit determination process, such as:

   a. Approximations involved in the method of orbit improvement and in the mathematical model,

   b. Errors in the observations,

   c. Errors in the computational procedure used in the solution process.

2. Errors in the numerical integration procedure caused by errors in the dynamical model and computer truncation and roundoff errors.

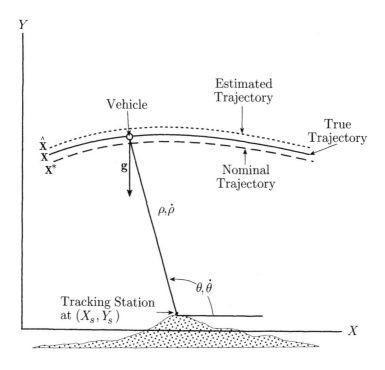

Figure 1.2.1: Uniform gravity field trajectory.

Consequently, the process of observation and estimation must be repeated continually as the vehicle's motion evolves. Furthermore, the orbit determination procedure may be used to obtain better estimates of the location of tracking stations, adjust the station clocks, calibrate radar biases, obtain a better estimate of geophysical constants, and so on. In fact, any observable quantity directly affecting either the motion of the vehicle or the observation-state relationship can be determined by appropriately using the observational data.

To formulate the orbit determination problem, we will use elements of probability and statistics as well as matrix theory. Appendixes A and B contain introductory material on these topics.

1.2 UNIFORM GRAVITY FIELD MODEL

To illustrate some of the basic ideas involved in the orbit determination process, consider the flight of a vehicle moving under the influence of a uniform gravitational field, as shown in Fig. 1.2.1. In the preliminary design of the trajectory, an initial state $\mathbf{X}_0^*$ (i.e., the initial position and the initial velocity vectors), is selected in such a way that the vehicle will follow a desired nominal (or design) trajectory. In practice, however, the design conditions are never realized. Hence, the *true initial state*, $\mathbf{X}_0$, will differ from the *nominal initial state*, $\mathbf{X}_0^*$, and consequently, the true trajectory followed by the vehicle will differ from the nominal trajectory. An indication of these two trajectories is given in Fig. 1.2.1. Here the true trajectory is denoted as $\mathbf{X}$, the nominal trajectory as $\mathbf{X}^*$, and the best estimate of the true trajectory is indicated by $\hat{\mathbf{X}}$. To determine an estimate, $\hat{\mathbf{X}}$, of the true trajectory of the vehicle, tracking information must be used.

Figure 1.2.1 shows a tracking station at a point whose coordinates are (X_s, Y_s). Assume that the range (linear distance along the line of sight), ρ, and the elevation, θ, of the line of sight to the satellite can be measured. The tracking information, or observations, ρ and θ, at any time depend on the true state of the satellite and the position of the tracking station. Since the true position of the satellite is unknown, the calculated or modeled values of the observations will depend on the nominal position of the satellite and the location of the tracking station at the time the observation is made. The difference between observed and calculated observations at the observation time provides the information used to obtain an improved estimate of the satellite's motion. Other observations such as range rate, $\dot{\rho}$, and elevation rate, $\dot{\theta}$, which depend on both the velocity and position of the satellite and tracking station, could also be used to determine $\hat{\mathbf{X}}$. Note that range rate is the line of sight velocity that is, the projection of the spacecraft's velocity vector on to the range vector

$$\dot{\rho} = \frac{\dot{\boldsymbol{\rho}} \cdot \boldsymbol{\rho}}{\rho}.$$

1.2.1 FORMULATION OF THE PROBLEM

For the uniform gravity field model, the differential equations of motion can be expressed as

$$
\begin{aligned}
\ddot{X}(t) &= 0 \\
\ddot{Y}(t) &= -g,
\end{aligned}
\tag{1.2.1}
$$

where g is the gravitational acceleration and is assumed to be constant. Integration of Eq. (1.2.1) leads to

$$
\begin{aligned}
X(t) &= X_0 + \dot{X}_0 t \\
Y(t) &= Y_0 + \dot{Y}_0 t - g\tfrac{t^2}{2}, \\
\dot{X}(t) &= \dot{X}_0 \\
\dot{Y}(t) &= \dot{Y}_0 - gt
\end{aligned}
\tag{1.2.2}
$$

where t represents the time, and the reference time, t_0, is chosen to be zero. The subscript 0 indicates values of the quantities at t_0. Now, providing the values of

X_0, Y_0, $\dot{X}_0$, $\dot{Y}_0$, and g are given, Eq. (1.2.2) can be used to predict the position and velocity of the vehicle at any time. However, as previously indicated, the values of these quantities are never known exactly in practice, and it is the task of the orbit determination procedure to estimate their values.

1.2.2 THE EQUATION OF THE ORBIT

The parameter t can be eliminated in the last two of Eq. (1.2.2) to obtain the curve followed by the vehicle in the (X, Y) plane; that is,

$$t = \frac{X - X_0}{\dot{X}_0}. \tag{1.2.3}$$

Hence,

$$Y(t) = Y_0 + \frac{\dot{Y}_0}{\dot{X}_0}(X - X_0) - \frac{1}{2}\frac{g}{\dot{X}_0^2}(X - X_0)^2. \tag{1.2.4}$$

Equation (1.2.4) can be recognized readily as the equation of a parabola. This equation is analogous to the conic curve obtained as the solution to motion in an inverse square force field (see Chapter 2). It should be noted that Eqs. (1.2.3) and (1.2.4) still involve the values of the four initial conditions. The relation indicates that if the vehicle was at X', then it must also have been at Y' obtained from Eq. (1.2.4). It is also obvious that any real value of X is possible, and it has associated with it a unique Y. However, the converse is not true. For each possible value of Y, there are usually two values of X possible. The time epoch, t', corresponding to the point (X', Y') can be determined from the relation $t' = (X' - X_0)/\dot{X}_0$. This relation is analogous to Kepler's equation for the motion in an inverse square force field. Since the epoch of time must be associated with the determination of the state of the vehicle if an ephemeris is to be generated and since the time is an integral part of all observations, the solution in the form of Eqs. (1.2.3) and (1.2.4) is usually not convenient for most orbit determination procedures.

In the subsequent discussions, either the differential equations, Eq. (1.2.1), or their solution in the form of Eq. (1.2.2) will be used.

1.2.3 THE ROLE OF THE OBSERVATION

As an example of the orbit determination procedure, consider the situation in which the state is observed at some time epoch, t_j. Then if X_j, Y_j, $\dot{X}_j$ and $\dot{Y}_j$ are given at t_j, Eq. (1.2.2) can be used to form four equations in terms of four unknowns. This system of equations can be used to determine the unknown components of the initial state. For example, from Eq. (1.2.2) it follows that

$$
\begin{bmatrix} X_j \\ Y_j + gt_j^2/2 \\ \dot{X}_j \\ \dot{Y}_j + gt_j \end{bmatrix}
=
\begin{bmatrix} 1 & 0 & t_j & 0 \\ 0 & 1 & 0 & t_j \\ 0 & 0 & 1 & 0 \\ 0 & 0 & 0 & 1 \end{bmatrix}
\begin{bmatrix} X_0 \\ Y_0 \\ \dot{X}_0 \\ \dot{Y}_0 \end{bmatrix}.
$$

Then, the initial state can be determined as follows:

$$
\begin{bmatrix} X_0 \\ Y_0 \\ \dot{X}_0 \\ \dot{Y}_0 \end{bmatrix}
=
\begin{bmatrix} 1 & 0 & t_j & 0 \\ 0 & 1 & 0 & t_j \\ 0 & 0 & 1 & 0 \\ 0 & 0 & 0 & 1 \end{bmatrix}^{-1}
\begin{bmatrix} X_j \\ Y_j + gt_j^2/2 \\ \dot{X}_j \\ \dot{Y}_j + gt_j \end{bmatrix}.
\tag{1.2.5}
$$

Providing that the matrix inverse in Eq. (1.2.5) exists, the true initial state $\mathbf{X}_0^T = [X_0, Y_0, \dot{X}_0, \dot{Y}_0]$ can be determined.

Unfortunately, in an actual trajectory or orbit determination process, the individual components of the state generally cannot be observed directly. Rather, the observations consist of nonlinear functions of the state; for example, range, elevation, range rate, and so on. In this case, the nonlinear observation-state relationship is

$$
\begin{aligned}
\rho &= \sqrt{(X - X_s)^2 + (Y - Y_s)^2} \\
\tan\theta &= (Y - Y_s)/(X - X_s) \\
\dot{\rho} &= \frac{1}{\rho}[(X - X_s)(\dot{X} - \dot{X}_s) + (Y - Y_s)(\dot{Y} - \dot{Y}_s)] \\
\dot{\theta} &= \frac{1}{\rho^2}[(X - X_s)(\dot{Y} - \dot{Y}_s) - (\dot{X} - \dot{X}_s)(Y - Y_s)],
\end{aligned}
\tag{1.2.6}
$$

where X_s, Y_s, $\dot{X}_s$, and $\dot{Y}_s$ are the position and velocity components, respectively, of the tracking station. Generally the station velocity components $\dot{X}_s$ and $\dot{Y}_s$ will be zero relative to an Earth-fixed frame unless we are accounting for tectonic plate motion. Now, if ρ, θ, $\dot{\rho}$ and $\dot{\theta}$ are given at t_j, then Eq. (1.2.2) can be substituted into Eq. (1.2.6) to obtain

$$
\begin{aligned}
\rho_j &= \sqrt{(X_0 - X_s + \dot{X}_0 t_j)^2 + (Y_0 - Y_s + \dot{Y}_0 t_j - gt_j^2/2)^2} \\
\theta_j &= \tan^{-1}[(Y_0 - Y_s + \dot{Y}_0 t_j - g\frac{t_j^2}{2})/(X_0 - X_s + \dot{X}_0 t_j)] \\
\dot{\rho}_j &= \frac{1}{\rho_j}[(X_0 - X_s + \dot{X}_0 t_j)(\dot{X}_0 - \dot{X}_s) \\
&\quad + (Y_0 - Y_s + \dot{Y}_0 t_j - g\frac{t_j^2}{2})(\dot{Y}_0 - gt_j - \dot{Y}_s)]
\end{aligned}
\tag{1.2.7}
$$

$$\dot{\theta}_j = \frac{1}{\rho_j^2}[(X_0 - X_s + \dot{X}_0 t_j)(\dot{Y}_0 - g t_j - \dot{Y}_s)$$

$$-(\dot{X}_0 - \dot{X}_s)(Y_0 - Y_s + \dot{Y}_0 t_j - g\frac{t_j^2}{2})].$$

Symbolically Eq. (1.2.7) represents four nonlinear algebraic equations in terms of the unknown components of the initial state. If $\mathbf{Y}^T$ represents the vector whose components are the observations $[\rho_j, \theta_j, \dot{\rho}_j, \dot{\theta}_j]$, then Eq. (1.2.7) can be expressed as

$$J(\mathbf{X}_0) \equiv \mathbf{Y} - G(\mathbf{X}_0, t) = 0, \qquad (1.2.8)$$

where $G(\mathbf{X}_0, t)$ is a 4×1 vector of nonlinear functions consisting of the right-hand side of Eq. (1.2.7). Eq. (1.2.8) can be solved iteratively by the Newton–Raphson iteration procedure. If $\mathbf{X}_0^n$ represents the nth approximation to the solution, then

$$J(\mathbf{X}_0^{n+1}) \cong J(\mathbf{X}_0^n) + \left[\frac{\partial J}{\partial \mathbf{X}_0^n}\right][\mathbf{X}_0^{n+1} - \mathbf{X}_0^n],$$

according to a 2-term Taylor series approximation. Now, setting $J(\mathbf{X}_0^{n+1}) = 0$ to comply with Eq. (1.2.8) and solving for $\mathbf{X}_0^{n+1}$ leads to

$$\mathbf{X}_0^{n+1} = \mathbf{X}_0^n - \left[\frac{\partial J}{\partial \mathbf{X}_0^n}\right]^{-1} J(\mathbf{X}_0^n), \qquad (1.2.9)$$

where

$$\frac{\partial J}{\partial \mathbf{X}_0^n} = -\frac{\partial G(\mathbf{X}_0^n, t)}{\partial \mathbf{X}_0^n}.$$

The process can be repeated until $||\mathbf{X}_0^{n+1} - \mathbf{X}_0^n|| \leq \epsilon$, where ϵ is a small positive number.

Several facts are obvious from this example:

1. The linear dependence of the state at time t on the initial conditions, as obtained in Eq. (1.2.2), usually does not occur.

2. In general, the equations that define the solution to the orbit determination problem are nonlinear, and the solution must be obtained by a numerical iteration procedure. Note that this iteration procedure requires an initial guess for $\mathbf{X}_0$. This initial guess must be close enough to the true solution for convergence to occur.

3. Since the equations are nonlinear, multiple solutions may exist; only one will correspond to the minimizing or correct solution. Generally the initial guess for $\mathbf{X}_0$ will be close to the truth so convergence to an incorrect solution is not a problem.

This situation represents a special case of the usual orbit determination problem; that is, the case where there are as many independent observations at one epoch of time as there are unknown components of the state vector. Usually, the number of observations available at any time will be smaller than the number of components in the state vector. For example, the radar may have the capability of measuring only range, ρ, and range rate, $\dot{\rho}$, or range and elevation, θ, at any one point in time. The observation set (ρ, θ) is not adequate to determine the state $(X_0, Y_0, \dot{X}_0, \dot{Y}_0)$, since using Eq. (1.2.6) will lead to a set of two nonlinear equations in terms of four unknowns. However, if observations are available at any two time points t_1 and t_2, then the set $(\rho_1, \theta_1, \rho_2, \theta_2)$ can be used to determine the trajectory. From Eq. (1.2.7), the following equations can be obtained:

$$
\begin{aligned}
\rho_1 &= \sqrt{(X_0 - X_s + \dot{X}_0 t_1)^2 + (Y_0 - Y_s + \dot{Y}_0 t_1 - g t_1^2/2)^2} \\
\theta_1 &= \tan^{-1}[(Y_0 - Y_s + \dot{Y}_0 t_1 - g t_1^2/2)/(X_0 - X_s + \dot{X}_0 t_1)] \\
\rho_2 &= \sqrt{(X_0 - X_s + \dot{X}_0 t_2)^2 + (Y_0 - Y_s + \dot{Y}_0 t_2 - g t_2^2/2)^2} \\
\theta_2 &= \tan^{-1}[(Y_0 - Y_s + \dot{Y}_0 t_2 - g t_2^2/2)/(X_0 - X_s + \dot{X}_0 t_2)].
\end{aligned}
\tag{1.2.10}
$$

Equation (1.2.10) represents four nonlinear equations in terms of four unknowns from which the unknowns can be determined as indicated by Eq. (1.2.9).

The approaches just outlined have two fundamental assumptions that cannot be realized in practice. These assumptions are:

1. Perfect knowledge of the differential equations that describe the motion.

2. Perfect observations.

The assumption of perfect knowledge of the differential equations implies that all the forces are modeled perfectly and that the values for the constant parameters used in the force model are exact. Usually, the true values for all of the parameters are not defined with complete numerical accuracy. But what is even more relevant, simplifying assumptions usually are made to reduce the computational requirements. Under such assumptions, certain unimportant or "negligible" components of the forces acting on the vehicle are omitted. As a consequence, the mathematical model used to describe the dynamical process will differ from the true process.

In general, the observations will be corrupted by random and systematic observation errors that are inherent in the measurement process. Hence, the observed value of the range, ρ, or elevation, θ, will differ from the true value because of these errors. In addition, the true and observed values for the range and the elevation will differ because of imperfect knowledge of the tracking station location, (X_s, Y_s).

Because of these factors, four observations will not be sufficient to determine the planar motion trajectory exactly. In the usual operational situation, many observations are made, and the trajectory is selected to give the best agreement with the observations obtained. In general the mechanism for obtaining the best estimate is to linearize the problem by expanding the equations of motion and the observation-state relationship about a reference trajectory. Deviations from the reference trajectory (e.g., position and velocity) are then determined to yield the "best" agreement with the observations. As we will see, the "best" agreement generally is based on the least squares criterion.

1.2.4 LINEARIZATION PROCEDURE

To illustrate the linearization procedure, return to the flat Earth example discussed previously. Assuming that there are errors in each of the initial position and velocity values and in g, the state vector is $X^T = [X, Y, \dot{X}, \dot{Y}, g]$. It is a straightforward procedure to replace X by $X^* + \delta X$, Y by $Y^* + \delta Y, \ldots,$ and g by $g^* + \delta g$ in Eq. (1.2.2) to obtain the equation for perturbation, or deviation, from the reference trajectory. The $(\ )^*$ values represent assumed or specified values. If the equations for the reference trajectory, designated by the $(\ )^*$ values, are subtracted from the result obtained by this substitution, the equations for the state deviations are obtained as

$$
\begin{aligned}
\delta X &= \delta X_0 + \delta \dot{X}_0 t \\
\delta Y &= \delta Y_0 + \delta \dot{Y}_0 t - \delta g \frac{t^2}{2} \\
\delta \dot{X} &= \delta \dot{X}_0 \\
\delta \dot{Y} &= \delta \dot{Y}_0 - \delta g t \\
\delta g &= \delta g.
\end{aligned}
\tag{1.2.11}
$$

Equation (1.2.11) can be expressed in matrix form in the following manner:

$$
\begin{bmatrix}
\delta X \\
\delta Y \\
\delta \dot{X} \\
\delta \dot{Y} \\
\delta g
\end{bmatrix}
=
\begin{bmatrix}
1 & 0 & t & 0 & 0 \\
0 & 1 & 0 & t & -t^2/2 \\
0 & 0 & 1 & 0 & 0 \\
0 & 0 & 0 & 1 & -t \\
0 & 0 & 0 & 0 & 1
\end{bmatrix}
\begin{bmatrix}
\delta X_0 \\
\delta Y_0 \\
\delta \dot{X}_0 \\
\delta \dot{Y}_0 \\
\delta g
\end{bmatrix}.
\tag{1.2.12}
$$

Equation (1.2.12) can be used to predict the deviation of the vehicle from the reference trajectory, $\mathbf{X}^*(t)$, at any time $t > 0$. Note that the deviations from the true state, $\delta \mathbf{X}(t) \equiv \mathbf{x}(t)$, will be caused by deviations in the original state or deviations in the gravitational acceleration, δg. However, the quantities $\delta X_0, \delta Y_0, \ldots,$ are not known, and it is the problem of the orbit determination procedure to estimate their values.

In this example, both Eqs. (1.2.2) and (1.2.12) are linear. In the case of central force field motion, the equations analogous to Eq. (1.2.2) are not linear, however, and the orbit determination process is simplified considerably by the linearization procedure. Care must be taken so that deviations from the reference conditions remain in the linear range.

As mentioned earlier, the observations ρ and θ, which are taken as the satellite follows the true trajectory, are nonlinear functions of X, Y, X_s, and Y_s. Furthermore, they contain random and systematic errors represented by ϵ_ρ and ϵ_θ; hence,

$$\rho = \sqrt{(X - X_s)^2 + (Y - Y_s)^2} + \epsilon_\rho \qquad (1.2.13)$$

$$\theta = \tan^{-1}\left(\frac{Y - Y_s}{X - X_s}\right) + \epsilon_\theta.$$

Our objective is to linearize the observations with respect to the reference trajectory. This can be accomplished by expanding Eq. (1.2.13) in a Taylor series about the reference or nominal trajectory as follows:

$$\rho \cong \rho^* + \left[\frac{\partial \rho}{\partial X}\right]^* (X - X^*) + \left[\frac{\partial \rho}{\partial Y}\right]^* (Y - Y^*) + \epsilon_\rho$$

$$\theta \cong \theta^* + \left[\frac{\partial \theta}{\partial X}\right]^* (X - X^*) + \left[\frac{\partial \theta}{\partial Y}\right]^* (Y - Y^*) + \epsilon_\theta. \qquad (1.2.14)$$

Note that the partials with respect to $\dot{X}$, $\dot{Y}$, and g are zero since they do not appear explicitly in Eq. (1.2.13). By defining

$$\begin{aligned}
\delta\rho &= \rho - \rho^* \\
\delta\theta &= \theta - \theta^* \\
\delta X &= X - X^* \\
\delta Y &= Y - Y^*,
\end{aligned} \qquad (1.2.15)$$

we can write Eq. (1.2.14) as

$$\delta\rho = \left[\frac{\partial \rho}{\partial X}\right]^* \delta X + \left[\frac{\partial \rho}{\partial Y}\right]^* \delta Y + \epsilon_\rho$$

$$\delta\theta = \left[\frac{\partial \theta}{\partial X}\right]^* \delta X + \left[\frac{\partial \theta}{\partial Y}\right]^* \delta Y + \epsilon_\theta. \qquad (1.2.16)$$

The symbol [ ]* indicates that the value in brackets is evaluated along the nominal trajectory. In Eq. (1.2.16), terms of order higher than the first in the state deviation values have been neglected assuming that these deviations are small. This requires that the reference trajectory and the true trajectory be close at all times in the interval of interest. Now, if the following definitions are used,

$$\mathbf{y}^T \equiv [\delta\rho \ \delta\theta]$$

$$\widetilde{H} \equiv \begin{bmatrix} \left[\frac{\partial\rho}{\partial X}\right]^* & \left[\frac{\partial\rho}{\partial Y}\right]^* & 0 & 0 & 0 \\ \left[\frac{\partial\theta}{\partial X}\right]^* & \left[\frac{\partial\theta}{\partial Y}\right]^* & 0 & 0 & 0 \end{bmatrix} \quad (1.2.17)$$

$$\mathbf{x}^T \equiv [\delta X \ \delta Y \ \delta\dot{X} \ \delta\dot{Y} \ \delta g]$$

$$\epsilon^T \equiv [\epsilon_\rho \ \epsilon_\theta],$$

then Eq. (1.2.16) can be expressed as the matrix equation,

$$\mathbf{y} = \widetilde{H}\mathbf{x} + \epsilon. \quad (1.2.18)$$

In Eq. (1.2.18), $\mathbf{y}$ is called the *observation deviation vector*, $\mathbf{x}$ is the *state deviation vector*, $\widetilde{H}$ is a mapping matrix that relates the observation deviation vector to the state deviation vector, and ϵ is a random vector that represents the noise or error in the observations. It is common practice to refer to $\mathbf{x}$ and $\mathbf{y}$ as the *state and observation vectors*, respectively, even though they really represent deviations from nominal values.

1.2.5 STATE TRANSITION MATRIX

The matrix multiplying the initial state vector in Eq. (1.2.12) is referred to as the *state transition matrix*, $\Phi(t, t_0)$. For the state deviation vector, $\mathbf{x}$, defined in Eq. (1.2.17), this matrix is given by

$$\Phi(t, t_0) = \begin{bmatrix} 1 & 0 & t & 0 & 0 \\ 0 & 1 & 0 & t & -t^2/2 \\ 0 & 0 & 1 & 0 & 0 \\ 0 & 0 & 0 & 1 & -t \\ 0 & 0 & 0 & 0 & 1 \end{bmatrix}. \quad (1.2.19)$$

The state transition matrix maps deviations in the state vector from one time to another. In this case, deviations in the state are mapped from t_0 to t. Note that we have arbitrarily set $t_0 = 0$. For this linear example, the mapping is exact, but in the general orbit determination case, the state equations will be nonlinear, and the state transition matrix is the linear term in a Taylor series expansion of the state vector at time t about the state vector at a reference time t_0. Hence, the state deviation vector and observation deviation vector at any time can be written in terms of x_0, as follows:

$$\mathbf{x}(t) = \Phi(t, t_0)\mathbf{x}_0 \tag{1.2.20}$$

$$\mathbf{y}(t) = \widetilde{H}(t)\mathbf{x}(t) + \boldsymbol{\epsilon}(t).$$

$\mathbf{y}(t)$ can be written in terms of $\mathbf{x}_0$ as

$$\mathbf{y}(t) = H(t)\mathbf{x}_0 + \boldsymbol{\epsilon}(t), \tag{1.2.21}$$

where

$$H(t) = \widetilde{H}(t)\Phi(t, t_0). \tag{1.2.22}$$

The problem that remains now is to determine the best estimate of $\mathbf{x}_0$ given the linearized system represented by Eqs. (1.2.20) and (1.2.21). Our problem can now be summarized as follows. Given an arbitrary epoch, t_k, and the state propagation equation and observation-state relationship

$$\mathbf{x}(t) = \Phi(t, t_k)\mathbf{x}_k \tag{1.2.23}$$

$$\mathbf{y} = H\mathbf{x}_k + \boldsymbol{\epsilon},$$

find the "best" estimate of $\mathbf{x}_k$. In Eq. (1.2.23), $\mathbf{y}$ is an $m \times 1$ vector, $\mathbf{x}$ is an $n \times 1$ vector, $\boldsymbol{\epsilon}$ is an $m \times 1$ vector, and H is an $m \times n$ mapping matrix, where n is the number of state variables and m is the total number of observations. If m is sufficiently large, the essential condition $m \geq n$ is satisfied.

In contrast to the algebraic solutions described earlier in this chapter, the system represented by the second of Eq. (1.2.23) is always underdetermined. That is, there are m-knowns (e.g., the observation deviations, $\mathbf{y}$) and $m + n$ unknowns (e.g., the m-observation errors, $\boldsymbol{\epsilon}$, and the n-unknown components of the state deviation vector, $\mathbf{x}_k$). In Chapter 4, several approaches for resolving this problem are discussed. The most straightforward is based on the method of least squares as proposed by Gauss (1809). In this approach, the best estimate for the unknown state vector, $\mathbf{x}_k$, is selected as the value $\hat{\mathbf{x}}_k$, which minimizes the sum of the squares of the calculated values of the observations errors. That is, if $\mathbf{x}_k^0$ is any value of $\mathbf{x}_k$, then $\boldsymbol{\epsilon}^0 = \mathbf{y} - H\mathbf{x}_k^0$ will be the m-calculated values of the observation residuals corresponding to the value $\mathbf{x}_k^0$. Then, the best estimate of $\mathbf{x}_k$ will be the value that leads to a minimal value of the performance index, $J(\mathbf{x}_k^0)$, where

$$J(\mathbf{x}_k^0) = 1/2(\boldsymbol{\epsilon}^{0T}\boldsymbol{\epsilon}^0) = 1/2(\mathbf{y} - H\mathbf{x}_k^0)^T(\mathbf{y} - H\mathbf{x}_k^0). \tag{1.2.24}$$

For a minimum of this quantity, it is necessary and sufficient that

$$\left.\frac{\partial J}{\partial \mathbf{x}_k^0}\right|_{\hat{\mathbf{x}}_k} = 0; \quad \delta \mathbf{x}_k^{0T} \left.\frac{\partial^2 J}{\partial \mathbf{x}_k^0 \partial \mathbf{x}_k^0}\right|_{\hat{\mathbf{x}}_x} \delta \mathbf{x}_k^0 > 0. \tag{1.2.25}$$

From the first of the conditions given by Eq. (1.2.25), it follows that

$$(\mathbf{y} - H\hat{\mathbf{x}}_k)^T H = 0$$

or by rearranging that

$$H^T H \hat{\mathbf{x}}_k = H^T \mathbf{y}. \tag{1.2.26}$$

If the $n \times n$ matrix $(H^T H)$ has an inverse, the solution can be expressed as

$$\hat{\mathbf{x}}_k = (H^T H)^{-1} H^T \mathbf{y}.$$

Chapter 4 expands and clarifies the procedure and techniques for determining, and assessing the accuracy of, the estimate $\hat{\mathbf{x}}_k$. Further details are given on computational methods for determining $\hat{\mathbf{x}}_k$ in Chapter 5.

1.3 BACKGROUND AND OVERVIEW

As noted in this introductory chapter, orbit determination methodology can be separated into two general classes, for example, classical (or deterministic) orbit determination and modern (or statistical based) orbit determination. In the classical approach, observational errors are not considered and the problem is reduced to making the necessary number of measurements to determine the orbit. There is an equivalence between the number of observations and the number of unknowns. The modern orbit determination problem recognizes the influence of observation errors and, to minimize the effects of the observational error, more observations are used than the number of parameters to be estimated. Modern orbit determination methodology dates back to the beginning of the 19th century, when Legendre (1806), in attempting to determine the orbits of the planets, recognized the importance of observational errors and proposed the least squares solution as a method for minimizing the effect of such errors. Gauss (1809) presented a probabilistic basis for the method of least squares, which was closely akin to the maximum likelihood method proposed by Fisher (1912). Gauss (1809) stated that he had used the method since 1795 and historians (Cajori, 1919) generally agree that Gauss originated the method. The notation of Gauss is still in use. Laplace (1812) added refinements to the computational approach and to the statistical foundations. Markov (1898), in his work on the mathematical foundations of probability theory, clarified many of the concepts associated with the method of least squares. Helmert (1900) extended the application to astronomy and geodesy. In the 20th century, Kolmogorov (1941) and Wiener (1949) linked the problem to modern systems theory and set the stage for the sequential filter formulation for processing tracking data. In 1959, Swerling gave the first conceptual approach to the sequential orbit determination problem. Kalman (1960) presented a mathematically rigorous approach for the sequential processing of observations of a linear dynamic system. In a later work, Kalman and Bucy (1961) gave an approach for the extended Kalman filter, which is the sequential approach for the general nonlinear orbit determination problem. Following the publication of Kalman's paper in 1960, there was an extensive series of publications related to this topic during the next three decades. Many of these investigations will be referenced in the subsequent chapters. A summary of the historical development of the least squares approach is given by Sorenson (1970).

The main thrust of this book will be aimed at developing computational data processing techniques for estimating the state of orbiting satellites. As noted in the previous discussion, the problem formulation will lead to a nonlinear relation between the observations and the epoch state of the satellite. Therefore, in modern system theory terms, we will be concerned with estimating the state of a nonlinear dynamic system, using observations of nonlinear functions of the system state, where the observations are corrupted by random and systematic observational errors. Although the main focus for the physical models will be the orbit determination problem, the problem will be cast in the state space notation defined by modern system theory. As such, the computational approaches presented in the treatment will be applicable to a wide range of problems, including missile guidance, attitude control, economic system analysis, chemical process control, and a wide variety of manufacturing processes. Once the defining equations from these processes have been reduced to the state space formulation, the computational techniques developed here will be applicable.

1.4   SUMMARY

Following the introduction of the basic concepts of nonlinear parameter estimation in Chapter 1, Chapter 2 provides a detailed discussion of the characteristics of the orbit for Earth-orbiting satellites. This is the primary dynamical system of interest in this treatment. Chapter 3 describes the various observations used in the Earth-orbiting satellite orbit determination problem. Chapter 4 describes the properties of the Weighted Least Squares and Minimum Variance Estimation algorithms as they are expressed in a batch estimation form and describes the transformation to the sequential estimation algorithm. The relation to the Kalman filter is noted. In Chapter 5, the computation algorithms for solving the linear system based on both the Cholesky Decomposition and the Orthogonal Transformation approach are presented. In addition, square root sequential estimation algorithms are given. Chapter 6 describes the error analysis approaches that evolve from the Consider covariance analysis. The algorithms developed here account for the effects of uncertainties on unestimated parameters that appear in either the dynamic or measurement equations.

As noted previously, a primary emphasis of this treatment will be devoted to developing computationally robust and efficient algorithms for processing the observations of Earth-orbiting satellites.

1.5 REFERENCES

Cajori, F., *A History of Mathematics*, MacMillan Co., New York, 1919.

Fisher, R. A., "On an absolute criteria for fitting frequency curves," *Mess. Math.*, Vol. 41, pp. 155–160, 1912.

Gauss, K. F., *Theoria Motus Corporum Coelestium*, 1809 (Translated into English: Davis, C. H., *Theory of the Motion of the Heavenly Bodies Moving about the Sun in Conic Sections*, Dover, New York, 1963).

Helmert, F. R., "Zur Bestimmung kleiner Flächenstücke des Geoids aus Lothabweichungen mit Rücksicht auf Lothkrümmung", Sitzungsberichte Preuss. Akad. Wiss., Berlin, Germany, 1900.

Kalman, R. E., "A New Approach to Linear Filtering and Prediction Theory," *J. Basic Eng.*, Vol. 82, Series E, No. 1, pp. 35–45, March 1960.

Kalman, R. E. and R. S. Bucy, "New Results in Linear Filtering and Prediction Theory," *J. Basic Eng.*, Vol. 83, Series D, No. 1, pp. 95–108, March 1961.

Kolmogorov, A. N., "Interpolation and Extrapolation of Stationary Random Sequences," *Bulletin of the Academy of Sciences of the USSR Math. Series*, Vol. 5, pp. 3–14, 1941.

Laplace, P. S., *Théorie Analytique de Probabilités*, Paris, 1812 (The 1814 edition included an introduction, *Essai Philosophique sur les Probabilités*, which has been translated into English: Dale, A. I., *Philosophical Essay on Probabilities*, Springer-Verlag, New York, 1995).

Legendre, A. M., *Nouvelles méthodes pour la détermination des orbites des comètes*, Paris, 1806.

Markov, A. A., "The law of large numbers and the method of Least Squares," (1898), *Izbr. Trudi., Izd. Akod. Nauk*, USSR, pp. 233–251, 1951.

Sorenson, H. W., "Least squares estimation: from Gauss to Kalman," *IEEE Spectrum*, Vol. 7, No. 7, pp. 63–68, July 1970.

Swerling, P., "First order error propagation in a stagewise differential smoothing procedure for satellite observations," *J. Astronaut. Sci.*, Vol. 6, pp. 46–52, 1959.

Wiener, N., *The Extrapolation, Interpolation and Smoothing of Stationary Time Series*, John Wiley & Sons, Inc., New York, 1949.

1.6   EXERCISES

1. Write a computer program that computes $\rho(t_i)$ for a uniform gravity field using Eq. (1.2.10). A set of initial conditions, X_0, $\dot{X}_0$, Y_0, $\dot{Y}_0$, g, X_s, Y_s, and observations, ρ, follow. With the exception of the station coordinates, the initial conditions have been perturbed so that they will not produce exactly the observations given. Use the Newton iteration scheme of Eq. (1.2.9) to recover the exact initial conditions for these quantities; that is, the values used to generate the observations. Assume that X_s and Y_s are known exactly. Hence, they are not solved for.

Unitless Initial Conditions

$X_0 = 1.5$
$Y_0 = 10.0$
$\dot{X}_0 = 2.2$
$\dot{Y}_0 = 0.5$
$g = 0.3$
$X_s = Y_s = 1.0$

Time	Range Observation, ρ
0	7.0
1	8.00390597
2	8.94427191
3	9.801147892
4	10.630145813

Answer

$X_0 = 1.0$
$Y_0 = 8.0$
$\dot{X}_0 = 2.0$
$\dot{Y}_0 = 1.0$
$g = 0.5$

2. In addition to the five state variables of Exercise 1, could X_s and Y_s be solved for given seven independent observations of ρ? Why or why not? Hint: Notice the relationship between X_0 and X_s and between Y_0 and Y_s in Eq. (1.2.7) for ρ_j. See also Section 4.12 on observability.

Chapter 2

The Orbit Problem

2.1 HISTORICAL BACKGROUND

Johannes Kepler, based on a study of planetary position data acquired by Tycho Brahe, formulated three laws of planetary motion (Plummer, 1918):

1. The orbit of each planet is an ellipse with the Sun at one focus;

2. The heliocentric radius vector of each planet sweeps over equal area in equal time;

3. The square of the orbital period is proportional to the cube of the ellipse semimajor axis.

But it was Isaac Newton who established the mathematical foundation from which Kepler's Laws can be derived. Newton's insightful analysis of the work of Kepler led to the formulation of the basic concepts of mechanics and gravitation at an early stage in his career. Not until he was in his forties, however, did Newton publish these concepts in the *Principia* (1687), as described in the extensive biography by Westfall (1980).

The *Principia* presented the Three Laws of Motion, quoted from p. 13 of Motte's 1729 English translation of Newton's Latin publication (with modern terminology in brackets):

> Law I: Every body continues in its state of rest, or of uniform motion in a right line [straight line] unless it is compelled to change that state by forces impressed upon it.

> Law II: The change of motion is proportional to the motive force impressed; and is made in the direction of the right line in which that force is impressed. [$\mathbf{F} = m\mathbf{a}$, where $\mathbf{F}$ is force, m is mass, and $\mathbf{a}$ is acceleration, which is equal to the time derivative of velocity, $\mathbf{V}$.]

Law III: To every action there is always opposed an equal reaction: or, the mutual actions of two bodies upon each other are always equal, and directed to contrary parts.

Newton deduced the Law of Gravitation as well, which states: the mutual force of gravitational attraction between two point masses is

$$F = \frac{GM_1 M_2}{d^2} \tag{2.1.1}$$

where G is the Universal Constant of Gravitation and d is the distance between the two point masses or particles, M_1 and M_2.

The Laws of Newton contain some important concepts. The concept of a *point mass* is embodied in these laws, which presumes that the body has no physical dimensions and all its mass is concentrated at a point. The concept of an *inertial reference frame* is crucial to Newton's Second Law; that is, the existence of a coordinate system that has zero acceleration. The concept of *force* (**F**) is introduced as the cause of the acceleration experienced by a point mass, and the mass (m) is the constant of proportionality implied by Newton's statement of the Second Law. Finally, the concept of a uniformly changing *time* is fundamental to Newton's Laws of Motion.

These laws are the foundation of the mathematical description of orbital dynamics. To this day, the laws enable a remarkably accurate description of the motion of both natural and artificial bodies. The work of Albert Einstein [1905] provided even greater insight into this description and allowed the explanation of small discrepancies between observation and Newtonian dynamics.

The description of orbital dynamics in this book is Newtonian in nature. Inclusion of relativistic corrections is important for those applications that require high accuracy, and the relevant terms are described in this chapter.

Subtle differences in terminology exist in the literature concerning the distinction between *reference frames* and *reference systems*, for example. A careful and precise set of definitions is discussed by Kovalevsky *et al.* (1989). For purposes of this book, the terminology *reference frame* will be used within a context that describes some specific characteristic (e.g., inertial reference frame). The terminology *coordinate system* generally will apply to a set of axes that is used to describe the motion of a body. In some instances, *reference frame* and *coordinate system* are used interchangeably where there is no ambiguity.

In the following sections, the description of satellite motion is developed, beginning with the two-body gravitational model of satellite motion. The fundamental description from the problem of two bodies serves as a useful reference, but it is inadequate for the accurate representation of the actual satellite motion. The more complete and realistic representation will be developed from the two-body case. It is not the intent of the development in this chapter to provide an

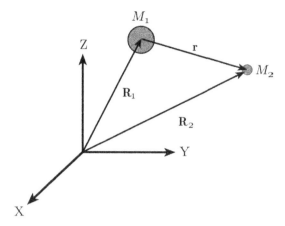

Figure 2.2.1: The problem of two bodies: M_1 and M_2 are spheres of constant, uniform density. The position vectors of the spheres refer to their respective geometrical center, which also coincide with the center of mass of the sphere.

exhaustive treatise on orbital mechanics, since excellent books already exist (e.g., Szebehely and Mark, 1998; Vallado, 2001; Bond and Allman, 1996; Chobotov, 1996; Prussing and Conway, 1993; Danby, 1988; Roy, 1988; Battin, 1999; and Schaub and Junkins, 2003). The philosophy in this chapter is to provide the key steps in the development and its interpretation, especially within the context of the main purpose related to the determination of the orbit. With this philosophy, many details are left to you, although the preceding list of references provides most of the details.

2.2 PROBLEM OF TWO BODIES: GENERAL PROPERTIES

The classic problem of two bodies was solved by Newton, which is repeated in a modern form by all treatises on celestial mechanics. The classic treatment describes the motion of two point masses. However, human experience demonstrates that the Earth has physical dimensions, and a point mass Earth is a serious stretch of the imagination. Nevertheless, a spherical body with constant, uniform mass distribution can be shown to be gravitationally equivalent to a point mass. Thus, as a more realistic approximation to the physical characteristics of a planet, a spherical body will be used in the following development.

Consider the motion of two spheres of mass M_1 and M_2, as illustrated in Fig. 2.2.1. No restrictions are placed on the magnitude of these masses, so for illustrative purposes, assume M_1 represents the Earth and M_2 represents a satellite, including the Moon. The coordinate system $(X,\ Y,\ Z)$ is inertial and nonrotating; that is, it has no acceleration, and there is no preferred orientation of the axes at this stage. The position vectors of the geometrical centers of the respective spheres are $\mathbf{R}_1$ and $\mathbf{R}_2$. The position of M_2 with respect to M_1 is described by $\mathbf{r}$. The equations of motion derived from Newton's Laws of Motion and Law of Gravitation are

$$M_1\ddot{\mathbf{R}}_1 \quad = \quad \frac{GM_1M_2\mathbf{r}}{r^3} \tag{2.2.1}$$

$$M_2\ddot{\mathbf{R}}_2 \quad = \quad -\frac{GM_1M_2\mathbf{r}}{r^3} \tag{2.2.2}$$

where $\ddot{\mathbf{R}}_1 = d^2\mathbf{R}_1/dt^2$ and t represents time. It can readily be shown that the equations of relative motion are

$$\ddot{\mathbf{r}} = -\frac{\mu\mathbf{r}}{r^3} \tag{2.2.3}$$

where $\mu = G(M_1 + M_2)$. In the case of an artificial satellite of the Earth, let M_1 represent the Earth and M_2 the satellite. Since the Earth is so massive, even a large satellite like a fully loaded space station results in

$$\mu = G(M_1 + M_2) \cong GM_1$$

to very high accuracy (better than one part in 10^{15}).

The general characteristics of the motion follow from manipulation of Eqs. (2.2.1) and (2.2.2):

1. Motion of Two-Body Center of Mass. The definition of the position of the center of mass of the two spheres, $\mathbf{R}_{cm}$, is

$$\mathbf{R}_{cm} = \frac{M_1\mathbf{R}_1 + M_2\mathbf{R}_2}{(M_1 + M_2)}. \tag{2.2.4}$$

By adding Eqs. (2.2.1) and (2.2.2) it follows that

$$\ddot{\mathbf{R}}_{cm} = 0 \quad \text{and} \quad \mathbf{R}_{cm} = \mathbf{C}_1(t - t_o) + \mathbf{C}_2 \tag{2.2.5}$$

where $\mathbf{C}_1$ and $\mathbf{C}_2$ are constants of the motion. Equation (2.2.5) states that the center of mass of the two spheres moves in a straight line with constant velocity. Note that the straight line motion of the center of mass is defined in an inertial coordinate system.

2. Angular Momentum. Taking the vector cross product of $\mathbf{r}$ with both sides of Eq. (2.2.3) gives

$$\mathbf{r} \times \ddot{\mathbf{r}} = 0 \quad \text{or} \quad \frac{d}{dt}(\mathbf{r} \times \dot{\mathbf{r}}) = 0\,.$$

Defining $\mathbf{h} = \mathbf{r} \times \dot{\mathbf{r}}$, the *angular momentum per unit mass* (or *specific angular momentum*), it follows that $\mathbf{h}$ is a vector constant:

$$\mathbf{h} = \mathbf{r} \times \dot{\mathbf{r}} = \textbf{constant}. \qquad (2.2.6)$$

An important interpretation of $\mathbf{h} = \textbf{constant}$ is that the orbital motion is planar, since $\mathbf{r}$ and $\dot{\mathbf{r}}$ must always be perpendicular to $\mathbf{h}$. The plane of the motion is referred to as the *orbit plane*, which is a plane that is perpendicular to the angular momentum vector. Note that the angular momentum in Eq. (2.2.6) is defined for the motion of M_2 with respect to M_1.

3. Energy. Taking the vector dot product of $\dot{\mathbf{r}}$ with Eq. (2.2.3), it can be shown that

$$\xi = \frac{\dot{\mathbf{r}} \cdot \dot{\mathbf{r}}}{2} - \frac{\mu}{r} = \text{constant} \qquad (2.2.7)$$

where ξ is the scalar *energy per unit mass* (also known as *specific energy* or *vis viva*) and $\dot{\mathbf{r}}$ is the velocity of M_2 with respect to M_1. Thus, the energy per unit mass is constant; that is, the sum of the kinetic and potential energies is constant. Note that this expression describes the energy in terms of the motion of M_2 with respect to M_1.

In summary, there are ten constants of the motion: $\mathbf{C}_1$, $\mathbf{C}_2$, $\mathbf{h}$, ξ. The following general properties of the orbital motion can be stated:

a. The motion of M_2 with respect to M_1 takes place in a plane, the orbit plane, defined as the plane that is perpendicular to the constant angular momentum vector ($\mathbf{h} = \mathbf{r} \times \dot{\mathbf{r}}$);

b. The center of mass of the two bodies moves in a straight line with constant velocity with respect to an inertial coordinate system;

c. The scalar energy per unit mass is constant, as given by Eq. (2.2.7).

2.2.1 MOTION IN THE PLANE

The previous section summarized some general characteristics of the two-body motion, but the differential equations of motion, Eq. (2.2.3), were not solved. The detailed characteristics of the orbital motion are still to be determined, as described in this section.

The previous section showed that the orbital motion of M_2 with respect to M_1 takes place in a plane perpendicular to the angular momentum vector. This fact of planar motion can be used to facilitate the solution of Eq. (2.2.3). The planar motion is illustrated in Fig. 2.2.2, described with respect to a nonrotating (X, Y, Z) coordinate system. The $\hat{X}$ axis is shown at the intersection of the orbit

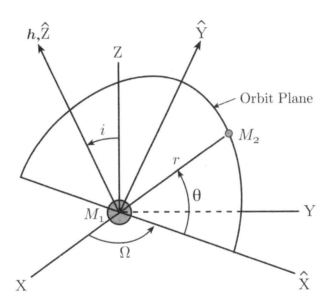

Figure 2.2.2: The orbit plane in space. Since the angular momentum is constant, the motion of M_2 with repect to M_1 takes place in a plane, known as the orbit plane. This plane is perpendicular to the angular momentum vector and the orientation is defined by the angles i and Ω.

plane and the (X, Y) plane, where the intersection between the planes is referred to as the *line of nodes*. The angles i and Ω are discussed in Section 2.2.2. The $\hat{Y}$ axis is perpendicular to $\hat{X}$, but it lies in the orbit plane, whereas $\hat{Z}$ is coincident with $\mathbf{h}$. Since the angular momentum vector, $\mathbf{h}$, is constant, the orientation of the orbit plane in space does not change with time, and the $(\hat{X}, \hat{Y}, \hat{Z})$ axes are, therefore, nonrotating. From the introduction of a unit vector, $\mathbf{u}_r$, defined by $\mathbf{r} = r\,\mathbf{u}_r$ and the unit vector, $\mathbf{u}_\theta$, as illustrated in Fig. 2.2.3, it follows that Eq. (2.2.3) can be written in polar coordinates as

$$\mathbf{u}_r \text{ component}: \quad \ddot{r} - r\dot{\theta}^2 = -\frac{\mu}{r^2} \tag{2.2.8}$$

$$\mathbf{u}_\theta \text{ component}: \quad 2\dot{r}\dot{\theta} + r\ddot{\theta} = 0. \tag{2.2.9}$$

Two cases exist: $\dot{\theta} = 0$ and $\dot{\theta} \neq 0$. The $\dot{\theta} = 0$ case is rectilinear motion, which is a special case of no interest in this discussion. For the orbital motion case, $\dot{\theta} \neq 0$, and it can be shown from Eq. (2.2.9) that

$$\frac{d}{dt}(r^2\dot{\theta}) = 0. \tag{2.2.10}$$

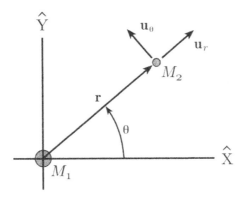

Figure 2.2.3: The planar motion described with polar coordinates.

Note that since $\dot{\mathbf{r}} = \dot{r}\mathbf{u}_r + r\dot{\theta}\mathbf{u}_\theta$, it follows from $\mathbf{h} = \mathbf{r} \times \dot{\mathbf{r}}$ that $\mathbf{h} = r^2\dot{\theta}\mathbf{u}_h$, where $\mathbf{u}_h$ is the unit vector in the $\mathbf{h}$ direction. It is evident that $h = r^2\dot{\theta}$ is constant. Furthermore, a change in both the dependent and independent variables in Eq. (2.2.8)

$$r \to 1/u, \quad t \to \theta \qquad (2.2.11)$$

gives the differential equation of motion

$$\frac{d^2u}{d\theta^2} + u = \frac{\mu}{h^2} \qquad (2.2.12)$$

which has the solution

$$u = \frac{\mu}{h^2} + A\cos(\theta - \omega) \qquad (2.2.13)$$

where A and ω are constants of integration.

From Eq. (2.2.7) for the energy, it can be shown that

$$A = \frac{\mu}{h^2}e$$

where

$$e = \left[1 + 2\frac{\xi h^2}{\mu^2}\right]^{\frac{1}{2}}, \qquad (2.2.14)$$

a positive quantity.

Table 2.2.1: Classes of Orbital Motion

Orbit Type	Eccentricity	Energy	Orbital Speed
Circle	$e = 0$	$\xi = -\frac{\mu}{2a}$	$V = \sqrt{\mu/r}$
Ellipse	$e < 1$	$\xi < 0$	$\sqrt{\mu/r} < V < \sqrt{2\mu/r}$
Parabola	$e = 1$	$\xi = 0$	$V = \sqrt{2\mu/r}$
Hyperbola	$e > 1$	$\xi > 0$	$V > \sqrt{2\mu/r}$

It follows from Eq. (2.2.13) that

$$r = \frac{h^2/\mu}{1 + e\cos(\theta - \omega)} .$$

It can be readily shown that

$$\frac{dr}{d\theta} = \frac{\mu e \sin(\theta - \omega)}{h^2(1 + \cos(\theta - \omega))^2}$$

from which it follows that

$$\frac{dr}{d\theta} = 0$$

when $\theta = \omega$ or $\theta - \omega = \pi$. Furthermore, since

$$\frac{d^2r}{d\theta^2} > 0$$

at $\theta = \omega$, it is evident that r is a minimum. This location with minimum r is referred to as the *perifocus*, or *perigee* in the case of a satellite of the Earth. As a consequence, the constant of integration ω is known as the *argument of perigee*.

Introducing the *true anomaly*, f defined by $f = \theta - \omega$, then the expression for r becomes

$$r = \frac{p}{1 + e\cos f} \tag{2.2.15}$$

where $p = h^2/\mu$.

The form of Eq. (2.2.15) is identical to the geometrical equation for a conic section (for example, see Danby [1988]). The geometrical quantities p and e are the *semilatus rectum* and *eccentricity* of the conic section, respectively. As shown in the preceding equations, these geometrical parameters are determined by dynamical parameters h, ξ, and μ.

The two-body orbit can be described as a conic section: circle, ellipse, parabola, or hyperbola. The specific conic section is determined by the eccentricity,

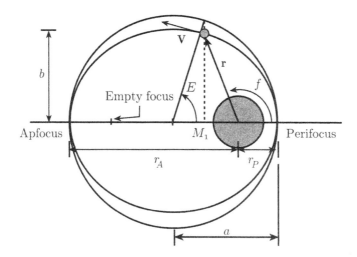

Figure 2.2.4: Characteristics of the elliptic orbit. The circle has radius equal to a and is used to illustrate the eccentric anomaly, E. The dashed line is perpendicular to the major axis of the ellipse.

e. Each conic section has a specific energy and velocity for a given r, as shown in Table 2.2.1.

Only the conic sections with $e < 1$ are closed paths. The parabola and hyperbola are often referred to as *escape trajectories*.

Equation (2.2.15) describes the position of M_2 with respect to M_1, where M_1 is located at a focus of the conic section, and the position of M_2 is described in terms of r and f. The ellipse conic section, as illustrated in Fig. 2.2.4, has two foci, where M_1 is located at one focus and the other is an *empty focus*. The ellipse is defined by p and e, but other parameters are commonly used, such as semimajor axis (a) and semiminor axis (b). Some useful geometrical relations are

$$
\begin{aligned}
p &= a(1 - e^2) \\
b &= a\sqrt{1 - e^2}.
\end{aligned}
$$

Note that $p = r$ when $f = \pm\pi/2$.

The minimum distance between M_1 and M_2 is the *perifocus* (or *perigee* if M_1 is the Earth), given by

$$r_p = a(1 - e) \tag{2.2.16}$$

and the maximum distance, *apfocus* (or *apogee* for an Earth satellite) is

$$r_a = a(1 + e). \tag{2.2.17}$$

Other relations between geometrical and dynamical parameters for the elliptical orbit can be shown to be

$$h^2 = \mu a(1 - e^2) = \mu p$$
$$\xi = -\frac{\mu}{2a} .$$

The M_2 velocity, $\mathbf{V}$, with respect to M_1 can be written as

$$\mathbf{V} = \dot{r}\mathbf{u}_r + r\dot{f}\mathbf{u}_\theta . \tag{2.2.18}$$

The magnitude of the M_2 velocity (with respect to M_1), or speed, is

$$V = \sqrt{\mu(2/r - 1/a)}$$

and it can be shown that the radial component of the velocity, $\dot{r}$, is

$$\dot{r} = \frac{he\sin f}{p} \tag{2.2.19}$$

from which it is evident that the radial velocity is positive (away from M_1) while traveling from perifocus to apfocus, and it is negative (toward M_1) from apfocus to perifocus. Furthermore, from $h = r^2\dot{f}$ (since $\dot{f} = \dot{\theta}$), the velocity component perpendicular to the position vector is

$$r\dot{f} = \frac{h}{r} . \tag{2.2.20}$$

As demonstrated by Vallado (2001), for example, the area (A) swept out by r as a function of time is

$$\frac{dA}{dt} = \frac{h}{2}$$

which is constant, thereby demonstrating Kepler's Second Law. If this equation is integrated over one orbital period, T, the time required for M_2 to complete one revolution of M_1 (for example, from perifocus and back to perifocus) is

$$T = \frac{2\pi a^{3/2}}{\mu^{1/2}} . \tag{2.2.21}$$

This equation verifies Kepler's Third law.

From the conic section equation, Eq. (2.2.14), if the angle f is known, the distance r can be determined. However, for most applications, the time of a particular event is the known quantity, rather than the true anomaly, f. A simple relation between time and true anomaly does not exist. The transformation between time and true anomaly uses an alternate angle, E, the *eccentric anomaly*, and is expressed by *Kepler's Equation*, which relates time and E as

$$E - e\sin E = n(t - t_p) \tag{2.2.22}$$

where n is the mean motion ($\sqrt{\mu/a^3}$) , t is time, and t_p is the time when M_2 is located at perifocus. The time t_p is a constant of integration. Since Kepler's Equation is a transcendental equation, its solution can be obtained by an iterative method, such as the Newton–Raphson method. If

$$g = E - e \sin E - M,$$

where $M = n(t - t_p)$, the *mean anomaly*, then the iterative solution gives

$$E_{k+1} = E_k - \left(\frac{g}{g'}\right)_k, \qquad (2.2.23)$$

where k represents the iteration number and

$$g' = \frac{dg}{dE} = 1 - e \cos E.$$

From known values of a and e and a specified t, the preceding iterative method will provide the eccentric anomaly, E. A typical value of E to start the iteration is $E_0 = M$. Some useful relations between the true anomaly and the eccentric anomaly are

$$\sin f = \frac{(1 - e^2)^{1/2} \sin E}{1 - e \cos E} \qquad \sin E = \frac{(1 - e^2)^{1/2} \sin f}{1 + e \cos f} \qquad (2.2.24)$$

$$\cos f = \frac{\cos E - e}{1 - e \cos E} \qquad \cos E = \frac{e + \cos f}{1 + e \cos f}$$

$$\tan \frac{f}{2} = \left(\frac{1 + e}{1 - e}\right)^{1/2} \tan \frac{E}{2} \qquad \tan \frac{E}{2} = \left(\frac{1 - e}{1 + e}\right)^{1/2} \tan \frac{f}{2}$$

$$M = E - e \sin E \quad \text{(Kepler's Equation)}$$

$$\frac{df}{dE} = \frac{(1 - e^2)^{1/2}}{1 - e \cos E} = \frac{1 + e \cos f}{(1 - e^2)^{1/2}}$$

$$\frac{dM}{dE} = 1 - e \cos E = \frac{1 - e^2}{1 + e \cos f}$$

$$\frac{dM}{df} = \frac{(1 - e \cos E)^2}{(1 - e^2)^{1/2}} = \frac{(1 - e^2)^{3/2}}{(1 + e \cos f)^2} .$$

It can be shown that

$$r = a(1 - e \cos E)$$

but in some cases it is useful to express r as a function of time, or mean anomaly. This relation for elliptic orbits, however, involves an infinite series because of the transcendental nature of Kepler's Equation; namely,

$$\frac{a}{r} = 1 + 2 \sum_{m=1}^{\infty} \hat{J}_m(me) \cos mM \qquad (2.2.25)$$

where $\hat{J}_m$ is the Bessel Function of the first kind of order m, with argument me (see Smart, 1961; Brouwer and Clemence, 1961). If the eccentricity is small, the following approximate equations may be useful:

$$\frac{a}{r} \cong 1 + e \cos M + \text{higher order terms} \qquad (2.2.26)$$

$$\cos f \cong \cos M + e(\cos 2M - 1) + \dots$$

$$\sin f \cong \sin M + e \sin 2M + \dots$$

$$f \cong M + 2e \sin M + \dots$$

$$\cos E \cong \cos M + \frac{e}{2}(\cos 2M - 1) + \dots$$

$$\sin E \cong \sin M + \frac{e}{2} \sin 2M + \dots$$

$$E \cong M + e \sin M + \dots .$$

2.2.2 MOTION IN SPACE

The preceding section has described the two-body orbital motion in the plane. The description of the complete three-dimensional motion requires two additional parameters. The angular momentum per unit mass, $\mathbf{h}$, is

$$\mathbf{h} = h_X \mathbf{u}_X + h_Y \mathbf{u}_Y + h_Z \mathbf{u}_Z \qquad (2.2.27)$$

where $\mathbf{u}_X$, $\mathbf{u}_Y$, and $\mathbf{u}_Z$ are unit vectors associated with the (X, Y, Z) axes. Two angles, i and Ω (see Fig. 2.2.2), can be used to describe the three-dimensional orientation of the orbit plane.

The *orbital inclination*, i, is the angle between the Z axis and $\mathbf{h}$. This angle can be determined from

$$\cos i = \frac{h_Z}{h}, \qquad (2.2.28)$$

where $0 \leq i \leq 180°$ and $h = |\mathbf{h}|$. If the (X, Y) axes are in the equator of the planet, then the orbit can be categorized by the inclination, as summarized in Table 2.2.2.

The orbit plane shown in Fig. 2.2.2 intersects the plane containing the (X, Y) axes at the *line of nodes*. Since the motion is planar, the satellite will move through the line of nodes at the *ascending node* (AN) where it moves from $-Z$ to $+Z$. With the (X, Y) axes in the equatorial plane, the satellite moves from the southern hemisphere to the northern hemisphere when it crosses the ascending node.

Table 2.2.2: Inclination Terminology

i	Orbit Type
$90°$	Polar
$0 \leq i < 90°$	Posigrade
$90° < i \leq 180°$	Retrograde
$i = 0°$	Posigrade equatorial
$i = 180°$	Retrograde equatorial

Similarly, the point where the satellite crosses the line of nodes from $+Z$ to $-Z$ is referred to as the *descending node* (DN). The angle, Ω, that defines the ascending node location is

$$\sin \Omega = \frac{h_X}{h_{XY}} \qquad (2.2.29)$$

$$\cos \Omega = -\frac{h_Y}{h_{XY}} \qquad (2.2.30)$$

where $h_{XY} = [h_X^2 + h_Y^2]^{1/2}$ and $0 \leq \Omega \leq 360°$. Note that both $\sin \Omega$ and $\cos \Omega$ are required to determine the quadrant of Ω. The specific terminology used for Ω depends on the (X, Y, Z) coordinate system. The term *right ascension of the ascending node* is used in cases where (X, Y, Z) is a celestial system, but *longitude of the ascending node* is used in the case of an Earth-fixed coordinate system.

2.2.3 BASIC COORDINATE SYSTEMS

The (X, Y, Z) coordinate system used in the preceding sections has no preferred orientation for the simple description of the two-body motion. Nevertheless, a specific, well-defined orientation of the (X, Y, Z) axes is required in practice.

Consider the motion of the Earth about the Sun. Since the mass of the Sun is 328,000 times the mass of the Earth, it is quite natural to describe the motion of this system as the motion of the Earth relative to the Sun. But from the description of the two-body motion, there is no requirement that M_1 be greater than M_2 or vice versa. The two-body motion of the Earth-Sun system is illustrated in Fig. 2.2.5 from a geocentric point of view. The sphere representing the Earth includes a body-fixed (Earth-fixed) coordinate system, with the (x, y) axes in the equatorial plane and with the x axis coincident with the intersection of the Greenwich

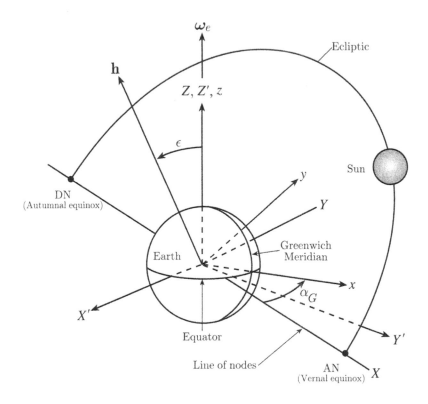

Figure 2.2.5: The Earth-Sun two-body motion. For convenience the motion can be described from the geocentric perspective to illustrate the relations between two-body parameters and astronomical conventions. The orbit plane defined by the angular momentum of the two-body system is the ecliptic, the ascending node is the vernal equinox, and the inclination is the obliquity of the ecliptic. The bodies are not drawn to scale since the diameter of the Sun is 100 times the diameter of the Earth. The (X', Y') axes have no preferred orientation, but the X axis is chosen to coincide with the vernal equinox.

meridian and the equator. The z axis of the Earth-fixed system coincides with the angular velocity vector of the Earth, ω_e. The nonrotating (X', Y') axes lie in the Earth's equator, but there is no preferred direction for these axes, except the Z' axis coincides with the angular velocity of the Earth in this model. The Earth-Sun orbit plane intersects the equator at the line of nodes. The point where the Sun moves from the southern hemisphere into the northern hemisphere is the ascending node, denoted by AN in Fig. 2.2.5 from a geocentric viewpoint. Similarly, the descending node is identified by DN.

The description given in the preceding paragraph and illustrated in Fig. 2.2.5 contains the essential descriptions of a well-defined and consistent celestial coordinate system. Since the two celestial bodies are modeled as rigid spheres with constant density, the node locations are fixed in space and the angular velocity vector, ω_e, of the rotating Earth is constant in magnitude and direction.

Special terms are assigned to the orbit parameters illustrated in Fig. 2.2.5. The orbit plane of the Sun about the Earth (or the Earth about the Sun) is referred to as the *ecliptic* and the inclination is the *obliquity of the ecliptic*, usually denoted by ϵ ($\simeq 23.5°$). The ascending node corresponds to the point known as the *vernal equinox* and the descending node is the *autumnal equinox*. The vernal equinox occurs about March 21 and the autumnal equinox occurs about September 21. The location of the Sun when it is 90° from the vernal equinox is the *summer solstice* and the location at 270° is the *winter solstice*, which occur about June 21 and December 21, respectively. It is somewhat prejudicial that these astronomical locations are identified with northern hemisphere seasons, since association with seasons is reversed in the southern hemisphere. Finally, the angle α_G defines the orientation of the Earth-fixed coordinate system (x, y, z) with respect to the vernal equinox (i.e., the x axis shown in Fig. 2.2.5). The angle α_G is known as the *Greenwich mean sidereal time* (GMST), defined by $\alpha_G = \omega_e(t - t_0) + \alpha_{G0}$. The period of Earth rotation with respect to a fixed direction in space is about 86,164 sec (23 hr 56 min 4 sec) (i.e., $\omega_e = 2\pi/86164$ rad/sec).

With the X axis directed toward the vernal equinox and with both the z and Z axes coincident with the angular velocity vector, ω_e, it follows that the relations between the nonrotating (X, Y, Z) axes and the Earth-fixed, rotating (x, y, z) axes as illustrated in Fig. 2.2.5 are

$$
\begin{aligned}
x &= X\cos\alpha_G + Y\sin\alpha_G \\
y &= -X\sin\alpha_G + Y\cos\alpha_G \\
z &= Z.
\end{aligned}
\tag{2.2.31}
$$

For the two-body problem that has been described up to this point, the orientation of the ecliptic defined in a consistent manner with the two-body dynamics will be fixed in space. Hence, the vernal equinox would be a fixed direction in space and the obliquity of the ecliptic would also be fixed. However, as described in Section 2.4, temporal variations exist in the location of the vernal equinox and in ϵ. With such variations, it is necessary to designate a specific epoch to be assigned to the equinox and obliquity. A commonly used reference is the *mean equator and equinox of 2000.0* (the vernal equinox on January 1, 2000, 12 hrs). The system defined for this date is referred to as the *J2000 system*. In some applications, the *M50 system* is used, which is defined by the *mean equator and equinox of*

1950.0. The nonrotating J2000 or M50 systems are sometimes referred to as *Earth-centered inertial* (ECI) and the Earth-fixed system (x, y, z) is referred to as Earth-centered, Earth-fixed (ECF or ECEF or ECF (Earth-centered, fixed)). We will use the notation J2000 to refer to the system and J2000.0 to refer to the date.

2.2.4   ORBIT ELEMENTS AND POSITION/VELOCITY

Assume that an artificial satellite orbits a planet whose mass, M_1, is known (i.e., $\mu = GM_1$ is known). Assume further that a technique and/or procedure has been utilized to determine the position and velocity of the satellite at a time, t_0, in a nonrotating coordinate system. The position and velocity at t_0 are

$$
\begin{aligned}
\mathbf{r}_0 &= X_0\mathbf{u}_X + Y_0\mathbf{u}_Y + Z_0\mathbf{u}_Z \\
\mathbf{V}_0 &= \dot{X}_0\mathbf{u}_X + \dot{Y}_0\mathbf{u}_Y + \dot{Z}_0\mathbf{u}_Z \ .
\end{aligned}
\tag{2.2.32}
$$

The angular momentum, $\mathbf{h}$, is

$$
\mathbf{h} = h_X\mathbf{u}_X + h_Y\mathbf{u}_Y + h_Z\mathbf{u}_Z
$$

where

$$
\begin{aligned}
h_X &= Y_0\dot{Z}_0 - \dot{Y}_0 Z_0 \\
h_Y &= Z_0\dot{X}_0 - \dot{Z}_0 X_0 \\
h_Z &= X_0\dot{Y}_0 - \dot{X}_0 Y_0 \ .
\end{aligned}
\tag{2.2.33}
$$

Also,

$$
h^2 = \mathbf{h} \cdot \mathbf{h}.
\tag{2.2.34}
$$

It follows that the inclination (i) and node location (Ω) can be determined from Eqs. (2.2.28), (2.2.29), (2.2.30), and (2.2.33).

The energy per unit mass can be determined from Eq. (2.2.7),

$$
\xi = \frac{V_0^2}{2} - \frac{\mu}{r_0}
$$

where

$$
V_0^2 = \mathbf{V}_0 \cdot \mathbf{V}_0
$$

and

$$
r_0 = [\mathbf{r}_0 \cdot \mathbf{r}_0]^{1/2} \ .
$$

From ξ, the semimajor axis is

$$
a = -\frac{\mu}{2\xi} \ .
$$

The eccentricity of the orbit is determined from Eq. (2.2.14),

$$e = \left[1 + \frac{2\xi h^2}{\mu^2}\right]^{1/2} .$$

The *semilatus rectum* is given by $p = h^2/\mu$ and the true anomaly is

$$\cos f = \frac{1}{r_0 e}[p - r_0]$$

$$\sin f = \frac{p}{he} \frac{\mathbf{r}_0 \cdot \dot{\mathbf{r}}_0}{r_0} .$$

The angle $(\omega + f)$ is

$$\cos(\omega + f) = \frac{X_0}{r_0} \cos \Omega + \frac{Y_0}{r_0} \sin \Omega$$

$$\sin(\omega + f) = \frac{Z_0}{r_0 \sin i}$$

where both cosine and sine expressions are required to properly determine the quadrant. The argument of perifocus, ω, can be determined from

$$\omega = (\omega + f) - f .$$

An alternate determination of the argument of perifocus can be obtained from the *eccentricity vector*, e, a vector directed toward perifocus with magnitude equal to the orbit eccentricity. The vector e is given by

$$\mathbf{e} = \mathbf{V}_0 \times \frac{\mathbf{h}}{\mu} - \frac{\mathbf{r}_0}{r_0} = e_X \mathbf{u}_X + e_Y \mathbf{u}_Y + e_Z \mathbf{u}_Z .$$

It can be shown that

$$\cos \omega = \frac{e_X \cos \Omega + e_Y \sin \Omega}{e}$$

$$\sin \omega = \frac{-\cos i \sin \Omega \, e_X + \cos i \cos \Omega \, e_Y + \sin i \, e_Z}{e} .$$

The eccentric anomaly, E_0, can be obtained from

$$\cos E_0 = \frac{r_0}{a} \cos f + e$$

$$\sin E_0 = \frac{r_0}{b} \sin f .$$

Furthermore, the mean anomaly, M_0, is given by

$$M_0 = E_0 - e \sin E_0$$

and the time of perifocus passage is

$$t_p = t_0 - \frac{M_0}{n} .$$

In summary, the position and velocity vectors at t_0 completely describe the two-body orbit, since they can be converted to an independent set of six orbit elements:

$$a$$
$$e$$
$$i$$
$$\Omega$$
$$\omega$$
$$t_p \text{ or } M_0 .$$

In other words, the position and velocity completely determine the nature of the two-body orbit, assuming that μ is known. Similarly, if the orbit elements are given, there are corresponding position and velocity vectors at a specified time.

Example 2.2.4.1

The ECI position and velocity of a Space Shuttle has been determined to be the following:

$$
\begin{aligned}
X &= & 5492000.34 & \quad \text{m} \\
Y &= & 3984001.40 & \quad \text{m} \\
Z &= & 2955.81 & \quad \text{m} \\
\dot{X} &= & -3931.046491 & \quad \text{m/sec} \\
\dot{Y} &= & 5498.676921 & \quad \text{m/sec} \\
\dot{Z} &= & 3665.980697 & \quad \text{m/sec}
\end{aligned}
$$

With $\mu = 3.9860044 \times 10^{14} \text{m}^3/\text{s}^2$ and the preceding equations, it follows that the corresponding orbit elements are

$$
\begin{aligned}
a &= & 6828973.232519\,m \\
e &= & 0.0090173388450585 \\
i &= & 28.474011884869° \\
\Omega &= & 35.911822759495° \\
\omega &= & -44.55584705279° \\
M_0 &= & 43.8860381032208°
\end{aligned}
$$

Also,

$$
\begin{aligned}
Period &= & 5616.2198\,\text{sec} \\
r_p &= & 6767394.07\,\text{m} \\
r_a &= & 6890552.40\,\text{m} \\
f &= & 44.608202°
\end{aligned}
$$

Example 2.2.4.2

A two-line element (TLE) format is commonly used by NORAD and NASA. An example for NOAA 14 is

```
NOAA 14
1 23455U  94089A  97229.90474114 .00000115  00000-0  88111-4  0 1530
2 23455 98.9964 181.3428 0010013 113.9737 246.2483 14.11685823135657
```

The interpretation of the TLE format is given in Table 2.2.3. Some components of the TLE pertain to the perturbing effects of additional forces, described in Section 2.3. Updated TLE sets can be found on the World Wide Web.

Convert the TLE for NOAA 14 into position and velocity at the epoch contained in the format. From the TLE set, the orbit elements on Day 229.90474114 of 1997 are

$$
\begin{aligned}
n &= 014.11685823 \text{ rev/day} \,(a = 7231745.57\text{m}) \\
e &= 0.0010013 \\
i &= 98.9964° \\
\Omega &= 181.3428° \\
\omega &= 113.9737° \\
M_0 &= 246.2483°
\end{aligned}
$$

It follows that

$$
\begin{aligned}
X &= -7232720.490 \text{ m} & \dot{X} &= -5.243469 \text{ m/s} \\
Y &= -167227.700 \text{ m} & \dot{Y} &= 1160.655450 \text{ m/s} \\
Z &= 14595.566 \text{ m} & \dot{Z} &= 7329.834189 \text{ m/s}
\end{aligned}
$$

2.2.5 POSITION/VELOCITY PREDICTION

With a specified set of orbit elements, the position and velocity at time t can be determined by solving Kepler's Equation for eccentric anomaly and determining true anomaly. From the orbit elements and the true anomaly at time t, the position and velocity are

$$
\begin{aligned}
X &= r \cos f Q_{11} + r \sin f Q_{12} \\
Y &= r \cos f Q_{21} + r \sin f Q_{22} \\
Z &= r \cos f Q_{31} + r \sin f Q_{32}
\end{aligned}
\qquad (2.2.35)
$$

$$
\begin{aligned}
\dot{X} &= \dot{X}^* Q_{11} + \dot{Y}^* Q_{12} \\
\dot{Y} &= \dot{X}^* Q_{21} + \dot{Y}^* Q_{22} \\
\dot{Z} &= \dot{X}^* Q_{31} + \dot{Y}^* Q_{32}
\end{aligned}
\qquad (2.2.36)
$$

Table 2.2.3: NORAD/NASA TWO-LINE ELEMENTS (TLE)

Line 1	
Column	Description
01–01	Line Number of Element Data
03–07	Satellite Number
10–11	International Designator (Last two digits of launch year)
12–14	International Designator (Launch number of the year)
15–17	International Designator (Component of launch)
19–20	Epoch Year (Last two digits of year)
21–32	Epoch (Day of year and fractional portion of the day)
34–43	First Time Derivative of the Mean Motion or Ballistic Coefficient
45–52	Second Time Derivative of Mean Motion (decimal point assumed)
54–61	Drag term or radiation pressure coefficient (decimal point assumed)
63–63	Ephemeris type
65–68	Element number
69–69	Check sum (Modulo 10)
Line 2	
Column	Description
01–01	Line Number of Element Data
03–07	Satellite Number
09–16	Inclination [Degrees]
18–25	Right Ascension of the Ascending Node [Degrees]
27–33	Eccentricity (decimal point assumed)
35–42	Argument of Perigee [Degrees]
44–51	Mean Anomaly [Degrees]
53–63	Mean Motion [Revs per day]
64–68	Revolution number at epoch [Revs]
69–69	Check Sum (Modulo 10)

Columns not defined are blank.

where

$$\dot{X}^* = V_r \cos f - V_\theta \sin f$$
$$\dot{Y}^* = V_r \sin f + V_\theta \cos f$$
$$V_r = \dot{r} = \frac{he}{p} \sin f$$
$$V_\theta = \frac{h}{r}$$
$$Q_{11} = \cos \omega \cos \Omega - \sin \omega \sin \Omega \cos i$$
$$Q_{12} = -\sin \omega \cos \Omega - \cos \omega \sin \Omega \cos i$$
$$Q_{21} = \cos \omega \sin \Omega + \sin \omega \cos \Omega \cos i$$
$$Q_{22} = -\sin \omega \sin \Omega + \cos \omega \cos \Omega \cos i$$
$$Q_{31} = \sin \omega \sin i$$
$$Q_{32} = \cos \omega \sin i .$$

Example 2.2.5

The position and velocity of the Space Shuttle at a later time can be determined from the state given in Example 2.2.4.1. For example, at 30 minutes after the epoch used in Example 2.2.4.1, the eccentric anomaly is 159.628138°. It follows that the position and velocity at 30 minutes is

$$X = -5579681.52m$$
$$Y = 2729244.60m$$
$$Z = 2973901.72m$$
$$\dot{X} = 3921.809270m/s$$
$$\dot{Y} = 6300.799313m/s$$
$$\dot{Z} = 1520.178404m/s.$$

An example of an *ephemeris* of positions, or table of positions, for the interval 30 min to 34 min is

t (min)	X (m)	Y (m)	Z (m)
30	−5579681.52	2729244.60	2973901.72
32	−5999982.83	1951421.98	2765929.81
34	−6315097.41	1139386.52	2509466.97

SHUTTLE GROUND TRACK

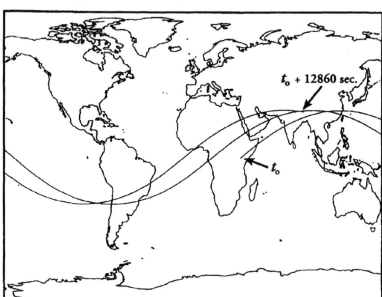

Figure 2.2.6: Shuttle ground track: the latitude-longitude ephemeris of Example 2.2.4.1. The latitude-longitude ephemeris is overlaid on a global map to illustrate the ground track of the satellite.

In some applications, the ephemeris is expressed in the body-fixed coordinate system. The transformation is given by Eq. (2.2.31), in which

$$\alpha_G = \omega_e(t - t_0) + \alpha_{G0}.$$

Assuming $t_0 = 0$ and $\alpha_{G0} = 0$, it follows that the preceding ephemeris expressed in ECF coordinates is:

t (min)	x (m)	y (m)	z (m)
30	−5174477.07	3436045.54	2973901.72
32	−5668947.18	2769635.28	2765929.81
34	−6076481.79	2062771.41	2509466.97

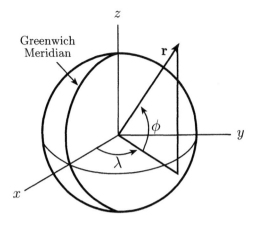

Figure 2.2.7: Definition of geocentric latitude (ϕ) and longitude (λ).

For other applications, these coordinates are further transformed into geocentric latitude (ϕ) and longitude (λ), illustrated in Fig. 2.2.7 and defined by

$$
\begin{aligned}
x &= r \cos \phi \cos \lambda \\
y &= r \cos \phi \sin \lambda \\
z &= r \sin \phi
\end{aligned}
\tag{2.2.37}
$$

or

$$
\sin \phi = \frac{z}{r} \qquad -90° \le \phi \le 90°
$$

$$
\cos \lambda = \frac{x}{r \cos \phi}
$$

$$
\sin \lambda = \frac{y}{r \cos \phi} \qquad 0 \le \lambda \le 360° \ .
$$

The altitude, h, is obtained from $h = r - r_e$, where r_e is the Earth radius. The geocentric latitude, longitude, and height ephemeris derived from the preceding ECF ephemeris for the interval is:

t(min)	ϕ(deg)	λ(deg)	h(m)
30	25.584	146.414	508495.95
32	23.672	153.962	510854.90
34	21.359	161.249	512151.92

Here, it has been assumed that the Earth is spherical with radius 6378137m. The latitude/longitude ephemeris can be plotted on a global or regional map to illustrate the path of the *subsatellite point*; that is, the point where the position vector of the satellite pierces the Earth's surface. The subsatellite locus of points produces a *ground track*. The ephemeris generated by the position and velocity in Example 2.2.4.1 for two orbital revolutions is illustrated in Fig. 2.2.6.

It is evident from Fig. 2.2.6 that, although Ω is constant, the longitude of the point where the satellite crosses the equator in a northward direction is not constant. The change in the *longitude of the ascending node* is caused solely by the rotation of the Earth in the two-body problem, but other factors contribute as well to the actual motion, as described in Section 2.3.

2.2.6 STATE TRANSITION MATRIX AND ERROR PROPAGATION

The initial state for a particular problem, such as the example state vectors in the Section 2.2.4 examples, can be expected to be in error at some level. If the error in position, for example, is $\Delta \mathbf{r}(t_0)$, this error will produce an error $\Delta \mathbf{r}(t)$ at a later time t, as well as an error in velocity, $\Delta \dot{\mathbf{r}}(t)$. The propagation of the error at t_0 to an error at t can be approximately expressed by

$$
\begin{bmatrix} \Delta X \\ \Delta Y \\ \Delta Z \\ \Delta \dot{X} \\ \Delta \dot{Y} \\ \Delta \dot{Z} \end{bmatrix}
=
\begin{bmatrix}
\frac{\partial X}{\partial X_0} & \frac{\partial X}{\partial Y_0} & \frac{\partial X}{\partial Z_0} & \frac{\partial X}{\partial \dot{X}_0} & \frac{\partial X}{\partial \dot{Y}_0} & \frac{\partial X}{\partial \dot{Z}_0} \\
\frac{\partial Y}{\partial X_0} & \frac{\partial Y}{\partial Y_0} & \frac{\partial Y}{\partial Z_0} & \frac{\partial Y}{\partial \dot{X}_0} & \frac{\partial Y}{\partial \dot{Y}_0} & \frac{\partial Y}{\partial \dot{Z}_0} \\
\frac{\partial Z}{\partial X_0} & \frac{\partial Z}{\partial Y_0} & \frac{\partial Z}{\partial Z_0} & \frac{\partial Z}{\partial \dot{X}_0} & \frac{\partial Z}{\partial \dot{Y}_0} & \frac{\partial Z}{\partial \dot{Z}_0} \\
\frac{\partial \dot{X}}{\partial X_0} & \frac{\partial \dot{X}}{\partial Y_0} & \frac{\partial \dot{X}}{\partial Z_0} & \frac{\partial \dot{X}}{\partial \dot{X}_0} & \frac{\partial \dot{X}}{\partial \dot{Y}_0} & \frac{\partial \dot{X}}{\partial \dot{Z}_0} \\
\frac{\partial \dot{Y}}{\partial X_0} & \frac{\partial \dot{Y}}{\partial Y_0} & \frac{\partial \dot{Y}}{\partial Z_0} & \frac{\partial \dot{Y}}{\partial \dot{X}_0} & \frac{\partial \dot{Y}}{\partial \dot{Y}_0} & \frac{\partial \dot{Y}}{\partial \dot{Z}_0} \\
\frac{\partial \dot{Z}}{\partial X_0} & \frac{\partial \dot{Z}}{\partial Y_0} & \frac{\partial \dot{Z}}{\partial Z_0} & \frac{\partial \dot{Z}}{\partial \dot{X}_0} & \frac{\partial \dot{Z}}{\partial \dot{Y}_0} & \frac{\partial \dot{Z}}{\partial \dot{Z}_0}
\end{bmatrix}
\begin{bmatrix} \Delta X_0 \\ \Delta Y_0 \\ \Delta Z_0 \\ \Delta \dot{X}_0 \\ \Delta \dot{Y}_0 \\ \Delta \dot{Z}_0 \end{bmatrix}
\tag{2.2.38}
$$

where the matrix of partial derivatives is referred to as the *state transition matrix*, also known as the *matrizant*. Although they are complicated, expressions for the partial derivatives, such as $\frac{\partial X}{\partial X_0}$, can be formed for the problem of two bodies (e.g., Goodyear, 1965). A general approach to the generation of the state transition matrix is given in Chapter 4. The origin of Eq. (2.2.38) can readily be seen by noting that a differential change in X at time t is related to differential changes in position and velocity at t_0 by

$$
dX = \frac{\partial X}{\partial X_0} dX_0 + \frac{\partial X}{\partial Y_0} dY_0 + \frac{\partial X}{\partial Z_0} dZ_0 + \frac{\partial X}{\partial \dot{X}_0} d\dot{X}_0
$$
$$
+ \frac{\partial X}{\partial \dot{Y}_0} d\dot{Y}_0 + \frac{\partial X}{\partial \dot{Z}_0} d\dot{Z}_0 .
\tag{2.2.39}
$$

For finite changes, let ΔX replace dX, ΔX_0 replace dX_0, and so on; then Eq. (2.2.38) becomes the equation produced by Eq. (2.2.39) for ΔX.

Example 2.2.6.1

Assuming an initial error in Example 2.2.4.1 given by

$$
\begin{aligned}
\Delta X_0 &= & 1\ \text{m} \\
\Delta Y_0 &= & 2\ \text{m} \\
\Delta Z_0 &= & 3\ \text{m} \\
\Delta \dot{X}_0 &= & 0 \\
\Delta \dot{Y}_0 &= & 0 \\
\Delta \dot{Z}_0 &= & 0,
\end{aligned}
$$

the Goodyear (1965) formulation of the state transition matrix can be used to map this error to 30 min, which produces

$$
\begin{aligned}
\Delta X &= & 0.65\ \text{m} \\
\Delta Y &= & 13.77\ \text{m} \\
\Delta Z &= & 4.78\ \text{m} \\
\Delta \dot{X} &= & -0.009953\ \text{m/s} \\
\Delta \dot{Y} &= & 0.011421\ \text{m/s} \\
\Delta \dot{Z} &= & 0.005718\ \text{m/s} \ \ .
\end{aligned}
$$

Note that the error at t_0 propagated to 30 min can also be obtained by adding the above initial error to the state, which results in

$$
\begin{aligned}
X &= & 5492001.34\ \text{m} \\
Y &= & 3984003.40\ \text{m} \\
Z &= & 2958.81\ \text{m}
\end{aligned}
$$

and no change in velocity. Following the procedure in Section 2.2.5, the predicted state at 1800 sec is

$$
\begin{aligned}
X &= & -5579680.87\ \text{m} \\
Y &= & 2729258.37\ \text{m} \\
Z &= & 2973906.50\ \text{m} \\
\dot{X} &= & -3921.819223\ \text{m/s} \\
\dot{Y} &= & -6300.787892\ \text{m/s} \\
\dot{Z} &= & -1520.172686\ \text{m/s} \ \ .
\end{aligned}
$$

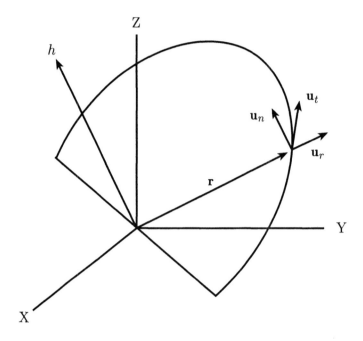

Figure 2.2.8: The RTN system: Radial (R), Transverse (T), and Normal (N).

Differencing this state with the predicted state in Section 2.2.5 produces the same error components obtained with the state transition matrix mapping. Since the differences are small in comparison to the total position magnitudes, the differential displacements in Eq. (2.2.38) are accurately approximated by the actual displacements. This aspect is discussed in Chapter 4 within the context of *linear approximations*.

For most applications, use of the (XYZ) coordinates to express the error is less revealing than other coordinate systems. An alternate system is shown in Fig. 2.2.8, in which unit vectors are defined in the radial (R) direction, transverse (T) direction, and normal (N) direction, referred to as the RTN system. To avoid ambiguity, these directions are specified by unit vectors $\mathbf{u}_r$, $\mathbf{u}_t$, and $\mathbf{u}_n$, which are defined by

$$\mathbf{r} = r\mathbf{u}_r$$
$$\mathbf{h} = h\mathbf{u}_n$$

and $\mathbf{u}_t$ completes the right-handed, orthogonal system.

A commonly used terminology for the RTN directions is: $\mathbf{u}_t$ may be referred to as the *along-track direction* and $\mathbf{u}_n$ may be referred to as the *cross-track direc-*

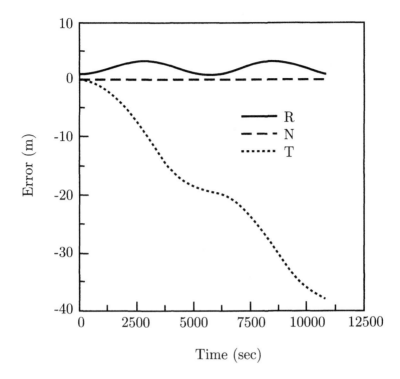

Figure 2.2.9: Orbit error evolution in RTN directions. A radial perturbation of 1 meter at $t = 0$ was applied to the state vector in Example 2.2.4.1. The plot illustrates the 1-m radially perturbed orbit minus the unperturbed orbit (Example 2.2.4.1).

tion. In some definitions, the along-track direction may refer to the direction of the velocity vector, which coincides with the $\mathbf{u}_t$ direction at every point only on a circular orbit. The ECI to RTN transformation is given by Eq. (4.16.21).

Example 2.2.6.2

To illustrate the nature of the orbit error, Fig. 2.2.9 shows the evolution of orbit error for two orbital revolutions resulting from a 1-meter error in the radial direction for the Shuttle described in Example 2.2.4.1. The 1-meter error in the radial component (increase in radial component) at the initial time produces the perturbed position of

$$
\begin{aligned}
X &= \quad 5492001.14945 \text{ m} \\
Y &= \quad 3984001.98719 \text{ m} \\
Z &= \quad\quad\ \ 2955.81044 \text{ m}
\end{aligned}
$$

and no change in velocity. Figure 2.2.9 shows the evolution of this 1-meter radial error in (RTN)-components, where the components are defined as the perturbed orbit position minus the reference orbit defined by the state vector in Example 2.2.4.1. With u_t defined by the reference orbit, the position on the perturbed orbit is about -40 m u_t at 11,000 sec; that is, it "trails" the reference orbit by about 40 m at this time. Since there is no out-of-plane perturbation of either position or velocity, the normal component is zero. Two significant characteristics are evident in the radial and transverse components. The transverse component exhibits a somewhat linear growth with time, or *secular* change. Furthermore, the existence of a *periodic* variation coinciding with the orbit period is evident in the radial and transverse components. This periodic variation is usually identified as a *once per revolution* variation. A more complete discussion is given by El´Yasberg (1967).

Further consideration of Fig. 2.2.9 explains the sign and periodicity in this figure. The increase in r with no change in initial velocity results in an increase in semimajor axis and an increase in orbital period. With the difference in orbital period, a satellite in the unperturbed orbit must move ahead of a satellite in the perturbed orbit. Since the transverse component is defined as perturbed minus unperturbed, the component will be negative. It appears linear because of the linear dependence in time exhibited by Kepler's Equation. The explanation of the periodic variation in this characteristic results from the difference in semimajor axis and eccentricity of the two orbits, which produces terms that are dependent on eccentric anomaly.

2.3 PERTURBED MOTION

The description of motion given in Section 2.2 is idealized. It has been obtained by assuming that only gravitational forces exist between two bodies, both of which have been further idealized by assuming them to be spheres that are gravitationally equivalent to point masses. The idealizations enable an analytical solution to the equations of motion and the motion can be interpreted as a simple geometrical figure (circle, ellipse, parabola, hyperbola). Because of the simple geometrical interpretation of the orbit, the idealized problem of two bodies provides a starting point for the interpretation of more realistic representations of the orbital motion.

Newton was well aware of the complexities surrounding the equations of motion that better represent the real world. He considered the gravitational problem of three bodies representing the Earth, Moon, and Sun, but he reportedly told his friend, the Royal Astronomer Halley, that the motion of the Moon "made his head ache and kept him awake so often that he would think of it no more" (Moulton, p. 363, 1914). The lunar problem (or the three-body problem) has no general, closed-form solution analogous to the conic section solution for the problem of two bodies (except for equilibrium solutions), but there are various ways of forming approximate solutions. Approximate analytical solutions use the problem of two bodies as a reference solution and expressions can be derived for the deviation, or perturbation, resulting from additional forces. These additional forces, that is, forces in addition to the point mass gravitation between two bodies, are referred to as *perturbing forces*. The analytical techniques used to obtain the approximate solutions are known as *general perturbations*.

With the advent of increasingly powerful digital computers, techniques that provide a numerical solution to the perturbed equations of motion have received increased attention as well. In these techniques, the equations of perturbed motion, represented by a set of ordinary differential equations, are solved numerically with specified initial conditions. These techniques are known as *special perturbations*. The procedures applied to the numerical integration of the equations of motion are described in most books on numerical analysis, but references such as Press *et al.* (1986) and Shampine and Gordon (1975) are good sources of applied information.

In the following sections, the contributions of various perturbing forces are summarized. These forces are separated into two categories: gravitational and nongravitational. The influence of these perturbing forces on the orbit is described in these sections.

2.3.1 CLASSICAL EXAMPLE: LUNAR PROBLEM

In spite of the notorious difficulty in solving the equations that describe the motion of the Moon, the problem can readily be solved by numerical integration. The results of the integration are instructive in describing the nature of perturbed motion and, additionally, the results are applicable to a further discussion of astronomical coordinate systems.

Following the approach used in the derivation of the equations of motion of two spheres given in Section 2.2, the equations of motion for three bodies that attract each other in accordance with Newtonian gravity can be derived in a similar manner. Let M_1, M_2, and M_3 represent the masses of three spheres, each with constant and uniform density. The equations of motion for each body can be expressed using an inertial coordinate system. But it is useful (and instructive) to describe the motion of two of the bodies with respect to the third body, say M_1. The resulting equations of motion for M_2 and M_3 are as follows, where the

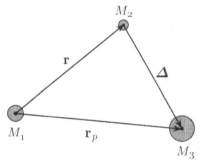

Figure 2.3.1: Vector definition for three bodies.

vectors are illustrated in Fig. 2.3.1:

$$\ddot{\mathbf{r}} = -\frac{\mu \mathbf{r}}{r^3} + GM_3 \left(\frac{\boldsymbol{\Delta}}{\Delta^3} - \frac{\mathbf{r}_p}{r_p^3} \right) \tag{2.3.1}$$

$$\ddot{\mathbf{r}}_p = -\frac{\mu' \mathbf{r}_p}{r_p^3} - GM_2 \left(\frac{\mathbf{r}}{r^3} + \frac{\boldsymbol{\Delta}}{\Delta^3} \right) \tag{2.3.2}$$

where $\mu = G(M_1 + M_2)$ and $\mu' = G(M_1 + M_3)$.

Let the bodies represent the Earth, Moon, and Sun, so that M_1 is the Earth's mass, M_2 is the mass of the Moon, and M_3 represents the Sun. This description is known as the *problem of three bodies* in celestial mechanics. The vector $\mathbf{r}$ represents the position vector of the Moon with respect to the Earth, $\boldsymbol{\Delta}$ is the position vector of the Sun with respect to the Moon, and $\mathbf{r}_p$ is the position vector of the Sun with respect to the Earth. Examination of the gravitational parameters of the Earth, Moon, and Sun as given in Appendix D shows that the mass of the Sun is more than 300,000 times greater than the Earth and the mass of the Moon is 81 times smaller than the Earth. An accurate approximation can be made in Eq. (2.3.2) by ignoring both the term with GM_2 and M_1 in μ'. With these approximations, the motion of the Sun about the Earth (or vice versa) is described by two-body dynamics. This approximation is known as the *restricted problem of three bodies*, described in more detail by Szebehely (1967).

Equations (2.3.1) and (2.3.2) can be integrated using a numerical method (e.g., see Shampine and Gordon, 1975) with specified initial conditions. Since the Sun is a dominant perturbation of the lunar motion, it is more appropriate to use a different orientation of the (X, Y, Z) axes. In this orientation, the Z axis is perpendicular to the ecliptic instead of the equator (as defined in Section 2.2.3), but the X axis is still directed to the vernal equinox. If the initial conditions in Table 2.3.1 are adopted, along with the mass parameters from Appendix D, the numer-

Table 2.3.1:
Ecliptic Position and Velocity of Sun and Moon
on January 1, 2000, 12 Hours

	MOON		
$X =$	-291608 km	$\dot{X} =$	0.643531 km/sec
$Y =$	-274979 km	$\dot{Y} =$	-0.730984 km/sec
$Z =$	36271 km	$\dot{Z} =$	-0.011506 km/sec

	SUN		
$X =$	26499000 km	$\dot{X} =$	29.794 km/sec
$Y =$	-144697300 km	$\dot{Y} =$	5.469 km/sec
$Z =$	0 km	$\dot{Z} =$	0 km/sec

ical integration of the equations for the Moon will produce an *ephemeris*, a table of time, position, and velocity. If the position and velocity vectors at each time are converted into orbit elements, referred to as *osculating elements*, using the equations in Section 2.2.4, the lunar inclination is illustrated in Fig. 2.3.2. The variation in the lunar ascending node is shown in Fig. 2.3.3 with respect to the vernal equinox over a period of one year. It is clearly evident from these figures that the orbit elements are not constant, which is the result of the influence of the gravitational perturbation of the Sun on the two bodies: Earth and Moon.

Figures 2.3.2 and 2.3.3 exhibit some important features of perturbed motion. The ascending node location in Fig. 2.3.3 exhibits a linear variation with time, referred to as a *secular variation*, as well as *periodic variations*. The inclination also shows periodic variations, but no apparent secular variation. Close examination of the periodic variations show the existence of approximately 180-day and 14-day periods, half the orbital period of the Sun about the Earth, and half the orbital period of the Moon, respectively. These periods appear in the plot for the node as well, but the 14-day periods are less evident. The secular node rate is negative, thus the change is referred to as a *regression of the node*, with a rate of about 19.4 degrees per year. The line of nodes of the Moon's orbit completes one revolution in 18.6 years, a fundamental period in the lunar motion that has been known since antiquity. Although less evident, other periodicities exist in the orbit element variations, including 28 days and 365 days, but the amplitudes of these terms are smaller and their effects are less evident in Figs. 2.3.2 and 2.3.3.

This example illustrates the influence that a perturbing force may have. An orbit element may exhibit secular variations, but it will also show periodic variations. The orbit elements that exhibit secular variations depend on the nature of the perturbing force, as do the specific periods associated with the periodic variations. The periodic variations are often loosely categorized as *short-period perturbations* or *long-period perturbations*. Depending on the nature of the perturbing forces, other periodicities may exist that do not fit these two categories, so other categories may be introduced to accommodate them. Furthermore, the period of a perturbation is not a definitive statement regarding how a perturbation should be categorized. For example, "short period" usually refers to perturbations in the frequency spectrum that are integer or near-integer multiples of the mean orbital angular rate. There are usually several specific frequencies that are categorized as short-period, but all are near the mean orbital angular rate.

2.3.2   VARIATION OF PARAMETERS

As illustrated in the previous section, the influence of a perturbing force on an orbit can be illustrated through time variations of the elements. The temporal variations illustrated for the Moon suggest the development of differential equations for the elements, analogous to the process known as *variation of parameters* in the solution of ordinary differential equations. The differential equations of motion for the perturbed satellite problem can be expressed as

$$\ddot{\mathbf{r}} = -\frac{\mu \mathbf{r}}{r^3} + \mathbf{f}, \tag{2.3.3}$$

where $\mathbf{r}$ is the position vector of the satellite center of mass expressed in an appropriate reference frame and $\mathbf{f}$ is the perturbing force. The perturbing force $\mathbf{f}$, or more correctly, the force per unit mass since $\mathbf{f}$ has acceleration units, can be further resolved into components in the RTN system (see Section 2.2.6),

$$\mathbf{f} = \hat{R}\mathbf{u}_r + \hat{T}\mathbf{u}_t + \hat{N}\mathbf{u}_n \tag{2.3.4}$$

where the unit vectors are defined by the RTN directions and $\hat{R}, \hat{T}, \hat{N}$ represent the force components. Note in Eq. (2.3.3) that if $\mathbf{f} = 0$, the resulting motion is two-body. Thus, $\mathbf{f}$ perturbs the two-body motion.

It can be shown that the equations of motion given by Eq. (2.3.3) can be equivalently expressed by a set of differential equations that describes the time rate of change of the orbit elements (Appendix D). For example, the rates of change of the node and inclination are

$$\frac{d\Omega}{dt} = \frac{r \sin(\omega + f)\hat{N}}{h \sin i} \tag{2.3.5}$$

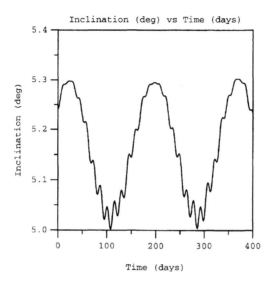

Figure 2.3.2: Time variation of the lunar inclination with respect to the ecliptic. Epoch: January 1, 2000, 12 hours.

$$\frac{di}{dt} = \frac{r\cos(\omega + f)\hat{N}}{h} \ .$$

The solutions to the full set of equations give the six orbit elements as a function of time, from which position and velocity can be obtained from the equations in Section 2.2.4.

In some cases, the perturbing force, $\mathbf{f}$, is derivable from a potential or disturbing function, D. The equations for time variation of the orbit elements can be also expressed as a function of D, as given in Appendix C. For example, the node and inclination are

$$\frac{d\Omega}{dt} = \frac{1}{h\sin i}\frac{\partial D}{\partial i} \tag{2.3.6}$$

$$\frac{di}{dt} = \frac{1}{h\sin i}\left[\cos i\frac{\partial D}{\partial \omega} - \frac{\partial D}{\partial \Omega}\right] \ .$$

The equations given by Eqs. (2.3.3), (2.3.5), and (2.3.6) are equivalent. That is, aside from errors introduced by the respective technique used to solve these equations, the solutions of the equations are identical for the same perturbing forces. Equation (2.3.6) is particularly useful in gaining insight into the nature of perturbed motion, whereas Eq. (2.3.3) is most often used for applications requiring high accuracy in the representation of $\mathbf{f}$.

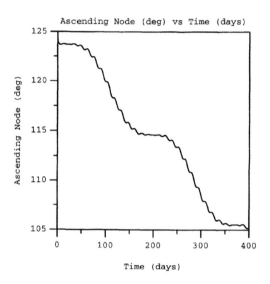

Figure 2.3.3: Time variation of the lunar node measured in the ecliptic. Epoch: January 1, 2000, 12 hours.

The perturbing force **f** in Eq. (2.3.3) can be categorized into two types of forces: gravitational and nongravitational. In the following sections, various sources that contribute to **f** are described.

2.3.3   GRAVITATIONAL PERTURBATIONS: MASS DISTRIBUTION

Consider the gravitational attraction between two point masses, M_1 and M_2, separated by a distance r. The gravitational potential between these two bodies, U, can be expressed as

$$U = \frac{GM_1M_2}{r}. \tag{2.3.7}$$

The gravitational force on M_2 resulting from the interaction with M_1 can be derived from the potential as the gradient of U; that is,

$$\mathbf{F} = \boldsymbol{\nabla}U = -\frac{GM_1M_2\mathbf{r}}{r^3} \tag{2.3.8}$$

where **r** is the position vector of M_2 with respect to M_1 given by

$$\mathbf{r} = x\mathbf{u}_x + y\mathbf{u}_y + z\mathbf{u}_z$$

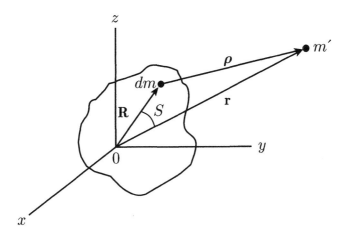

Figure 2.3.4: Definition of position vectors and differential mass for a body with arbitrary mass distribution.

and

$$\mathbf{\nabla} = \frac{\partial}{\partial x}\mathbf{u}_x + \frac{\partial}{\partial y}\mathbf{u}_y + \frac{\partial}{\partial z}\mathbf{u}_z \, .$$

If a body of total mass M has an arbitrary distribution of mass, it can be modeled as the collection of a large number of point masses. The gravitational potential sensed by a point mass, m', at an external location is

$$U = m' \int \int \int \frac{G\gamma\, dx\, dy\, dz}{\rho} \qquad (2.3.9)$$

where γ is the mass density associated with an element of mass dm, $dx\, dy\, dz$ is the differential volume, and ρ is the distance between the differential mass dm and the external mass m' (see Fig. 2.3.4). For convenience, the external mass is usually taken to be unity ($m' = 1$), and the integral expression is written in a more compact notation

$$U = \int_M \frac{G\, dm}{\rho} \qquad (2.3.10)$$

where the integral notation represents integration over the entire mass of the body. Equation (2.3.10) can be directly integrated for some bodies, such as a sphere of constant density γ. This integration (see Danby, 1988) shows that the constant density sphere is gravitationally equivalent to a point mass.

The position vector of m' with respect to the coordinate system (x, y, z) is $\mathbf{r}$. The location of the origin, O, and the orientation of (x, y, z) is arbitrary; however, for convenience, the (x, y, z) system is assumed to be body-fixed. Hence, the (x, y, z) axes are fixed in the body and rotate with it. For the Earth, as an example, the (x, y) plane would coincide with the equator, with the x axis also coinciding with the Greenwich meridian.

Equation (2.3.10) can be expanded into the infinite series

$$U = \frac{G}{r} \int_M \sum_{\ell=0}^{\infty} \left(\frac{R}{r}\right)^{\ell} P_{\ell}(\cos S)dm \qquad (2.3.11)$$

where R is the distance between the origin O and the differential mass dm, and P_{ℓ} is the Legendre polynomial of *degree* ℓ, with argument equal to the cosine of the angle between the two vectors, $\mathbf{R}$ and $\mathbf{r}$. Using the decomposition formula (Heiskanen and Moritz, 1967, p. 33), the Legendre polynomial can be expanded into spherical harmonics and terms that are dependent on the mass distribution collected into coefficients to yield

$$
\begin{aligned}
U &= \frac{\mu}{r} + U' \\[2mm]
U' &= -\frac{\mu^*}{r} \sum_{\ell=1}^{\infty} \left(\frac{a_e}{r}\right)^{\ell} P_{\ell}(\sin\phi)J_{\ell} \\[2mm]
&\quad + \frac{\mu^*}{r} \sum_{\ell=1}^{\infty} \sum_{m=1}^{\ell} \left(\frac{a_e}{r}\right)^{\ell} P_{\ell m}(\sin\phi)[C_{\ell m}\cos m\lambda + S_{\ell m}\sin m\lambda]
\end{aligned}
\qquad (2.3.12)
$$

where the coordinates of m' are now expressed with spherical coordinates (r, ϕ, λ). Scale factors involving a reference mass ($\mu^* = GM^*$) and a reference distance (a_e) have been introduced to nondimensionalize the mass property coefficients, $C_{\ell m}$ and $S_{\ell m}$. The decomposition formula has introduced Legendre's Associated Functions, $P_{\ell m}$, of degree ℓ and *order m*. The spherical coordinate ϕ represents a *geocentric latitude* and λ represents a *longitude angle*. The relations between the spherical coordinates and (x, y, z) are given by Eq. (2.2.37).

The coefficients J_{ℓ}, $C_{\ell m}$, and $S_{\ell m}$ are referred to as *spherical harmonic coefficients*, or *Stokes coefficients*, representing mass properties of the body. For example, J_{ℓ} is given by

$$J_{\ell} = -\left(\frac{1}{M^* a_e^{\ell}}\right) \int_M R^{\ell} P_{\ell}(\sin\phi')dm \qquad (2.3.13)$$

where ϕ' is the geocentric latitude of the differential mass dm, M^* is usually taken to be the mass of the body, and a_e is its mean equatorial radius. The Legendre polynomial P_{ℓ} equals $P_{\ell,0}$. The similar expressions for $C_{\ell m}$ and $S_{\ell m}$ are dependent on the longitude of the differential mass. Note that if the body is rigid and the (x, y, z) system is fixed in the body, then the J_{ℓ}, $C_{\ell m}$, and $S_{\ell m}$ coefficients will be constant. On the other hand, if (x, y, z) is not body-fixed or if temporal redistribution of mass takes place, the coefficients will be time-dependent.

The J_ℓ coefficients are referred to as *zonal harmonics*, which are related to $C_{\ell m}$ by the relation $J_\ell = -C_{\ell,0}$. The $C_{\ell m}$ and $S_{\ell m}$ terms are named according to the values of ℓ and m. If $\ell \neq m$, the $C_{\ell m}$ and $S_{\ell m}$ coefficients are referred to as *tesseral harmonics* and if $\ell = m$, they are referred to as *sectoral harmonics*. The terminology is derived from the fact that the J_ℓ terms are independent of longitude and the zeroes of P_ℓ can be visualized as dividing the body into *zones of latitude*. Similarly, when $\ell = m$, the coefficients divide the body into sectors defined by longitude. The number of sectors is determined by m. Additional descriptions are given by Heiskanen and Moritz (1967).

An alternate formulation for the gravitational potential uses *normalized* Legendre functions and coefficients. The Legendre functions for high degree acquire large numerical values and the mass coefficients are very small. The normalization process results in a set of mass coefficients, usually identified as $\overline{C}_{\ell m}$ and $\overline{S}_{\ell m}$, with the same order of magnitude over a wide range of ℓ and m. For the Earth, a set of mass coefficients is given in Appendix D along with the relations between normalized coefficients and conventional coefficients. Note that if normalized coefficients are used, a normalized set of Legendre functions must be used also.

Recursive expressions for the generation of both conventional and normalized expressions are given by Lundberg and Schutz (1988). The recursive expression for the conventional Legendre polynomial, $P_\ell(\sin \phi)$, is

$$P_\ell(\sin \phi) = \left[(2\ell - 1) \sin \phi P_{\ell-1}(\sin \phi) - (\ell - 1) P_{\ell-2}(\sin \phi) \right] / \ell$$

where $P_0(\sin \phi) = 1$ and $P_1(\sin \phi) = \sin \phi$. The recursions for the conventional Legendre Associated Functions are

$$P_{\ell,\ell}(\sin \phi) = (2\ell - 1) \cos \phi P_{\ell-1,\ell-1}(\sin \phi)$$

where $P_{1,1}(\sin \phi) = \cos \phi$. Furthermore, for $l \neq m$

$$P_{\ell,m}(\sin \phi) = \left[(2\ell - 1) \sin \phi P_{\ell-1,m}(\sin \phi) - (\ell + m - 1) P_{\ell-2,m} \right] / (\ell - m)$$

where $P_{i,j} = 0$ if $j > i$. In addition, note the following recursions:

$$\begin{aligned} \sin m\lambda &= 2 \cos \lambda \sin(m-1)\lambda - \sin(m-2)\lambda \\ \cos m\lambda &= 2 \cos \lambda \cos(m-1)\lambda - \cos(m-2)\lambda. \end{aligned}$$

It can be shown that the degree one terms (J_1, $C_{1,1}$, and $S_{1,1}$) are proportional to the distance that the origin O is offset from the center of mass of the body. If the origin O coincides with the center of mass, the degree one terms will be zero.

In the case of the Earth, the center of mass origin usually is referred to as the *geocenter*. Furthermore, it can be shown that the degree two terms are proportional to the moments and products of inertia of the body and their combinations:

$$J_2 = \frac{2C - B - A}{2Ma_e^2}$$

$$C_{2,1} = \frac{I_{xz}}{Ma_e^2} \quad S_{2,1} = \frac{I_{yz}}{Ma_e^2} \tag{2.3.14}$$

$$C_{2,2} = \frac{B - A}{4Ma_e^2} \quad S_{2,2} = \frac{I_{xy}}{2Ma_e^2}$$

where A and B are equatorial moments of inertia (x and y axes); C is the polar moment of inertia (z axis); I_{xy}, I_{xz}, and I_{yz} are the products of inertia; M is the mass of the body; and a_e is the mean radius.

$\mathbf{F}^*$, experienced by m', can be expressed as

$$\mathbf{F}^* = m'\nabla U \tag{2.3.15}$$

and the equations of motion for m' can be shown to be

$$\ddot{\mathbf{r}} = \left(1 + \frac{m'}{M}\right)\nabla U . \tag{2.3.16}$$

However, if m' represents an artificial satellite of the Earth, m'/M is very small, and this equation becomes

$$\ddot{\mathbf{r}} = \nabla U = -\frac{\mu\mathbf{r}}{r^3} + \mathbf{f}_{NS} \tag{2.3.17}$$

where $\mathbf{f}_{NS}$ denotes the contribution from the *nonspherical U'* terms; that is, $\nabla U'$. Integration of these equations, even with a numerical method, requires some careful consideration. The acceleration term, $\ddot{\mathbf{r}}$, can be expressed in either a body-fixed, rotating system (x, y, z) or a nonrotating system (X, Y, Z). However, the gravitational potential is expressed using spherical coordinates, presumed to be defined using a body-fixed coordinate system so that the mass property coefficients are constant (at least for a rigid body). Thus,

$$\nabla U = \frac{\partial U}{\partial r}\mathbf{u}_r + \frac{1}{r}\frac{\partial U}{\partial \phi}\mathbf{u}_\phi + \frac{1}{r\cos\phi}\frac{\partial U}{\partial \lambda}\mathbf{u}_\lambda . \tag{2.3.18}$$

Although Eq. (2.3.18) contains a singularity at the poles, which poses a problem for exactly polar orbits, alternate formulations exist that remove this singularity (e.g., Pines, 1973). If the body-fixed system is chosen to represent $\mathbf{r}$, then $\ddot{\mathbf{r}}$ will include the Coriolis, centripetal, tangential, and relative acceleration terms. Alternatively, expressing $\ddot{\mathbf{r}}$ in the nonrotating expression is simpler. Since the gradient of U given by Eq. (2.3.18) provides force components in spherical coordinates, a coordinate transformation matrix $T_{r\phi\lambda}^{xyz}$ is required to rotate the $(\mathbf{u}_r, \mathbf{u}_\phi, \mathbf{u}_\lambda)$ components of $\mathbf{f}$ into (x, y, z) components, as follows:

$$T_{r\phi\lambda}^{xyz} = \begin{bmatrix} \cos\phi\cos\lambda & -\sin\phi\cos\lambda & -\sin\lambda \\ \cos\phi\sin\lambda & -\sin\phi\sin\lambda & \cos\lambda \\ \sin\phi & \cos\phi & 0 \end{bmatrix} \tag{2.3.19}$$

where the latitude ϕ and longitude λ are related to (x, y, z) through Eq. (2.2.36). In addition, if $\ddot{\mathbf{r}}$ is expressed in the nonrotating system, then a further transformation, T_{xyz}^{XYZ}, is needed between the rotating axes (x, y, z) and the nonrotating axes (X, Y, Z). The details of this transformation are dependent on variations in both the magnitude and direction of the angular velocity vector of the rotating axes. For simplicity in this discussion, if the (x, y, z) axes rotate about the Z axis and the z axis is coincident with the Z axis, it follows that the transformation is

$$T_{xyz}^{XYZ} = \begin{bmatrix} \cos\alpha & -\sin\alpha & 0 \\ \sin\alpha & \cos\alpha & 0 \\ 0 & 0 & 1 \end{bmatrix} \tag{2.3.20}$$

where α represents the angle of rotation.

If the (x, y, z) axes are fixed in the Earth, with the x axis coinciding with the intersection between the Greenwich meridian and the equator and the X axis coincides with the vernal equinox, the angle α corresponds to α_G, described in Section 2.2.3. Thus, the contribution of the mass distribution of the central body to the perturbing force in Eq. (2.3.3) is expressed by

$$\mathbf{f}_{NS} = T_{xyz}^{XYZ} \, T_{r\phi\lambda}^{xyz} \, \nabla U' \tag{2.3.21}$$

assuming $\mathbf{r}$ is expressed in the nonrotating system (X, Y, Z) in the equations of motion. Hence, Eq. (2.3.17) becomes

$$\ddot{\mathbf{r}} = -\frac{\mu}{r^3}\mathbf{r} + T_{xyz}^{XYZ} \, T_{r\phi\lambda}^{xyz} \, \nabla U'. \tag{2.3.22}$$

The alternate expression for the equations of motion use the rotating system to express $\mathbf{r}$. In this case,

$$\begin{aligned} \ddot{\mathbf{r}}_{\text{Rel}} = \ & -\boldsymbol{\omega}_c \times (\boldsymbol{\omega}_c \times \mathbf{r}) - 2\boldsymbol{\omega}_c \times \dot{\mathbf{r}}_{\text{Rel}} - \dot{\boldsymbol{\omega}}_c \times \mathbf{r} \\ & -\frac{\mu\mathbf{r}}{r^3} + T_{r\phi\lambda}^{xyz}\nabla U' \end{aligned} \tag{2.3.23}$$

where ω_c represents the angular velocity vector of (x, y, z) with respect to (X, Y, Z), the position vector $\mathbf{r}$ is expressed in terms of (x, y, z) components, and "Rel" means that the time derivative does not include derivatives of the unit vectors.

If ω_c is constant such that $\omega_c = \omega_c \mathbf{u}_z$, it can be shown that an energy-like integral exists, known as the *Jacobi Integral*:

$$\frac{v^2}{2} - \frac{\omega_c^2(x^2 + y^2)}{2} - \frac{\mu}{r} - U' = K \qquad (2.3.24)$$

where K is the Jacobi Constant and v is the speed of the satellite with respect to the rotating system (x, y, z). Another form of this integral is given by Bond and Allman (1996), which is suitable for the nonrotating formulation:

$$\frac{V^2}{2} - \omega_c \cdot \mathbf{h} - \frac{\mu}{r} - U' = K \qquad (2.3.25)$$

where V is the ECI speed, ω_c and $\mathbf{h}$ are also expressed in the ECI coordinate system. The relation between ECI velocity ($\mathbf{V}$) and ECF velocity ($\mathbf{v}$) is

$$\mathbf{V} = \mathbf{v} + \omega_c \times \mathbf{r}$$

For the same U' and the same assumptions about reference frames, the various differential equations (Eqs. (2.3.22), (2.3.23)) and the equations in Appendix C for orbit elements give equivalent results. In practice, the solution method applied to the respective equations may produce a different error spectrum. The nonrotating coordinate system and the equations represented by Eq. (2.3.22) are widely used in precision orbit determination software.

2.3.4   GRAVITATIONAL PERTURBATIONS: OBLATENESS AND OTHER EFFECTS

Consider an ellipsoid of revolution with constant mass density. It is apparent that $C_{\ell m}$ and $S_{\ell m}$ will be zero for all nonzero m because there is no longitudinal dependency of the mass distribution. Furthermore, it can readily be shown that because of symmetry with respect to the equator there will be no odd-degree zonal harmonics. Hence, the gravitational potential of such a body will be represented by only the even-degree zonal harmonics. Furthermore, for a body like the Earth, more than 95% of the gravitational force (other than μ/r^2) will be derived from the degree-2 term, J_2.

Assuming that the entire gravitational potential for the ellipsoid of revolution can be represented by the J_2 term, it follows from Eq. (2.3.12) that

$$U' = -\frac{\mu}{r}\left(\frac{a_e}{r}\right)^2 J_2 P_2(\sin\phi) \qquad (2.3.26)$$

Table 2.3.2:

Node Motion as a Function of i $(J_2 > 0)$

Inclination	$\dot{\Omega}_s$	Terminology
Posigrade	Negative	Regression of Nodes
Retrograde	Positive	Progression of Nodes
Polar	Zero	

where $\mu^* = \mu$ and $P_2(\sin\phi) = (3\sin^2\phi - 1)/2$.

Noting that $\sin\phi = \sin i \sin(\omega + f)$ from spherical trigonometry, it can be shown that the gravitational potential can be expressed in terms of orbit elements with the help of the expansions in eccentricity given in Eq. (2.2.26). By ignoring terms of order e and higher, it follows that

$$U' = -\frac{\mu}{a}\left(\frac{a_e}{a}\right)^2 J_2\{3/4\sin^2 i[1 - \cos(2\omega + 2M)] - 1/2\}$$
$$+ \quad \text{higher order terms}.$$

To a first order approximation, the orbit elements have a small variation with time, except M, which is dominated by nt. If U' is separated into

$$U' = U_s + U_p \tag{2.3.27}$$

where

$$U_s = -\frac{GM}{a}\left(\frac{a_e}{a}\right)^2 J_2(3/4\sin^2 i - 1/2) \tag{2.3.28}$$

$$U_p = \frac{GM}{a}\left(\frac{a_e}{a}\right)^2 J_2\left(3/4\sin^2 i\cos(2\omega + 2M)\right) \tag{2.3.29}$$

it can be shown from Eq. (2.3.6) that

$$\dot{\Omega}_s \cong -\frac{3}{2}J_2 n\left(\frac{a_e}{a}\right)^2 \cos i \tag{2.3.30}$$

if the disturbing function D is taken to be U_s and terms with e are ignored. For convenience the notation $\dot{\Omega}_s$ has been introduced to identify the node rate caused by U_s. It can be further shown using U_s that a, e, and i do not change with time. Hence, for a given set of a, e, and i, $\dot{\Omega}_s$ will be constant, referred to as the *secular node rate*. Furthermore, it is evident that the sign of $\dot{\Omega}_s$ is determined by the inclination. Specifically, if J_2 is positive (oblate spheroid, such as Earth), then Table 2.3.2 summarizes the characteristics and terminology associated with various inclinations.

From $\dot{\Omega}_s$ = constant, it is evident that the ascending node location changes linearly with time at a constant rate, referred to as a *secular perturbation*. Similarly, the term U_p is responsible for *periodic perturbations*. It is evident from Eq. (2.3.30) that the magnitude of the node rate for a given inclination will increase as the altitude decreases.

An important application of the secular perturbation in the node is the *solar-synchronous* satellite. As the name implies, such a satellite is synchronized with the Sun, such that the line of nodes remains aligned with the Earth-Sun direction. This orbit design is adopted for various reasons; for example, lighting conditions on the Earth's surfaces. For a solar-synchronous satellite $\dot{\Omega}_s = 360°/365.25$ days $\cong 1°$/day. For a near-circular orbit with a semimajor axis of about 7000 km, the inclination would be about 98°.

A more complete expression with no eccentricity truncation for the expansion of the potential as a function of orbit elements can be found in Kaula (1966). This expansion enables the following expressions for the secular rates of Ω, ω, and M to be readily obtained:

$$\dot{\Omega}_s = -\frac{3}{2}J_2\frac{n}{(1-e^2)^2}\left(\frac{a_e}{a}\right)^2\cos i \tag{2.3.31}$$

$$\dot{\omega}_s = \frac{3}{4}J_2\frac{n}{(1-e^2)^2}\left(\frac{a_e}{a}\right)^2(5\cos^2 i - 1) \tag{2.3.32}$$

$$\dot{M}_s = \bar{n} + \frac{3}{4}J_2\frac{n}{(1-e^2)^{3/2}}\left(\frac{a_e}{a}\right)^2(3\cos^2 i - 1). \tag{2.3.33}$$

The values of a, e, and i in these expressions represent the *mean elements*; that is, the elements with periodic variations averaged out. In Eq. (2.3.33), $\bar{n} = \mu^{1/2}/\bar{a}^{3/2}$, where $\bar{a}$ is the mean value. Table 2.3.3 provides the characteristics and motion expressed by $\dot{\omega}_s$. As shown in this table, two inclinations exist where $\dot{\omega}_s = 0$, referred to as the *critical inclinations*.

Considering only the effect of U_s, the following orbit elements exhibit linear variations with time: Ω, ω, M. It can be shown that a, e, and i do not exhibit such change. As a consequence, a simple model of the perturbed motion is a *secularly precessing ellipse* (SPE). In this model, a, e, and i are constant, and Ω, ω, and M change linearly with time at the rates derived from U_s. With the SPE as a reference, the effect of U_p, Eq. (2.3.29), can be introduced to show that, to first order,

$$a(t) = \bar{a} + 3\bar{n}\,\bar{a}J_2\left(\frac{a_e}{\bar{a}}\right)^2\sin^2 i\frac{\cos(2\omega + 2M)}{2\dot{\omega}_s + 2\dot{M}_s} \tag{2.3.34}$$

Table 2.3.3:

Perigee Motion as a Function of Inclination ($J_2 > 0$)

Inclination i	$\dot{\omega}_s$	Terminology
$i < 63.435°$	Positive	Progression of perigee
$i = 63.435°$	Zero	Critical inclination
$63.435° < i < 116.656°$	Negative	Regression of perigee
$i = 116.565°$	Zero	Critical inclination
$i > 116.565$	Positive	Progression of perigee

where ($^-$) denotes the mean value of the element. Recall that this expression is approximate since terms with order e^1 and higher powers, as well as J_2^2, have been ignored. If the eccentricity is small, this expression is a good approximation. It should also be noted that mean elements can be replaced by osculating elements, except in the computation of $\bar{n}$ in the first term of $\dot{M}$, since terms of J_2^2 are ignored. Note that $\bar{n} = \mu^{1/2}/\bar{a}^{3/2}$. If J_2^2 terms are ignored, $\bar{a}$ may be computed from Eq. (2.3.34) by using an osculating value for $a(t)$ and in the second term for $\bar{a}$.

The preceding discussion illustrates the concepts of general perturbations; that is, an analytical solution of Lagranges planetary equations. A well-known solution for all the orbit elements is given by Brouwer for the J_2 problem (Appendix E). The equations in this appendix have truncated Brouwer's solution so that terms of order eJ_2 are ignored.

The linear solution of Kaula (1966) is an example of a solution that enables categorization of the influence of all gravity coefficients on the orbit perturbation. As shown by Kaula, the general frequency of a perturbation is given by

$$\dot{\psi}_{lmpq} = (l - 2p)\dot{\omega}_s + (l - 2p + q)\dot{M}_s + m(\dot{\Omega}_s - \omega_e) \qquad (2.3.35)$$

where p is an index that ranges from 0 to l, q is an index associated with a function that is dependent on the orbit eccentricity, and ω_e is the rotation rate of the primary body. The orbit element change with time is dependent on eccentricity and inclination functions. The eccentricity-dependent function has a leading term that is dependent on $e^{|q|}$; thus, for small eccentricity orbits, q is usually considered to range between -2 and $+2$. The characteristics of the orbit perturbations caused by individual degree and order terms in the gravity field are summarized in Table 2.3.4, where the categories are defined by the nature of $\dot{\psi}_{lmpq}$. Note

Table 2.3.4:

Perturbation Categories and Sources

Category	Source	Comments
Secular	Even-degree zonals	$l-2p=0;\ l-2p+q=0;\ m=0$
Long period	Odd-degree zonals	$l-2p\neq0;\ l-2p+q=0;\ m=0$
Short period	All gravity coefficients	$l-2p+q\neq0$
m-daily	Order m terms	$l-2p=0;\ l-2p+q=0;\ m\neq0$
Resonance	Order m terms	$(l-2p)\dot{\omega}+(l-2p+q)\dot{M}\cong-m(\dot{\Omega}-\omega_e)$

that the period of the perturbation is $2\pi/\dot{\psi}_{lmpq}$. An example of *resonance* is the geostationary satellite, where $\dot{M}=\omega_e$.

A special class of perturbed orbits, known as *frozen orbits*, have proven useful for some applications. For these orbits, even though the inclination is not critical, J_2 and J_3 interact to produce $\dot{\omega}_s=0$ (Cook, 1966; Born *et al.*, 1987) and the mean perigee remains fixed at about 90° or 270° in an average sense. Frozen orbits are useful in applications where it is desired that the mean altitude have small variations for a given latitude. They are required if it is necessary to have an exactly repeating ground track. Otherwise, the time to go between two latitudes will vary as the perigee rotates, and the ground tracks will not repeat.

The *nodal period* is the time required for a satellite to complete successive equatorial crossings, such as ascending node to ascending node. The nodal period, P_n, is

$$P_n = \frac{2\pi}{\dot{\omega}_s + \dot{M}_s} \tag{2.3.36}$$

or

$$P_n = \frac{2\pi}{\bar{n}}\left[1 - 3/2J_2\left(\frac{a_e}{\bar{a}}\right)^2\left(4\cos^2 i - 1\right)\right] \tag{2.3.37}$$

which is obtained using Eqs. (2.3.31) and (2.3.32). Using osculating elements, the nodal period for a near-circular orbit is

$$P_n = \frac{2\pi}{n}\left[1 - 3/4\,J_2\left(\frac{a_e}{a}\right)\right.$$

$$\left.\left\{1 + 5\cos^2 i - 6\sin^2 i\sin^2\left(\omega + M\right)\right\}\right]. \tag{2.3.38}$$

Note that either mean or osculating elements can be used in any term that is multiplied by J_2, since the approximation used in the analytic theory is accurate only to order J_2. However, unless a mean value of semimajor axis is used in Eq. (2.3.37), an error of order J_2 will be introduced in the lead term of the nodal period equation. Note also that Eq. (2.3.38) requires an osculating value for a to compute n.

Another useful concept in mission design is that of the nodal day, which is the time required for the Earth to make one revolution relative to the satellite's line of nodes. It is given by

$$D_n = \frac{2\Pi}{\dot{\Omega}_s + \omega_e} .$$

Hence, a nodal day will be greater than a siderial day for retrograde orbits and less than a siderial day for posigrade orbits.

Example 2.3.1

For the Space Shuttle example, Example 2.2.4.1, use the state vector as initial conditions for the numerical integration of the equations of motion given by Eqs. (2.3.17) and (2.3.26). For the Earth, $J_2 = 0.001082636$ (the normalized value $C_{2,0} = -0.0004841695$). The position-velocity ephemeris, converted into orbit elements, produces the results shown in Fig. 2.3.5 for a, e, i, Ω, and ω for a one-day interval. Evaluation of Eq. (2.3.31) for $\dot{\Omega}_s$ with the approximate mean orbit elements ($a = 6827$ km, $e = 0.008$, $i = 28.455$) gives $-6.91°$/day. The approximate slope of Ω in Fig. 2.3.5 is $-6.93°$/day. Note that a and i exhibit no secular rate (as expected), only a dominant twice-per-revolution change, as suggested by Eq. (2.3.34) for semimajor axis. Note that e exhibits no secular variation, although its periodic variations includes a combination of frequencies (see Appendix E).

2.3.5 GRAVITATIONAL PERTURBATIONS: THIRD-BODY EFFECTS

An artificial satellite of the Earth experiences the gravitational attraction of the Sun, the Moon, and to a lesser extent the planets and other bodies in the solar system. Equation (2.3.1), which was derived for the perturbed lunar motion based on the gravitational problem of three bodies, can be adapted to represent the gravitational effect of celestial bodies on the motion of a satellite. Since the description is based on the problem of three bodies, the perturbation is often referred to as the *third-body effect*, even when more than three bodies are considered. In some literature, the terminology *n-body effects* may be used to accommodate the case when many perturbing bodies are included. With the third-body effects, it

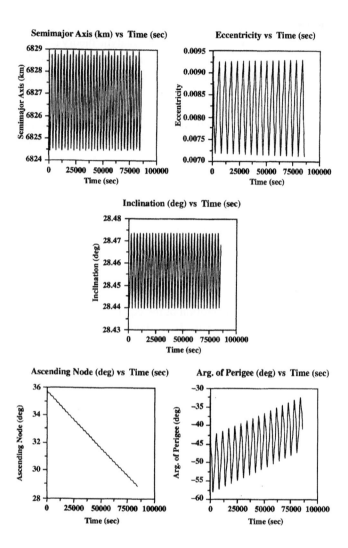

Figure 2.3.5: Orbit element variations produced by Earth oblateness, J_2. The mean anomaly is not shown since it is a straight line with a slope approximately equal to n. The ascending node and the argument of perigee exhibit significant linear variation, whereas a, e, and i do not. These results were obtained from numerical integration of the equations of motion.

can be shown that the perturbing force contribution to the equations of motion, Eq. (2.3.3), is

$$\mathbf{f}_{3B} = \sum_{j=1}^{n_p} \mu_j \left(\frac{\mathbf{\Delta}_j}{\Delta_j^3} - \frac{\mathbf{r}_j}{r_j^3} \right) \tag{2.3.39}$$

where n_p is the number of additional bodies, j represents a specific body (Sun, Moon, or planet), $\mu_j = GM_j$ is the gravitational parameter of body j, $\mathbf{\Delta}_j$ is the position vector of M_j with respect to the satellite, and $\mathbf{r}_j$ is the position vector of M_j with respect to the Earth (see Fig. 2.3.1). The first term $(\mathbf{\Delta}_j/\Delta_j^3)$ of the third-body perturbation is known as the *direct effect* and the second term $(\mathbf{r}_j/r_j^3)$ is the *indirect effect*. The notation for the latter term arises from the observation that it is independent of the artificial satellite coordinates. Since Eq. (2.3.39) describes the satellite motion with respect to the Earth center of mass, which is a noninertial point, the indirect term accounts for the inertial acceleration of the geocenter.

The third-body terms in Eq. (2.3.39) are based on the assumption that the satellite, Earth, and perturbing body (M_j) are point masses (or spheres of constant density). In reality, the Moon and Sun should be represented by a gravitational potential, such as Eq. (2.3.12). The satellite equations of motion will be complicated by the inclusion of these terms, but the accuracy required for most applications does not warrant such modeling of the luni-solar gravitational fields. For some high-accuracy applications, the dominant contribution is the interaction of the third body with the Earth oblateness through the indirect term, sometimes referred to as the *indirect oblateness*.

The nature of the third-body perturbation is similar to the Earth oblateness effect; namely, the Sun and Moon produce a secular variation in the satellite node, argument of perigee, and mean anomaly. Expressing the third-body perturbation with a gravitational potential and following a treatment similar to the approach used in Section 2.3.4, it can be shown that the secular node rate for an Earth satellite is

$$\dot{\Omega}_{\mathrm{sec}} = \frac{3}{4} \frac{\mu_j}{a_j^3} \frac{\cos i}{n} \left(\frac{3}{2} \sin^2 \epsilon - 1 \right) \tag{2.3.40}$$

where the 5° inclination of the Moon's orbit with respect to the ecliptic has been ignored, the subscript j identifies the Moon or the Sun, and ϵ is the obliquity of the ecliptic.

It is evident from Eq. (2.3.40) that the third-body effect on the nodal motion increases as the altitude increases, since n gets smaller with increasing semimajor axis. This characteristic is consistent with the expectation that the effect would increase as the satellite altitude approaches the source of the perturbation. Note the dependency of $\dot{\Omega}_{\mathrm{sec}}$ on $\cos i$, which is similar to the J_2 effect. Since the quantity in parentheses is negative, $\dot{\Omega}_{\mathrm{sec}}$ is retrograde for a posigrade orbit, for example.

In addition to the secular effect contributed by the Moon and Sun, significant periodic effects exist. The largest periodic effects have periods of 183 days and 14 days, caused by the Sun and Moon, respectively. The semiannual solar effects are greater than the lunar semimonthly effects. The coordinates of the Sun, Moon, and planets required for the third-body perturbation are available in numerical form (e.g., Standish *et al.*, 1997). These coordinates are expressed in the J2000.0 system.

2.3.6   GRAVITATIONAL PERTURBATIONS: TEMPORAL CHANGES IN GRAVITY

If the mass distribution within a body is constant over time in an appropriate coordinate system that is fixed in the body, the mass property coefficients, C_{lm} and S_{lm}, will be constant. Such is the case if the body is perfectly rigid. However, all celestial bodies deform at some level and this deformation redistributes mass so that the mass property coefficients vary with time. The ocean tides are well known and familiar mass-redistribution phenomena, but the solid Earth also responds to the same luni-solar gravitational forces that are responsible for the ocean tides. The previous formulation for the gravitational potential, Eq. (2.3.12), is valid even if the C_{lm} and S_{lm} coefficients are time dependent. The detailed formulation for the temporal changes in these coefficients produced by solid Earth and ocean tides is given by McCarthy (1996). Further discussion of the topic is given by Lambeck (1988).

The dominant tide effect appears in temporal changes of the degree two coefficients; namely,

$$
\begin{array}{llll}
\Delta J_2 & = & -K_T\, q_{2,0} & \\
\Delta C_{2,1} & = & K_T\, q_{2,1} & \quad \Delta S_{2,1} = K_T\, u_{2,1} \\
\Delta C_{2,2} & = & K_T\, q_{2,2} & \quad \Delta S_{2,2} = K_T\, u_{2,2}
\end{array}
\qquad (2.3.41)
$$

$$
K_T = k_2 \frac{a_e^3}{\mu}.
$$

Here, k_2 = second-degree Love Number (dependent on elastic properties of the Earth, with an approximate value of 0.3), thus:

$$
q_{2,0} = \sum_j C_T P_{2,0}(\sin \phi_j)
$$

$$
q_{2,1} = \frac{1}{3} \sum_j C_T P_{2,1}(\sin \phi_j) \cos \lambda_j
$$

$$q_{2,2} = \frac{1}{12}\sum_j C_T P_{2,2}(\sin\phi_j)\cos 2\lambda_j$$

$$u_{2,1} = \frac{1}{3}\sum_j C_T P_{2,1}(\sin\phi_j)\sin\lambda_j$$

$$u_{2,2} = \frac{1}{12}\sum_j C_T P_{2,2}(\sin\phi_j)\sin 2\lambda_j$$

$$C_T = \frac{\mu_j}{r_j^3}$$

where

j is determined by the Moon or Sun,

r_j is the distance from the geocenter to the center of mass of the Moon or Sun, and

$\phi_j \lambda_j$ are the geocentric latitude and longitude of the Moon and Sun, respectively.

The magnitude of ΔJ_2 is about 10^{-8}, compared to the J_2 magnitude of 10^{-3}. Care should be taken with applying ΔJ_2, since the average value produced by Eq. (2.3.41) is nonzero and this average may already be accommodated into J_2.

Other phenomena contribute to temporal changes in gravity. Among these are meteorologically driven redistributions of atmospheric mass as well as redistribution of solid Earth and ocean mass in response to changes in the magnitude and direction of the Earth's angular velocity vector. The latter phenomena are referred to as *pole tides*.

2.3.7 GRAVITATIONAL PERTURBATIONS: GENERAL RELATIVITY

Although the preceding developments are well served by the foundations of Newtonian mechanics, some high-accuracy applications will require additional terms introduced by relativistic effects (Einstein, 1905; Huang *et al.* 1990; Ries *et al.*, 1991). The dominant dynamical effect from general relativity is a small, but observable, advance in the argument of perigee (Ries *et al.*, 1991). For example, a satellite such as BE-C ($a = 7500$ km, $e = 0.025$, $i = 41.2°$) experiences an

advance of 11 arcsec/yr in the location of perigee. The relativistic term that should
be included in the equations of motion is

$$\mathbf{f}_g = \frac{\mu}{c^2 r^3} \left\{ \left[2(\beta + \gamma)\frac{\mu}{r} - \gamma(\dot{\mathbf{r}} \cdot \dot{\mathbf{r}}) \right] \mathbf{r} + 2(1 + \gamma)(\mathbf{r} \cdot \dot{\mathbf{r}})\dot{\mathbf{r}} \right\} \qquad (2.3.42)$$

where

  c is the speed of light,

  $\mathbf{r}, \dot{\mathbf{r}}$ are the position and velocity vectors of the satellite in geocentric nonrotat-
      ing coordinates,

  β, γ are parameters in the isotropic Parameterized Post-Newtonian metric, which
      are exactly equal to one in General Relativity.

As previously noted, the dominant effect of this contribution is a secular motion
of perigee.

2.3.8 NONGRAVITATIONAL PERTURBATIONS: ATMOSPHERIC RESISTANCE

For a low-altitude Earth satellite, atmospheric *drag* can be a dominant perturbing force. The drag force is derived from aerodynamic considerations and the
representation of the drag force on an orbiting vehicle is the same as an atmospheric flight vehicle:

$$\mathbf{F}_{\text{Drag}} = -1/2\rho C_D A V_r \mathbf{V}_r$$

where ρ is the atmospheric density, C_D is the *drag coefficient*, $\mathbf{V}_r$ is the vehicle
velocity relative to the atmosphere, A is the satellite cross-sectional area and the
minus sign denotes the direction opposite to $\mathbf{V}_r$. When this force is included in
the derivation of the equations of satellite motion with respect to the geocenter,
the following perturbing force per unit mass results, which will be added to the
other forces in Eq. (2.3.3):

$$\mathbf{f}_{\text{Drag}} = -1/2\rho \left(\frac{C_D A}{m} \right) V_r \mathbf{V}_r \qquad (2.3.43)$$

where the satellite mass, m, now appears as a divisor in the drag term. A commonly used parameter in the drag term is $(C_D A/m)$, sometimes referred to as the
ballistic coefficient. At low altitudes, such as those where the Space Shuttle operates (~ 350 km), the atmospheric density is about 10^{-11} that of sea level density.
Such low densities have mean molecular distances measured in meters. Nevertheless, atmospheric drag removes energy from the orbit and produces a secular

decay in semimajor axis and eccentricity. This decay eventually brings the satellite into the higher density regions of the atmosphere, and it may be destroyed by the heating produced from atmospheric friction generated by the high orbital speeds. Some satellites have been designed to accommodate the high heat loads produced by atmospheric entry, such as Space Shuttle, but most satellites will not survive intact.

The drag term in Eq. (2.3.43) is dependent on several factors. Most notably, the atmospheric density exhibits an approximately exponential reduction with increasing altitude, $\rho = \rho_0 e^{-\beta(h-h_0)}$, where ρ_0 is the density at a reference altitude h_0, h is the altitude of interest, and β is a scale parameter. But the specific density at any location will be influenced by the day-night cycle and the location of the subsolar point, solar activity (e.g., solar storms), geomagnetic activity, and seasonal variations. A satellite at a constant altitude can experience an order of magnitude change in density over an orbital revolution. Acting over many orbital revolutions, the decay in semimajor axis will determine the *lifetime* of a satellite. A very low-altitude satellite with a high ballistic coefficient will have a short lifetime, perhaps a few weeks or a few months in duration. Prediction of satellite lifetime is a challenging problem, but it is made particularly difficult by uncertainties in predicting atmospheric density and other factors. Various models of atmospheric density exist, such as those developed by Barlier, *et al.* (1978) and Jacchia (1977).

To account for the variations in drag produced by changes in atmospheric density, as well as the area-to-mass ratio (A/m), other parameterizations are introduced. These parameterizations may include piecewise constant values of C_D, which are defined over specific intervals of time, or more complex representations of C_D.

Some satellites have been designed to reduce the effects of atmospheric drag. Most notably, these are satellites where high-accuracy knowledge of the orbital position is required and it is desirable to minimize the influence of forces with significant uncertainty such as drag. The French Starlette satellite, launched in 1975, is spherical with a diameter of 14 cm, but it contains a core of depleted uranium to increase the mass. The area-to-mass ratio for Starlette is 0.001 m^2/kg. By contrast, the Echo balloon satellite had an A/m of about 10 m^2/kg. It is obvious that the balloon satellite will have a much shorter lifetime than Starlette at the same altitude.

If it is assumed that $V_r = V$, it can be shown (Danby, 1988) that

$$\frac{da}{dt} = -\left(\frac{C_D A}{m}\right)\frac{\rho a^2}{\mu}V^3 . \qquad (2.3.44)$$

Since V_r is almost entirely in the orbit plane, drag has only a very small out-of-plane force and, hence, there is no significant drag effect on the orbit node or inclination. In addition to the secular change in semimajor axis, e and ω also exhibit secular change because of drag.

2.3.9 NONGRAVITATIONAL PERTURBATIONS: RADIATION PRESSURE

For an Earth satellite, the Sun is the source for perturbations referred to as *solar radiation pressure*. This force results from the transfer of momentum through impact, reflection, absorption, and re-emission of photons. The Earth is an additional perturbation source, through both reflected solar radiation and emitted radiation, both of which exhibit considerable variation over an orbital revolution. The solar radiation pressure force contribution ($\mathbf{f}_{SRP}$) to Eq. (2.3.3) is given by

$$\mathbf{f}_{SRP} = -P\frac{\nu A}{m}C_R\mathbf{u} \qquad (2.3.45)$$

where P is the momentum flux from the Sun at the satellite location, A is the cross-sectional area of the satellite normal to the Sun, $\mathbf{u}$ is the unit vector pointing from the satellite to the Sun, C_R is the reflectivity coefficient with a value of approximately one, and ν is an eclipse factor such that $\nu = 0$ if the satellite is in the Earth's shadow (umbra), $\nu = 1$ if the satellite is in sunlight, and $0 < \nu < 1$ if the satellite is in partial shadow or penumbra. The passage of the satellite from sunlight to shadow is not abrupt, but the interval of time spent in partial shadow will be very brief for near-Earth satellites. In the vicinity of Earth, P is approximately 4.56×10^{-6} Newtons/m^2.

The modeling of the Earth radiation components is commonly performed with spherical harmonics to represent latitudinal and longitudinal variations of Earth albedo and emissivity. This force can be expressed using Eq. (2.3.45), except the momentum flux parameter would account for the albedo and emissivity and $\mathbf{u}$ would be directed toward the Earth. It is known from observations that the simple model given by Eq. (2.3.45) is inadequate for applications with high accuracy requirements. Alternate models for GPS are described by Fliegel *et al.* (1992) and Beutler *et al.* (1994).

2.3.10 NONGRAVITATIONAL PERTURBATIONS: OTHER

Other nongravitational forces exist that may influence the orbit in complex ways. Aside from forces that involve the magnetic field of the planet (e.g., if the satellite has a charge), there are other forces that have been studied because of their anomalous nature.

In addition to solar radiation pressure modeled by Eq. (2.3.45), heating of satellite components by the Sun and the Earth, as well as internal heat sources, will produce an unsymmetrical thermal distribution. Such a thermal imbalance will generate a *thermal force*, which may produce drag-like characteristics. The modeling of a spacecraft with more complex shapes, such as TOPEX/Poseidon and GPS, has required analysis by finite element methods for high-accuracy applications. A review of nongravitational force modeling is given by Ries *et al.* (1993). For purposes of this book, these forces will be covered by the term f_{Other} in the satellite equations of motion.

Satellite-dependent models have been developed to represent some nongravitational components. The GPS satellites experience a force parallel to the solar panel rotation axis, referred to as the *y-bias force*, thought to result from solar panel misalignment and/or thermal force effects (Vigue, *et al.*, 1994). index[aut]Schutz, B. E.

Additional forces exist, usually with short time duration, associated with mass expulsion from a spacecraft. The resulting force, or thrust, may be difficult to model because the time at which the expulsion occurred and magnitude of the event may be unknown.

For some applications, *empirical forces* are included in the equations of motion to accommodate forces of unknown origin or errors that result from mismodeling of forces. As described in this chapter, a typical error signature has a periodicity approximately equal to the orbit period; that is, a *once per revolution* characteristic. A convenient modeling takes the form of $A \cos nt + B \sin nt + C$, where n is the mean motion and A, B, C are empirical parameters to be estimated in the orbit determination process. The empirical model may be applied to the radial, transverse, and/or normal components.

2.3.11 PERTURBED EQUATIONS OF MOTION: SUMMARY

Based on Eq. (2.3.3), the satellite equations of motion are

$$\ddot{\mathbf{r}} = -\frac{\mu \mathbf{r}}{r^3} + \mathbf{f} \tag{2.3.46}$$

where $\mathbf{r}$ is the position vector of the satellite center of mass in an appropriate reference frame. The perturbing force $\mathbf{f}$ consists of the following

$$\mathbf{f} = \mathbf{f}_{NS} + \mathbf{f}_{3B} + \mathbf{f}_g + \mathbf{f}_{\text{Drag}} + \mathbf{f}_{\text{SRP}} + \mathbf{f}_{\text{ERP}} + \mathbf{f}_{\text{Other}} \tag{2.3.47}$$

where

$\mathbf{f}_{NS}$ is given by Eq. (2.3.21), including the tide contribution in Eq. (2.3.41),

$\mathbf{f}_{3B}$ is given by Eq. (2.3.39),

$\mathbf{f}_g$ represents the contributions of general relativity, Eq. (2.3.42),

$\mathbf{f}_{\text{Drag}}$ is given by Eq. (2.3.43),

f_{SRP} is given by Eq. (2.3.45),

f_{ERP} is Earth radiation pressure,

f_{Other} represents other forces, such as thermal effects, magnetically induced forces, and so on.

The various contributions to the equations of motion must use a consistent coordinate system. If, for example, the acceleration $\ddot{r}$ is expressed in J2000, all individual components of f must be expressed in the same system. As a result, some of these components may require coordinate transformations similar to those given in Eq. (2.3.21) for the nonspherical mass distribution effects. Advanced aspects of the relevant transformation between, for example, Earth-fixed and J2000 are discussed in the next section.

As noted previously, Eq. (2.3.46) can be solved with the use of a numerical method or through analytical approaches. Both introduce different levels of approximation. Various high-accuracy numerical programs have been developed in the community based on different numerical methods (single-step integrators, multistep integrators, etc.).

The advantage of speed offered by analytical techniques may be traded for force model accuracy (see Section 2.5). In these cases, a simplified geopotential consisting of a few zonal harmonics plus the effects of luni-solar gravity and drag may be used. NORAD, for example, has adopted such a general perturbation method, based on Brouwer's Theory (see Appendix E).

Example 2.3.11.1

The node rate of the GPS satellite known as PRN-5 was observed to have a dominant secular rate of -0.04109 deg/day during 1995 and 1996. We wish to determine the most significant factors that contribute to this rate. The following mean elements for the satellite also were observed:

$$
\begin{aligned}
a &= 26560.5 \text{ km} \\
e &= 0.0015 \\
i &= 54.5 \text{ deg}.
\end{aligned}
$$

The following contributions can readily be obtained:

J_2 Eq. (2.3.31)	-0.03927 deg/day
Moon Eq. (2.3.40)	-0.00097
Sun Eq. (2.3.40)	-0.00045
Total	-0.04069 deg/day

The difference between the observed value and this total is 0.0004 deg/day, or 1%, caused mainly by secular contributions from the other even-degree zonal harmonics and J_2^2.

2.4 COORDINATE SYSTEMS AND TIME: INTRODUCTION

Time and coordinate systems are an inherent component of the orbit determination problem. As described in Chapter 3, measurements of some aspect of a satellite's position or motion must be indexed in time. The time index of the measurements must be related to the time used in the equations of motion, Eq. (2.3.46), but different time systems are used, all of which attempt to represent the concept of Newtonian time. The relation between the various time systems must be known.

Measurements collected by some instrumentation located on the Earth's surface suggest the use of coordinate systems attached to the Earth at the location of the instrument. Such topocentric coordinate systems rotate since they are fixed to the planet's surface. On the other hand, the equations of motion use an inertial system. It is evident that transformation equations between the various coordinate systems are required.

The coordinate description given in Section 2.2.3, based on the problem of two bodies, assumed that the bodies were rigid spheres with constant density. In this model, the equinoxes remained fixed in direction, and the angular velocity vectors of the rotating bodies maintain the same direction in space. The following sections relax these conditions and examine the effects resulting from more realistic models of the celestial bodies. The resulting transformations and relationships are essential in the modern orbit determination problem. More complete discussions of the characteristics of the rotation of the Earth are given by Lambeck (1980) and Moritz and Mueller (1987).

2.4.1 PRECESSION AND NUTATION

Consider the motion of a spherical Moon orbiting the nonspherical Earth. Based on the discussion in Section 2.3, the gradient of the gravitational potential given by Eq. (2.3.12), multiplied by the mass of the Moon, yields the gravitational force experienced by the Moon. By Newton's Third Law, the Earth will experience an equal and opposite force; however, this force will not, in general, be directed through the center of mass of the Earth. As a consequence, the Earth will experience a gravitational torque, $\mathbf{T}$, given by

$$\mathbf{T} = -\mathbf{r}_p \times M_P \boldsymbol{\nabla}_P U' \qquad (2.4.1)$$

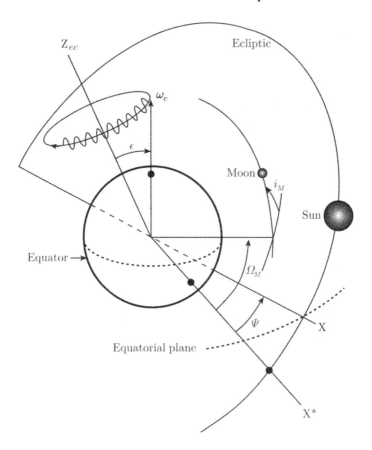

Figure 2.4.1: Precession and nutation. The Sun and Moon interact with the oblateness of the Earth to produce a westward motion of the vernal equinox along the ecliptic (Ψ) and oscillations of the obliquity of the ecliptic (ϵ).

where U' is the Earth's gravitational potential, ∇_P is the gradient operator with respect to the coordinates of the Moon ($\mathbf{r}_p$), and M_P is the mass of the Moon. The torque acting on the Earth produces changes in the inertial orientation of the Earth.

In the case of the Earth, both the Moon and the Sun are significant contributors to this gravitational torque. A dominant characteristic of the changes in inertial orientation is a conical motion of the Earth's angular velocity vector in inertial space, referred to as *precession*. Superimposed on this conical motion are small oscillations, referred to as *nutation*.

Consider the two angles in Fig. 2.4.1, Ψ and ϵ. The angle Ψ is measured in the ecliptic from a fixed direction in space, X^*, to the instantaneous intersection of the Earth's equator and the ecliptic. As discussed in Section 2.2.3, the intersection between the equator and the ecliptic where the Sun crosses the equator from the southern hemisphere into the northern hemisphere is the vernal equinox. For this discussion, the ecliptic is assumed to maintain a fixed orientation in space, although this is not precisely the case because of gravitational perturbations from the planets. The angle ϵ represents the angle between the angular velocity vector and an axis Z_{ec}, which is perpendicular to the ecliptic. The gravitational interaction between the Earth's mass distribution, predominantly represented by the oblateness term, J_2, and the Sun or Moon produces a change in Ψ, which can be shown to be

$$\dot{\Psi} = -\frac{3}{2} \left(\frac{\mu J_2 a_e^2}{C \omega_e} \right) \frac{M_p}{a_p^3} \cos \bar{\epsilon} \qquad (2.4.2)$$

where the subscript p represents the Sun or the Moon, the parameters in the parentheses refer to Earth, C is the polar moment of inertia of Earth, and $\bar{\epsilon}$ is the mean obliquity of the ecliptic. This motion, known as the *precession of the equinoxes*, has been known since antiquity. The combined effect of the Sun and the Moon produce $\dot{\Psi} = -50$ arcsec/yr; that is, a westward motion of the vernal equinox amounting to $1.4°$ in a century. Thus, the conical motion of ω_e about Z_{ec} requires about 26,000 years for one revolution.

The obliquity of the ecliptic, ϵ, is $\epsilon = \bar{\epsilon} + \Delta\epsilon$, where $\bar{\epsilon}$ represents the mean value, approximately $23.5°$. The nutations are represented by $\Delta\epsilon$, which exhibits periodicities of about 14 days (half the lunar period, or semimonthly) and 183 days (half the solar period, or semiannual), plus other long period variations. The $5°$ inclination of the Moon's orbit with respect to the ecliptic, i_M, produces a long period nutation given by:

$$\Delta\epsilon = \frac{3}{2} \left(\frac{\mu J_2 a_e^2}{C \omega_e} \right) \frac{M_M}{a_M^3} \cos i_M \sin i_M \cos \epsilon \frac{\cos(\Omega_M - \Psi)}{(\dot{\Omega}_M - \dot{\Psi})} \qquad (2.4.3)$$

where the subscript M refers to the Moon, and the amplitude of the coefficient of $\cos(\Omega_M - \Psi)$ in Eq. (2.4.3) is about 9 arcsec with a period of 18.6 years. The 9 arcsec amplitude of this term is known as the *constant of nutation*. The amplitudes of the semimonthly and semiannual terms are smaller than the constant of nutation, but all are important for high-precision applications.

The precession/nutation series for a rigid body model of the Earth, including the ecliptic variations, were derived by Woolard (1953). Extensions that include the effects of nonrigidity have been developed. Nevertheless, small discrepancies between observation and theory exist, believed to be associated with geophysical effects in the Earth's interior (Herring *et al.*, 1991; Wahr, 1981). Such observational evidence will lead, in time, to the adoption of new series representing the precession and nutation. The current 1980 International Astronomical Union

(IAU) series are available with the lunar, solar, and planetary ephemerides in an efficient numerical form based on results generated by the Jet Propulsion Laboratory (Standish *et al.*, 1997).

As described in Section 2.2.3, the J2000.0 mean equator and equinox specifies a specific point on the celestial sphere on January 1, 2000, 12 hrs. The *true of date* (TOD) designation identifies the *true* equator and equinox at a specified date. The J2000.0 equinox is fixed in space, but the location of the TOD equinox changes with time.

2.4.2 EARTH ROTATION AND TIME

In the simplified model of Earth rotation described in Section 2.2.3, the angular velocity vector, ω_e, is fixed in direction and magnitude. It has been known since the late 1800s that the direction of this vector does not remain fixed in direction with respect to an Earth-fixed frame of reference (Seidelmann, 1992). The vector describes a conical counter-clockwise motion, making a complete revolution in about 430 days (known as the Chandler period), but exhibiting an annual variation as well. The combination of Chandler and annual periods results in a six-year variation. Over a decadal time scale, this *polar motion* is confined to a region around the z axis with a diameter of less than 0.6 arcsec (or 18.5 m at the Earth's surface). The components of the angular velocity vector in the (x, y) plane are usually expressed in terms of angular components (x_p, y_p) using a left-handed coordinate system. The x axis of this system is directed along the Greenwich meridian and the y axis is directed toward 90° west. These components (x_p, y_p) provide the direction of ω_e but they cannot be predicted accurately over long periods of time, so they are determined from modern observational techniques (see Chapter 3) and reported by the International Earth Rotation Service (IERS).

In addition to changes that take place in the direction of ω_e, the magnitude of the vector changes with time. These changes in the magnitude are caused by transfer of angular momentum from atmospheric winds to the solid Earth and by tidal deformations that produce changes in the Earth's moments of inertia. A change in the angular velocity vector produces a corresponding change in the length of day (LOD), the period required for a complete revolution of the Earth with respect to the stars (but not the Sun). This change in LOD usually is expressed with a parameter using units of time, a form of Universal Time known as $\Delta(UT1)$. The variations in LOD, as well as $\Delta(UT1)$, are difficult to predict so they are observed regularly and reported by the IERS. In cases that rely on the use of satellite observations, the quantities x_p, y_p, and LOD may be estimated simultanelously with the satellite state.

Although $\Delta(UT1)$ can, in principle, be estimated with polar motion and satellite state, the separation of $\Delta(UT1)$ from the orbit node, Ω, complicates such estimation. Although satellite observations from satellites such as GPS and Lageos have been used to determine both Δ(UT1) and Ω over time scales of a few days, it should be noted that errors in modeling Ω will be absorbed in the estimation of Δ(UT1). Since modeling the satellite Ω over very long periods, such as semiannual and longer, is complicated by force model errors from, for example, radiation pressure, artificial satellite observations cannot provide reliable long-term estimates of Δ(UT1).

Time is expressed in different ways. An event can uniquely be identified by the calendar date and time of day, expressed in local time or Universal Time (discussed in a later paragraph). For example, the epoch *J2000.0* is January 1, 2000, 12 hours. The event may be identified by the *Julian Date* (JD), which is measured from 4713 b.c., but the day begins at noon. The epoch J2000.0 is represented by JD2451545.0. However, January 1, 2000, 0 hours (midnight) is JD 2451544.5. A *modified Julian date* (MJD) is represented by the JD minus 2400000.5 and the *Julian day number* is the integral part of the Julian date. Another commonly used expression of time is the *day of year* (DOY), in which January 1, 2000, would be day 1 of 2000 and February 1, 2000, would be day 32. The general format for the system is DOY followed by hours, minutes, and seconds past midnight.

Time is the independent variable of the equations of motion, Eq. (2.3.46), and the ability to relate the time tag associated with some observation of an aspect of satellite motion to the independent variable is essential for the problem of orbit determination. Observations such as the distance from an instrument to a satellite are described in Chapter 3, but each observation will be *time tagged* using a reference clock at the tracking station.

A clock requires the repeated recurrence of an event with a known interval. Up until the early part of the twentieth century, the rotation of the Earth served as a clock, but this use was abandoned when evidence mounted that the rotation rate was not uniform. In the modern context, clocks are a particular application of devices that oscillate with specific frequencies. By counting the number of cycles completed over a specified time interval, the oscillator becomes a clock. For example, crystal oscillators abound in modern life (e.g., wristwatches, computers, etc.) The reference frequency of crystal oscillators will change with temperature, for example. Hence, the frequency will exhibit temporal characteristics, usually represented by an offset from "true" time, a drift term and perhaps higher order terms, including stochastic components.

Modern atomic frequency standards, also used as clocks, are classified as rubidium, cesium, and hydrogen maser, given in the order of increasing accuracy and cost. In the case of a cesium frequency standard, a tunable quartz oscillator is used to excite the cesium at the natural frequency of 9,192,631,770 Hz. This frequency is used to define the System International (SI) second. Cesium clocks exhibit high stability, losing about 1 second in 1.5 million years. Hydrogen maser clocks exhibit about an order of magnitude even higher stability.

Various time systems are in use. For Earth satellite applications, the Terrestrial

Table 2.4.1:

Difference Between GPS-Time and UTC[a]

Date (00:00:01 UTC)			(GPS-T) minus UTC seconds
July	1,	1981	1
July	1,	1982	2
July	1,	1983	3
July	1,	1985	4
January	1,	1988	5
January	1,	1990	6
January	1,	1991	7
July	1,	1992	8
July	1,	1993	9
July	1,	1994	10
January	1,	1996	11
July	1,	1997	12
January	1,	1999	13

[a]Dates shown are leap second adjustments.

Dynamical Time (TDT) is used as the independent variable in the equations of motion, although Barycentric Dynamical Time (TDB) is used for solar system applications. Both TDT and TDB were related to International Atomic Time (TAI) at a specified epoch. TAI is a time scale based on cesium atomic clocks operated by numerous international agencies. In principle, by averaging over numerous atomic clocks, stochastic variations can be diminished and TAI will be a close approximation of a uniform time scale. TDT has come to be known as Terrestrial Time (TT).

Universal Time is the measure of time that is the theoretical basis for all civil time-keeping. Aside from the previously defined $\Delta(UT1)$, Coordinated Universal Time (UTC) is derived from TAI, where UT1 = UTC $+\Delta(UT1)$. UTC is maintained by the atomic clocks of various agencies, such as the National Institute of Standards and Time (NIST) and the U.S. Naval Observatory (USNO). The specific time systems are designated UTC(NIST) or UTC(USNO), for example. UTC(USNO) and TAI are based on the ensemble average of several cesium oscillators and hydrogen masers (about 30 cesium and 5–10 hydrogen masers in early 2000). The primary difference between UTC and TAI is the introduction of *leap seconds* to maintain synchronization between UTC and Δ(UT1) to within ± 0.9 seconds.

The difference between TAI and TT is, by definition, a constant. The transformation is given by

$$TT - TAI = 32.184 \text{ sec}. \tag{2.4.4}$$

The transformation between TDB and TT is a purely periodic function of the position and velocity of the Earth, Moon, planets, and Sun, a relativistic effect. The largest term in this transformation is an annual term with an amplitude of 1.658 milliseconds.

With the introduction of the Global Positioning System (GPS), another time system was introduced. To avoid discontinuities that occur with UTC, GPS-Time (GPS-T) is related to atomic time (maintained by USNO), but the leap second adjustments applied to UTC are not introduced in GPS-T. One of the hydrogen maser clocks at the USNO is used to provide the physical realization of both UTC(USNO) and GPS-T, which is determined from the ensemble average. This clock is steered to UTC(USNO) and is referred to as the *Master Clock* (MC). The MC is used for both UTC(USNO) and GPS-T. The offsets between GPS-T and UTC are shown in Table 2.4.1. As of January 1, 1999, and until the next leap second, TAI-UTC = 32 seconds.

In GPS operations, time is measured by *GPS-week* and time within the week, measured from Sunday midnight. The GPS-week is an integer with the initial week assigned a value of zero, corresponding to January 6, 1980. Since 10 bits were allocated in the GPS broadcast message to the GPS-week, the week number is limited to 1023, which occurred the week of August 15, 1999. After reaching the value of 1023, the week "rolled over" to zero for the week of August 22. Although an ambiguity exists with the uniqueness of the GPS-week as the result of the rollover, resolution of this ambiguity is not a problem for most applications.

References containing more detailed information include Seidelmann (1992) and McCarthy (1996). The latter reference documents the definitions, algorithms, models and constants adopted by the IERS. The document on the "IERS Conventions" is updated every few years.

2.4.3 EARTH-FIXED AND TOPOCENTRIC SYSTEMS

Since the Earth is not a rigid body, the concept of an Earth-fixed reference frame is not precisely defined. As previously noted, the Earth deforms in response to luni-solar gravity, which produces a redistribution of mass. The deformations associated with this redistribution, or tides, produce not only changes in gravity but changes in the coordinates of points on the Earth's surface. In addition, the Earth's crust consists of several moving tectonic plates. As a consequence, a relative motion exists between points on different plates. With these considerations, a true Earth-fixed reference frame does not exist. The selected reference frame must be consistent with the description of the Earth's rotational characteristics and deformation.

A Terrestrial Reference Frame (TRF) for precision orbit determination is "attached" to the Earth with an origin that is coincident with the center of mass. The (x, y, z) axes are orthogonal and the z axis of this coordinate system is approximately coincident with ω_e, but it cannot remain coincident because of polar motion. The x axis is usually chosen to be approximately coincident with the Greenwich meridian and the y axis is mutually perpendicular. Specific realizations of the TRF include the IERS TRF, or *ITRF*, and the TRF commonly used with many GPS applications, referred to as the WGS-84 (World Geodetic System) system. WGS-84 is closely aligned with the ITRF.

With the ITRF, for example, velocities of the observing sites are determined to account for plate motion, including local crustal deformations. To maintain a consistent TRF, these site velocities are determined within a process that defines the ITRF as well as the Earth rotation parameters. ITRF-97 (Boucher *et al.*, 1999) provides the coordinates and velocities for stations participating in the International Space Geodetic Network. The IERS Conventions (McCarthy, 1996) are a useful source of information for the effects of tides on station coordinates.

As defined in Section 2.3.3, spherical coordinates are used to describe the planetary gravitational potential. These spherical coordinates are related to the (x, y, z) coordinates by the following relations:

$$
\begin{aligned}
x &= r \cos \phi \cos \lambda \\
y &= r \cos \phi \sin \lambda \\
z &= r \sin \phi.
\end{aligned}
\tag{2.4.5}
$$

Here the angle ϕ is the geocentric latitude, λ is the longitude, and r is the magnitude of the position vector, $\mathbf{r}$. Since an ellipsoid of revolution is a better approximation to the physical topography of the Earth than a sphere, an alternate set of coordinates is used to describe positions on the surface of the ellipsoid, as illustrated in Fig. 2.4.2. This alternate set uses a *geodetic latitude*, ϕ', *longitude*, λ, and *height above the ellipsoid*, h. The relationships between these coordinates and (x, y, z) are

$$
\begin{aligned}
x &= (N_h + h) \cos \phi' \cos \lambda \\
y &= (N_h + h) \cos \phi' \sin \lambda
\end{aligned}
\tag{2.4.6}
$$

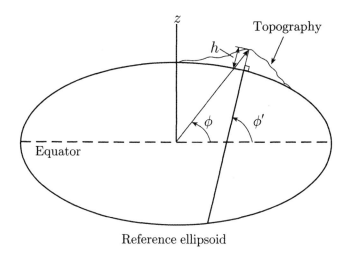

Reference ellipsoid

Figure 2.4.2: Geodetic coordinates. Coordinates on the surface of the Earth are usually defined in terms of geodetic latitude, ϕ', which differs from geocentric latitude, ϕ.

$$z = (N_h + h - \tilde{e}^2 N_h) \sin \phi'$$

where eccentricity of the elliptical cross-section of the ellipsoid is

$$\tilde{e}^2 = \tilde{f}(2 - \tilde{f}),$$

$$N_h = \frac{R_e}{(1 - \tilde{e}^2 \sin^2 \phi')^{1/2}} \qquad (2.4.7)$$

and $\tilde{f}$ represents the flattening parameter. The parameter $\tilde{f}$ is defined by

$$\tilde{f} = \frac{R_e - R_p}{R_e},$$

where R_e is the equatorial radius of the ellipsoid, and R_p is its polar radius. Another useful equation for the ellipsoid is

$$x^2 + y^2 + \left(\frac{R_e}{R_p}\right)^2 z^2 = R_e^2 \qquad (2.4.8)$$

where (x, y, z) are coordinates of a point on the ellipsoid surface. Further geodetic details can be found in Torge (1980). This equation can be a useful constraint to determine the location, for example, of the point where a space-borne instrument boresight pierces the ellipsoid for a specified spacecraft orientation in the ECF.

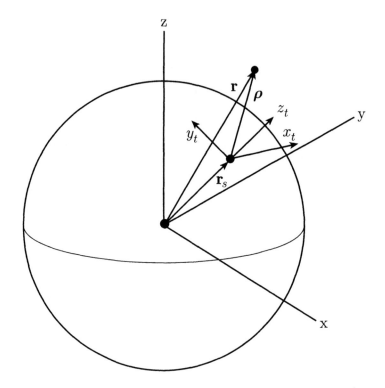

Figure 2.4.3: Topocentric coordinates. A useful description of satellite position can be made with topocentric coordinates. The origin is an observer's position on the surface of the Earth and the topocentric coordinates are defined with x_t directed eastward, y_t northward, and z_t is radial.

For some applications, a coordinate system attached to a point on the surface of the Earth is useful in the description of satellite motion from a ground-based observer's viewpoint. A topocentric coordinate system can be defined such that z_t is along the local vertical, x_t is eastward, and y_t is northward, as illustrated in Fig. 2.4.3. The position vector of a satellite, for example, in an Earth-fixed system, $\mathbf{r}$, can be expressed in the topocentric system as $\mathbf{r}_t$:

$$\mathbf{r}_t = T_t(\mathbf{r} - \mathbf{r}_s) = T_t \boldsymbol{\rho}$$

$$T_t = \begin{bmatrix} -\sin\lambda & \cos\lambda & 0 \\ -\sin\phi\cos\lambda & -\sin\phi\sin\lambda & \cos\phi \\ \cos\phi\cos\lambda & \cos\phi\sin\lambda & \sin\phi \end{bmatrix}. \qquad (2.4.9)$$

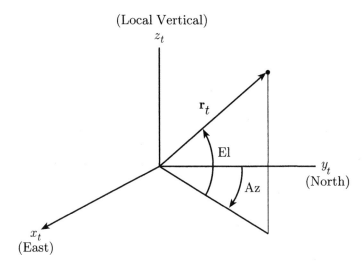

Figure 2.4.4: Azimuth and elevation. These angles are defined in a topocentric system with azimuth measured eastward from north and elevation measured above the plane tangent to the Earth at the observer's location.

Where geocentric latitude has been used for simplicity, $\mathbf{r}$ and $\mathbf{r}_s$ are the position vectors of the satellite, and the observer is expressed in the Earth-fixed system. With the components of satellite position expressed in the topocentric system, $(x, y, z)_t$, two commonly used angles can be introduced: *azimuth* (Az) and *elevation* (El). These angles are illustrated in Fig. 2.4.4 and can be determined from

$$\sin(\text{El}) = \frac{z_t}{r_t} \qquad -90° \leq \text{El} \leq 90°$$

$$\sin(\text{Az}) = \frac{x_t}{r_{xy}} \qquad \qquad (2.4.10)$$
$$0 \leq \text{Az} < 360°.$$

$$\cos(\text{Az}) = \frac{y_t}{r_{xy}}$$

Where (Az) is measured from North in an eastward direction, (El) is measured above (or below) the $(x, y)_t$ plane (that is, the local tangent plane), and $r_{xy} = [x_t^2 + y_t^2]^{1/2}$. Both sin (Az) and cos (Az) are needed to determine the correct quadrant.

2.4.4   TRANSFORMATION BETWEEN ECF AND ECI

If the equations of motion given by Eqs. (2.3.46) and (2.3.47) are expressed in an ECI system, such as the J2000 system, the gravitational force given by Eq. (2.3.18) must be transformed from the Earth-fixed system used to describe the potential into the J2000 system. For the simplified case of simple ECF rotation about the body-fixed z axis, the necessary transformation was given in Eq. (2.3.20). In reality, with consideration given to precession, nutation, polar motion, and UT1, the transformation is considerably more complex. If T^{xyz}_{XYZ} represents the transformation matrix from J2000 coordinates to Earth-fixed, then

$$\begin{bmatrix} x \\ y \\ z \end{bmatrix}_{ECF} = T^{xyz}_{XYZ} \begin{bmatrix} X \\ Y \\ Z \end{bmatrix}_{ECI} \qquad (2.4.11)$$

where the vector components $(X, Y, Z)_{ECI}$ represent Earth-centered inertial coordinates (e.g., J2000) and $(x, y, z)_{ECF}$ represents the Earth-fixed components (e.g., ITRF). The transformation matrix can be further decomposed into

$$T^{xyz}_{XYZ} = WS'NP \qquad (2.4.12)$$

where the specific components of these matrices and additional details are given in Appendix H. The general purpose of these matrices is

P: applies precession, from a specified epoch to the current time,

N: applies nutation at the current time,

S′: applies the rotation to account for true sidereal time,

W: applies polar motion to align the z axis (true pole) with the pole of the ECF system.

2.5   ORBIT ACCURACY

Different conditions and considerations must be applied to a discussion about orbit accuracy. The term *orbit accuracy* may refer to the accuracy of the solution of the equations of motion with specified parameters in the model of the forces acting on the satellite. In this case, little consideration is given to the accuracy of the parameters in the equations of motion. Instead the focus is simply on the accuracy of the technique used to solve the nonlinear ordinary differential equations given by Eqs. (2.3.46) and (2.3.47). If a general perturbation technique is used, the concern is usually with the small parameters, such as eccentricity, that are used to expand the forces and facilitate the solution. On the other hand, if a special perturbation technique is used, the concern will be with the step size of the numerical method chosen to solve the equations of motion. In either case, the solution technique is an approximation and will always introduce some error in the solution. To clarify this aspect, the terminology *solution technique accuracy* refers to the error introduced in the solution of the equations of motion by the solution technique, with no consideration given to the accuracy of the parameters in the equations. The solution technique accuracy can be controlled at different levels. For example, a numerical integration technique can be controlled by the step size and order of the technique.

Since the errors introduced by the solution technique can be controlled at some level, they are seldom the dominant error source. The parameters in the force models as well as the modeling of the forces are usually the most significant error sources. For example, all parameters in the force model have been determined by some means, usually as part of the orbit determination process. As a consequence, all parameters in the equations of motion contain some level of error. Such errors are crucial in the comparison of the solution to the equations of motion to physical measurements of some characteristics of the motion. In some cases, the error source may be embedded in the force modeling. In other words, the force model may be incorrect or incomplete. These errors will be described by the terminology *force model accuracy*.

In some cases, intentional mismodeling may be introduced to simplify the equations of motion. For example, the simple evaluation of the spherical harmonic terms in the gravitational force will require significantly greater computation time if the representation includes terms to degree and order 180 compared to degree and order 8. In this case, it may be acceptable to introduce a level of spherical harmonic truncation to reduce the computation time, though some error will be introduced into the solution by the omitted terms.

The most important consideration for the orbit accuracy is the specification of a *requirement* for the accuracy in each application. The requirement may be motivated by a wide variety of considerations, such as sensor accuracy, but the requirement will determine the selection of parameters for the solution technique, the accuracy of the force model parameters, the selection of force models, and the selection of a solution technique. The equations of motion are distinctly different if the orbit of a satellite must be known (and hence *determined*) with centimeter accuracy versus kilometer accuracy. Depending on these requirements, intentional model simplifications may be introduced. Identification of the accuracy to which the orbit must be determined is a vital aspect of the orbit determination problem.

2.6 REFERENCES

Barlier, F., C. Berger, J. Falin, G. Kockarts, and G. Thuiller, "A thermospheric model based on satellite drag data," *Annales de Géophysique*, Vol. 34, No. 1, pp. 9–24, 1978.

Battin, R., *An Introduction to the Mathematics and Methods of Astrodynamics*, American Institute of Aeronautics and Astronautics, Reston, VA, 1999.

Beutler, G., E. Brockmann, W. Gurtner, U. Hugentobler, L. Mervart, M. Rothacher, and A. Verdun, "Extended orbit modeling techniques at the CODE Processing Center of the International GPS Service for Geodynamics (IGS): theory and initial results," *Manuscripta Geodaetica*, Vol. 19, No. 6, pp. 367–385, April 1994.

Bond, V., and M. Allman, *Modern Astrodynamics*, Princeton University Press, Princeton, NJ, 1996.

Born, G., J. Mitchell, and G. Hegler, "GEOSAT ERM – mission design," *J. Astronaut. Sci.*, Vol. 35, No. 2, pp. 119–134, April–June 1987.

Boucher, C., Z. Altamini, and P. Sillard, *The International Terrestrial Reference Frame (ITRF97)*, IERS Technical Note 27, International Earth Rotation Service, Observatoire de Paris, May 1999.

Brouwer, D., "Solutions of the problem of artificial satellite theory without drag," *Astron. J.*, Vol. 64, No. 9, pp. 378–397, November 1959.

Brouwer, D., and G. Clemence, *Methods of Celestial Mechanics*, Academic Press, New York, 1961.

Chobotov, V. (ed.), *Orbital Mechanics*, American Institute of Aeronautics and Astronautics, Inc., Reston, VA, 1996.

Cook, G. E., "Perturbations of near-circular orbits by the Earth's gravitational potential," *Planetary and Space Science*, Vol. 14, No. 5, pp. 433–444, May 1966.

Danby, J. M. A., *Fundamentals of Celestial Mechanics*, Willmann-Bell, Inc., Richmond, VA, 1988.

Einstein, A., "Zur Elektrodynamik bewegter Körper," *Annalen der Physik*, Vol. 17, No. 10, pp. 891–921, 1905 (translated to English: Perrett, W., and G. Jeffery, *The Principle of Relativity*, Methuen and Co., 1923; republished by Dover, New York).

El'Yasberg, P. E., *Introduction to the Theory of Flight of Artificial Earth Satellites*, translated from Russian, Israel Program for Scientific Translations, 1967.

Fliegel, H., T. Gallini, and E. Swift, "Global Positioning System radiation force models for geodetic applications," *J. Geophys. Res.*, Vol. 97, No. B1, 559–568, January 10, 1992.

Goodyear, W. H., "Completely general closed form solution for coordinates and partial derivatives of the two-body problem," *Astron. J.*, Vol. 70, No. 3, pp. 189–192, April 1965.

Heiskanen, W., and H. Moritz, *Physical Geodesy*, W. H. Freeman and Co., San Francisco, 1967.

Herring, T., B. Buffett, P. Mathews, and I. Shapiro, "Free nutations of the Earth: influence of inner core dynamics," *J. Geophys. Res.*, Vol. 96, No. B5, pp. 8259–8273, May 10, 1991.

Huang, C., J. C. Ries, B. Tapley, and M. Watkins, "Relativistic effects for near-Earth satellite orbit determination," *Celest. Mech.*, Vol. 48, No. 2, pp. 167–185, 1990.

Jacchia, L., *Thermospheric temperature, density and composition: new models*, Special Report 375, Smithsonian Astrophysical Observatory, Cambridge, MA, 1977.

Kaula, W. M., *Theory of Satellite Geodesy*, Blaisdell Publishing Co., Waltham, 1966 (republished by Dover, New York, 2000).

Kovalevsky, J., I. Mueller, and B. Kolaczek (eds.), *Reference Frames in Astronomy and Geophysics*, Kluwer Academic Publishers, Dordrecht, 1989.

Lambeck, K., *The Earth's Variable Rotation*, Cambridge University Press, Cambridge, 1980.

Lambeck, K., *Geophysical Geodesy*, Clarendon Press, Oxford, 1988.

Lundberg, J., and B. Schutz, "Recursion formulas of Legendre functions for use with nonsingular geopotential models," *J. Guid. Cont. Dyn.*, Vol. 11, No. 1, pp. 31–38, January–February 1988.

McCarthy, D. (ed.), *IERS Conventions (1996)*, IERS Technical Note 21, International Earth Rotation Service, Observatoire de Paris, July 1996.

Moritz, H., and I. Mueller, *Earth Rotation: Theory and Observation*, Ungar Publishing Company, New York, 1987.

Moulton, F. R., *An Introduction to Celestial Mechanics*, MacMillan Co., New York, 1914.

Newton, I., *Philosophiae Naturalis Principia Mathematica*, 1687 (translated into English: A. Motte, 1729; revised by F. Cajori, University of California Press, 1962).

Pines, S., "Uniform representation of the gravitational potential and its derivatives," *AIAA J.*, Vol. 11, No. 11, pp. 1508–1511, November 1973.

Plummer, H. C., *An Introductory Treatise on Dynamical Astronomy*, Cambridge University Press, 1918 (republished by Dover Publications, New York, 1966).

Pollard, H., *Mathematical Introduction to Celestial Mechanics*, Prentice-Hall, Englewood Cliffs, NJ, 1966.

Press, W., B. Flannery, S. Teukolsky, and W. Vetterling, *Numerical Recipes*, Cambridge University Press, Cambridge, 1986.

Prussing, J., and B. Conway, *Orbit Mechanics*, Oxford University Press, New York, 1993.

Ries, J. C., C. K. Shum, and B. Tapley, "Surface force modeling for precision orbit determination," *Geophysical Monograph Series, Vol. 73*, A. Jones (ed.), American Geophysical Union, Washington, DC, 1993.

Ries, J. C., C. Huang, M. M. Watkins, and B. D. Tapley, "Orbit determination in the relativistic geocentric reference frame," *J. Astronaut. Sci.*, Vol. 39, No. 2, pp. 173–181, April–June 1991.

Roy, A. E., *Orbital Motion*, John Wiley & Sons Inc., New York, 1988.

Schaub, H., and J. Junkins, *Analytical Mechanics of Space Systems*, American Institute of Aeronautics and Astronautics, Reston, VA, 2003.

Seidelmann, P. K. (ed.), *Explanatory Supplement to the Astronomical Almanac*, University Science Books, Mill Valley, CA, 1992.

Shampine, L., and M. Gordon, *Computer Solution of Ordinary Differential Equations, The Initial Value Problem*, W. H. Freeman and Co., San Francisco, 1975.

Smart, W. M., *Celestial Mechanics*, John Wiley & Sons Inc., New York, 1961.

Standish, E. M., X. Newhall, J. Williams, and W. Folkner, *JPL Planetary and Lunar Ephemerides* (CD-ROM), Willmann-Bell, Inc., Richmond, VA, 1997.

Szebehely, V., *Theory of Orbits*, Academic Press, New York, 1967.

Szebehely, V., and H. Mark, *Adventures in Celestial Mechanics*, John Wiley & Sons, Inc., New York, 1998.

Torge, W., *Geodesy*, Walter de Gruyter, Berlin, 1980 (translated to English: Jekeli, C.).

Vallado, D., *Fundamentals of Astrodynamics and Applications*, Space Technology Library, Microcosm Press, El Segundo, CA, 2001.

Vigue, Y., B. Schutz, and P. Abusali, "Thermal force modeling for the Global Positioning System satellites using the finite element method," *J. Spacecr. Rockets*, Vol. 31, No. 5, pp. 866–859, 1994.

Wahr, J. M., "The forced nutations of an elliptical, rotating, elastic, and oceanless Earth," *Geophys. J. of Royal Astronom. Soc.*, 64, pp. 705–727, 1981.

Westfall, R., *Never at Rest: A Biography of Isaac Newton*, Cambridge University Press, Cambridge, 1980.

Woolard, E., *Theory of the rotation of the Earth around its center of mass*, Astronomical Papers—American Ephemeris and Nautical Almanac, Vol. XV, Part I, U.S. Naval Observatory, Washington, DC, 1953.

2.7 EXERCISES

1. By differentiating $\mathbf{r} = r\mathbf{u}_r$, derive Eqs. (2.2.8) and (2.2.9). Recall that the derivative of a unit vector in a rotating frame is given by $\dot{\mathbf{u}} = \boldsymbol{\omega} \times \mathbf{u}$, where ω is the angular velocity of the rotating frame. In this case, $\boldsymbol{\omega} = \dot{\theta}\mathbf{u}_h$.

2. Determine the orbital period of the following satellites:

    (a) Shuttle: $a = 6,778$ km

    (b) Earth-observing platform (EOS/Terra): $a = 7,083$ km

    (c) Geodetic satellite (Lageos): $a = 12,300$ km

3. Determine the semimajor axis from the period for the following:

    (a) Geosynchronous ($T = 86,164$ sec)

    (b) GPS ($T = 86,164/2$ sec)

4. Given the following position and velocity of a satellite expressed in a non-rotating, geocentric coordinate system:

	Position (m)	Velocity (m/sec)
X	7088580.789	−10.20544809
Y	−64.326	−522.85385193
Z	920.514	7482.07514112

Determine the six orbit elements (a, e, i, Ω, ω, M_0). Express the angles in degrees.

5. Given the position and velocity in Exercise 4 at $t = 0$, predict the position and velocity of the satellite at $t = 3,000$ sec, assuming two-body motion. Determine the radial and horizontal components of velocity and the flight path angle at $t = 3,000$ sec. The flight path angle is the angle in the orbit plane between the normal to the position vector and the velocity vector.

6. Determine the latitude and longitude of the subsatellite point for $t = 3,000$ sec in Exercise 5 if α_G at $t = 0$ is zero. Assume the Z axis of the nonrotating system is coincident with the z axis of the rotating system.

7. Determine the node rate and perigee rate due to J_2 of the Earth for the following satellites:

Satellite	Altitude	Eccentricity	Inclination
Mir	400 km	0	51 deg
EOS/AM	705 km	0	98 deg
GPS	20,000 km	0	55 deg
Geosync	36,000 km	0.1	63 deg

(Assume semimajor axis is altitude + mean radius of Earth.)

8. Determine the inclination required for a "solar-synchronous" satellite with an altitude of 600 km (add Earth radius to obtain semimajor axis) and zero eccentricity.

9. Determine the node rate caused by the Sun and Moon for the set of satellites in Exercise 7. Determine the individual luni-solar contributions and the combined effect.

10. In November, 1996, a payload known as Wakeshield (WSF-03) was deployed from Columbia on STS-80. This payload carried a GPS receiver. Using an on-board navigation method, the receiver measurements were processed into position in an Earth-centered, Earth-fixed coordinate system. The GPS-determined positions for selected times follow.

WSF-03 Earth-fixed Position Determined by GPS

WGS-84 Positions in Meters

November 24, 1996

GPS-T (hrs:min:sec)	00:00:0.000000	00:00:03.000000
x	3325396.441	3309747.175
y	5472597.483	5485240.159
z	-2057129.050	-2048664.333

November 25, 1996

GPS-T (hrs:min:sec)	00:00:0.000000	00:00:03.000000
x	4389882.255	4402505.030
y	-4444406.953	-4428002.728
z	-2508462.520	-2515303.456

(a) Use these positions to demonstrate that the WSF-03 node location is not fixed in space and determine an approximate rate of node change (degrees/day) from these positions. Compare the node rate with the value predicted by Eq. (2.3.31). (Hint: Determine the node location in the first 3 sec and the last 3 sec. Also, recall that for two-body dynamics, the position vectors at two times will define the orbit plane.)

(b) Determine the inclination of WSF-03 during the first 3-sec interval and the last 3-sec interval. (Hint: Assume two-body dynamics applies during these respective 3-sec intervals.)

Comment: The position vectors determined by GPS in this case are influenced at the 100-meter level by Selective Ability, but the error does not significantly affect this problem.

11. Assuming an Earth satellite is influenced by J_2 only, derive the equations of motion in nonrotating coordinates. Assume the nonrotating Z axis coincides with the Earth-fixed z axis.

12. Using the result of Exercise 11, numerically integrate the equations of motion for one day with the following initial conditions for a high-alititude, GLONASS-like satellite:

a	25500.0 km
e	0.0015
i	63 deg
Ω	-60 deg
ω	0 deg
M_0	0 deg

(a) During the integration, compute the Jacobi constant and the Z component of the angular momentum. Are these quantities constant?

(b) Plot the six orbit elements as a function of time.

(c) Identify features that are similar to and different from Fig. 2.3.5 for the Shuttle.

(d) Compare the node rate predicted by Eq. (2.3.31) with a value estimated from (b).

(e) Compare the amplitude of the semimajor axis periodic term with Eq. (2.3.34).

(f) Plot the ground track. Does the ground track repeat after one day?

13. Using Fig. 2.3.5, estimate the mean values of a, e, and i and the linear rates of node and perigee change. Compare the estimated node and perigee rates with the rates predicted by Eqs. (2.3.31) and (2.3.32).

14. For the orbit of the TOPEX/Poseidon, assume the ECF frame rotates with a constant rate ω_e about the z axis and the initial osculating elements are:

a	7712 km
e	0.0001
i	63.4°
Ω	135°
ω	90°
M_0	0

(a) Use the J_2 result of Exercise 11 to numerically integrate the equations of motion for one day.

(b) Compare the numerical results with results from the analytical theory in Appendix E. Note that errors in the analytical theory are of the order eJ_2 and J_2^2.

15. Using the gravitational potential in Eq. (2.3.12), derive

$$\frac{\partial U}{\partial r}, \quad \frac{\partial U}{\partial \phi}, \quad \text{and} \quad \frac{\partial U}{\partial \lambda}.$$

Derive a recursive relation for

$$\frac{\partial P_{\ell m}}{\partial \phi}.$$

16. With the initial conditions from Exercise 12 or 14, derive the equations of motion in both the ECF and ECI with all gravity coefficients through degree and order 3. Numerically integrate for 10 days both ECF (Eq. 2.3.22) and ECI (Eq. 2.3.23) and compute the Jacobi integral for both cases. Assume the Earth rotates with constant angular velocity (neglect polar motion, Δ (LOD), precession, and nutation). Assume the Greenwich meridian coincides with the vernal equinox at $t = 0$.

    (a) Assess the numerical integration accuracy from the Jacobi integral.

    (b) Compare the ECF and ECI position results by applying an appropriate transformation into a common coordinate system.

17. Determine the contributions from the Moon and the Sun to the precession of the equinoxes using Eq. 2.4.2. Compare the combined luni-solar rate with the observed rate of -50.25 arcsec/yr. Note that the polar moment of inertia of the Earth is 8.12×10^{44} g cm^2, the semimajor axis of the Moon's orbit is 384,000 km, and for the Sun $a = 149,599,000$ km (~ 1 Astronomical Unit).

18. Determine the secular change in semimajor axis (in m per day) for a balloon-like satellite ($A/m = 10$ m^2 kg^{-1}) at an altitude of 600 km, with atmospheric density $\rho = 5 \times 10^{-13}$ kg m^{-3}. Compare the result to a geodetic satellite with $A/m = 0.001$ m^2 kg^{-1}. Assume $C_D = 2$ for both cases.

Chapter 3

Observations

3.1 INTRODUCTION

In the late 1950s, a ground-based tracking system known as *Minitrack* was installed to track the Vanguard satellites. This system, which was based on radio interferometry, provided a single set of angle observations when the satellite passed above an instrumented station's horizon. These angle measurements, collected from a global network of stations over several years, were used to show that the Earth had a pear-shape. They also provided the means to investigate the nature of forces acting on Vanguard and their role in the multiyear orbit evolution of the satellite.

A variety of observations are now available to support the determination of a satellite orbit. For the most part, these observations are collected by instrumentation that measures some scalar quantity related to the satellite's position or velocity. Modern systems such as the *Global Positioning System* (GPS) provide the position of a GPS receiver, but even GPS requires the use of scalar measurements to determine the orbits of the individual satellites that comprise the GPS constellation. The dependence of scalar measurements on the satellite state (position and velocity) is a key element in the orbit determination process.

Section 3.2 describes an ideal representation of the commonly used measurements. This ideal representation not only serves as an introduction to the measurement type, it also can be used for simulations and other analyses of the orbit determination process. A more realistic description of those measurements is provided in Section 3.3, but from a conceptual viewpoint. The hardware realization of the measurements is discussed in Section 3.4, including other effects, such as those associated with propagation of signals through the atmosphere. The hardware realization must be modeled in terms of the orbit parameters to be determined; hence, this modeling produces a *computed observation*, which is an essential part of the orbit determination process. Examples of measurements obtained by selected instrumentation are discussed in Section 3.5. The use of differenced measurements to remove errors with specific characteristics is discussed in Section 3.6. The use of a satellite ephemeris and angles are discussed in the final sections of this chapter. Supplementary material can be found in Montenbruck and Gill (2001) and Seeber (1993).

3.2   OBSERVATIONS

3.2.1   IDEAL RANGE

A common measurement in orbit determination is the distance between an Earth-based instrument performing the measurement and the satellite. If the position vector of an instrument is $\mathbf{r}_I$ and the position vector of the satellite is $\mathbf{r}$, the ideal range ρ is the scalar magnitude of the position vector of the satellite with respect to the instrument,

$$\rho = [(\mathbf{r} - \mathbf{r}_I) \cdot (\mathbf{r} - \mathbf{r}_I)]^{1/2}. \qquad (3.2.1)$$

In this ideal representation, the range should be more precisely described as the *geometric range* or *instantaneous range*, since it represents the instantaneous, geometric distance between the measuring instrument and the satellite. This ideal representation has ignored subtle issues, such as the finite speed of light and the fact that ρ is the distance between a specific point within the instrument and a specific point on the satellite. Furthermore, if ρ_{obs} represents the measured range at time t, $\mathbf{r}$ and $\mathbf{r}_I$ represent the true position vectors at this time, then ρ and ρ_{obs} are related by

$$\rho_{\mathrm{obs}} = \rho + \epsilon \qquad (3.2.2)$$

where ϵ represents instrumental errors and propagation delays. Note also that if the position vectors $\mathbf{r}$ and $\mathbf{r}_I$ are in error, then ϵ must contain terms that would be necessary to make Eq. (3.2.2) an equality.

The geometric range, ρ, is invariant under rotation of axes used to describe $\mathbf{r}$ and $\mathbf{r}_I$. That is, if (X, Y, Z) represents an inertial system and (x, y, z) represents an Earth-fixed system, the geometric range can be expressed as

$$\rho = [(X - X_I)^2 + (Y - Y_I)^2 + (Z - Z_I)^2]^{1/2} \qquad (3.2.3)$$

or

$$\rho = [(x - x_I)^2 + (y - y_I)^2 + (z - z_I)^2]^{1/2} \qquad (3.2.4)$$

where ρ is identical for both representations.

3.2.2 IDEAL RANGE RATE

In some cases, the time rate of change of the range, or range-rate, may be the measured quantity. From the expression of range in the (X, Y, Z) nonrotating system, Eq. (3.2.3), differentiation with respect to time yields

$$\dot{\rho} = \left[(X - X_I)(\dot{X} - \dot{X}_I) + (Y - Y_I)(\dot{Y} - \dot{Y}_I) \right.$$
$$\left. + (Z - Z_I)(\dot{Z} - \dot{Z}_I) \right] \Big/ \rho \tag{3.2.5}$$

or

$$\dot{\rho} = \frac{\boldsymbol{\rho} \cdot \dot{\boldsymbol{\rho}}}{\rho} \tag{3.2.6}$$

where $\boldsymbol{\rho} = (X - X_I)\mathbf{u}_X + (Y - Y_I)\mathbf{u}_Y + (Z - Z_I)\mathbf{u}_Z$, the position vector of the satellite with respect to the instrument. The relative velocity is

$$\dot{\boldsymbol{\rho}} = (\dot{X} - \dot{X}_I)\mathbf{u}_X + (\dot{Y} - \dot{Y}_I)\mathbf{u}_Y + (\dot{Z} - \dot{Z}_I)\mathbf{u}_Z. \tag{3.2.7}$$

Note that the position and velocity of a ground-based instrument expressed in the nonrotating (X, Y, Z) system will be dependent on the rotation of the Earth.

Equation (3.2.6) can readily be interpreted as the component of the relative velocity in the direction defined by the relative position vector, $\boldsymbol{\rho}$. In other words, the range-rate is the component of the relative velocity between the observing instrument and the satellite in the line-of-sight direction—the direction defined by $\boldsymbol{\rho}$. As described in the preceding section, if $\dot{\rho}_{\rm obs}$ is the observed parameter, then $\dot{\rho}_{\rm obs}$ is equal to $\dot{\rho}$ plus the inclusion of instrumental, media, and other errors.

3.2.3 IDEAL AZIMUTH AND ELEVATION ANGLES

A topocentric system was defined in Section 2.4.3. If the relative position vector, $\boldsymbol{\rho}$, is expressed in this system, then the angles, azimuth and elevation, can be defined, as noted in Section 2.4.3. The angles are illustrated in Fig. 2.4.4.

3.2.4 EXAMPLES: IDEAL OBSERVATIONS

In Example 2.2.4.1, the initial conditions for a Shuttle-like orbit were given and the state was predicted forward in time with Example 2.2.5. This ephemeris was transformed into Earth-fixed coordinates and the ground track can be found in Fig. 2.2.7. If the initial conditions differ by 1 meter in the radial component with respect to those in Example 2.2.4.1, the difference between the two ephemerides is illustrated in Fig. 2.2.9 for about two orbital revolutions. If the "perturbed" initial conditions in Example 2.2.6.1 are assumed to represent the "truth," then a set of *simulated observations* can be generated. For this discussion, simulated

Table 3.2.1:

Coordinates of Simulated Ranging Instruments

Site	Earth-Fixed Coordinates			Geocentric	
	x (m)	y (m)	z (m)	Latitude	Longitude
FZ	4985447.872	−3955045.423	−428435.301	3.8516°S	38.4256°W
EI	−1886260.450	−5361224.413	−2894810.165	26.9919°S	109.3836°W

FZ: Fortaleza

EI: Easter Island

Note: x,y,z computed from latitude, longitude, and a spherical Earth with radius 6378137.0 m

observations are represented by geometric ranges (Section 3.2.1) and range-rate (Section 3.2.2). Simulated observations are useful to characterize different error sources in a controlled experiment.

Consider two hypothetical sites capable of making range measurements: Easter Island and Fortaleza, Brazil. Assume the true coordinates of these two sites are given in Table 3.2.1. The ground tracks of the orbit in the vicinity of these sites are illustrated in Fig. 3.2.1, extracted from the global ground tracks shown in Fig. 2.2.6. The interval while the satellite is above the station's horizon is termed a *pass*. In some cases, the measurements below a specified elevation may be ignored, usually referred to as the *elevation mask*. The first candidate pass from Fortaleza rises less than 1° in elevation, so with a 5° elevation mask, for example, this pass would be ignored. An elevation mask usually is adopted to edit low-elevation passes because of poor signal-to-noise ratios, excessive atmospheric refraction, or multipath effects. For the case shown in Fig. 3.2.1, two passes are observed from Easter Island and one pass from Fortaleza, ignoring the very low-elevation pass at Fortaleza.

The geometric ranges to the satellite observed by each station over three passes are shown in Fig. 3.2.2, where time is measured from the time of the initial state given in Example 2.2.4.1. The minimum range point in each plot is the *point of closest approach* of the satellite to the station. If the ranging instrument had a measurement noise at the one meter level, it is clear that this characteristic would not be visually obvious in Fig. 3.2.2.

The range-rates observed from Easter Island are shown in Fig. 3.2.3. Since range-rate is the time derivative of range, the point of minimum range corresponds to zero range-rate.

A *sky plot* is a useful way to visualize the geometry of a satellite pass above an

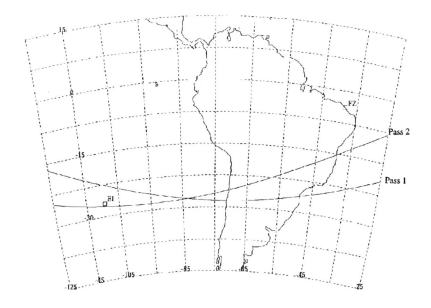

Figure 3.2.1: Portion of shuttle ground track from Fig. 2.2.6. The two orbital revolutions in the vicinity of simulated tracking stations at EI (Easter Island) and FZ (Fortaleza) produce two passes that are within view of the stations. However, the satellite in Pass 1 at Fortaleza rises above the horizon by less than 5° in elevation.

observing station's horizon. The sky plot for each of the three passes illustrated in Fig. 3.2.2 is shown in Fig. 3.2.4. The concentric circles represent an elevation increment of 30°, where the outermost circle is the station's horizon (0° elevation), and the center point is 90° elevation (directly overhead the station). The plot shows the azimuth, measured eastward from north. For example, for Pass 1 from Easter Island, the satellite rises above the station horizon at an azimuth of −59° (i.e., 59° west of north) and sets at 96° azimuth (96° east of north). The highest elevation for this pass is about 40° at an azimuth of 23°.

If the initial conditions in Example 2.2.6.2 represent the true initial conditions, and the set given by Example 2.2.4.1 are used as a nominal or reference set, a *residual* can be formed as the *observed minus computed*, or *O minus C* $(O - C)$. Within the context of orbit determination applications, the true initial conditions are unknown, but the observations represented by the ranges in Fig. 3.2.2 are the result of the true state and the measuring instrument characteristics. Usually, a nominal set of initial conditions are available but they deviate from the true values. Nevertheless, based on the nominal initial conditions, a *computed range* can be formed. The resulting residuals for the three passes are shown in Fig.

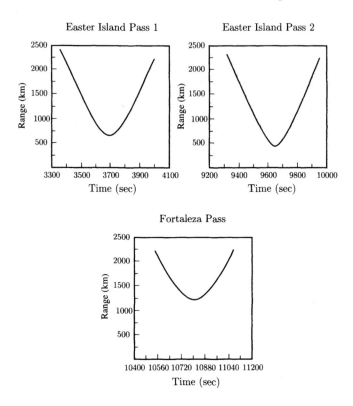

Figure 3.2.2: Simulated range vs time. These plots show the geometric range vs time measured by the two sites, Easter Island and Fortaleza, shown in Fig. 3.2.1.

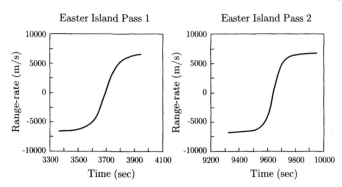

Figure 3.2.3: Simulated range-rate vs time.

3.2.5. It has been assumed that all ranges are instantaneous and no atmospheric delays have been included in these simulations. In essence, the residuals in Fig. 3.2.5 illustrate how the orbit differences of Fig. 2.2.9 are exhibited in terms of range residuals from ground-based stations.

Consider the residual, $O - C$, for the second pass from Easter Island. As shown in the sky plot for this pass in Fig. 3.2.4, the satellite passes nearly overhead, with a maximum elevation of about $78°$. When the satellite rises above the Easter Island horizon in the west, the measured range reflects a significant component in the along-track direction. As described in Section 2.2.6.2, the position on the true orbit trails the position on the nominal orbit in the along-track direction. It follows that the true range will be longer than the range to the reference position. As a consequence, the orbit differences shown in Fig. 2.2.9 will result in $(O - C) > 0$ at satellite rise, but $(O - C) < 0$ when the satellite sets at the observing station's horizon. At some point during the pass over the station, the satellite motion will be perpendicular to the observer's viewing direction, so none of the along-track differences will be "seen"; that is, "O" goes to zero. If the pass has a lower elevation, the basic characteristics are the same, as shown for the low-elevation Fortaleza pass in Fig. 3.2.5.

3.3 CONCEPTUAL MEASUREMENT SYSTEMS

3.3.1 RANGE

All measurements of range are based on the *time-of-flight* principle. Simply stated, an instrument transmits a signal, usually with some appropriate time duration, which is reflected by a passive target or retransmitted by an active target. A *two-way range* may originate with an instrument located on the surface of the Earth, or *ground-based* instrument, and the signal will travel to the satellite and back to the transmitting site. One example of a two-way instrument is the satellite laser ranging (SLR) hardware, which is described in Section 3.5.2. In the SLR measurement, a laser pulse travels on an *uplink path* to the satellite, where it is reflected. The *downlink path* takes the signal from the satellite back to the transmitting site. The uplink and downlink paths constitute the two-way measurement. Some two-way range measurements may be transmitted and received at the same location by essentially the same instrumentation, but others may use a transmitter at one location and a receiver at another. This latter case is sometimes referred to as a *three-way measurement*, in spite of only uplink and downlink paths in the measurement. In some communities, the term *bi-static measurement* may be used.

The *one-way range* is based on the transmittal of a signal by a satellite or a ground-based transmitter that is received by separate instrumentation. If the transmit source is within the satellite, the signal may travel only a downlink path.

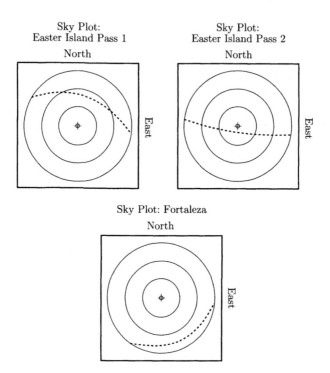

Figure 3.2.4: Sky plots. These plots illustrate the azimuth and elevation for the three passes shown in the ground track of Fig. 3.2.1. The circles represent increments of 30° elevation, with the outermost circle at 0° elevation and the center of the circles at 90° elevation. Azimuth is measured clockwise from north.

Alternatively, the transmitter could be ground-based, and the signal might be received by satellite-borne instrumentation on the uplink path. An important example of the satellite-borne transmitter is the Global Positioning System (GPS), discussed in Section 3.5.1. Even in the case of GPS, the receiver could be located on a satellite, rather than being ground-based.

One-Way Range

Assume a signal of very short duration is transmitted with a specific electromagnetic frequency at time t_T. The transmitted signal propagates along a path and arrives at a point instrumented to receive the transmitted signal. The signal arrives at the receiver at time t_R. Since the signal travels at the speed of light, denoted by c, the signal has traveled a one-way distance given by

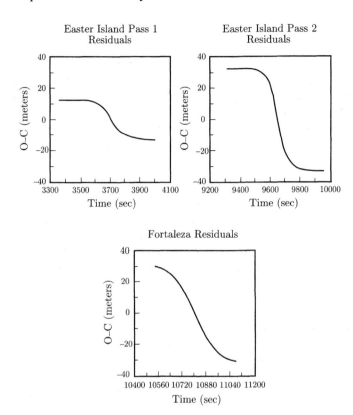

Figure 3.2.5: Range residuals vs time. The range residuals, $(O - C)$, for the three passes illustrate the effect of the orbit error shown in Fig. 2.2.9 on range residuals. The orbit used to generate O is Example 2.2.6.1 and the orbit used to generate C is Example 2.2.4.1.

$$\tilde{\rho} = c(t_R - t_T).\tag{3.3.1}$$

Different clocks are used to register the respective times, t_R and t_T. If the clocks used to record t_R and t_T are precisely synchronized, then $\tilde{\rho}$ represents the true range, aside from other effects such as atmospheric delays of the signal propagation. For the discussion in this section, c will represent the speed of light in a vacuum. Unfortunately, the clocks used are generally not well-synchronized. In fact, if the clock used to determine the transmit time is running significantly faster than the receiver clock, the quantity $\tilde{\rho}$ could be negative! With the synchronization aspect in mind, the quantity $\tilde{\rho}$ generally will not represent the true range, but it is related to the true range. As a consequence, $\tilde{\rho}$ is designated the *pseudorange*.

Using the GPS case as an example, a signal with special characteristics is transmitted. The time when this signal is transmitted is predetermined, but the actual time when the signal is emitted will be controlled by the satellite clock. The time of signal arrival at a reception site is measured by an independent clock. Let t represent *clock time*—time as registered by a clock—but let T represent true time. Everyday experience suggests that a reasonable model for the relation between clock time and true time is the linear relation

$$t = T + a + b(T - T_0) + \epsilon' \tag{3.3.2}$$

where a represents a constant offset of clock time from true time, b represents a linear clock drift, T_0 is a reference time, and ϵ' represents other errors, such as nonlinear terms and stochastic components. In essence, the linear clock model is assumed to be valid over the interval from T_0 to T. Applying Eq. (3.3.2) to both the transmitter clock and the receiver clock, the pseudorange becomes

$$\tilde{\rho} = c(T_R - T_T) + c(a_R - a_T) + c(b_R - b_T)(T - T_0) + \epsilon \tag{3.3.3}$$

where the subscript T refers to the transmitter clock, R denotes the receiver clock, and ϵ is the difference between the other error sources. Since the term $c(T_R - T_T)$ is the true range, it is evident that

$$\tilde{\rho} = \rho(T_T, T_R) + c(a_R - a_T) + c(b_R - b_T)(T - T_0) + \epsilon \tag{3.3.4}$$

where $\rho(T_T, T_R)$ is the distance between the transmit point at time T_T and the receive point at T_R; that is, the true range. Recall that the true range is distinctly different from the ideal range, which does not account for the finite speed of light. It is further evident that pseudorange differs from true range by the clock-related terms. Furthermore, if the time interval $(T - T_0)$ and the difference in drift terms are sufficiently small, then

$$\tilde{\rho} = \rho + c(a_R - a_T) + \epsilon \tag{3.3.5}$$

which illustrates that pseudorange is a *biased range*; the pseudorange differs from true range by a "constant" term (plus noise). Letting ρ_b represent $c(a_R - a_T)$, a range bias, it follows that

$$\tilde{\rho} = \rho + \rho_b + \epsilon. \tag{3.3.6}$$

This discussion has ignored the influence of other delays that contribute to the signal arrival time. In particular, atmospheric refraction will delay the signal (see Section 3.4.2) and contribute an additional term, $\delta\rho_{\text{atm}}$, to Eq. 3.3.5 to yield

$$\tilde{\rho} = \rho + \rho_b + \delta\rho_{\text{atm}} + \epsilon. \tag{3.3.7}$$

The one-way range instrumentation measures $\tilde{\rho}$ by determining t_T and t_R. Equation (3.3.7) shows that the measured quantity is a biased range. A useful alternate form of Eq. (3.3.7) is

$$\tilde{\rho} = \rho + c\delta t_R - c\delta t_T + \delta\rho_{\text{atm}} + \epsilon \qquad (3.3.8)$$

where δt_R is the receiver clock correction and δt_T is the transmitter clock correction. For example, from Eq. (3.3.4), $\delta t_R = a_R + b_R(T - T_0)$ plus higher order and random components.

From Eq. (3.3.7) the measured quantity, $\tilde{\rho}$, is related to the satellite position at time T_T and the receiver position at time T_R, or

$$\tilde{\rho} = [(X - X_I)^2 + (Y - Y_I)^2 + (Z - Z_I)^2]^{1/2} + \rho_b + \delta\rho_{\text{atm}} + \epsilon \qquad (3.3.9)$$

where (X, Y, Z) represents the true position of the satellite at time T_T and the true instrument components $(X, Y, Z)_I$ are determined at T_R.

A computed pseudorange, $\tilde{\rho}_c$, would be formed with an assumed, or nominal, ephemeris for the satellite and coordinates of the instrument and other parameters in Eq. (3.3.9). A *residual* would be obtained from the difference, $\tilde{\rho} - \tilde{\rho}_c$. Such a residual is required in orbit determination.

It is significant to note that the true range, ρ, is formed as the magnitude of the difference between two position vectors, each of which has a different time attached to it. As a consequence, these two vectors must be expressed in the same reference frame. If the reference frame is nonrotating, the resulting path is simply the straight line distance between the two vectors. If an Earth-fixed frame is used, for example, the path appears to be curved and it will be necessary to account for this curvature. Unless otherwise stated, it will be assumed that $\rho(T_T, T_R)$ will be formed using a nonrotating reference frame.

Table 3.3.1 shows the quantity $(\delta T = T_R - T_T)$ for transmitting satellites at different altitudes, assuming the satellite is at the zenith of a ground-based receiver. With $c = 299,792.458$ km/sec, the GPS case is $20,000 \div c = 0.08673$ sec.

Two-Way Range

The two-way range includes both an uplink and a downlink path, where the order will be dependent on the transmitter source. For this discussion, assume the transmitter to be located on the Earth's surface and the same instrumentation will be used to receive the return signal. For simplicity, assume no time delay exists when the signal arrives at the satellite and is retransmitted. This description matches the ground-based satellite laser ranging (SLR) instrumentation, where the transmitted laser pulse is reflected by the satellite back to the transmitting station. Another important example is a satellite altimeter, where the satellite-borne

Table 3.3.1:

One-Way $\delta T = T_R - T_T$ at Ground Observer Zenith

Transmitter	Altitude (km)	δT (millisec)
Shuttle	400	1.33
ERS-1	700	2.33
TOPEX/Poseidon	1336	4.46
GPS	20,000	86.73
Geosynchronous	36,000	140.10

altimeter emits a pulse in the satellite nadir direction and the pulse is reflected by the Earth's surface (e.g., the ocean) back to the satellite. For simplicity, assume the transmitted pulse has a Gaussian shape. A commonly used point on the pulse for determination of t_T is the centroid, but other points may be used that are determined by the hardware characteristics. The return pulse is usually characterized by lower energy, but if the return pulse is also Gaussian, the receive time, t_R, can be determined for the centroid. In the SLR case, the instrumentation records the time when the laser pulse was transmitted by the ground-based hardware, t_T, and the time, t_R, represents the time when some portion of that pulse returns to the transmitting site after being reflected by the satellite. Hence, the round-trip distance traversed by the pulse is

$$\rho_{rt} = c(t_R - t_T) \qquad (3.3.10)$$

where the transmit and receive times are measured by the same clock, denoted by clock time, t. Since the path length is the round-trip distance, the two-way ranges corresponding to the examples in Table 3.3.1 will be double those shown.

As in the one-way example, assume the instrumentation clock can be characterized by a linear clock model, Eq. (3.3.2). It follows that

$$\rho_{rt} = c(T_R - T_T) + b(T_R - T_T) + \epsilon \qquad (3.3.11)$$

where the offset term, a, has cancelled since it is assumed to be based on the same clock. If the clock drift is sufficiently small over the round-trip time interval and other clock error terms plus atmospheric delays can be ignored, then the round-trip range is essentially the true round trip distance. The removal of the major clock term, a, in the two-way range is a significant advantage for this measurement type.

As in the one-way range case, atmospheric delays must be considered. Assuming the drift term can be ignored, it follows that a more accurate representation of Eq. (3.3.11) is

$$\rho_{rt} = c(T_R - T_T) + \delta\rho_{\text{atm}} + \epsilon \tag{3.3.12}$$

where $\delta\rho_{\text{atm}}$ is the atmospheric delay contribution.

The orbit determination process will compare the measured range, ρ_{rt}, to a computed value based on t_T and t_R, a nominal satellite ephemeris and nominal ground station coordinates, plus modeled atmospheric contributions and other effects. A computed value of ρ_{rt} with this information is not straightforward, since the computation requires the time when the signal arrives at the satellite, but this parameter is not directly measured.

An iterative process can be invoked for the computed two-way range to determine the unknown time of arrival of the signal at the satellite. Two sets of iterations are required. For this discussion, assume that errors in clock time, t, are insignificant. Further assume that an ECI nominal satellite ephemeris is available and that an appropriate procedure is available to evaluate the ephemeris for any arbitrary time. It is also assumed that nominal coordinates of a ground-based ranging station are given in ECF, but the transformation between ECF and ECI is known. The following description of the iteration assumes that the signal propagates in a straight line path in the ECI frame (e.g., a very narrow laser beam); hence the description of the procedure is based on ECI.

If the measured signal transmit time, t_T, is taken as the starting point, then the computed two-way range can be determined as follows:

1. Determine the instantaneous range, ρ, between the station and the satellite at t_T using Eq. (3.2.3). This step requires evaluation of the ephemeris at t_T and the station coordinates in ECI, so the ECF coordinates are transformed into ECI using the ECF-to-ECI transformation matrix, Eq. (2.4.11).

2. Compute the approximate signal arrival time at the satellite as $t_a = t_T + \rho/c$.

3. Evaluate the ephemeris at t_a and compute the corresponding range, ρ_{new}, between the ECI satellite position at t_a and the ECI position of the station at t_T.

4. Compare ρ and ρ_{new}; if the difference is less than some specified criteria, such as 1 micrometer, then the process has converged and no more iterations are required. Otherwise, rename ρ_{new} to be ρ and go to step 2.

After converging on the computed signal arrival time, t_a, and the corresponding uplink range, ρ, the next set of computations determine the downlink range and time of arrival at the station based on the models. A procedure similar to the one given earlier can be followed. In this case, the arrival time at the satellite will be assumed to be known from the first iteration, and the arrival time at the station is unknown. This arrival time will most likely differ from the measured arrival time because of errors in the nominal orbit, errors in the station coordinates, and mismodeling of the media corrections.

It turns out, however, that the preceding iterative process can be remarkably simplified. From the measured round-trip range, ρ_{rt}, the average range can be determined as $\rho_{avg} = \rho_{rt}/2$ and the approximate pulse arrival time at the satellite is $\rho_{rt}/(2c) = \delta\tau$. It follows that the satellite pulse arrival time is $t_a = t_T + \delta\tau$. If the instantaneous range is computed at t_a (evaluate the ephemeris and the ECI station coordinates at this t_a), this range is approximately equal to ρ_{avg}, at least to the submillimeter level. As a consequence, the measured round-trip range can be modeled by the instantaneous range (one-way) determined at the time t_a.

In the case of a satellite-borne altimeter, the two-way measurement originates with an instrument in the satellite, but the signal propagation is earthward, usually in the nadir direction defined by either the local geodetic vertical or the radial direction to the geocenter. In the altimeter case, the signal is reflected by the Earth's surface (e.g., the ocean). The preceding discussion in this section for the ground-based instrument applies as well to the altimeter. Let h_{rt} represent the round-trip altitude, and

$$h_{rt} = c(t_R - t_T) \qquad (3.3.13)$$

where t_R is the signal transmit time and t_T is the receive time. Note that the measurement consists of a downlink path (satellite to surface) and an uplink path for the echo (surface to satellite). Even in the ECI coordinate system, the motion of the Earth has a small contribution to h_{rt} during the round-trip interval.

As in the case of the previous two-way range, the computed altitude requires an iterative process to determine the surface time of arrival. In the general case that allows for an off-nadir pointed altimeter, the process is similar to the preceding discussion for the ground-based two-way ranging station. The average altitude is

$$h_{avg} = h_{rt}/2 \qquad (3.3.14)$$

and the average time is

$$t_{avg} = t_T + h_{avg}/c. \qquad (3.3.15)$$

As in the preceding case, h_{avg} is a good approximation (submillimeter) to the instantaneous altitude formed at t_{avg}. The time of signal arrival at the surface is closely represented by t_{avg}.

Example

Consider a satellite in an equatorial posigrade, circular orbit with an altitude of 600 km above a spherical Earth. Assume the satellite is $20°$ in true anomaly past the zenith direction of a two-way ranging station, which places the satellite at 4.3° elevation with respect to the station. Assume a signal is transmitted by the station at $t = 0$. The uplink iteration to determine the computed range gives 2393433.99356 meters, with signal arrival at the satellite 0.007983636445 sec after transmission. The downlink iteration gives 2393426.58799 meters and a signal arrival at the ground station 0.015967248187 sec ($\delta\tau$) after transmission. The average range (sum of uplink and downlink, divided by two) is 2393430.290775 meters. Determination of the instantaneous range at $\delta\tau/2$ (0.007983624094 sec) gives 2393430.290689 meters.

3.3.2 RANGE-RATE

Most *range-rate* systems in current use are based on a single propagation path, either uplink or downlink. The following discussion treats the problem from two viewpoints: an instrument transmitting a short duration pulse at a known interval and a beacon that transmits a signal with a known frequency.

Repeated Pulse Transmission

Assume a satellite-borne instrument transmits a pulse at a specified and fixed interval, δt_T. Hence, the pulses are transmitted at a sequence of times (e.g., t_{T1}, t_{T2}, etc.). The transmitted pulses are received by a ground-based instrument and the arrival times are recorded as t_{R1}, t_{R2}, and so on, where t_{R1} is the arrival of the pulse transmitted at t_{T1}. The transmit and receive times are related by

$$
\begin{aligned}
t_{R1} &= t_{T1} + \rho_1/c \\
t_{R2} &= t_{T2} + \rho_2/c,
\end{aligned}
\tag{3.3.16}
$$

and so on, where the relation to Eq. (3.3.1) is obvious. If t_{R1} is subtracted from t_{R2}, it follows that

$$
t_{R2} - t_{R1} = \delta t_T + (\rho_2 - \rho_1)/c
\tag{3.3.17}
$$

or

$$
\delta t = \delta t_T + \delta\rho/c
\tag{3.3.18}
$$

where $\delta t = t_{R2} - t_{R1}$ and $\delta\rho$ represents the *range change* between t_{T2} and t_{T1}. Consider the following cases:

- If $\rho_2 > \rho_1$: the length of the path traversed by the signal is getting longer (the transmitter and receiver are getting farther apart), then $\delta t > \delta t_T$. In other words, the time between successive pulse arrivals is longer than the fixed interval between their transmission, δt_T.

- If $\rho_2 < \rho_1$: the length of the signal path is getting shorter (the transmitter and receiver are moving toward each other), then $\delta t < \delta t_T$. In this case, the time between successive pulse arrivals is shorter than δt_T.

- If $\rho_2 = \rho_1$: there is no change in the signal path length and the pulse arrival interval equals the pulse transmit interval.

Equation (3.3.18) can be rewritten as

$$\delta t = \delta t_T(1 + (\delta\rho/\delta t_T)/c) \tag{3.3.19}$$

where $(\delta\rho)$ is the range change in the time interval δt_T. Hence, $\delta\rho/\delta t_T$ is a form of *range-rate*. If the ground-based instrument measures δt, it is evident that this measured quantity is dependent on the range change during the interval δt_T. Note that the measured quantity is not the instantaneous range-rate, although the shorter the interval δt_T, the closer the measured quantity will be to the instantaneous range-rate. In a sense, the previous description is a form of measurement differences, which is discussed later in this chapter. Since the same clock is used to determine the pulse transmit times and a separate clock is used to measure the pulse receive times, the quantities δt and δt_T are time differences based on the time recorded by separate clocks. As in the case of two-way ranging, the dominant clock model term, a, will cancel in the differencing process while the drift and higher order clock effects will remain at some level.

Transmitter Beacon

Assume a satellite contains a radio beacon that transmits a signal with a known frequency, f_T. The transmitted signal arrives at a ground-based site with an apparent frequency f_R, but this arrival signal is usually mixed with a reference frequency standard at the site. The arriving signal is multiplied with the reference frequency, f_G, which yields a signal that contains frequencies of $(f_G + f_R)$ and $(f_G - f_R)$. The sum of frequencies is filtered and if f_G is essentially f_T, the resulting $(f_G - f_R)$ signal represents the apparent change in transmitted frequency. The receiver is designed to count the number of cycles at this frequency between two times, t_{R_1} and t_{R_2}, thus

$$N_{1,2} = \int_{t_{R_1}}^{t_{R_2}} (f_G - f_R)dt \tag{3.3.20}$$

where $N_{1,2}$ represents the number of integer cycles plus the fractional part of a cycle in the measured frequency $(f_G - f_R)$ during the time interval between t_{R_1} and t_{R_2}. Let

$$t_{R_1} = t_{T_1} + \Delta t_1$$

and

$$t_{R_2} = t_{T_2} + \Delta t_2$$

where $\Delta t_1 = \rho_1/c$ and ρ_1 is the distance from the transmit point to the receive point at t_{R_1}. Assuming that f_G is constant,

$$N_{1,2} = f_G \left[t_{T_2} - t_{T_1} + (\rho_2 - \rho_1)/c \right] - \int_{t_{R_1}}^{t_{R_2}} f_R dt. \qquad (3.3.21)$$

However, the last integral is

$$\int_{t_{R_1}}^{t_{R_2}} f_R dt = \int_{t_{T_1}}^{t_{T_2}} f_T dt = f_T \left[t_{T_2} - t_{T_1} \right]. \qquad (3.3.22)$$

Thus,

$$N_{1,2} = (f_G - f_T)(t_{T_2} - t_{T_1}) + f_G(\rho_2 - \rho_1)/c. \qquad (3.3.23)$$

Let $\delta t = t_{T_2} - t_{T_1}$, $\delta \rho = \rho_2 - \rho_1$, and f_G equal f_T, then

$$N_{1,2}/\delta t = (f_T/c)(\delta \rho/\delta t). \qquad (3.3.24)$$

It follows that

$$f_R = f_T - N_{1,2}/\delta t, \qquad (3.3.25)$$

so the apparent received frequency is $f_T(1 - (\delta\rho/\delta t)/c)$, and it is evident that this frequency depends on the range-rate. In other words, Eq. (3.3.25) illustrates the *Doppler effect*. If the relative motion between the transmitter and receiver is positive (the two are moving apart), then $f_T - f_R$ is less than f_T; that is, the apparent frequency of the transmitter beacon is lower than the actual frequency. But if the relative motion is negative, the apparent frequency of the transmitter beacon is higher than f_T. These results are consistent with those in the preceding section for the repeated pulse transmission, as they should be since both cases are based on the Doppler effect. An important element of the preceding discussion is the fact that instantaneous range-rate cannot be measured. Instead, the measured quantity $(N_{1,2})$ is related to the range change $(\delta\rho)$ in a specified interval of time (δt). Depending on the realization of the previously described concept, the instrument design may measure $N_{1,2}$ by allowing the zero crossings of $(f_G - f_T)$ to increment a counter, thereby generating an integer count, or Doppler count. In some designs, the counter may reset at the end of the count interval (t_{R_2} in the previous discussion), or it may simply record the content of the counter at the interval end. In the continuous count case, the count is made with respect to some interval when the signal from the transmitter first appeared, where the start count may be nonzero.

Another interpretation of $N_{1,2}$ can be obtained by rearranging the terms in Eq. (3.3.23) to give

$$\rho_2 = \rho_1 + N_{1,2} \ c/f_G - c/f_G(f_G - f_T)(t_{T_2} - t_{T_1}), \qquad (3.3.26)$$

which suggests that a range measurement can be formed from the measured $N_{1,2}$. In this representation, ρ_1 is not known so ρ_2 can be regarded to be a biased range, similar to the previously described pseudorange. If t_2 is any time after t_1, Eq. (3.3.26) demonstrates that $N_{1,2}$ exhibits a range-like variation with time. For these applications, $N_{1,2}$ is usually the accumulated integer cycles plus the fractional part of the cycle. Although the fractional part is the *phase*, the term is commonly used to describe the sum of the integer cycles plus the fractional part. In the case when $f_G = f_T$, it follows that Eq. (3.3.26) can be written as

$$\rho_2 = \rho_1 + \tilde{\rho}_{PH} \qquad (3.3.27)$$

where $\tilde{\rho}_{PH} = \lambda N_{1,2}$, $\lambda = c/f_T$, the signal wavelength. The quantity $\tilde{\rho}_{PH}$ is a pseudorange formed from the phase, *phase pseudorange*.

3.4   REALIZATION OF MEASUREMENTS

3.4.1   CONSIDERATIONS

The preceding section provided a conceptual discussion of the commonly used observations with little consideration given to the actual instrumentation applied to the realization of those observations. Although the discussion referred to the transmission of signals, the characteristics of those signals and the effects of the atmosphere were not treated. All signals are transmitted using a region of the electromagnetic spectrum, spanning from radio frequencies to visible regions of the spectrum. The commonly used regions of the spectrum are identified using terminology established during the 1940s global conflict and are shown in Table 3.4.1. Depending on their frequency, the signals will be influenced differently by the atmosphere, as discussed in the next section. Current technologies used in orbit determination are summarized in the following sections.

3.4.2   ATMOSPHERIC EFFECTS

Measurements used for orbit determination that propagate through the Earth's atmosphere will experience refractive effects. These effects delay the signals and, for example, lengthen the apparent range. Account must be taken of these delays in most cases, though the overall effect is unimportant in cases with coarse accuracy requirements (e.g., km-level). Since the atmosphere consists of distinctive layers, the propagation effects can be discussed within the context of the two most influential layers: troposphere and ionosphere.

Table 3.4.1:

Electromagnetic Spectrum Regions for Satellite Transmissions

Band Category	Approx. Frequency Range (MHz)	Wavelength Range (approx. cm)
L–Band	1000–2000	30–15
S–Band	2000–4000	15–7.5
C–Band	4000–8000	7.5–3.75
X–Band	8000–12500	3.75–2.4
K–Band	12500–40000	2.4–0.75
Infrared	3×10^8	0.00010
Green	5.6×10^8	0.0000532

(Skolnik, 1990)

Troposphere

The troposphere, which extends from the surface to about 10 km, is the region of the atmosphere where most meteorological effects that influence the surface take place. A transition region known as the tropopause lies between the troposphere and the next layer, the stratosphere. The dominant refractive contributions from these regions are associated with the troposphere, so the effect is referred to as the *tropospheric delay*, even though the stratosphere accounts for about 20% of the total. For radio frequencies, the troposphere and stratosphere are electrically neutral, or a *nondispersive medium*. The propagation delay, $\delta\rho$, is

$$\delta\rho = 10^{-6} \int N ds \qquad (3.4.1)$$

where ds is a differential length along the path, N is the *refractivity*

$$N = (n - 1)10^6 \qquad (3.4.2)$$

and n is the *index of refraction*. The refractivity is usually written as the sum of the two components

$$N = N_d + N_w \qquad (3.4.3)$$

where the subscripts denote *dry* (d) and *wet* (w) components. The *dry component*, or *hydrostatic component*, accounts for about 90% of the total effect. It assumes the atmosphere behaves consistently with the ideal gas law and that it is in hydro-static equilibrium so that only the surface pressure needs to be known. Modeling of the *wet component* is more difficult because of variability in the partial water vapor pressure, so this correction is estimated from the tracking data or may be obtained from other instrumentation, such as *water vapor radiometers*. A commonly applied expression for frequencies less than about 30 GHz was given by Smith and Weintraub (1953)

$$
\begin{aligned}
N_d &= 77.6(P/T) \\
N_w &= 3.73 \times 10^5 (\tilde{e}/T^2)
\end{aligned}
\tag{3.4.4}
$$

where P is pressure in millibars (mb), T is temperature in degrees Kelvin, and partial water vapor pressure, $\tilde{e}$, is in mb, all measured at the surface. Evaluation of Eq. (3.4.1) has been treated by Saastamoinen (1972).

In the zenith direction, the tropospheric delay is about 2 meters at sea level, but the effect is dependent on the elevation angle of the viewing direction through the atmosphere. For example, at 10° elevation, the delay is about 12 meters. Mapping functions are usually adopted to map the zenith effect to any elevation angle. These mapping functions are fundamentally dependent on the cosecant of the elevation angle, but the complexity of the mapping functions varies.

Various methods are used to generate the propagation delay, including ray-tracing. The signal delay is predominantly dependent on the cosecant of the satellite elevation seen at the station. A frequently used expression for the tropospheric delay, $\Delta \rho_t$, is

$$
\Delta \rho_t(El) = \tau_d m_d(El) + \tau_w m_w(El)
\tag{3.4.5}
$$

where τ represents the zenith delay, $m(El)$ represents a mapping function with elevation dependency and the subscript d is the dry component and w is the wet component. The simplest mapping function is

$$
m(El) = 1/(\sin(El)).
\tag{3.4.6}
$$

For optical wavelengths, such as those used with lasers, the troposphere behaves dispersively. As a consequence, the delay is dependent on the wavelength of the signal. For laser ranging systems, a commonly used correction for troposphere delay is given by Marini and Murray (1973):

$$
\Delta \rho_t = \frac{f(\lambda)}{f(\phi, H)} \frac{A + B}{\sin(El) + \frac{B/(A+B)}{\sin(El)+0.01}}
\tag{3.4.7}
$$

where

$$
A = 0.002357 P_0 + 0.000141 e_0
$$

$$B = (1.084 \times 10^{-8})P_0 T_0 K + (4.734 \times 10^{-8})\frac{P_0^2}{T_0}\frac{2}{(3 - 1/K)}$$

$$f(\phi, H) = 1 - 0.0026\cos 2\phi - 0.00031H$$

$$K = 1.163 - 0.00968\cos(2\phi) - 0.00104T_0 + 0.00001435P_0$$

$$e_0 = \frac{R_h}{100} \times 6.11 \times 10^a \qquad\qquad (3.4.8)$$

$$a = \frac{7.5(T_0 - 273.15)}{237.3 + (T_0 - 273.15)}.$$

El is the true elevation, P_0 is the atmospheric pressure at the laser site (mb), T_0 is the atmospheric temperature at the laser site (degrees Kelvin), e_0 is the water vapor pressure at the site (millibars), R_h is relative humidity (%), ϕ is the site geodetic latitude, and H is the height of the site above the ellipsoid (km). Note that $f(\lambda)$ is

$$f(\lambda) = 0.9650 + \frac{0.0164}{\lambda^2} + \frac{0.000228}{\lambda^4} \qquad\qquad (3.4.9)$$

where λ is the laser wavelength in microns. For example, a ruby laser has a wavelength of 0.6943 microns and $f(\lambda)$ is 1.0000, whereas a green laser ($\lambda = 0.532$ microns) has $f(\lambda) = 1.02579$. At optical wavelengths, the wet component is small and usually ignored.

Other mapping functions have been developed for radio frequency measurements. For example, the MIT Thermospheric mapping function, known as MTT, has close functional similarity to the Marini-Murray function (Herring *et al.*, 1992):

$$m(El) = \frac{1 + \dfrac{a}{1 + \frac{b}{1+c}}}{\sin(El) + \dfrac{a}{\sin(El) + \frac{b}{\sin(El) + \frac{b}{\sin(El)+c}}}} \qquad\qquad (3.4.10)$$

where the coefficients for the dry part, m_d, are

$$
\begin{aligned}
a &= [1.2320 + 0.0130\cos\phi - 0.0209H_s \\
&\quad + 0.00215(T_s - T_0)] \times 10^{-3} \\
b &= [3.1612 - 0.1600\cos\phi - 0.0331H_s \\
&\quad + 0.00206(T_s - T_0)] \times 10^{-3} \\
c &= [71.244 - 4.293\cos\phi - 0.149H_s \\
&\quad - 0.0021(T_s - T_0)] \times 10^{-3}
\end{aligned}
\qquad (3.4.11)
$$

where T_s is surface temperature in °C, $T_0 = 10$°C, ϕ is the site geodetic latitude, and H_s is the height of the site above the reference ellipsoid. The coefficients for the wet part, m_w, are

$$\begin{aligned}
a &= [0.583 - 0.011 \cos\phi - 0.052 H_s \\
&\quad +0.0014(T_s - T_0)] \times 10^{-3} \\
b &= [1.402 + 0.102 \cos\phi - 0.101 H_s \\
&\quad +0.0020(T_s - T_0)] \times 10^{-3} \\
c &= [45.85 - 1.91 \cos\phi - 1.29 H_s \\
&\quad +0.015(T_s - T_0)] \times 10^{-3}
\end{aligned}$$

(3.4.12)

In very high-accuracy cases, azimuthal dependencies should be accounted for in the mapping function or measured with water vapor radiometers.

Ionosphere

The ionosphere is the portion of the atmosphere that is characterized by the presence of ions and free electrons. The ionosphere is most distinctive at altitudes above 80 km and consists of several layers, but not all of the layers are completely distinct. The ionosphere has no significant effect on signals with optical wavelengths (e.g., lasers). For radio frequencies, the ionosphere is dispersive and the signal time delay is dependent on the frequency, as well as the electron content along the signal path. Thus, the time delay δt at a frequency f is, to order $1/f^3$,

$$\delta t = \frac{\alpha}{f^2} \qquad (3.4.13)$$

where α is related to the *TEC (total electron content)* along the path. The term α is positive for group delays (e.g., pseudorange) and negative for carrier phase. Since the ionospheric delay is dependent on frequency, transmission of signals at different frequencies allows removal of the ionosphere, at least to an acceptable level for orbit determination applications. Given a ranging code transmitted at two different frequencies, f_1 and f_2, this equation can be used to eliminate the TEC and obtain an ionosphere-free range, ρ_c. Using the linear combination of range measurements made with f_1 and f_2, ρ_c can be shown to be

$$\rho_c = \gamma_1 \rho_1 - \gamma_2 \rho_2 \qquad (3.4.14)$$

where ρ_1 is the range measurement from f_1 and ρ_2 is the measurement from f_2, and

$$\begin{aligned}
\gamma_1 &= \frac{f_1^2}{f_1^2 - f_2^2} \\
\gamma_2 &= \frac{f_2^2}{f_1^2 - f_2^2} \, .
\end{aligned}$$

(3.4.15)

Significant changes in total electron content take place both spatially and temporally. Predictive ionosphere models may be used to correct single-frequency ranging data, but the highest accuracy is achieved with dual-frequency measurements. For measurement systems that operate in the L-band of the frequency spectrum (e.g., GPS with $f_1 = 1575.42$ MHz and $f_2 = 1227.60$ MHz), the ionosphere delay can be meter-level to tens of meters, depending on the TEC. Although the linear combination increases the measurement noise compared to the single-frequency measurement, the removal of the systematic effects associated with ionospheric variability is essential for high-accuracy applications. Appropriate estimation techniques can aid in smoothing the increased measurement noise produced by the linear combination.

3.4.3 GENERAL RELATIVITY

As noted in Chapter 2, an Earth orbiter experiences a small relativistic precession of the perigee. In this section, the relativistic effects associated with various measurements (e.g., range or time delay) are described. These effects include the constancy of the speed of light, time dilation, gravitational frequency shifts, and the Sagnac effect. Further details for GPS are given by Ashby (2002).

Depending on the altitude, a clock (oscillator) in a circular orbit may beat fractionally faster or slower than a clock with identical characteristics on the surface of the Earth. This effect applies to systems such as GPS, Transit, and DORIS. In the case of GPS, this relativistic effect is usually accounted for by slightly lowering the satellite oscillator frequency to 10.229999999543 MHz to achieve the effective frequency of 10.23 MHz when the satellite is in orbit. It is evident that this effect does not apply to passive satellites, such as Lageos, with no on-board oscillators.

If the clock's orbit is not perfectly circular, the clock will vary about the mean frequency with the same period as the orbit. For a GPS satellite with an orbital eccentricity of 0.01, for example, this could lead to a navigation error of up to 7 m. The correction to the time offset is easily calculated from knowledge of the GPS satellite position and velocity from

$$\Delta t_r = \frac{2\mathbf{r} \cdot \mathbf{v}}{c^2}. \tag{3.4.16}$$

In a system of several massive bodies, the gravitational fields of these bodies will influence the arrival time of a signal traveling at the speed of light from a source. For Earth orbiters, a nonrotating inertial reference system tied to the center of mass of the Earth usually is used, and the relativistic effect of the Earth's mass is the only contributor. If the signal is transmitted from a position at a distance r_1 from the geocenter and received at a position that is a distance r_2, and ρ is the distance between transmitter and receiver, then the relativistic correction to the time delay caused by the Earth is:

$$\Delta t_{\text{delay}} = \frac{2GM}{c^3} \ln \left[\frac{r_1 + r_2 + \rho}{r_1 + r_2 - \rho} \right]. \tag{3.4.17}$$

For low-altitude Earth satellites, the contribution from other celestial bodies is negligible. In the case of GPS, this correction is a few cm, but the effect mostly cancels in differenced measurements (see Section 3.6).

An additional relativistic effect is known as the Sagnac effect. When working in the ECI frame, the Sagnac effect is not applicable; the iteration to calculate the light-time accounts for the motion of the receiver clock during the time of propagation. The Sagnac effect is generally a concern only for high-precision time transfer between ground-based clocks via a satellite. Ries *et al.* (1991) describe the relativistic effects for orbit determination.

3.5   MEASUREMENT SYSTEMS

3.5.1   ONE-WAY RANGE

GPS

The basic concept of a one-way range was described in Section 3.3.1. As noted in that description, the one-way range is based on the time required for a signal to travel from a transmitter to a receiver. The wide use of GPS makes it a suitable example of the one-way ranging technique. Further details of GPS are given by Hofmann-Wellenhof *et al.* (1997) and Leick (2003).

The GPS consists of a constellation of satellites, developed by the U.S. Department of Defense. Each satellite broadcasts signals to support one-way ranging at radio frequencies using the following basic elements: a radio transmitter, a computer, a frequency standard (provides both time and frequency reference for radio transmissions), a transmit antenna, and a power source. The characteristics of the constellation in January 2000 are summarized in Table 3.5.1. The GPS satellites are organized into six orbit planes ($i = 55°$) with four or five satellites in each plane. Each satellite has an orbital period of 11 hours 58 minutes.

All GPS satellites use the same frequencies in the L-band: 1575.42 MHz (known as L_1) and 1227.60 (known as L_2). These frequencies are termed the *carrier frequencies*, which are 154 f_0 and 120 f_0, where $f_0 = 10.23$ MHz, the GPS fundamental frequency. A third civil frequency (known as L_5 at 1176.45 MHz) is being planned for future GPS satellites starting with Block IIF. A thorough discussion of the GPS signals can be found in Parkinson *et al.* (1996).

Table 3.5.1

GPS Constellation Status: 1 January 2000 00:00:00.000 GPS-Time

Orb./Plane & Position	PRN	S/C type	Clock	a (km)	e	i (deg.)	Ω (deg.)	ω (deg.)	$\omega+M$ (deg.)	λ_{AN} (deg.)
A-1	09	IIA	Cs	26559.3	0.0098	54.0	175.7	32.3	103.0	127.0	-53.0
A-2	25	IIA	Cs	26561.7	0.0078	53.6	173.5	232.9	356.0	71.9	-108.1
A-3	27	IIA	Cs	26559.9	0.0137	53.9	174.7	198.3	212.2	1.1	-178.9
A-4	19	II	Rb	26560.0	0.0053	53.1	172.4	203.6	238.8	12.0	-168.0
A-5	08	IIA	Rb	26539.7	0.0086	54.8	178.2	102.1	148.2	152.2	-28.0
B-1	22	IIA	Rb	26559.9	0.0129	53.5	233.7	30.5	278.4	92.6	-87.4
B-2	30	IIA	Cs	26562.8	0.0056	54.1	235.7	83.8	3.5	137.2	-42.8
B-3	02	II	Cs	26561.3	0.0194	53.6	232.9	236.7	140.7	24.3	-155.7
B-4	05	IIA	Cs	26562.0	0.0019	53.7	234.0	9.5	38.6	153.3	-26.7
C-1	06	IIA	Cs	26558.9	0.0068	54.5	297.2	222.1	306.4	170.8	-9.2
C-2	03	IIA	Cs	26561.2	0.0010	54.1	294.8	72.8	207.5	118.6	-61.4
C-3	31	IIA	Cs	26561.7	0.0092	54.6	295.4	45.7	175.0	102.7	-77.3
C-4	07	IIA	Cs	26559.5	0.0109	54.6	295.4	239.7	75.1	53.6	-126.4
D-1	24	IIA	Rb	26561.1	0.0090	56.5	358.4	261.4	325.4	61.7	-118.3
D-2	15	II	Cs	26555.5	0.0073	56.3	0.2	85.8	118.6	139.2	-40.7
D-3	17	II	Cs	26558.9	0.0113	56.4	2.5	167.5	224.5	14.7	-165.3
D-4	04	IIA	Rb	26562.0	0.0053	56.0	357.8	323.0	0.5	78.4	-101.6
D-5	11	IIR	Cs	26559.3	0.0029	53.0	355.2	189.8	101.0	125.8	-54.2
E-1	14	II	Cs	26562.4	0.0005	56.1	59.3	129.5	31.4	155.0	-25.0
E-2	21	II	Cs	26559.9	0.0160	55.7	56.9	211.5	130.0	22.5	-157.5
E-3	10	IIA	Cs	26557.3	0.0038	55.8	56.5	353.0	256.6	84.7	-95.3
E-4	23	IIA	Cs	26562.3	0.0145	55.9	59.2	249.3	163.6	41.8	-138.2
E-5	16	II	Cs	26562.3	0.0044	55.9	59.5	19.6	355.5	137.2	-42.8
F-1	01	IIA	Cs	26568.4	0.0048	55.0	117.9	258.8	335.0	5.7	-174.4
F-2	26	IIA	Rb	26562.3	0.0116	55.2	116.9	2.1	180.0	106.9	-73.1
F-3	18	II	Cs	26559.2	0.0076	54.4	114.0	107.0	117.9	162.7	-17.4
F-4	29	IIA	Rb	26558.6	0.0073	55.0	115.3	248.2	81.0	56.3	-123.7
F-5	13	IIR	Rb	26558.7	0.0022	55.2	116.5	322.6	244.1	138.7	-41.3

The osculating orbit elements are expressed in J2000. PRN is the GPS identifier, Cs denotes cesium and Rb is rubidium, S/C Type identifies the satellite design type, and λ_{AN} is longitude of the ascending node.

The discussion in this section also applies to the current Russian navigation satellite system, known as GLONASS (Global Navigation Satellite System) and the future European Space Agency GALILEO (planned for operation by 2008).

Both satellite constellations use three orbital planes with about 10 satellites (including spares) in each plane (GALILEO). The GLONASS satellites use a 63° inclination and orbit periods of 11 hours 15 minutes. GALILEO satellites are expected to use a 56° inclination with an orbit period of 14 hours 22 minutes. Each GLONASS satellite broadcasts on a different frequency in the L-band, but GALILEO will have broadcast characteristics similar to GPS.

A Block II GPS satellite is illustrated in Fig. 3.5.1. The transmit antenna is the array shown on the main body of the spacecraft with the helical windings. The large panels on both sides are the solar panels used to generate power. Each satellite carries standard quartz frequency standards, as well as multiple atomic frequency standards (two cesium and two rubidium). In fact, a tunable quartz standard is used to excite the cesium standard, for example, at the natural frequency of $9, 192, 631, 770$ Hz. It is this frequency that, in fact, defines the SI second.

The carrier frequencies are derived from the frequency standard in use (cesium, rubidium, or quartz), but additional information is superimposed. A simple analogy can be drawn with a common radio where a carrier frequency is used (e.g., 100 MHz) to carry the audio ($< 20, 000$ Hz). In this example, the transmitter superimposes audio on the carrier, and the receiver extracts the audio signal when the radio is properly tuned to the carrier frequency. In the case of GPS (or GLONASS), the information superimposed on the carrier includes ranging codes and other data necessary to perform the navigation function offered by the satellite constellation.

GPS uses several ranging codes, though they have many similarities. In concept, the ranging codes are generated as *pseudo-random noise* (PRN). Consider the PRN code to be a series of binary digits (bits)—001101011100—for example, superimposed on the carrier. Each bit, known in GPS terminology as a *chip*, has a specific time duration depending on the code. The PRN bit sequence is determined by a documented algorithm. Each bit in the sequence will be transmitted by the satellite at a specific time determined by the satellite clock (which is derived from the frequency standard). Hence, as the receiver extracts the bit sequence from the carrier, it will assign receive times to each bit based on its own clock. With the ability to replicate the PRN code, the receiver will cross-correlate, or align, the received bit sequence with the sequence it is able to generate. Since each bit has a known transmit time, the difference between receive time and transmit time is obtained by this cross-correlation; that is, the quantity $t_R - t_T$ in Eq. (3.3.1) is determined, as well as the individual times, t_R and t_T. Note that the time when each bit in the code is transmitted is determined by the satellite clock, so the time t_T is based on the satellite clock. Similarly, the time when the bit is

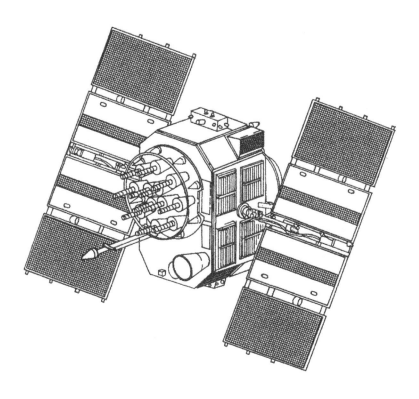

Figure 3.5.1: Block II GPS satellite. The satellite solar panels rotate about the axis mounted perpendicular to the main body (spacecraft bus) and the transmit antenna array, shown with the helical windings in the center, is directed toward the Earth's center. The antenna transmit pattern encompasses the visible Earth. A body-fixed set of axes includes a y axis coincident with the solar panel axes and a z axis directed toward the Earth's center. The spacecraft can present the maximum cross-sectional area of the solar panels to the Sun by rotating the bus about the z axis (yaw) and rotating the solar panels about the y axis.

received, t_R, is determined by the clock in the receiver.

The PRN codes currently transmitted by the GPS satellites are:

- C/A (Coarse Acquisition): This code uses 1023 bits and repeats every 1 ms. The algorithm for generating the sequence is described in detail by Hofmann-Wellenhof *et al.* (1997). Each bit requires about 1 microsec for transmission or about 300 meters in distance. One major purpose of this code is to facilitate acquisition of the P-code, which is a much longer bit sequence. Since the C/A code repeats every millisecond, an ambiguity exists between each millisecond interval. In other words, there is no information about absolute time within the C/A code. Resolving this ambiguity to determine the correct time interval requires additional information (e.g., information broadcast by the GPS satellites about their position).

- P (Precise): This code has a much longer duration of 37 weeks before it repeats. But this long repeat interval is divided into one week segments and each segment is assigned to a specific GPS satellite. Each satellite, in turn, repeats its assigned code each week. The duration for each bit is the equivalent of 30 meters in distance, corresponding to a transmission rate of 10.23×10^6 bits per sec. All information about the P-code is readily available. Most receivers use the C/A code for initial acquisition, then transition to the P-code. Since the P-code within each satellite does not repeat for one week, direct cross-correlation without use of the C/A code is challenging. Direct P-code acquisition is easier if the receiver has an accurate clock.

- Y: This code is generated from the P-code, but a classified code (W-code) is used to mix with the P-code. This mixing produces an encrypted P-code. When the Y-code is being transmitted, it is said that *Anti-Spoofing* (AS) has been invoked. The terminology arises from the military consideration that an adversary could transmit false P-code signals to confuse, or spoof, a receiver. When AS is used, the classified Y-code avoids this possibility.

In the GPS satellites known as Block II, including Block IIA and Block IIR, the C/A code is transmitted only on L_1 and the P/Y codes are transmitted on both L_1 and L_2. As a consequence, receivers capable of operating with only C/A are single-frequency receivers. Without the second frequency, the ionosphere correction cannot be made as accurately as measurements obtained with two frequencies, since it must rely on less accurate ionosphere models. Modern dual-frequency receivers are available that may be capable of correlating directly with the Y-code or they may use signal processing techniques, such as cross-correlation of the Y-code on the two frequencies to obtain the ionosphere delay. The method based on advanced signal processing effectively provides a dual-frequency measurement without knowledge of the Y-code. A pseudo-measurement usually is created by adding the measured ionosphere delay to the C/A pseudorange.

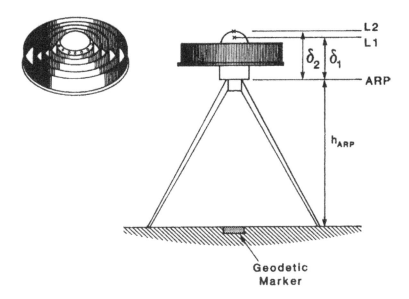

Figure 3.5.2: Typical GPS antenna setup. A choke-ring antenna is shown on the left and the antenna set up with a tripod over a geodetic marker is shown on the right. The height of the antenna reference point (ARP) above the geodetic marker is usually made to high accuracy with a geodetic-quality graduated ruler. The antenna phase center locations, denoted by L1 and L2, are separately calibrated.

A typical GPS receiver antenna installation is shown in Fig. 3.5.2, which shows a *choke-ring antenna* on the left. The choke ring is designed to reduce the effects of *multipath*, a phenomenon that occurs when the incoming GPS signal is reflected from nearby objects, such as buildings for a ground receiver. The reflected signal travels a longer path to the antenna, thereby introducing another error source into the measured signal. The choke-ring antenna is used for both ground-based and space-borne installations.

The relationship between the measured ranges and the satellite state requires the specification, or estimation, of the receiver antenna location. For precision orbit determination, the coordinates of the antenna shown in Fig. 3.5.2 must be known or determined in an appropriate reference frame. Furthermore, the reference point to which those coordinates refer and the specific point to which the range measurements are made are required. The antenna usually is described with a *phase center*, the point to which range measurements refer. This phase center usually is obtained by empirical testing, including testing in anechoic chambers. Such testing determines the phase center pattern for a particular antenna. Experi-

ence has shown that such patterns are common to a particular antenna type so that
the test results of a subset may be applied to other nontested antennas.

For some antennas, the phase center pattern may be azimuthally symmetric,
but may exhibit change as a function of elevation. For precision applications,
the characteristics of the phase center variation must be known, as well as the
location of the phase center even if it is invariant. Furthermore, in some cases, the
coordinates published for a particular antenna may refer to a specific point other
than the phase center, usually referred to as the *antenna reference point* (ARP).
In these cases, the correction to the phase center must be separately applied. It is
important that the specific point where *a priori* coordinates for the antenna should
be applied is understood. To further complicate matters, the phase center for L_1
is usually different from the location for L_2. As shown in Fig. 3.5.2, the location
for the L_1 phase center is δ_1 with respect to the ARP and the location of the L_2
phase center is δ_2.

The precision of GPS pseudorange measurements is receiver dependent, but
most modern receivers are at the level of about 1 meter for P-code pseudorange.
Note that the combination of two measurements made at different frequencies to
remove ionosphere effects will produce a noisier measurement (Eq. 3.4.14). Although the precision of the pseudorange is meter level, the accuracy of the measurement is usually not comparable. For example, pseudorange measurements
may be negative because of unsynchronized clocks, a clear indication that a measurement may be precise, but not accurate. Nevertheless, the correction for clock
errors can be determined through appropriate estimation strategies and the corrections obtained from them will render the measurement accurate, as well as
precise. As a means to control the accuracy of pseudorange measurements, the
GPS satellite clocks may be *dithered*. This dithering produces clock errors that
cannot be accounted for without access to Department of Defense classified information. When clock dithering is activated, it is said that *Selective Availability*
(SA) has been invoked. Unclassified techniques to remove SA will be described
in Section 3.6, but SA was deactivated in May, 2000.

The pseudorange measurement given by Eq. (3.3.9) can be expanded for either
L_1 or L_2 to give

$$\tilde{\rho} = \rho + c(\delta t_R - \delta t_T) + \delta\rho_{\text{trop}} + \delta\rho_{\text{ion}} + \epsilon \qquad (3.5.1)$$

where

$\tilde{\rho}$ is the measured pseudorange,

ρ is the true range between the true transmit time and the true receive time,

δt_R is the receiver clock difference with true time,

δt_T is the transmitter clock difference with true time,

$\delta\rho_{\text{trop}}$ is the troposphere delay,

$\delta\rho_{\text{ion}}$ is the ionosphere contribution, and

ϵ represents remaining errors, such as instrument noise.

Computation of the true range, ρ, would require knowledge of the true GPS satellite position and the true receiver antenna coordinates, as well as the true transmit and receive times. In practice, these quantities may not be known with high accuracy. In the orbit determination problem, for example, the satellite position may be known *a priori* with an accuracy of several meters. In this instance, the error term ϵ will represent the error between the true range and the computed range formed from the *a priori* known position. The receiver ability to measure pseudorange is characterized by the instrument's precision, usually at the meter level.

A more precise GPS measurement is based on the carrier phase. With the previously described atmospheric effects, the usual form of the phase, expressed as range and similar to Eq. (3.5.1), can be obtained from Eq. (3.3.26). For either L_1 or L_2, it can be shown that

$$\Phi = \rho + c(\delta t_R - \delta t_T) + \lambda \widetilde{N} + \delta\rho_{\text{trop}} - \delta\rho_{\text{ion}} + \epsilon \qquad (3.5.2)$$

where

Φ is the measured phase range for the specified frequency,

λ is the wavelength of the signal (L_1 or L_2),

$\widetilde{N}$ is the integer phase ambiguity,

and the other terms were defined with Eq. (3.5.1).

Note that the raw phase measurement, ϕ, provided by a receiver consists of the accumulated integer cycles since a reference time, plus the fractional part of a cycle. The measured phase range is

$$\Phi = \lambda \phi \qquad (3.5.3)$$

but in some receivers the expression may require $\Phi = -\lambda\phi$.

For comparison with pseudorange, the precision of phase range is usually characterized at the several millimeter level. If the GPS receiver is carried on a satellite, the term $\delta\rho_{\text{trop}}$ is zero and $\delta\rho_{\text{ion}}$ may be sufficiently small to neglect. Even at 1000 km altitude, the ionosphere contribution is at the decimeter level. The phase range from L_1 and L_2 can be combined to remove the ionosphere contribution using the same approach applied to pseudorange (Hofmann-Wellenhof *et al.*, 1997).

In applications of GPS measurements to the determination of an orbit of a low Earth orbiter (LEO), such as described in Section 3.7, the positions of the GPS satellites must be known or determined. One option is to apply the techniques of estimation described in the following chapters to the determination of the orbits of the GPS satellites using a network of ground stations. In some cases, the GPS satellite orbits may be determined simultaneously with the orbit of a LEO, but in others the GPS satellite orbits may be fixed to orbits determined by other sources.

The GPS ephemerides can be recreated using information broadcast by the satellites in near real time. These *broadcast ephemerides* can be generated from 16 parameters (*navigation message* or *ephemeris parameters*) based on Keplerian orbit elements and coefficients of terms that represent time dependency (see Hofmann-Wellenhof *et al.*, 1997) . The set of broadcast parameters applies to a specific interval of time, typically two hours, and new sets of parameters are broadcast automatically by a GPS satellite as the applicable time interval changes. The accuracy of the broadcast ephemerides generally is regarded to be at the 10-meter level. The primary intent is for the broadcast ephemerides to support real-time or near real-time applications. The information used to create the parameters broadcast by the GPS satellites is based on determination of the GPS satellite orbits by the Department of Defense using a set of six monitor stations. These monitor stations record pseudorange and carrier phase measurements between the ground-based receiver and the GPS satellite, which are then used to determine the orbits through a process that makes use of the methodologies in the following chapters. The orbits determined by this process are then extrapolated forward in time and the extrapolated ephemerides are approximated with a model based on the 16 ephemeris parameters. The ephemeris parameters are uploaded to the GPS satellites for broadcast during the appropriate time interval.

Precise GPS ephemerides are available with a latency of about one day or longer. These ephemerides are intended to support applications with high accuracy requirements, such as those related to space geodesy. Depending on the application, the position accuracy of a satellite carrying a GPS receiver may approach the centimeter level, while other applications may require an accuracy of ten meters. Two sources of ephemerides are available: *National Imagery and Mapping Agency* (NIMA) and the *International GPS Service* (IGS). In both cases, the respective agency operates a ground network of GPS receivers to support the determination of the GPS orbits. In the case of the IGS, an international collaboration of agencies supports a ground network of 200 receivers, somewhat uniformly distributed around the Earth. Seven Analysis Centers of the IGS use a subset of measurements from these receivers to determine GPS ephemerides and the IGS combines these products into a single official IGS product. The IGS final product is available with a latency of two to three weeks with an accuracy expected to be at the decimeter level, but a rapid product is available with a one-day latency.

Satellite-to-Satellite Tracking (SST)

Various forms of *satellite-to-satellite tracking* (SST) are in use. This terminology usually applies to measurements collected between a pair of satellites, but the common terminology enables identification of the respective satellite altitude. If a GPS receiver is carried on a LEO satellite, then the previously described GPS measurements would be categorized as *high-low* SST measurements. For some GPS satellites, an inter-satellite range measurement is made, known as *cross-link ranging*, that would be a *high-high* SST measurement.

A recent example of *low-low* SST measurements is represented by the *Gravity Recovery And Climate Experiment* (GRACE). The SST measurements are primarily used to detect components of the Earth's gravitational field and especially gravity variations associated with redistribution of mass. Two low-altitude satellites are used and each satellite transmits signals at two frequencies in the K-band (24 GHz) and Ka-band (32 GHz), but the actual frequencies used by each satellite are not identical. The satellites have approximately the same 500-km altitude and are in the same orbit plane, but separated in the along-track direction by about 200 km. Each satellite carries a GPS receiver, but these receivers have been modified to track both the GPS signals and the K-band signals transmitted by the other GRACE satellite.

The GRACE SST measurements in the K-band are similar to GPS one-way measurements of carrier phase made in the L-band. Two K-band measurements are made at different frequencies on each satellite to enable a correction for the ionosphere using the technique discussed in Section 3.4.2. The measurements made and recorded by each GRACE satellite are a form of one-way carrier phase similar to GPS. Each GRACE satellite carries an ultra-stable oscillator as the frequency reference ($\sim$4.8 MHz) that is used to generate the transmitted signal and to mix with the signal that arrives from the other satellite. Simply stated, each satellite measures the carrier phase signal it receives from the other satellite relative to the signal that it transmits. In an approach similar to measurement differencing (Section 3.6), the measurements collected on each satellite can be added to obtain a measurement that is proportional to the range between the satellites, while at the same time removing long-term oscillator instability. This form of SST measurement has been termed *dual-one-way-ranging*. The measurement and instrumentation has been described by Dunn *et al.* (2003).

3.5.2 TWO-WAY RANGE

SLR

The basic concept of a two-way range was described in Section 3.3.1. Examples of this measurement type include ground-based *satellite laser ranging* (SLR) systems and *satellite-borne altimeters*. SLR was applied for the first time by

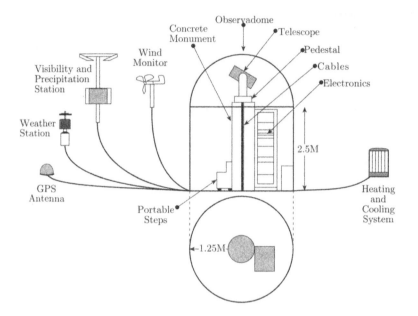

Figure 3.5.3: The illustrated SLR-2000 is designed for autonomous operation, whereas previous designs have required operators. The SLR concept is based on the measurement of time-of-flight between laser pulse transmission and receipt of echo photons. The telescope collects the echo photons and the electronics provide the time-of-flight (Degnan and McGarry, 1997).

NASA Goddard Space Flight Center in 1965 to Explorer 22 (Beacon Explorer-B).

Modern SLR stations use a Nd:YAG (neodymium yttrium aluminum garnet) laser to illuminate a satellite, usually at 0.532 micrometers (green). SLR systems rely on specially designed cube corners to reflect the incoming laser light back to the source, where a telescope collects the returning photons to measure the round-trip time from pulse transmission to pulse reception. Since lasers operate in the optical wavelengths, their operation will be dependent on local weather conditions. Nevertheless, the optical cube corners carried on a satellite to support SLR are simple and completely passive.

After conversion into a distance, the question arises for purposes of the computed range: what does the measured distance represent? Clearly, the range is measured to the cube corner on the satellite that reflected the arriving photons, but the originating point in the SLR may be more complex. In some cases, the distances of the laser paths within the instrument are carefully measured and a correction to some adopted reference point is made. In other cases, the instrument may be used to range to a calibration target at a very precisely (and independently) measured distance to derive a correction to an adopted reference point. As a consequence, a range measurement represents the distance from some adopted reference point within the SLR system to a cube corner on the satellite.

The collimated laser beam has a divergence with a diameter of a few hundred meters at satellite altitude; hence, the satellite must be within the beam to achieve a successful echo. Errors that contribute to the satellite's position with respect to the laser beam include (a) satellite position prediction and (b) the instrument's telescope mount.

Position prediction errors result from the generation of a predicted ephemeris, usually obtained from the solution of the satellite equations of motion (e.g., Eq. 2.3.46) with a specified initial state. The initial state usually is determined from the application of the estimation techniques described in later chapters. The resulting ephemeris can be expected to contain errors from uncertainties in the force parameters in the equations of motion (e.g., the value of the drag coefficient), errors in the force modeling, and errors in the initial conditions.

In some systems, the position prediction errors are exceeded by errors in the telescope mount, which contributes to the laser pointing direction. In such cases, an adaptive search technique may be applied during the tracking pass to compensate for all of these errors.

A concept under development by NASA for an autonomous system is illustrated in Fig. 3.5.3, known as SLR-2000, which is expected to be deployed around 2005 (Degnan and McGarry, 1997). In the SLR system, the predicted ephemeris is used to drive the laser/telescope pointing. A clock is an essential element, although different clocks may be used. A time-interval counter or an event timer is used to measure time over the short duration between the emission of the transmit, t_T, pulse, and the reception of the echo pulse, t_R. A second clock, usually synchronized to GPS-Time using a time-transfer receiver, provides the absolute time required for time tagging the measurement.

The *International Laser Ranging Service* (ILRS) coordinates the global operation of independent SLR stations. The global network of SLR stations is supported by national agencies, and the recorded data from these stations are made available with a few hours latency through the ILRS data centers. The range data are regularly used to monitor geocenter motion and Earth orientation, and to determine SLR station velocities.

Several satellites have been designed to support SLR geodetic science. These satellites are spherical with a small area-to-mass ratio to reduce the effects of nongravitational forces. The relation of target cross-section to altitude is evident by the characteristics of the three satellite pairs summarized in Table 3.5.2. The solution of the satellite equations of motion (Eq. 2.3.46) provide the position of the satellite center of mass in an appropriate reference frame, but the SLR measure-

Table 3.5.2:

Geodetic Satellite "Twins"

Satellite	Launch	Diameter (cm)	Mass (kg)	a (km)	e	i (deg)
Starlette	1975 France	24	47	7326	0.020	50
Stella	1993 France	24	48	7184	0.001	99
Lageos-I	1976 U.S.	60	411	12254	0.005	110
Lageos-II	1992 U.S./Italy	60	405	12145	0.010	53
Etalon-I	1989 Soviet Union	129	1415	25514	0.001	65
Etalon-II	1989 Soviet Union	129	1415	25490	0.001	65

Lageos: LAser GEOdynamics Satellite

Etalon satellites are in GLONASS orbit planes

ments are made to the outer surface of the satellite, so a correction to the measured range for satellite center of mass is required. This correction is essentially the radius of the sphere, though in a precise sense it will account for the effective radius based on the laser pulse interaction with the cube corners.

Nonspherical satellites, such as TOPEX/Poseidon and ICESat, carry arrays of cube corners to support SLR tracking. These arrays usually mount the cube corners in some hemispheric-like pattern to enable ranging returns from the expected SLR directions. The array carried on ICESat is shown in Fig. 3.5.4. The necessary corrections from the SLR array optical center to the satellite center of mass are obviously more complicated than those for a spherical satellite.

The single shot precision of modern SLR systems is better than 1 cm, which is usually representative of the measurement accuracy from the hardware viewpoint. Another interpretation of accuracy, however, involves the application of all corrections in the orbit determination process. Orbits determined for Lageos-I from SLR, for example, usually are characterized with an accuracy at the centimeter level, in an RMS sense.

PRARE

Another modern two-way ranging system was developed in Germany and carried for the first time on the European Earth Resources Satellites (*ERS*), known as ERS-1 and ERS-2. *PRARE* (Precise Range and Range Rate Equipment) places the primary transmitter and receiver in the satellite, but ground stations have both transmit/receive functions as well. Two signals are transmitted by the satellite-borne PRARE, one at 2.2 GHz (S-band) and the other at 8.5 GHz (X-band), and

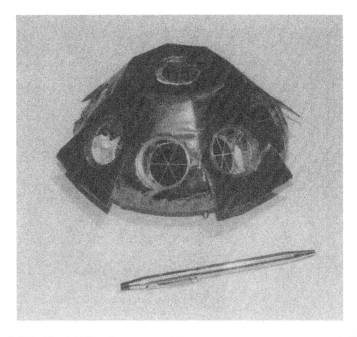

Figure 3.5.4: The ICESat SLR array. The circular components on each flat face contain a cube corner with the optical property of reflecting incoming light back to the source. There are nine corner cubes, each with a diameter of 3.2 cm. (Photo courtesy of ITE, Inc.)

both signals have a PRN code imposed. The time delay between the two frequencies is determined by the ground receiver and transmitted to the satellite. The received X-band signal is retransmitted to the satellite on a 7.2-GHz carrier frequency with the PRN code. A two-way range is obtained in X-band by measurement of the transmit and receive times of the X-band signal. The ionosphere correction is applied with the ground-transmitted determination based on the downlink signal. A one-way Doppler measurement is also made using the X-band.

Altimeter

A radar altimeter, such as those carried on TOPEX/Poseidon (T/P), operates similarly to the SLR system as a two-way measurement. The altimeter sends a chirp and measures the round-trip time. The TOPEX altimeter (U.S.) operates with two frequencies to enable a correction for the ionosphere, whereas the Poseidon altimeter (France) uses a single frequency and relies on other data to pro-

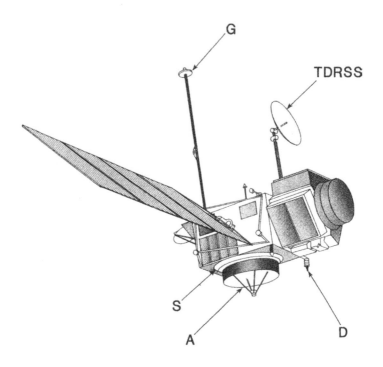

Figure 3.5.5: The TOPEX/Poseidon satellite. The radar altimeter antenna (A) is directed toward the Earth (nadir direction), the DORIS receiver antenna is D, the SLR array (S) consists of corner cubes mounted on the circumference of the altimeter antenna, the GPS receiver choke-ring antenna (G) is mounted at the end of a long boom to reduce multipath effects, and the TDRSS antenna is shown.

vide the ionosphere correction. The TOPEX frequencies are 13.6 GHz and 5.3 GHz, and the Poseidon frequency is 13.65 GHz. The altimeters share a common antenna, but only one altimeter is operated at a time. The TOPEX ionosphere correction can be obtained from Eq. (3.4.12), and the antenna is shown in Fig. 3.5.5. As shown, TOPEX carried several tracking systems that supported determination of the TOPEX orbit.

Laser altimeters, such as the *Mars Orbiter Laser Altimeter* (MOLA) and Earth-orbiting versions (such as the *Geoscience Laser Altimeter System*, or GLAS, carried on ICESat), function similarly to SLR. Both MOLA and GLAS operate with a near-infrared Nd:YAG wavelength of 1064 nanometers.

TDRSS

It is apparent from the discussion in Section 3.2.4 that tracking measurements collected by ground-based instrumentation will be limited to those time periods when the satellite is above the horizon at the ground location. For a low-altitude satellite, say 1000 km, such viewing opportunities will be limited to a few passes per day, with an individual pass duration of about 10 minutes or less. On the other hand, an appropriately positioned geosynchronous satellite always remains above a ground station's horizon and the geosynchronous satellite will have a very broad region to view the low-altitude satellite. The *NASA Tracking and Data Relay Satellite System* (TDRSS) takes advantage of this latter characteristic by enabling range and range-rate measurements collected between a satellite in the TDRSS constellation and a low-altitude satellite. In principle, a constellation of three satellites in equatorial, geosynchronous orbit separated by 120 degrees can provide continuous tracking coverage of low Earth orbiters. In addition to the satellite constellation, a ground station provides the crucial link back to the ground.

The basic TDRSS constellation consists of two operational satellites with an angular separation of 130°, which provides 85%–100% coverage for low-altitude satellites. For redundancy, on-orbit satellite spares are included in the constellation. Most of the operational satellites and stored spares are clustered near 41° W and 171° W, with one satellite at 275° W. The system is supported by a ground complex located at White Sands, New Mexico, with a remote extension in Guam. The constellation consisted of eight satellites in mid-2002, including some of the original satellites and two of a new generation of satellites, known as TDRS-H, -I, and -J in the prelaunch naming convention. TDRS-H was launched on June 30, 2000, and it was renamed TDRS-8 when it was accepted in October, 2001. The position of TDRS-8 is 171° W. TDRS-I and -J were launched in 2002 and were stationed at 150° W and 150.7° W.

With the definitions applied previously, the TDRSS employs *multiway* measurements. For example, a signal is transmitted from the White Sands complex to a TDRS in the K-band with a ranging code. This signal is retransmitted at the TDRS to a user satellite on S-band or K-band and, in turn, retransmitted back to the TDRS, where it is transmitted back to White Sands. Although the signal travels multiple paths, the basic observation equation is given by Eq. (3.3.10) for two-way range. In the case of TDRSS, however, corrections to this equation are required to account for delays at the respective instrumentation for signal retransmital, or transponder delays. Furthermore, the problem of determining the orbit of a user satellite is now linked to the problem of determining the orbit of the TDRS satellite used in the transmission links. It is evident that any errors in the TDRS orbit will map into errors for the user satellite, but the TDRS orbit errors may be accommodated by estimating the TDRS orbit parameters as well as those for the user satellite. For detailed discussion on orbit determination using TDRSS, see papers by Rowlands *et al.* (1997) , Marshall *et al.* (1996), and Visser and Ambrosius (1997).

3.5.3 DOPPLER SYSTEMS

One-Way

A classic example of this technique has been the Navy Navigation Satellite System (NNSS) or Transit, which consisted of a constellation of several satellites in polar orbit. The NNSS began operation in the 1960s with the purpose that the constellation was to support navigation of U.S. Navy submarines. Operations in support of Navy applications ceased in the 1990s with the operational use of GPS. Each satellite carried a beacon that broadcasted signals with two frequencies for ionosphere correction: 150 MHz and 400 MHz. As described in Section 3.3.2, the receiver measured the Doppler count over specified intervals of time (e.g., 3.8 sec), thereby providing a measure of the Doppler shift or range-rate as the basic measurement for the navigation process.

In the 1970s, another spaceborne beacon was used for geodetic applications. For example, the satellite GEOS-3 carried an SLR reflector and beacons that broadcast signals at 162 MHz and 324 MHz, as well as a single-frequency radar altimeter operating at 13.9 GHz.

The French space agency, Centre National d'Etudes Spatiales (CNES), operates a Doppler system that utilizes ground-based beacons. In this system, the less expensive beacons are placed on the ground and a single receiver is carried on the satellite. This system, known as *DORIS (Doppler Orbitography and Radio positioning Integrated by Satellite)*, was used for the first time on the French imaging satellite, SPOT, in the 1980s, but this application was followed by use on TOPEX/Poseidon (see Fig. 3.5.5) in the early 1990s. DORIS will be used on several satellites to be operated in the early years of the 21st century. DORIS uses two frequencies for ionosphere correction: 401.25 MHz and 2036.25 MHz. DORIS measurements are usually processed as range-rate and the precision is better than 0.5 mm/sec.

The GPS carrier phase is another form of Doppler, although it is typically converted into a range measurement for processing. The basic measurement is usually extracted from the instrumentation as an integer number of cycles, measured since some start time, plus the fractional part of a cycle. The accumulated phase from the start time, expressed by Eq. (3.3.23), usually is converted into a distance measurement by multiplying with the wavelength of the signal (Eq. (3.5.3)), but the phase measurement can be converted into Doppler as an alternate measurement.

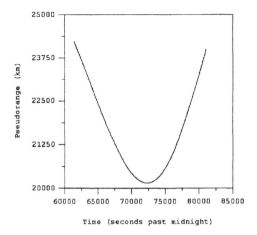

Figure 3.5.6: McDonald Observatory (MDO1) L1 pseudorange from PRN-6 on 19 June 1999. The uncorrected pseudorange as recorded by the receiver at a 30 second interval is shown. No pseudoranges were recorded for 200 seconds at about 67000 seconds.

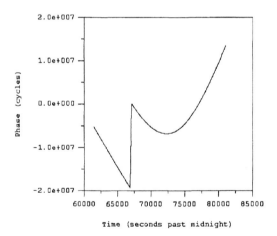

Figure 3.5.7: McDonald Observatory (MDO1) L1 carrier phase from PRN-6 on 19 June 1999. The uncorrected carrier phase as recorded by the receiver at a 30 second interval is shown. A 200 second interval where the receiver loses count of cycles is evident as the discontinuity.

3.5.4 EXAMPLES

GPS

Figures 3.5.6 and 3.5.7 illustrate measurements acquired by a GPS receiver (known as MDO1) at McDonald Observatory, near Ft. Davis, Texas. The receiver was a TurboRogue SNR8000 with a choke-ring antenna. The data were obtained on June 19, 1999, and recorded at a 30-second interval. Figure 3.5.6 shows the L1 pseudorange, expressed in kilometers, obtained from PRN-6. The measurements were recorded by the receiver at a 30-second interval. The SNR8000 receiver performs a *clock-steering*, meaning that the receiver reference oscillator is changed (a disciplined oscillator) so that the clock is steered toward GPS-Time, thereby closely synchronizing the receiver with GPS-Time to better than 100 ns. However, the MDO1 receiver uses an externally controlled high-quality quartz reference oscillator that is also used by the SLR system (known as *MLRS*), which is adjusted to GPS-Time separately.

The SNR8000 is an example of a class of geodetic quality receivers available from several vendors (e.g., Trimble Navigation, Ashtech, and Leica). These receivers simultaneously track the signals from up to 12 GPS satellites. The receiver used at MDO1 could track up to eight satellites. In spite of Y-code transmission (encrypted P-code) by the GPS satellites, the receivers provide measurements of pseudorange and carrier phase on L1 and L2 at a selected time interval.

The L1 pseudorange measurement for PRN-6 as recorded by the receiver is illustrated in Fig. 3.5.6, which simply illustrates the expected pattern. Comparing this figure to the simulated case in Fig. 3.2.2 shows the expected effect of the higher altitude GPS satellite on the magnitude of the range and the point of closest approach.

The L1 carrier phase measurement for PRN-6 is illustrated in Fig. 3.5.7, where the units are in cycles. Aside from the difference in units, the carrier phase exhibits similar characteristics to the pseudorange, except for the obvious discontinuity at 67000 seconds that does not appear in the pseudorange. This discontinuity results from an event that caused the receiver to lose lock on all GPS satellites for about 200 seconds and when tracking resumes, the carrier phase count was reinitialized to zero. On the other hand, since the pseudorange is based on code-correlation, there is no discontinuity. A discontinuity in carrier phase is referred to as *cycle slip*, which may produce dramatic discontinuities such as the Fig. 3.5.7 case, but they commonly produce cases where a small number of cycles are lost. Various techniques have been developed to fix cycle slips, which amount to using the carrier phase data before and after the discontinuity to estimate the integer number of cycles to adjust the data to create a continuous change in carrier phase. The data sampling of 30 seconds and the 200-second data gap make this particular cycle slip difficult to fix, but cases where the break may be a few cycles within the 30-second recording are straightforward to repair. If the cycle slip cannot be fixed,

Table 3.5.3:

ITRF-2000 Coordinates and Velocities for MDO1 and PIE1 Reference Points

ID	P/V	x	y	z
MDO1[1]	P	-1329998.6780	-5328393.3870	3236504.1990
	V	-0.0125	-0.0001	-0.0065
PIE1[2]	P	-1640916.7930	-5014781.2040	3575447.1420
	V	-0.0147	-0.0006	-0.0084

[1]Location: McDonald Observatory, Texas
[2]Location: Pie Town, New Mexico

Position (P) for January 1, 1997, in meters; velocity (V) in meters per year. The station velocity is caused primarily by tectonic plate motion. Since both stations are fixed to the North American Plate, which is rotating as a rigid plate, their velocities are similar. (xyz) is the ECF realized by ITRF-2000.

the single pass may be divided into two passes where the pass division occurs at the break.

The L_1 carrier phase can be converted to phase-range, a form of pseudorange, by multiplying by the signal wavelength (19 cm). This form of the range is given in Eq. (3.3.23), where the linear term disappears when f_G equals f_T, which is the case for this receiver. Direct comparison of the phase-range to the pseudorange derived from the transmitted code (Figs. 3.5.6 and 3.5.7) shows that there is a large bias between the two forms of pseudorange. This bias is caused by an *ambiguity* in the phase measurement (the term $\tilde{N}$ in Eq. (3.5.2)), which results from the lack of any unique feature in individual cycles that can be correlated with time. In the pseudorange measurement based on a code, the correlation with time is performed because of the unique character of the code sequence. With carrier phase measurements, various techniques have been developed for *ambiguity resolution*; that is, determination of $\tilde{N}$. Such techniques are attractive since the precision of phase-range is a few millimeters, whereas code pseudorange is decimeter-level. By resolving the ambiguity in the carrier phase measurement, the phase-range is significantly more precise than code pseudorange.

The quality of the measurements cannot be ascertained from Figs. 3.5.6 and 3.5.7, since the character is dominated by the orbital motion. For this example, the orbit position will be assumed to be known, thereby allowing a computed range, C, to be obtained and the residual O-C to be generated. By removing the dominant orbital motion effect, the measurement quality (including the quality of

Table 3.5.4:
Contributing Errors to Figs. 3.5.8 and 3.5.9

(O-C) Contributing Errors	Expected Error Magnitude
Use of geometric range	~ 100m
Satellite and receiver clock corrections	~ 100m
Selective Availability (SA)	~ 50m
Ionosphere and troposphere corrections	~ 10m
Transmit phase center	< 1m
Receiver phase center	< 0.1m
Receiver noise	0.5m
IGS ephemeris and ITRF coordinates	0.1m

the computed range) can be more readily exhibited.

Using the IGS final ephemeris and ECF coordinates of the receiver, the O-C can be computed for both code pseudorange and phase-range. For the cases described in the following paragraphs, the ITRF-2000 coordinates for the MDO1 site are given in Table 3.5.3. The time tag of each measurement recorded by this receiver is the *receive time* and the IGS ephemeris for the satellite is provided at an interval of 15 min. The MDO1 coordinates refer to the location of a specific reference point expressed with respect to the Earth center of mass as implied by the use of ITRF-2000. The IGS ephemeris for the PRN-6 center of mass actually refers to ITRF-97, but the difference between ITRF-97 and ITRF-2000 ephemerides is small.

The residual O-C for MDO1 is shown in Fig. 3.5.8, after removal of the computed range from the recorded pseudorange shown in Fig. 3.5.6. For this example, the computed range, C, was formed in the ECF using Eq. (3.2.4), with no corrections applied. Since no corrections have been made and geometric range was used for C, the error sources contributing to these residuals are summarized in Table 3.5.4.

When the residuals are computed for an IGS receiver located at Pie Town, New Mexico (known as PIE1 and located about 550 km northwest of MDO1), strong similarities are observed. Figure 3.5.9 shows the PIE1 residuals, with the computed range formed under the same assumptions as those for MDO1 in Fig. 3.5.8. Both Figs. 3.5.8 and 3.5.9 are dominated by two characteristics: an average linear trend and periodic variations. Close inspection of the periodic variations

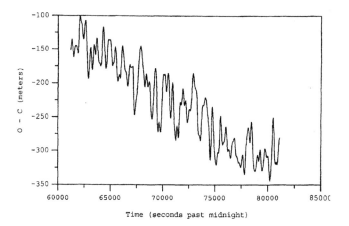

Figure 3.5.8: MDO1 L_1 pseudorange residuals (O-C) from 19 June 1998 for PRN6. The computed pseudorange (C) was obtained from the geometric range at signal receive time with no corrections. The one meter measurement noise is not evident at this scale since the O-C characteristics are dominated by an overall drift and high-frequency variations.

shows that this effect is essentially identical at both receivers. Using terminology introduced later in this book, one would say that the two characteristics have a high *correlation*. Of the sources listed in Table 3.5.4, only *Selective Availability* (SA) is expected to exhibit the somewhat periodic variations shown. In fact, repeating the same experiment one year later after SA has been "turned off" shows no evidence of these variations. The direct removal of SA is limited by the fact that information about the SA characteristics is classified, which eliminates the direct removal for nonclassified applications. The usual alternate approach, which is described in a later section, is to difference Figs. 3.5.8 and 3.5.9, which removes errors that are common to both receivers, such as SA.

After SA, the next dominant characteristics in the residuals are the nonzero mean and the linear slope that remains after "removal" of the mean and the periodic variations. Both the nonzero mean and the linear slope are caused by uncorrected clock errors, with some contribution from the generation of the computed range using the geometric range.

With the expectation that the receiver coordinates and the IGS ephemerides for the GPS satellites have errors at the decimeter level (see Table 3.5.4), the plots for (O-C) would show only random noise at the 50-cm level if the one-way range were properly computed from the available information, the clock errors were corrected, and the atmospheric delays were accounted for. The two receivers used

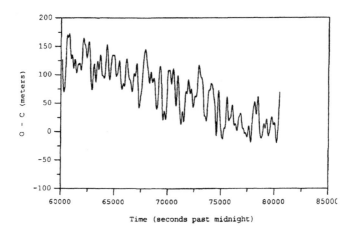

Figure 3.5.9: PIE1 $L - 1$ pseudorange residuals (O-C) for PRN6 from 19 June 1998. These residuals were formed with the same procedure used in Fig. 3.5.8. The high-frequency variations have a high correlation with those observed for the same time period at MDO1; therefore, these variations are not produced by the receiver. The variations are associated with Selective Availability.

in this example record pseudorange and carrier phase at a 30-second interval.

If carrier phase is converted to range, the plots for (O-C) would be dominated by a very large bias: the phase ambiguity term. However, removal of the ambiguity results in residual plots with characteristics very similar to Figs. 3.5.8 and 3.5.9. Since phase-range noise is usually a few millimeters, removal of the effects described in the preceding paragraph would illustrate the low noise level. The instrument noise statistics are usually used to describe the *measurement precision*.

DORIS

An example of DORIS measurements obtained from TOPEX/Poseidon is shown in Fig. 3.5.10. The data in this figure are based on a DORIS beacon located at Fairbanks, Alaska. Both DORIS frequencies were used to remove the ionosphere, but there would be no noticeable difference on the scale shown if the plot had been made using a single-frequency measurement.

Altimeter

As previously described, the altimeter measurement is based on two-way time of flight. The altimeter antenna is directed toward the nadir and a one-way range

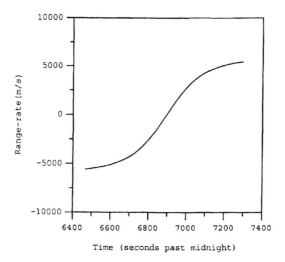

Figure 3.5.10: DORIS Range-rate from TOPEX/Poseidon on June 8, 1999.

(altitude) is formed by dividing the round-trip time-of-flight by two. The TOPEX altimeter data collected during a long pass over the Pacific (TOPEX/Poseidon Repeat Cycle 303) are shown in Fig. 3.5.11. For this example, the dual-frequency correction for the ionosphere was made, along with other corrections, at the meter level. The T/P orbit is frozen (see Chapter 2) with argument of perigee at 90°, which occurs just past 35500 seconds. As illustrated, the minimum altitude occurs near 34500 seconds, approximately corresponding to a latitude of 0°. The apparent contradiction of the perigee location and the minimum altitude is explained by the ellipsoidal character of the Earth, which plays a dominant role in Fig. 3.5.11, since the eccentricity of the T/P orbit is small (< 0.001).

Using an ephemeris of high accuracy for T/P, a computed altitude (C) to the Earth ellipsoid can be formed. The altitude residual, O-C, can be generated and is shown in Fig. 3.5.12, using satelllite geodetic latitude for the abscissa. The geodetic latitude at $-65.7°$ corresponds to the initial time of 32759 seconds. The residuals illustrate the profile of the ocean surface, expressed with respect to a reference ellipsoid, since this pass crosses mostly ocean.

3.6   DIFFERENCED MEASUREMENTS

In some cases, the measurements made by a particular technique are differenced in special ways. The differencing of the measurements removes or dimin-

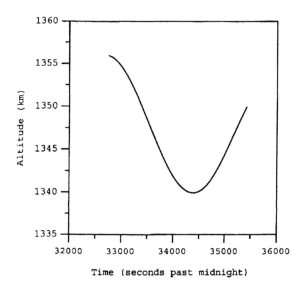

Time (seconds past midnight)

Figure 3.5.11: The TOPEX altimeter data collected during a long pass over the Pacific (TOPEX/Poseidon Repeat Cycle 303) are shown here.

ishes one or more error sources. The most common use of such differencing is with GPS pseudorange and carrier phase measurements and with altimeter measurements.

3.6.1 DIFFERENCED GPS MEASUREMENTS

Figures 3.5.8 and 3.5.9 show a very clear correlation in high-frequency variations in the PRN-6 pseudorange residuals collected with two receivers separated by 550 km, as discussed in Section 3.5.4. These variations are caused by *Selective Availability* (SA), an intentional dithering of the broadcasting satellite clock. SA was turned off in May 2000, but provides an excellent example of the application of differenced data. An obvious approach to removing this effect is to difference the residuals (or the measurements) shown in Figs. 3.5.8 and 3.5.9, which is referred to as a *single difference (SD)*. In this case, the pseudorange measurements (or carrier phase) from a specific GPS satellite recorded at two different receivers can be differenced to produce:

$$SD_{jk}^i = \rho_j^i - \rho_k^i \qquad (3.6.1)$$

where i identifies the GPS satellite, j identifies one receiver, and k identifies the other receiver. Alternatively, the single difference could be formed using mea-

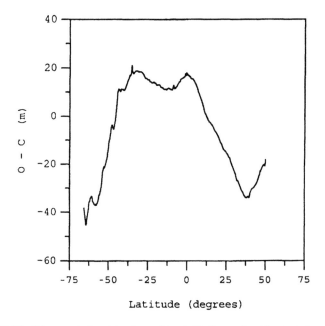

Figure 3.5.12: The geodetic altitude residual, O-C, can be illustrated using satellite geodetic latitude for the abscissa, as shown here. The O-C is generated from the observed geodetic altitude (Fig. 3.5.11) and a computed altitude formed from a TOPEX ephemeris determined form DORIS and SLR. The residual illustrates the along-track ocean surface profile with respect to the reference ellipsoid.

surements collected by one receiver and two different satellites.

An SD residual can be formed from the *a priori* receiver coordinates and the given coordinates for the broadcasting satellites to generate the *computed SD*. Hence, the SD residual is

$$SD(O - C)^i_{jk} = (O - C)^i_j - (O - C)^i_k, \qquad (3.6.2)$$

where O can be either pseudorange or carrier phase and $SD(O - C)$ represents the single difference residual using satellite i and receivers j and k. Using Eq. (3.6.2), if the residuals shown in Fig. 3.5.8 are differenced with those in Fig. 3.5.9, the result is shown in Fig. 3.6.1, where it is evident that the SA high-frequency variations have been removed.

The single-difference residuals formed using a different satellite, PRN-30, are shown in Fig. 3.6.2. Although they are not shown, the residuals formed in the same way as Fig. 3.5.8 or 3.5.9, but using PRN-30 instead of PRN-6, exhibit similar high-frequency variations. Since the SA effect is satellite-dependent, the

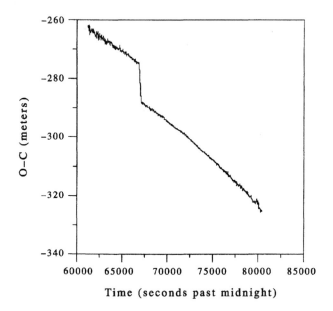

Figure 3.6.1: Single difference L1 pseudorange residuals formed with PRN-6, MDO1, and PIE1.

difference in the specific characteristics of the high-frequency variations is expected.

The SD behavior shown in Figs. 3.6.1 and 3.6.2 warrants further discussion. The formation of the SD has removed most of the high-frequency variations that are evident in Figs. 3.5.8 and 3.5.9. The remaining SD residuals exhibit a linear slope, a discontinuity at about 67000 sec, and some very small amplitude variations with an apparent high frequency. The remaining linear slope suggests that the effect is caused by sources that are not common in the data collected by the two receivers. Such effects include errors in the respective receiver clocks, errors contributed by atmosphere delays, and errors in the approximation of the computed range by the geometric range. Although the receiver separation of 550 km perhaps seems large, it is not when compared to the GPS orbit radius of 26000 km. Since both PRN-6 and PRN-30 are within view of each receiver, the line-of-sight to each satellite from the respective receiver is similar, which means that the atmospheric delays will be very similar and some portion (but not all) of the atmospheric delay will be removed in the SD formation. Errors associated with use of geometric range will partially cancel in the SD as well. However, the clock

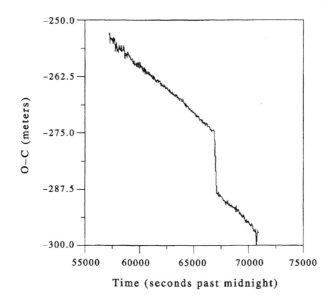

Figure 3.6.2: Single difference L1 pseudorange residuals formed with PRN-30, MDO1, and PIE1.

behavior of each receiver will be different and cannot be expected to cancel. As a consequence, receiver clock errors are expected to be the dominant error source that produces the linear variation in SD residuals shown in Figs. 3.6.1 and 3.6.2.

The discontinuity in Figs. 3.6.1 and 3.6.2 is another matter. This discontinuity occurs when the MDO1 receiver did not record data, but the PIE1 receiver does not show a gap in the data record. Although the precise cause of the discontinuity is not known, it appears to be associated with some event at MDO1, such as a clock anomaly related to the data gap.

The third characteristic evident in the SD residuals is an apparently small amplitude variation, which exhibits larger amplitude at the endpoints. These endpoints correlate with satellite rise and satellite set of one or both satellites. Clearly, SD measurements require that data from two satellites be available, so the difference between satellites requires that both satellites are located above the receiver horizon. The nature of these apparent high-frequency variations will become clearer in the following discussion.

The single differences for PRN-6 and PRN-30 can be further differenced to remove additional common errors. The resulting double difference (DD) is

$$DD_{jk}^{im} = SD_j^i - SD_k^m \qquad (3.6.3)$$

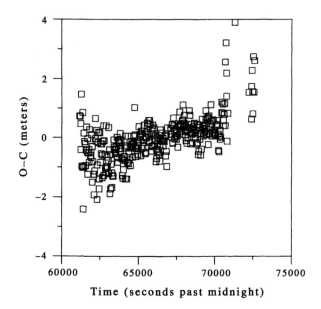

Figure 3.6.3: Double difference L1 pseudorange residuals formed from PRN-6, PRN-30, MDO1, and PIE1.

where DD represents the DD residual, formed between satellites i and m, and receivers j and k. The DD residual can be formed by differencing the single difference residuals, for example.

Double difference residuals can be formed by differencing the data used to create the plots of Figs. 3.6.1 and 3.6.2. The corresponding pseudorange DD residuals are shown in Fig. 3.6.4. It is evident that the dominant characteristics observed in the SD residuals have disappeared and the residuals are now scattered near zero. These residuals now exhibit a more random character and the only systematic characteristic is a slight linear slope. The fact that the Fig. 3.6.4 residuals are near zero demonstrates that the coordinates of the receivers given in Table 3.5.3 and the IGS ephemerides are quite accurate. If the receiver coordinates or the satellite ephemerides contained significant errors, the resulting DD residuals would reflect those errors, though not necessarily in a one-to-one ratio.

Note these additional points from Fig. 3.6.4. The obvious scatter is referred to as a *measurement noise* or *measurement precision*. The scatter seems to increase near the initial times and the final times, which is likely caused by low elevation effects at one or both receivers where the signal-to-noise ratio is degraded.

If the SD residuals had been computed using carrier phase (converted to phase

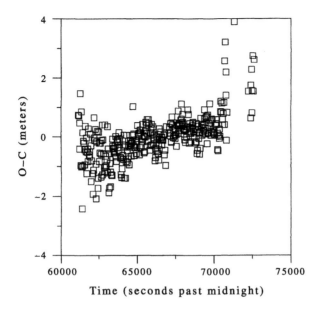

Figure 3.6.4: Double difference L1 pseudorange residuals formed from PRN-6, PRN-30, MDO1, and PIE1.

range) and if the computations had been performed in the same way as those for pseudorange described earlier, the residuals would have a large bias caused by the phase ambiguity. Without any correction for this ambiguity, the ambiguities in each of the phase range measurements would combine to produce a constant bias in the phase range DD residuals. Determination of the respective phase ambiguities is referred to as ambiguity resolution. In many cases the ambiguity may be estimated with other parameters in the estimation process. Once the phase ambiguity has been removed, either in the individual receiver/satellite measurements or as a combined term in the DD, the residuals formed after ambiguity removal would exhibit characteristics similar to Fig. 3.6.4, with the major exception that the residuals would be at the level of a few millimeters instead of the meter level evident in Fig. 3.6.4.

3.6.2 DIFFERENCED ALTIMETER DATA

Although a satellite-borne altimeter has a distinct advantage in its global distribution of data, consideration of possible error sources when altimeter data are directly used for orbit determination suggests several disadvantages. In particular, long wavelength oceanographic features and nontemporal ocean topography can be absorbed into the orbit when altimeter data are directly used for orbit and geodetic parameter determination. The nontemporal ocean topography, mostly due to error in the global geoid, has meter-level uncertainty and is significant when decimeter radial orbit accuracy of certain satellites, such as TOPEX/Poseidon, is desired.

A technique that eliminates the altimeter dependence on the nontemporal ocean topography is the use of altimeter measurements at the points where the orbit ground track intersects. These points are referred to as *crossover points*, and the differenced altimeter measurements at these points are referred to as *crossover measurements*.

Crossovers have been valuable for the evaluation of radial orbit ephemeris error. Although the nontemporal portion of the ocean topography can be eliminated at the crossover point, the remaining temporal changes, such as ocean tide, unmodeled orbit error, short wavelength phenomena, as well as altimeter time tag error, can still be aliased into the radial orbit error on a global basis. With the exception of the geographically correlated error due to inaccuracy in the Earth's gravity field (Christensen *et al.*, 1994) , global computation and analysis of crossover residuals can provide valuable information about radial orbit error sources.

In order to use crossover measurements, several procedures that are unique to these measurements must be employed. First the crossover times, t_i and t_j, must be generated from a nominal orbit. This nominal orbit may have been previously determined from a least squares estimate using GPS data or ground-based tracking data such as laser range. The crossover times t_i and t_j are computed using a ground track corresponding to the nominal orbit. The final results of this step are t_i and t_j for all crossover points in the nominal ephemeris.

The crossover measurements are generated using the crossover times. Since the altimeter measurements are usually recorded at a fixed time interval, it is rare that a measurement will exactly correspond to the crossover times (t_i and t_j). As a consequence, the altimeter measurements in the vicinity of each crossover time are used to create a pseudo-measurement, by the application of an interpolating function evaluated at the crossover time. Candidate interpolating functions include a least squares cubic spline, but other functions may be used. Depending on the nature of the surface (such as laser altimeter data on rough land surfaces), which may produce altimeter data with more noise than on a smooth surface, care must be taken in the application of an interpolating function to minimize the effect of the noise. With pseduo-measurements at each crossover time, the crossover measurement is obtained by differencing the two measurements and the crossover is assigned two time tags, t_i and t_j. For most applications, the crossovers should be formed consistently, such as the ascending track pseudo-measurement minus the descending track pseudo-measurement. To facilitate subsequent application in the estimation process, these crossover measurements may be sorted in chrono-logical order of the ascending track time. Additional information on altimeter crossover measurements can be found in Born *et al.* (1986) , Shum *et al.* (1990), and Lemoine *et al.* (2001).

3.7 SATELLITE POSITIONS

In some cases, the observational set may be an ephemeris of satellite posi-tions. The estimation techniques of the following chapters can be applied to such observations. Depending on the source of the ephemeris, it may exhibit a variety of error sources that depend on the methodology used to derive the ephemeris.

A commonly used ephemeris can be readily obtained if the satellite carries a GPS receiver. The GPS receiver usually provides its position in near real time from pseudorange measurements and the ephemeris information broadcast by the GPS satellites. As described in Section 3.5.1, a GPS receiver extracts the broad-cast ephemeris and clock parameters from the information superimposed on the carrier signal. Equation (3.5.1) gives the relationship for pseudorange as

$$\tilde{\rho} = \rho + c(\delta t_R - \delta t_T) + \delta\rho_{\text{trop}} + \delta\rho_{\text{ion}} + \epsilon.$$

With a receiver carried on a LEO satellite, the GPS signals arriving at the antenna do not pass through the troposphere unless the antenna is oriented at a significant angle with respect to the zenith direction, so the term $\delta\rho_{\text{trop}}$ usually can be ignored. The ionosphere contribution, $\delta\rho_{\text{ion}}$, may be small at LEO satellite altitude (e.g., < 10 cm at TOPEX altitude). As a consequence, the term can be ig-nored for some applications, but it is required for the most demanding applications with high accuracy. For those applications where the ionosphere term is required, some GPS receiver provides two frequency measurements and the correction for the ionosphere can readily be made with Eq. (3.4.14). Another approach is to apply a model for the ionosphere if a single-frequency receiver is available, but the accuracy of the correction will be degraded with respect to the dual-frequency measurement.

The transmitter clock correction in Eq. (3.5.1), δt_T, can be determined from the set of parameters broadcast by the satellite (or from other sources) for its clock and ρ is dependent on the GPS satellite position at the transmit time and the receiver position at the receive time. In the preceding equation, the unknowns are the receiver clock correction (δt_R) and the position vector of the GPS receiver carried on a LEO, which defines four unknowns. If the receiver measures the pseudorange from a minimum of four or more GPS satellites, then four equations similar to the preceding equation can be formed for each of the observed GPS satellite, which defines a system of four nonlinear algebraic equations. If more than four GPS satellites are visible, then an overdetermined system of equations exists. Such a system of equations can be solved using, for example, the Newton Raphson method. This method is described and applied to the solution of an overdetermined system of least squares equations in Section 4.3.4.

The solution for receiver position and clock offset is referred to as the *navigation solution* and it is commonly computed in near real time by GPS receivers, especially those designed for use on LEO satellites. Depending on the characteristics of the receiver, the accuracy of the position in the absence of Selective Availability is about 10 meters, where the error in the solution is dominated by the error in the GPS broadcast ephemerides. Nevertheless, an ephemeris formed by the navigation solution could be used as a set of observations in the application of an estimation algorithm (e.g., Thompson *et al.*, 2002).

The navigation solution is one form of *kinematic solution*. As the term implies, the solution is not dependent on the dynamical description of the satellite and the equations of motion (Eq. 2.3.46) are not required. A significant advantage for some applications is the fact that this solution can be generated in near real time. Another form of kinematic solution makes use of carrier phase measurements and the resolution of ambiguities to obtain a more accurate solution (Hofmann-Wellenhof *et al.*, 1997), but these solutions typically make use of high-accuracy GPS ephemerides, which usually preclude real-time LEO applications.

3.8   ANGLES

At the beginning of the space age, tracking of artificial satellites included an extension of astronomical techniques based on the use of photographic plates. With a camera mounted on a telescope, a photographic plate was produced for a region of the sky during a specified exposure time. Since the mount would compensate for the rotation of the Earth, stars would appear as dots but celestial bodies and artificial satellites would produce a streak on the plate since they move at non-sidereal rates. The Baker-Nunn system was introduced to track satellites using photographic techniques. In the modern era, the photographic plates have been replaced by a *Charge Coupled Device* (CCD) camera, but the methodology remains much the same. Calibration of the camera/mount is an important part of the operation of these instruments.

A photograph of a satellite streak against a background of stars provides a set of observations that can be deduced from the location of the satellite with respect to the nearby stars. The stellar coordinates can be determined from a star catalog, such as the catalog derived from the European Space Agency Hipparcos astrometry satellite. The basic set of satellite observations derived from the stellar coordinates are right ascension and declination, typically expressed with respect to J2000. One modern application of right ascension and declination observations of a satellite is in space surveillance applications. For example, the *Air Force Maui Optical and Supercomputing* (AMOS) site operates several telescopes on Mt. Haleakala to support surveillance.

The remainder of the Minitrack system built to track the Vanguard satellites is now confined to the continental U.S., but this space fence (operated by the U.S. Navy to support space surveillance) provides angular measurements in terms of direction cosines. The space fence consists of a network of transmitters and receivers located along a great circle. When a satellite (including space debris) crosses the electronic fence, a set of measurements given by time and direction cosines are determined, which is introduced into a process to determine both the orbit of the object and its identity.

3.9 REFERENCES

Ashby, N., "Relativity and the Global Positioning System," *Physics Today*, Vol. 55, No. 5, pp. 41–47, May 2002.

Born, G. H., B. D. Tapley, and M. L. Santee, "Orbit determination using dual crossing arc altimetry," *Acta Astronautica*, Vol. 13, No. 4, pp. 157–163, 1986.

Christensen, E., B. Haines, K. C. McColl, and R. S. Nerem, "Observations of geographically correlated orbit errors for TOPEX/Poseidon using the Global Positioning System," *Geophys. Res. Ltrs.*, Vol. 21, No. 19, pp. 2175–2178, Sept. 15, 1994.

Degnan, J., and J. McGarry, "SLR2000: Eyesafe and autonomous satellite laser ranging at kilohertz rates," *SPIE Vol. 3218, Laser Radar Ranging and Atmospheric Lidar Techniques*, pp. 63–77, London, 1997.

Dunn, C., W. Bertiger, Y. Bar-Sever, S. Desai, B. Haines, D. Kuang, G. Franklin, I. Harris, G. Kruizinga, T. Meehan, S. Nandi, D. Nguyen, T. Rogstad, J. Thomas, J. Tien, L. Romans, M. Watkins, S. C. Wu, S. Bettadpur, and J. Kim, "Instrument of GRACE," *GPS World*, Vol. 14, No. 2, pp. 16–28, February 2003.

Herring, T., "Modeling atmospheric delays in the analysis of space geodetic data," in *Refraction of Transatmospheric Signals in Geodesy*, J. C. DeMunck and T. A. Th. Spoelstra (eds.), Netherlands Geodetic Commission Publications in Geodesy, 36, pp. 157–164, 1992.

Hofmann-Wellenhof, B., H. Lichtenegger, and J. Collins, *Global Positioning System: Theory and Practice*, Springer-Verlag, Wien-New York, 1997.

Leick, A., *GPS Satellite Surveying*, J. Wiley & Sons, Inc., New York, 2003.

Lemoine, F., D. Rowlands, S. Luthcke, N. Zelensky, D. Chinn, D. Pavlis, and G. Marr, "Precise orbit determination of GEOSAT follow-on using satellite laser ranging and intermission altimeter crossovers," NASA/CP-2001-209986, *Flight Mechanics Symposium*, John Lynch (ed.), NASA Goddard Space Flight Center, Greenbelt, MD, pp. 377–392, June 2001.

Marini, J. W., and C. W. Murray, "Correction of laser range tracking data for atmospheric refraction at elevations above 10 degrees," *NASA GSFC X591-73-351*, Greenbelt, MD, 1973.

Marshall, J. A., F. J. Lerch, S. B. Luthcke, R. G. Williamson, and C. Chan, "An Assessment of TDRSS for Precision Orbit Determination," *J. Astronaut. Sci.*, Vol. 44, No. 1, pp. 115–127, January–March, 1996.

Montenbruck, O., and E. Gill, *Satellite Orbits: Models, Methods, and Applications*, Springer-Verlag, Berlin, 2001.

Parkinson, B., J. Spilker, P. Axelrad, and P. Enge (eds.), *Global Positioning System: Theory and Applications*, Vols. 1–3, American Institute of Aeronautics and Astronautics, Inc., Washington DC, 1966.

Ries, J. C., C. Huang, M. M. Watkins, and B. D. Tapley, "Orbit determination in the relativistic geocentric reference frame," *J. Astronaut. Sci.*, Vol. 39, No. 2, pp. 173–181, April–June 1991.

Rowlands, D. D., S. B. Luthcke, J. A. Marshall, C. M. Cox, R. G. Williamson, and S. C. Rowton, "Space Shuttle precision orbit determination in support of SLA-1 using TDRSS and GPS tracking data," *J. Astronaut. Sci.*, Vol. 45, No. 1, pp. 113–129, January–March 1997.

Saastamoinen, J., "Atmospheric correction for the troposphere and stratosphere in radio ranging of satellites," *Geophysical Monograph Series, Vol. 15*, S. Henriksen, A. Mancini, B. Chovitz (eds.), American Geophysical Union, Washington, DC, pp. 247–251, 1972.

Seeber, G., *Satellite Geodesy: Foundations, Methods & Applications*, Walter de Gruyter, New York, 1993.

Shum, C. K., B. Zhang, B. Schutz, and B. Tapley, "Altimeter crossover methods for precise orbit determination and the mapping of geophysical parameters," *J. Astronaut. Sci.*, Vol. 38, No. 3, pp. 355–368, July–September 1990.

Skolnik, M.I. (ed.), *Radar Handbook*, McGraw-Hill, New York, 1990.

Thompson, B., M. Meek, K. Gold, P. Axelrad, G. Born, and D. Kubitschek, "Orbit determination for the QUIKSCAT spacecraft," *J. Spacecr. Rockets*, Vol. 39, No. 6, pp. 852–858, November–December 2002.

Visser, P., and B. Ambrosius, "Orbit determination of TOPEX/Poseidon and TDRSS satellites using TDRSS and BRTS tracking," *Adv. Space Res.*, Vol. 19, pp. 1641–1644, 1997.

Wells, D., *Guide GPS Positioning*, Canadian GPS Associates, Fredericton, 1987.

3.10 EXERCISES

1. Determine the Doppler shift in frequency over a pass for:

   a. TRANSIT: 150 MHz

   b. GPS: 1575 MHz

   To simplify this problem, assume the satellite is in a posigrade equatorial orbit and the observer is located on the equator. Determine the Doppler shift when the satellite rises and sets on the observer's local horizon and when the satellite is at closest approach to the observer.

2. The TOPEX/Poseidon spacecraft, launched in August 1992, carried a GPS receiver. At the following date and time, the receiver made the following measurements:

   Date/time: 31 March 1993 02:00:00.000 (receiver time)

GPS PRN	L1 pseudorange (m)
21	-16049870.249
28	-14818339.994

   From other sources, the T/P position at this time was determined to be

   $$x \quad -2107527.21\text{m}$$
   $$y \quad 6247884.75$$
   $$z \quad -4010524.01$$

   where (x, y, z) is ECF. The positions of the GPS satellites at this time were

	PRN21	PRN28
x	10800116.93m	10414902.30
y	23914912.70	13538107.48
z	1934886.67	-20329185.40

where these positions are expressed in the same reference frame as T/P.

Determine the T/P GPS receiver clock offset using both satellites. Ignore the GPS transmitter clock correction. State and justify any assumptions.

3. Repeat the example in Section 3.3.1 but use orbit altitudes of

   a. Lageos (5900 km)

   b. GPS (20,000 km)

   c. Moon (380,000 km)

The following data apply to Exercises 4–13:

GPS Ephemerides

The following ephemerides were taken from the "IGS final ephemerides" and the format is similar to the SP-3 format used by IGS:

*YR/MO/DA HR:MN:SEC					
P	PRN	x	y	z	δt_T
	$\vdots$				
P	PRN	x	y	z	δt_T

where time is given in GPS-Time as

   YR = Year

   MO = Month of year

   DA = Day of month

   HR = Hour of day

   MN = Minutes

   SEC = Seconds

The lines following the epoch line are

   P = Position ephemeris

   PRN = GPS Satellite identifier

   x,y,z = ITRF-2000 position of GPS PRN, in km

   δt_T = IGS-determined clock offset for PRN, in microseconds

* 2003/7/3 5:45:0.00000000					
P	1	4566.504088	−17778.663761	−19019.826900	292.783269
P	4	−10978.897783	−10863.301011	−21544.271426	44.796690
P	7	−16743.206878	−20415.491050	−3315.879862	556.798655
P	8	−4587.665320	−17960.987250	18706.552158	378.701406
P	11	14431.377269	−21405.604003	6070.147167	54.319645
P	13	−4652.983492	−23260.723995	−12070.391803	−29.105842
P	27	1214.284685	−23377.459743	12442.855885	377.394607
P	28	−14937.324209	−12545.945359	18318.739786	−28.452370
P	29	−17278.539907	−1224.933899	20384.228178	155.457984
P	31	13278.146508	−8114.211495	21225.174835	300.884498
* 2003/7/3 6:0:0.00000000					
P	1	5861.745350	−15901.667555	−20276.875963	292.785440
P	4	−9777.977569	−12987.742895	−20912.838681	44.796231
P	7	−16651.970525	−20805.209726	−525.893274	556.811719
P	8	−3449.706143	−19727.772705	17096.293643	378.709764
P	11	14464.812694	−20477.667957	8654.011503	54.321275
P	13	−3700.036312	−22089.587975	−14378.853276	−29.106054
P	27	1969.580616	−24408.106473	10001.540814	377.440654
P	28	−12810.474492	−12772.984277	19707.048914	−28.448653
P	29	−18461.999653	−3089.021101	19091.163769	155.460061
P	31	14725.051413	−6093.976687	20978.562271	300.903599
* 2003/7/3 6:15:0.00000000					
P	1	7328.259794	−13996.483360	−21179.722738	292.788770
P	4	−8727.133318	−15090.465960	−19918.643118	44.795835
P	7	−16383.736913	−20953.296813	2273.086980	556.824434
P	8	−2489.551621	−21368.525118	15184.136954	378.718629
P	11	14449.564065	−19284.658838	11088.302121	54.322742
P	13	−2536.235846	−20774.358288	−16441.402012	−29.106257
P	27	2578.663020	−25207.610015	7386.006812	377.487071
P	28	−10560.925449	−13127.209192	20763.433991	−28.445536
P	29	−19645.300961	−4758.653980	17476.041504	155.462272
P	31	16233.527710	−4186.595815	20362.080900	300.922651

Observations

The following data are given in a format that is similar to the Receiver IN-
dependent EXchange (RINEX) format for GPS receiver observations (refer
to IGS for RINEX details):

YR/MO/DA HR:MN:SEC	EF	N	PRN-List
L1 L2 P2 P1			
⋮			
L1 L2 P2 P1			

where time is given in GPS receiver time (close to GPS-Time) and

YR = Year

MO = Month of year

DA = Day of month

HR = Hour of day

MN = Minutes

SEC = Seconds

EF = Epoch Flag (0 means no problem)

N = Number of tracked GPS satellites

PRN-List = list of PRN-identifiers of tracked GPS satellites

The lines following the epoch line contain the observations for the PRN
satellites, given in the same order as the PRN-list. In the observations:

L1 = L1 carrier phase (cycles)

L2 = L2 carrier phase (cycles)

P2 = L2 pseudorange (meters)

P1 = L1 pseudorange (meters)

(Note that fourth and fifth digits to right of decimal for carrier phase pertain
to signal strength.)

The following data were recorded by a Black Jack receiver carried on ICE-
Sat:

2003/7/3 6:0:0.0000000	0 8 1 4 7 8 11 13 27 28		
13313150.27606	10373859.94706	25276179.66000	25276168.63500
3638113.46106	2834889.84706	25214111.21600	25214104.18700
−8930705.47408	−6958991.53008	20768986.07800	20768979.92300
−6413864.84708	−4997816.68208	20025186.23300	20025180.71900
−10979121.08307	−8555158.14008	22308126.25500	22308121.45400
−2786774.49707	−2171514.86607	22773866.90400	22773861.60000
−23434105.64308	−18260340.20309	19833911.31900	19833905.82300
−7593037.70208	−5916651.79208	21846321.76500	21846317.42700

2003/7/3 6:0:10.0000000	0 8 1 4 7 8 11 13 27 28		
13739083.01006	10705752.39405	25357229.60300	25357220.91900
3975221.46206	3097571.22406	25278260.02000	25278253.89400
−8829945.83208	−6880477.50608	20788159.91600	20788153.76600
−6629467.22309	−5165818.45908	19984158.47600	19984153.00800
−10987968.90807	−8562052.48708	22306442.29400	22306437.67400
−2390872.85807	−1863020.13207	22849204.79500	22849199.34400
−23466816.37608	−18285829.03209	19827686.48700	19827681.02000
−7876627.23108	−6137630.55108	21792356.26200	21792351.82000

2003/7/3 6:4:0.0000000	0 8 7 8 11 13 27 28 29 31		
−5513215.18608	−4296011.91608	21419312.62400	21419306.89400
−9854402.24209	−7678754.03309	19370472.81400	19370467.60200
−10546995.52107	−8218436.17807	22390355.62500	22390351.48800
7267464.77706	5662956.04106	24687125.70900	24687120.66100
−22384357.99608	−17442354.31309	20033670.90000	20033665.56700
−13406848.48208	−10446892.85208	20739989.40200	20739985.24200
−4806438.03607	−3745274.75407	22815283.89300	22815278.35000
−1904315.82106	−1483881.67706	23910972.26500	23910967.00200

2003/7/3 6:4:10.0000000	0 8 7 8 11 13 27 28 29 31		
−5327807.74108	−4151538.49708	21454594.42700	21454588.71500
−10500588.28407	−8182274.62907	22399186.76600	22399182.59700
7703560.72106	6002770.96806	24770114.33000	24770107.20000
−22259480.22808	−17345046.90309	20057434.39800	20057428.92100
−13598356.03208	−10596119.48608	20703546.53400	20703542.38700
−5057074.29907	−3940575.68207	22767589.44100	22767583.83300
−2109142.85306	−1643487.08006	23871994.45100	23871989.67700

The following data were collected by a Turbo Rogue GPS receiver at Pie Town (NM):

2003/7/3 6:0:0.0000000	0 8 8 27 26 11 29 28 31 7		
−24388891.10649	−19004308.54347	19950321.918	19950319.447
−23612111.74249	−18399018.28147	20633681.740	20633678.842
−354035.84848	−275871.64045	23938856.683	23938853.490
−6053270.88448	−4716827.20645	22881055.819	22881053.542
−11945499.34248	−9308173.13145	22918548.923	22918545.380
−21453687.95449	−16717124.87047	21107802.248	21107800.600
12464204.51048	9712367.70644	23823735.311	23823729.233
−4481731.59449	−3492253.10746	22002748.155	22002743.741
2003/7/3 6:4:0.0000000	0 8 8 27 26 11 29 28 31 7		
−24416407.84949	−19025750.07047	19945085.523	19945082.850
−23327037.21349	−18176882.46347	20687929.760	20687926.774
−898114.97748	−699828.98244	23835323.854	23835317.574
−6258015.55248	−4876368.11345	22842093.804	22842091.419
−12296146.77548	−9581404.44246	22851822.101	22851818.825
−21614641.52349	−16842543.10147	21077173.381	21077172.008
13324935.88948	10383067.35145	23987526.176	23987521.928
−5099063.91549	−3973290.68546	21885273.341	21885269.023

The following data were collected by a Turbo Rogue GPS receiver at Mc-Donald Observatory (TX):

2003/7/3 6:0:0.0000000	0 8 29 27 11 8 13 26 28 7		
−9944212.12448	−7748726.63345	23401416.592	23401412.281
−24875552.81149	−19383519.73947	20395733.147	20395730.434
−11346725.95449	−8841595.17445	22528013.534	22528011.897
−28075193.49449	−21876748.09447	19982685.387	19982683.005
−4164302.39448	—	—	24437530.876
−2296772.12848	—	—	24480165.844
−18590371.83649	−14485969.70246	21419438.945	21419437.517
−16001981.31449	−12469051.09846	21932274.397	21932270.319

2003/7/3 6:4:0.0000000	0 8 29 27 11 8 13 26 28 7		
−10328700.49948	−8048327.52545	23328248.166	23328245.981
−24679310.88249	−19230604.03647	20433076.681	20433074.090
−11480544.84549	−8945869.42745	22502548.387	22502546.906
−28204694.64049	−21977658.03247	19958042.069	19958039.765
−3297123.18548	—	—	24602548.472
−2850548.61348	−2221193.01544	24374788.863	24374785.477
−18774076.61449	−14629116.05246	21384480.700	21384479.692
−16569301.13749	−12911117.94346	21824316.532	21824312.371

4. Use the ICESat data for any of the given epochs to determine:

   a. ionosphere-free pseudoranges for each satellite

   b. ionosphere correction to the L1 pseudoranges

   c. convert carrier phase to "phase range"

   d. ionosphere-free phase ranges for all satellites

5. Use the ICESat data separated by 10 sec and by 4 min to determine:

   a. range change over the respective time interval from pseudorange and phase range

   b. an approximate value of phase range ambiguity for each observed GPS. satellite

6. a. Determine an ICESat position at 06:00:00.000000 (and any other relevant parameters). Provide justification for the solved for parameter set and identify the measurements used. Strive for highest accuracy. Use all observed satellites, but provide justification for any satellites that should be edited from the solution. Identify and justify assumptions.

   b. Provide the O-C for each of the satellites used in a.

   c. Determine the azimuth and elevation of each GPS satellite with respect to the ICESat local horizontal plane and vertical direction using true north as a reference direction.

7. Repeat Exercise 6, but use the epoch 06:04:00.000000.

8. a. Form pseudorange single differences with the ICESat data at 06:00:00.000000. Identify the measurements used and strive for highest accuracy. Identify and justify assumptions.

    b. Determine an ICESat position and any other relevant parameters using the single differences of a. Provide justification for the solved for parameter set.

    c. Provide the O-C for all measurements used in b.

9. Repeat Exercise 8, but use the epoch 06:04:00.000000.

10.   a. Form pseudorange double-difference measurements between ICESat and Pie Town at 06:00:00.000000. The coordinates for Pie Town and McDonald are given in Table 3.5.3.

    b. Determine an ICESat position and any other relevant parameters using double differences of a. Provide justification for the solved for parameter set. Are the double differences in your solution independent (see Hoffman-Wellenhof *et al.*, 1997)? Strive for high accuracy. Identify and justify assumptions.

    c. Provide the O-C for all measurements used in b.

11. Repeat Exercise 10, but use the epoch 06:04:00.000000.

12. Repeat Exercise 10, but use McDonald instead of Pie Town. State any assumptions.

13. Form double-difference pseudorange measurements between Pie Town and McDonald and compute the O-C using the available ephemerides and information from Table 3.5.3.

Chapter 4

Fundamentals of Orbit Determination

4.1 INTRODUCTION

During the decade of the 1960s, the accuracy of the radio frequency and optical measurement systems, the force models that govern the satellite motion, and the analysis techniques, combined with relatively primitive computing technology, restricted the positioning of Earth-orbiting satellites to accuracies of hundreds of meters. With the decade of the 1970s, improvements in all of these areas, particularly in mathematical force models and computing capability, facilitated orbit determination accuracy at the level of a few tens of meters by mid-decade and to a few meters by the end of the decade. During the 1980s, significant improvements were made in the models for the Earth's gravity field, including solid body and ocean tides, and in the models for the surface force effects. All these improvements, which were greatly advanced by the significant developments in computer technology, allowed orbit determination accuracies to increase to the tens-of-centimeter level by the end of the decade.

This improvement in orbit determination accuracy was motivated by the ever-increasing demands of scientists in the oceanographic and geodetic communities. In particular, the need for centimeter-level accuracy in global ocean topography obtained from altimetric satellites spurred extensive and unprecedented model improvements during the past two decades. These studies have led to even further improvements in the technology; today the orbit of the oceanographic satellite TOPEX/Poseidon, launched in 1992, is routinely computed with an accuracy approaching 2 cm RMS in the radial component and 8 cm RSS for all components (Tapley *et al.*,1994; Schutz *et al.*, 1994; Marshall *et al.*, 1995; Bertiger *et al.*, 1995). Orbits for the Jason-1 altimetric satellite, launched in 2001, are rou-

159

tinely computed with an accuracy of 1 cm in the radial component and 4 cm RSS in all components in a post-processing mode using GPS, SLR, and DORIS data (Lutchke *et al.*, 2003). In near-real time (3–5 hours) the Jason-1 radial component is computed to better than 2.5 cm RMS using GPS data (Desai and Haines, 2003).

The discussion in the following sections will focus on the techniques used to estimate the orbits. The approach adopted here will follow that given by Tapley (1973, 1989). The role of the estimation process in improving the force and measurement models as an integral step in the process will be illustrated.

4.2   LINEARIZATION OF THE ORBIT DETERMINATION PROCESS

In Chapter 1, the general formulation of the orbit determination problem was discussed using a dynamical system governed by a simple linear force model. The role of the measurement model in introducing nonlinearity to the process was described. In the general orbit determination problem, both the dynamics and the measurements involve significant nonlinear relationships. For the general case, the governing relations involve the nonlinear expression

$$\dot{\mathbf{X}} = F(\mathbf{X}, t), \qquad \mathbf{X}(t_k) \equiv \mathbf{X}_k \tag{4.2.1}$$

$$\mathbf{Y}_i = G(\mathbf{X}_i, t_i) + \boldsymbol{\epsilon}_i; \qquad i = 1, \ldots, \ell \tag{4.2.2}$$

where $\mathbf{X}_k$ is the unknown n-dimensional state vector at the time t_k, and $\mathbf{Y}_i$ for $i = 1, \ldots, \ell$, is a p-dimensional set of observations that are to be used to obtain a best estimate of the unknown value of $\mathbf{X}_k$ (i.e., $\hat{\mathbf{X}}_k$). In general, $p < n$ and $m = p \times \ell \gg n$. The formulation represented by Eqs. (4.2.1) and (4.2.2) is characterized by: (1) the inability to observe the state directly, (2) nonlinear relations between the observations and the state, (3) fewer observations at any time epoch than there are state vector components ($p < n$), and (4) errors in the observations represented by $\boldsymbol{\epsilon}_i$. The problem of determining the trajectory of a space vehicle in the presence of these effects is referred to as the nonlinear estimation (or orbit determination) problem. If the state vector and the observation vector can be related in a linear manner, then several powerful techniques from the field of linear estimation theory can be applied to the orbit determination problem.

If a reasonable reference trajectory is available and if $\mathbf{X}$, the true trajectory, and $\mathbf{X}^*$, the reference trajectory, remain sufficiently close throughout the time interval of interest, then the trajectory for the actual motion can be expanded in a Taylor's series about the reference trajectory at each point in time. If this expansion is truncated to eliminate higher order terms, then the deviation in the state from the reference trajectory can be described by a set of linear differential equations with time-dependent coefficients. A linear relation between the observation deviation and the state deviation can be obtained by a similar expansion procedure. Then, the nonlinear orbit determination problem in which the complete state vector is to be estimated can be replaced by a linear orbit determination problem in which the deviation from some reference solution is to be determined.

To conduct this linearization procedure, let the $n \times 1$ state deviation vector, $\mathbf{x}$, and the $p \times 1$ observation deviation vector, $\mathbf{y}$, be defined as follows:

$$\mathbf{x}(t) = \mathbf{X}(t) - \mathbf{X}^*(t), \qquad \mathbf{y}(t) = \mathbf{Y}(t) - \mathbf{Y}^*(t). \tag{4.2.3}$$

It follows that

$$\dot{\mathbf{x}}(t) = \dot{\mathbf{X}}(t) - \dot{\mathbf{X}}^*(t). \tag{4.2.4}$$

Expanding Eqs. (4.2.1) and (4.2.2) in a Taylor's series about the reference trajectory leads to

$$
\begin{aligned}
\dot{\mathbf{X}}(t) &= F(\mathbf{X}, t) = F(\mathbf{X}^*, t) + \left[\frac{\partial F(t)}{\partial \mathbf{X}(t)} \right]^* [\mathbf{X}(t) - \mathbf{X}^*(t)] \\
&\quad + O_F \left[\mathbf{X}(t) - \mathbf{X}^*(t) \right] \\
\mathbf{Y}_i &= G(\mathbf{X}_i, t_i) + \epsilon_i = G(\mathbf{X}_i^*, t_i) + \left[\frac{\partial G}{\partial \mathbf{X}} \right]_i^* [\mathbf{X}(t_i) - \mathbf{X}^*(t_i)]_i \\
&\quad + O_G \left[\mathbf{X}(t_i) - \mathbf{X}^*(t_i) \right] + \epsilon_i
\end{aligned} \tag{4.2.5}
$$

where $[\ \ ]^*$ indicates that the partial derivative matrix is evaluated on the reference solution, $\mathbf{X}^*(t)$, which is obtained by integrating Eq. (4.2.1) with the specified initial conditions, $\mathbf{X}^*(t_0)$. The symbols O_F and O_G indicate terms in the expansion containing products of the difference, $\mathbf{X}(t) - \mathbf{X}^*(t)$, higher than the first order. If the terms of order higher than the first in Eq. (4.2.5) are neglected, under the assumption that the higher order products are small compared to the first order terms, and if the condition $\dot{\mathbf{X}}^* = F(\mathbf{X}^*, t)$ and $\mathbf{Y}_i^* = G(\mathbf{X}_i^*, t_i)$ are used, Eq. (4.2.5) can be written as

$$\dot{\mathbf{x}}(t) = A(t)\mathbf{x}(t) \tag{4.2.6}$$
$$\mathbf{y}_i = \widetilde{H}_i \mathbf{x}_i + \epsilon_i \qquad (i = 1, \ldots, \ell)$$

where

$$A(t) = \left[\frac{\partial F(t)}{\partial \mathbf{X}(t)} \right]^* \qquad \widetilde{H}_i = \left[\frac{\partial G}{\partial \mathbf{X}} \right]_i^*.$$

Hence, the original nonlinear estimation problem is replaced by the linear estimation problem described by Eq. (4.2.6), where

$$
\begin{aligned}
\mathbf{x}(t) &= \mathbf{X}(t) - \mathbf{X}^*(t), \\
\mathbf{x}_i &= \mathbf{X}(t_i) - \mathbf{X}^*(t_i)
\end{aligned}
$$

and

$$\mathbf{y}_i = \mathbf{Y}_i - G(\mathbf{X}_i^*, t_i).$$

Notice that if the original system of differential equations $\dot{\mathbf{X}} = F(\mathbf{X}, t)$ is linear, the second and higher order partial derivatives of $F(\mathbf{X}, t)$ are zero (i.e., $\frac{\partial^i F}{\partial \mathbf{X}^i} = 0, i \geq 2$). The same statements apply to $G(\mathbf{X}_i, t_i)$ in Eq. (4.2.5). Hence, for a linear system there is no need to deal with a state or observational deviation vector or a reference solution. However, for the orbit determination problem, $F(\mathbf{X}, t)$ and $G(\mathbf{X}_i, t_i)$ will always be nonlinear in $\mathbf{X}(t)$, thus requiring that we deal with deviation vectors and a reference trajectory in order to linearize the system.

Generally in this text, uppercase $\mathbf{X}$ and $\mathbf{Y}$ will represent the state and the observation vectors and lowercase $\mathbf{x}$ and $\mathbf{y}$ will represent the state and observation deviation vectors as defined by Eq. (4.2.3). However, this notation will not always be adhered to and sometimes $\mathbf{x}$ and $\mathbf{y}$ will be referred to as the state and observation vectors, respectively.

Example 4.2.1

Compute the A matrix and the $\widetilde{H}$ matrix for a satellite in a plane under the influence of only a *central force*. Assume that the satellite is being tracked with range observations, ρ, from a single ground station. Assume that the station coordinates, (X_S, Y_S), and the gravitational parameter are unknown. Then, the state vector, $\mathbf{X}$, is given by

$$\mathbf{X} = \begin{bmatrix} X \\ Y \\ U \\ V \\ \mu \\ X_S \\ Y_S \end{bmatrix}$$

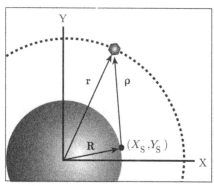

where U and V are velocity components and X_S and Y_S are coordinates of the tracking station. From Newton's Second Law and the law of gravitation,

$$\ddot{\mathbf{r}} = -\frac{\mu \mathbf{r}}{r^3}$$

or in component form,

$$\ddot{X} = -\frac{\mu X}{r^3}$$

$$\ddot{Y} = -\frac{\mu Y}{r^3}.$$

If these equations are expressed in first-order form, the following expression is obtained:

$$\dot{\mathbf{X}} = \begin{bmatrix} \dot{X} \\ \dot{Y} \\ \dot{U} \\ \dot{V} \\ \dot{\mu} \\ \dot{X}_S \\ \dot{Y}_S \end{bmatrix} = \begin{bmatrix} F_1 \\ F_2 \\ F_3 \\ F_4 \\ F_5 \\ F_6 \\ F_7 \end{bmatrix} = \begin{bmatrix} U \\ V \\ -\dfrac{\mu X}{r^3} \\ -\dfrac{\mu Y}{r^3} \\ 0 \\ 0 \\ 0 \end{bmatrix}.$$

Then,

$$A(t) = \frac{\partial F(\mathbf{X}^*, t)}{\partial \mathbf{X}} = \begin{bmatrix} \dfrac{\partial F_1}{\partial X} & \dfrac{\partial F_1}{\partial Y} & \dfrac{\partial F_1}{\partial U} & \dfrac{\partial F_1}{\partial V} & \dfrac{\partial F_1}{\partial \mu} & \dfrac{\partial F_1}{\partial X_S} & \dfrac{\partial F_1}{\partial Y_S} \\ \dfrac{\partial F_2}{\partial X} & \cdots & \cdots & \cdots & \cdots & \cdots & \dfrac{\partial F_2}{\partial Y_S} \\ \vdots & \vdots & \vdots & \vdots & \vdots & \vdots & \vdots \\ \vdots & \vdots & \vdots & \vdots & \vdots & \vdots & \vdots \\ \vdots & \vdots & \vdots & \vdots & \vdots & \vdots & \vdots \\ \vdots & \vdots & \vdots & \vdots & \vdots & \vdots & \vdots \\ \dfrac{\partial F_7}{\partial X} & \cdots & \cdots & \cdots & \cdots & \cdots & \dfrac{\partial F_7}{\partial Y_S} \end{bmatrix}^*$$

$$
= \begin{bmatrix}
0 & 0 & 1 & 0 & 0 & 0 & 0 \\
0 & 0 & 0 & 1 & 0 & 0 & 0 \\
-\dfrac{\mu}{r^3} + \dfrac{3\mu X^2}{r^5} & \dfrac{3\mu XY}{r^5} & 0 & 0 & -\dfrac{X}{r^3} & 0 & 0 \\
\dfrac{3\mu XY}{r^5} & -\dfrac{\mu}{r^3} + \dfrac{3\mu Y^2}{r^5} & 0 & 0 & -\dfrac{Y}{r^3} & 0 & 0 \\
0 & 0 & 0 & 0 & 0 & 0 & 0 \\
0 & 0 & 0 & 0 & 0 & 0 & 0 \\
0 & 0 & 0 & 0 & 0 & 0 & 0
\end{bmatrix}
$$

The $\tilde{H}$ matrix is given by

$$
\tilde{H} = \frac{\partial \rho}{\partial \mathbf{X}} = \begin{bmatrix} \dfrac{\partial \rho}{\partial X} & \dfrac{\partial \rho}{\partial Y} & \dfrac{\partial \rho}{\partial U} & \dfrac{\partial \rho}{\partial V} & \dfrac{\partial \rho}{\partial \mu} & \dfrac{\partial \rho}{\partial X_S} & \dfrac{\partial \rho}{\partial Y_S} \end{bmatrix}^*
$$

where

$$
\rho = \left[(X - X_S)^2 + (Y - Y_S)^2 \right]^{1/2}.
$$

It follows then that

$$
\tilde{H} = \begin{bmatrix} \dfrac{X - X_S}{\rho} & \dfrac{Y - Y_S}{\rho} & 0 & 0 & 0 & -\dfrac{(X - X_S)}{\rho} & -\dfrac{(Y - Y_S)}{\rho} \end{bmatrix}^*.
$$

4.2.1 THE STATE TRANSITION MATRIX

The first of Eq. (4.2.6) represents a system of linear differential equations with time-dependent coefficients. The symbol $[\,]^*$ indicates that the values of $\mathbf{X}$ are derived from a particular solution to the equations $\dot{\mathbf{X}} = F(\mathbf{X}, t)$ which is generated with the initial conditions $\mathbf{X}(t_0) = \mathbf{X}_0^*$. The general solution for this system, $\dot{\mathbf{x}}(t) = A(t)\mathbf{x}(t)$, can be expressed as

$$
\mathbf{x}(t) = \Phi(t, t_k)\mathbf{x}_k \tag{4.2.7}
$$

where $\mathbf{x}_k$ is the value of $\mathbf{x}$ at t_k; that is, $\mathbf{x}_k = \mathbf{x}(t_k)$. The matrix $\Phi(t_i, t_k)$ is called the state transition matrix and was introduced in Chapter 1, Section 1.2.5. The state transition matrix for the two-body problem was discussed in Chapter 2, Section 2.2.6. $\Phi(t, t_k)$ has the following useful properties, which can be demonstrated from Eq. (4.2.7).

$$
\begin{array}{llll}
1. & \Phi(t_k, t_k) & = & I \\
2. & \Phi(t_i, t_k) & = & \Phi(t_i, t_j)\Phi(t_j, t_k) \\
3. & \Phi(t_i, t_k) & = & \Phi^{-1}(t_k, t_i).
\end{array}
\tag{4.2.8}
$$

The differential equation for $\Phi(t_i, t_k)$ can be obtained by differentiating Eq. (4.2.7) (noting that $\mathbf{x}_k$ is a constant). This yields

$$
\dot{\mathbf{x}}(t) = \dot{\Phi}(t, t_k)\mathbf{x}_k.
\tag{4.2.9}
$$

Substituting Eq. (4.2.9) into the first of Eq. (4.2.6) and using Eq. (4.2.7) yields

$$
\dot{\Phi}(t, t_k)\mathbf{x}_k = A(t)\Phi(t, t_k)\mathbf{x}_k.
$$

Since this condition must be satisfied for all $\mathbf{x}_k$, it follows that

$$
\dot{\Phi}(t, t_k) = A(t)\Phi(t, t_k)
\tag{4.2.10}
$$

with initial conditions

$$
\Phi(t_k, t_k) = I.
$$

By differentiating $\mathbf{x}_k = \Phi^{-1}(t, t_k)\mathbf{x}(t)$ from Eq. (4.2.7) and using the first of Eq. (4.2.6) it can be shown that

$$
\dot{\Phi}^{-1}(t, t_k) = -\Phi^{-1}(t, t_k)\, A(t),
\tag{4.2.11}
$$

with initial conditions

$$
\Phi^{-1}(t_k, t_k) = I.
$$

Under certain conditions on $A(t)$, the state transition matrix may be inverted analytically (Battin, 1999).

If the matrix $A(t)$ can be partitioned in the form

$$
A(t) = \left[
\begin{array}{c|c}
A_1 & A_2 \\
\hline
A_3 & A_4
\end{array}
\right]
\tag{4.2.12}
$$

where the submatrices have the properties that

$$
A_1^T = -A_4, \quad A_2^T = A_2 \ \text{and} \ A_3^T = A_3.
\tag{4.2.13}
$$

Then $\Phi(t, t_k)$ can be similarly partitioned as

$$
\Phi(t, t_k) = \left[
\begin{array}{c|c}
\Phi_1 & \Phi_2 \\
\hline
\Phi_3 & \Phi_4
\end{array}
\right]
$$

and $\Phi^{-1}(t, t_k)$ may be written as

$$\Phi^{-1}(t, t_k) = \left[\begin{array}{c|c} \Phi_4^T & -\Phi_2^T \\ \hline -\Phi_3^T & \Phi_1^T \end{array}\right]. \tag{4.2.14}$$

The proof follows: Define

$$J = \left[\begin{array}{cc} 0 & I \\ -I & 0 \end{array}\right] \tag{4.2.15}$$

where I is the identity matrix and 0 is the null matrix. Then

$$\begin{aligned} \frac{d}{dt}(J\Phi(t, t_k)J) &= J\dot{\Phi}(t, t_k)J \\ &= -(JA(t)J)(J\Phi(t, t_k)J), \end{aligned} \tag{4.2.16}$$

where we have used the fact that $J^2 = -I$ and $\dot{\Phi}(t, t_k) = A(t)\Phi(t, t_k)$.
   Define

$$V(t, t_k) \equiv -(J\Phi(t, t_k)J)^T. \tag{4.2.17}$$

Taking the transpose of Eq. (4.2.16) and using Eq. (4.2.17),

$$\begin{aligned} (J\dot{\Phi}(t, t_k)J)^T &= -\dot{V}(t, t_k) \\ &= -(J\Phi(t, t_k)J)^T(JA(t)J)^T \end{aligned}$$

or

$$\dot{V}(t, t_k) = -V(t, t_k)(JA(t)J)^T. \tag{4.2.18}$$

Using Eq. (4.2.12) for $A(t)$ yields

$$(JA(t)J)^T = \left[\begin{array}{c|c} -A_4^T & A_2^T \\ \hline A_3^T & -A_1^T \end{array}\right]. \tag{4.2.19}$$

Consequently, if $A(t)$ satisfies Eq. (4.2.13) and

$$(JA(t)J)^T = A(t), \tag{4.2.20}$$

then

$$\dot{V}(t, t_k) = -V(t, t_k)A(t) \tag{4.2.21}$$

and

$$V(t_0, t_0) = -(J\Phi(t_0, t_0)J)^T = I.$$

Hence, from Eq. (4.2.11) and Eq. (4.2.21),

$$V(t, t_k) = \Phi^{-1}(t, t_k)$$

or

$$\begin{aligned}
\Phi^{-1}(t, t_k) &= -(J\Phi(t, t_k)J)^T \\
&= \left[\begin{array}{c|c} \Phi_4^T & -\Phi_2^T \\ \hline -\Phi_3^T & \Phi_1^T \end{array} \right]
\end{aligned} \tag{4.2.22}$$

which is identical to Eq. (4.2.14).

An even dimensional matrix, B, which has the property that $B^T J B = J$ (where J is defined by Eq. (4.2.15)) is called *symplectic* (Battin, 1999). It easily is shown that $\Phi(t, t_k)$ has this property when A satisfies Eq. (4.2.13). An important case where $\Phi(t, t_k)$ is symplectic arises when the acceleration can be written as the gradient of a potential function; that is,

$$\ddot{\mathbf{r}} = \nabla U. \tag{4.2.23}$$

4.2.2 SOLUTION FOR THE STATE TRANSITION MATRIX

A linear differential equation of the type $\dot{\mathbf{x}}(t) = A(t)\mathbf{x}(t)$ or $\dot{\Phi}(t, t_0) = A(t)\Phi(t, t_0)$ has an infinite number of solutions in terms of arbitrary constants. However, when initial conditions, $\mathbf{x}(t_0)$ and $\Phi(t_0, t_0)$, are specified and the elements of $A(t)$ are continuous functions of time, the solution becomes unique. One could ask the question "Why bother to solve for the state transition matrix when the state deviation vector can be determined directly by solving $\dot{\mathbf{x}}(t) = A(t)\mathbf{x}(t)$?" The answer is that the computational algorithms for determining the best estimate of $\mathbf{x}$ and for mapping the associated error covariance matrices are most easily formulated in terms of the state transition matrix. Since $\mathbf{x}_k$ in Eq. (4.2.7) is unknown in the orbit determination problem, the state transition matrix allows the solution, $\mathbf{x}(t)$, to be expressed in terms of the unknown initial state, $\mathbf{x}_k$. Hence, it is essential in relating observations made at different times.

The solution for $\Phi(t, t_0)$ is facilitated by noting that the individual columns of the differential equation for $\dot{\Phi}(t, t_0)$ are uncoupled and independent. To illustrate this, consider a one-dimensional case where the state vector consists of a single position and velocity coordinate. Equation (4.2.10) can be written in terms of the individual elements of the state transition matrix as follows:

$$\dot{\Phi}(t, t_0) = \left[\begin{array}{cc} \dot{\phi}_{11} & \dot{\phi}_{12} \\ \dot{\phi}_{21} & \dot{\phi}_{22} \end{array} \right] = \left[\begin{array}{cc} A_{11} & A_{12} \\ A_{21} & A_{22} \end{array} \right] \left[\begin{array}{cc} \phi_{11} & \phi_{12} \\ \phi_{21} & \phi_{22} \end{array} \right] \tag{4.2.24}$$

subject to the following initial conditions at t_0

$$\left[\begin{array}{cc} \phi_{11} & \phi_{12} \\ \phi_{21} & \phi_{22} \end{array} \right] = \left[\begin{array}{cc} 1 & 0 \\ 0 & 1 \end{array} \right]. \tag{4.2.25}$$

Equation (4.2.24) expands to

$$
\begin{bmatrix} \dot{\phi}_{11} & \dot{\phi}_{12} \\ \dot{\phi}_{21} & \dot{\phi}_{22} \end{bmatrix} = \begin{bmatrix} A_{11}\phi_{11} + A_{12}\phi_{21} & A_{11}\phi_{12} + A_{12}\phi_{22} \\ A_{21}\phi_{11} + A_{22}\phi_{21} & A_{21}\phi_{12} + A_{22}\phi_{22} \end{bmatrix}. \qquad (4.2.26)
$$

Recall that the A_{ij} are known quantities obtained by evaluating

$$
A_{ij}(t) = \left[\frac{\partial F_i(t)}{\partial \mathbf{X}_j(t)} \right]^*
$$

on the reference trajectory. From Eq. (4.2.26) we see that the columns of $\dot{\Phi}(t, t_0)$ are independent; for example, the first column of $\dot{\Phi}(t, t_0)$ does not contain elements of $\Phi(t, t_0)$ from the second column. Hence, we can solve for $\Phi(t, t_0)$ by integrating independently two 2×1 vector differential equations. For any practical orbit determination application, the solution for $\Phi(t, t_0)$ will be obtained via numerical integration. Hence, we can supply a vector of derivative values for the differential equation of the nominal state vector and $\dot{\Phi}(t, t_0)$ to the numerical integration routine. For this 1D case we would supply the integrator with the following vector at each time point:

$$
\begin{bmatrix} \dot{X} \\ \dot{U} \\ \dot{\phi}_{11} \\ \dot{\phi}_{21} \\ \dot{\phi}_{12} \\ \dot{\phi}_{22} \end{bmatrix}. \qquad (4.2.27)
$$

The first two elements would provide the reference orbit, $\mathbf{X}^*(t)$, and the next four would yield the elements of $\Phi(t, t_0)$. The reference orbit is used to evaluate $A(t)$, which is needed to evaluate $\dot{\Phi}(t, t_0)$ in Eq. (4.2.26).

Notice that when $\dot{\Phi} = A\Phi$ is numerically integrated for the two-body case given in Example 4.2.1 we need to concern ourselves only with the upper 4×5 portion of $\dot{\Phi}$. The final three rows and two columns of Φ remain zero except for the values of unity in the last three diagonal elements. Hence, the numerical integration algorithm should be structured to take advantage of this fact.

Another approach to solving the linearized equations of motion is given in Appendix F. Additional clarification on the structure of Φ and associated matrices also is given.

Generally all orbit determination problems will result in $A(t)$ being a function of time. For example, the simple two-body Example 4.2.1 involves a time-varying matrix for $A(t)$. However, it is instructive to consider the case where A is a constant matrix because of the insight gained about the state transition matrix. Such a case is illustrated in the following example.

Example 4.2.2

Consider a system of linear first-order differential equations with constant coefficients

$$\dot{x}_1 = x_1$$
$$\dot{x}_2 = \beta x_1 \qquad (4.2.28)$$

where β is a constant. Initial conditions are $x_1(t_0) = x_{10}$ and $x_2(t_0) = x_{20}$.

a) Write Eq. (4.2.28) in state space form; that is, $\dot{\mathbf{X}} = A\mathbf{X}$,

$$\dot{\mathbf{X}} = \begin{bmatrix} \dot{x}_1 \\ \dot{x}_2 \end{bmatrix} = \begin{bmatrix} 1 & 0 \\ \beta & 0 \end{bmatrix} \begin{bmatrix} x_1 \\ x_2 \end{bmatrix} = A\mathbf{X}. \qquad (4.2.29)$$

b) Solve Eq. (4.2.29) by determining the state transition matrix; that is, $\mathbf{X}(t) = \Phi(t, t_0)\mathbf{X}(t_0)$, where $\dot{\Phi}(t, t_0) = A(t)\Phi(t, t_0)$. Because $A(t)$ is a constant matrix there are a number of ways to solve for $\Phi(t, t_0)$. These include but are not limited to the following:

   1. Because Eq. (4.2.28) is linear with constant coefficients it can be solved directly. From the first equation

$$\frac{dx_1}{x_1} = dt$$
$$x_1(t) = ce^t; \qquad \text{at } t = t_0, \ x_1 = x_{10}$$
$$c = x_{10}e^{-t_0}$$
$$x_1(t) = x_{10}e^{(t-t_0)}, \qquad (4.2.30)$$

from the second of Eq. (4.2.28)

$$dx_2 = \beta x_{10}e^{(t-t_0)} dt$$
$$x_2(t) = \beta x_{10}e^{(t-t_0)} + c; \qquad \text{at } t = t_0, \ x_2 = x_{20}.$$

Hence,

$$x_2(t) = x_{20} + \beta x_{10}(e^{(t-t_0)} - 1). \qquad (4.2.31)$$

The state transition matrix can be determined from a direct differentiation of the solution

$$\Phi(t, t_0) \equiv \frac{\partial \mathbf{X}(t)}{\partial \mathbf{X}(t_0)} = \begin{bmatrix} \dfrac{\partial x_1(t)}{\partial x_1(t_0)} & \dfrac{\partial x_1(t)}{\partial x_2(t_0)} \\[3mm] \dfrac{\partial x_2(t)}{\partial x_1(t_0)} & \dfrac{\partial x_2(t)}{\partial x_2(t_0)} \end{bmatrix}. \qquad (4.2.32)$$

The elements of Eq. (4.2.32) are obtained by differentiating Eqs. (4.2.30) and (4.2.31) to yield

$$\Phi(t, t_0) = \begin{bmatrix} e^{(t-t_0)} & 0 \\ \beta(e^{(t-t_0)} - 1) & 1 \end{bmatrix}. \qquad (4.2.33)$$

2. We may integrate $\dot{\Phi} = A\Phi$ directly:

$$\begin{bmatrix} \dot{\phi}_{11} & \dot{\phi}_{12} \\ \dot{\phi}_{21} & \dot{\phi}_{22} \end{bmatrix} = \begin{bmatrix} 1 & 0 \\ \beta & 0 \end{bmatrix} \begin{bmatrix} \phi_{11} & \phi_{12} \\ \phi_{21} & \phi_{22} \end{bmatrix} ; \qquad \Phi(t_0, t_0) = I.$$

The equations to be integrated are

$$\begin{aligned} \dot{\phi}_{11} &= \phi_{11} \\ \dot{\phi}_{21} &= \beta\phi_{11} \\ \dot{\phi}_{12} &= \phi_{12} \\ \dot{\phi}_{22} &= \beta\phi_{12}. \end{aligned} \qquad (4.2.34)$$

Solving Eq. (4.2.34) yields

$$\begin{aligned} \phi_{11} &= ce^t ; & \phi_{11}(t_0) = 1 \\ c &= e^{-t_0} \\ \phi_{11} &= e^{(t-t_0)}. \end{aligned}$$

Next,

$$\begin{aligned} \dot{\phi}_{21} &= \beta e^{(t-t_0)} \\ \phi_{21} &= \beta e^{(t-t_0)} + c; & \phi_{21}(t_0) = 0. \end{aligned}$$

If this expression is evaluated at t_0, it follows that

$$c = -\beta$$

and

$$\phi_{21} = \beta(e^{(t-t_0)} - 1).$$

Finally, the second two of Eq. (4.2.34) along with the initial conditions lead to

$$\begin{aligned} \phi_{12} &= 0 \\ \phi_{22} &= 1. \end{aligned}$$

These solutions are identical to the elements given by Eq. (4.2.33).

3. We may use Laplace transforms to solve for $\Phi(t, t_0)$. In this case

$$
\begin{aligned}
\Phi(t, t_0) &= \mathcal{L}^{-1} [SI - A]^{-1} = \mathcal{L}^{-1} \begin{bmatrix} S - 1 & 0 \\ -\beta & S \end{bmatrix}^{-1} \\
&= \mathcal{L}^{-1} \begin{bmatrix} \frac{1}{S-1} & 0 \\ \frac{\beta}{S(S-1)} & \frac{1}{S} \end{bmatrix}.
\end{aligned}
$$

Using a table of Laplace transforms yields

$$
\Phi(t, t_0) = \begin{bmatrix} e^{(t-t_0)} & 0 \\ \beta(e^{(t-t_0)} - 1) & 1 \end{bmatrix}.
$$

4. Another solution, whenever A is a constant matrix, uses the eigenvalues and eigenvectors of A to yield the solution

$$
\Phi(t, t_0) = V e^{\lambda(t-t_0)} V^{-1}, \tag{4.2.35}
$$

where V is the matrix of normalized eigenvectors of A. Also,

$$
e^{\lambda(t, t_0)} = \begin{bmatrix} e^{\lambda_1(t-t_0)} & & & \\ & e^{\lambda_2(t-t_0)} & & 0 \\ & & \ddots & \\ 0 & & & e^{\lambda_n(t-t_0)} \end{bmatrix} \tag{4.2.36}
$$

where $\lambda_1, \lambda_2 \ldots \lambda_n$ are the eigenvalues of the $n \times n$ matrix, A.

This method requires that A have a complete set of linearly independent eigenvectors. Otherwise there is no invertible matrix of eigenvectors, V, and the algorithm fails.

For the example we are considering, the matrix

$$
A = \begin{bmatrix} 1 & 0 \\ \beta & 0 \end{bmatrix}
$$

has eigenvalues (see Appendix B, Section B.6)

$$
\begin{aligned}
\lambda_1 &= 0 \\
\lambda_2 &= 1
\end{aligned}
$$

and normalized eigenvectors

$$V_1 = \begin{bmatrix} 0 \\ 1 \end{bmatrix}, \qquad V_2 = \begin{bmatrix} \dfrac{1}{\sqrt{1+\beta^2}} \\[2ex] \dfrac{\beta}{\sqrt{1+\beta^2}} \end{bmatrix}.$$

Hence,

$$V = \begin{bmatrix} 0 & \dfrac{1}{\sqrt{1+\beta^2}} \\[2ex] 1 & \dfrac{\beta}{\sqrt{1+\beta^2}} \end{bmatrix}, \qquad V^{-1} = \begin{bmatrix} -\beta & 1 \\[1ex] \sqrt{1+\beta^2} & 0 \end{bmatrix}$$

$$e^{\lambda(t,t_0)} = \begin{bmatrix} 1 & 0 \\ 0 & e^{(t-t_0)} \end{bmatrix}$$

and from Eq. (4.2.35)

$$\Phi(t,t_0) = \begin{bmatrix} e^{(t-t_0)} & 0 \\ \beta(e^{(t-t_0)} - 1) & 1 \end{bmatrix}.$$

Moler and Van Loan (1978) describe 19 ways to compute the exponential of a matrix, including the methods presented here.

4.2.3 RELATING THE OBSERVATIONS TO AN EPOCH STATE

Note from the second of Eq. (4.2.6) that there is an unknown state vector x_i corresponding to each observation y_i. Hence, it is desirable to use the state transition matrix to express all observations in terms of the state at a single epoch in order to reduce the number of unknown state vectors from $\ell \times n$ to n. Using Eq. (4.2.7), the second of Eq. (4.2.6) may be written in terms of the state at t_k as

$$\begin{aligned} y_1 &= \tilde{H}_1 \Phi(t_1, t_k) x_k + \epsilon_1 \\ y_2 &= \tilde{H}_2 \Phi(t_2, t_k) x_k + \epsilon_2 \\ &\;\;\vdots \\ y_\ell &= \tilde{H}_\ell \Phi(t_\ell, t_k) x_k + \epsilon_\ell. \end{aligned} \qquad (4.2.37)$$

Equation (4.2.37) now contains $m = p \times \ell$ observations and only n unknown components of the state. If ϵ_i, $i = 1, \ldots, \ell$ is zero, any linearly independent n of Eq. (4.2.37) can be used to determine x_k.

If the following definitions are used

$$\mathbf{y} \equiv \begin{bmatrix} y_1 \\ \vdots \\ y_\ell \end{bmatrix}; \quad H \equiv \begin{bmatrix} \widetilde{H}_1 \Phi(t_1, t_k) \\ \vdots \\ \widetilde{H}_\ell \Phi(t_\ell, t_k) \end{bmatrix}; \quad \epsilon \equiv \begin{bmatrix} \epsilon_1 \\ \vdots \\ \epsilon_\ell \end{bmatrix} \qquad (4.2.38)$$

and if the subscript on $\mathbf{x}_k$ is dropped for convenience, then Eq. (4.2.37) can be expressed as follows:

$$\mathbf{y} = H\mathbf{x} + \epsilon \qquad (4.2.39)$$

where $\mathbf{y}$ is an $m \times 1$ vector, $\mathbf{x}$ is an $n \times 1$ vector, ϵ is an $m \times 1$ vector, H is an $m \times n$ mapping matrix, where $m = p \times \ell$ is the total number of observations. If p or ℓ is sufficiently large, the essential condition $m > n$ is satisfied. However, we are still faced with m unknown observation errors resulting in $m + n$ total unknowns and only m equations. The least squares criterion provides us with conditions on the m observation errors that allow a solution for the n state variables, $\mathbf{x}$, at the epoch time t_k.

4.3 THE LEAST SQUARES SOLUTION

The least squares solution selects the estimate of $\mathbf{x}$ as that value that minimizes the sum of the squares of the calculated observation residuals. That is, $\mathbf{x}$ is selected to minimize the following *performance index* (Lawson and Hanson, 1974; Björck, 1997):

$$J(\mathbf{x}) = 1/2 \epsilon^T \epsilon. \qquad (4.3.1)$$

The least squares criterion was first proposed by Gauss (1809) and is commonly used today. The sum of the squares of the calculated observation errors is a logical choice for the performance index. A criterion defined, for example, by the sum of the calculated observation errors could be identically zero with very large observation errors having plus and minus signs that cancel each other. Whether the observation error is positive or negative, its square will be positive and the performance index defined by Eq. (4.3.1) can vanish only if each of the observation errors is identically zero. If ϵ, as defined by Eq. (4.2.39), is substituted into Eq. (4.3.1), the following expression is obtained:

$$J(\mathbf{x}) = 1/2 \epsilon^T \epsilon = \sum_{i=1}^{\ell} 1/2 \epsilon_i^T \epsilon_i = 1/2 (\mathbf{y} - H\mathbf{x})^T (\mathbf{y} - H\mathbf{x}). \qquad (4.3.2)$$

Note that Eq. (4.3.2) is a quadratic function of $\mathbf{x}$, and as a consequence the expression will have a unique minima when (see Appendix B, Eq. (B.8.2))

$$\frac{\partial J}{\partial \mathbf{x}} = 0, \quad \text{and} \quad \delta \mathbf{x}^T \frac{\partial^2 J}{\partial \mathbf{x}^2} \delta \mathbf{x} > 0$$

for all $\delta \mathbf{x} \neq 0$. The second condition implies that the symmetric matrix

$$\frac{\partial^2 J}{\partial \mathbf{x}^2}$$

is positive definite.

Carrying out the first operation on Eq. (4.3.2) (see Appendix B, Eq. (B.7.3)) yields

$$\frac{\partial J}{\partial \mathbf{x}} = 0 = -(\mathbf{y} - H\mathbf{x})^T H = -H^T(\mathbf{y} - H\mathbf{x}). \qquad (4.3.3)$$

The value of $\mathbf{x}$ that satisfies Eq. (4.3.3) will be the best estimate of $\mathbf{x}$, which we will call $\hat{\mathbf{x}}$. Hence,

$$(H^T H)\hat{\mathbf{x}} = H^T \mathbf{y}. \qquad (4.3.4)$$

Also, from Eqs. (4.3.3) and (B.7.3) it follows that

$$\frac{\partial^2 J}{\partial \mathbf{x}^2} = H^T H \qquad (4.3.5)$$

which will be positive definite if H is full rank.

Equation (4.3.4) is referred to as the normal equation, and $(H^T H)$ is referred to as the *normal matrix*. Note that the matrix $H^T H$ is an $n \times n$ symmetric matrix, and if this matrix is positive definite (H is rank n) then the solution for the best estimate of $\mathbf{x}$ is given by

$$\hat{\mathbf{x}} = (H^T H)^{-1} H^T \mathbf{y}. \qquad (4.3.6)$$

Equation (4.3.6) is the well-known least squares solution for the best estimate of $\mathbf{x}$ given the linear observation state relationship expressed by Eq. (4.2.39). With the observations, $\mathbf{y}$, and a specified value of $\hat{\mathbf{x}}$, the value for the best estimate of the observation errors, $\hat{\epsilon}$, can be computed from Eq. (4.2.39) as

$$\hat{\epsilon} = \mathbf{y} - H\hat{\mathbf{x}}. \qquad (4.3.7)$$

4.3.1   THE MINIMUM NORM SOLUTION

For the solution given by Eq. (4.3.6) to exist, it is required that $m \geq n$ and H have rank n. Consider the case where $m < n$; that is, H is of rank $< n$.

In other words, there are more unknowns than linearly independent observations. One could choose to specify any $n - m$ of the n components of $\mathbf{x}$ and solve for the remaining m components of $\mathbf{x}$ using the observation equations with $\epsilon = 0$. However, this leads to an infinite number of solutions for $\hat{\mathbf{x}}$. As an alternative, to obtain a unique solution, one can use the minimum norm criterion to determine $\hat{\mathbf{x}}$. Generally, a nominal or initial guess for $\mathbf{x}$ exists. Recall that the differential equations have been linearized and $\mathbf{x} = \mathbf{X} - \mathbf{X}^*$. The *minimum norm* criterion chooses $\mathbf{x}$ to minimize the sum of the squares of the difference between $\mathbf{X}$ and $\mathbf{X}^*$, subject to the constraint that $\epsilon = 0$; that is, $\mathbf{y} = H\mathbf{x}$. Hence, the performance index becomes

$$J(\mathbf{x}, \boldsymbol{\lambda}) = 1/2\mathbf{x}^T\mathbf{x} + \boldsymbol{\lambda}^T(\mathbf{y} - H\mathbf{x}) \tag{4.3.8}$$

where the constraint has been adjoined with an m-dimensional vector of Lagrange multipliers (see Appendix B, Section B.8). Since both $\mathbf{x}$ and $\boldsymbol{\lambda}$ are unknown, the necessary condition for a minimum of $J(\mathbf{x}, \boldsymbol{\lambda})$ is that its derivative with respect to $\mathbf{x}$ and $\boldsymbol{\lambda}$ vanish. This leads to

$$\frac{\partial J(\mathbf{x}, \boldsymbol{\lambda})}{\partial \mathbf{x}} = 0 = \mathbf{x} - H^T\boldsymbol{\lambda} \tag{4.3.9}$$

$$\frac{\partial J(\mathbf{x}, \boldsymbol{\lambda})}{\partial \boldsymbol{\lambda}} = 0 = \mathbf{y} - H\mathbf{x}. \tag{4.3.10}$$

Substituting the expression for $\hat{\mathbf{x}}$ from Eq. (4.3.9) into Eq. (4.3.10) yields

$$\mathbf{y} = HH^T\boldsymbol{\lambda}, \tag{4.3.11}$$

and solving for $\boldsymbol{\lambda}$,

$$\boldsymbol{\lambda} = (HH^T)^{-1}\mathbf{y}. \tag{4.3.12}$$

Substituting Eq. (4.3.12) into Eq. (4.3.9) yields

$$\hat{\mathbf{x}} = H^T(HH^T)^{-1}\mathbf{y} \tag{4.3.13}$$

where HH^T is an $m \times m$ matrix of rank m. The quantities $H^T(HH^T)^{-1}$ of Eq. (4.3.13) and $(H^TH)^{-1}H^T$ of Eq. (4.3.6) are called the *pseudo inverses* of H in the equation $H\hat{\mathbf{x}} = \mathbf{y}$. They apply when there are more unknowns than equations or more equations than unknowns, respectively. Eq. (4.3.13) is the solution for $\mathbf{x}$ of minimum length. In summary,

$$\begin{aligned}
\hat{\mathbf{x}} &= (H^TH)^{-1}H^T\mathbf{y}, & \text{if } m > n \\
\hat{\mathbf{x}} &= H^{-1}\mathbf{y}, & \text{if } m = n \\
\hat{\mathbf{x}} &= H^T(HH^T)^{-1}\mathbf{y}, & \text{if } m < n.
\end{aligned} \tag{4.3.14}$$

4.3.2 Shortcomings of the Least Squares Solution

Three major shortcomings of the simple least squares solution are:

1. Each observation error is weighted equally even though the accuracy of observations may differ.

2. The observation errors may be correlated (not independent), and the simple least squares solution makes no allowance for this.

3. The method does not consider that the errors are samples from a random process and makes no attempt to utilize any statistical information.

The first of these objections is overcome through the use of the weighted least squares approach.

4.3.3 Weighted Least Squares Solution

Equation (4.3.14) has no means of preferentially ordering one observation with respect to another. A more general expression can be obtained by considering the following formulation. Given a vector sequence of observations $\mathbf{y}_1$, $\mathbf{y}_2, \ldots,$ $\mathbf{y}_\ell$ related through the state transition matrix to the state at some epoch time, $\mathbf{x}_k$, and an associated weighting matrix, w_i, for each of the observation vectors, one can write

$$
\begin{aligned}
\mathbf{y}_1 &= H_1\mathbf{x}_k + \epsilon_1; && w_1 \\
\mathbf{y}_2 &= H_2\mathbf{x}_k + \epsilon_2; && w_2 \\
&\quad \vdots \quad \vdots \quad \vdots \\
\mathbf{y}_\ell &= H_\ell\mathbf{x}_k + \epsilon_\ell; && w_\ell
\end{aligned}
\qquad (4.3.15)
$$

where

$$
H_i = \widetilde{H}_i \Phi(t_i, t_k).
$$

In Eq. (4.3.15) the weighting matrices, w_i, are assumed to be diagonal with their elements normalized to a range between zero and one. Observations weighted with a one would be given the highest possible weight and those weighted with zero would be neglected. To reduce Eq. (4.3.15) to an expression similar to

(4.2.14), the following definitions can be used:

$$\mathbf{y} = \begin{bmatrix} \mathbf{y}_1 \\ \mathbf{y}_2 \\ \vdots \\ \mathbf{y}_\ell \end{bmatrix}; \quad H = \begin{bmatrix} H_1 \\ H_2 \\ \vdots \\ H_\ell \end{bmatrix};$$

$$\boldsymbol{\epsilon} = \begin{bmatrix} \epsilon_1 \\ \epsilon_2 \\ \vdots \\ \epsilon_\ell \end{bmatrix}; \quad W = \begin{bmatrix} w_1 & 0 & \cdots & 0 \\ 0 & w_2 & \cdots & 0 \\ & & \ddots & \\ 0 & 0 & \cdots & w_\ell \end{bmatrix}. \tag{4.3.16}$$

Each observation $\mathbf{y}_i$ is assumed to be a p-vector and $\mathbf{x}_k$ is an n-vector. Equation (4.3.15) now can be expressed as

$$\mathbf{y} = H\mathbf{x}_k + \boldsymbol{\epsilon}; \quad W. \tag{4.3.17}$$

One can then pose the weighted least squares problem as follows. Given the linear observation state relationship expressed by (4.3.17), find the estimate of $\mathbf{x}_k$ to minimize the weighted sum of the squares of the calculated observation errors. The performance index is

$$J(\mathbf{x}_k) = 1/2\boldsymbol{\epsilon}^T W \boldsymbol{\epsilon} = \sum_{i=1}^{\ell} 1/2\epsilon_i^T w_i \epsilon_i. \tag{4.3.18}$$

Using Eq. (4.3.17), $J(\mathbf{x}_k)$ can be expressed as

$$J(\mathbf{x}_k) = 1/2(\mathbf{y} - H\mathbf{x}_k)^T W(\mathbf{y} - H\mathbf{x}_k). \tag{4.3.19}$$

A necessary condition for a minimum of $J(\mathbf{x}_k)$ is that its first derivative with respect to $\mathbf{x}_k$ vanishes (see Eq. B.7.4),

$$\frac{\partial J}{\partial \mathbf{x}_k} = 0 = -(\mathbf{y} - H\mathbf{x}_k)^T W H = -H^T W(\mathbf{y} - H\mathbf{x}_k). \tag{4.3.20}$$

This expression can be rearranged to obtain the normal equations analogous to Eq. (4.3.6) in the least squares formulation as

$$(H^T W H)\mathbf{x}_k = H^T W \mathbf{y}. \tag{4.3.21}$$

If the normal matrix $H^T W H$ is positive definite, it will have an inverse and the solution to (4.3.21) is

$$\hat{\mathbf{x}}_k = (H^T W H)^{-1} H^T W \mathbf{y}. \tag{4.3.22}$$

The value of $\hat{\mathbf{x}}_k$ given by Eq. (4.3.22) is the weighted least squares estimate and is the estimate that minimizes the sum of squares of the weighted observation errors. Note that Eq. (4.3.22) can be expressed as

$$\hat{\mathbf{x}}_k = P_k H^T W \mathbf{y},$$

where

$$P_k = (H^T W H)^{-1}. \tag{4.3.23}$$

The $n \times n$ matrix P_k is symmetric, as can be seen from the definition. Furthermore, if it exists, it must be positive definite, since it is computed as the inverse of the positive definite matrix, $H^T W H$. The parameter observability is related to the rank of this matrix. If all the parameters in $\mathbf{x}_k$ are observable (i.e., can be uniquely determined with the observation set $\mathbf{y}$), then P_k will be full rank and P_k will have an inverse. The number of independent observations must be greater than or equal to the number of parameters being estimated if P_k is to be invertible. Furthermore, P_k is related to the accuracy of the estimate, $\hat{\mathbf{x}}_k$. In general, the larger the magnitude of the elements of the matrix, P_k, the less accurate the estimate. Since the weighting matrix, W, usually results from an initial judgment on the accuracy of the observations followed by a normalization procedure to scale the weights to values between zero and one, this interpretation is not strictly valid in the statistical sense. Hence, some caution should be used when attempting to infer the accuracy of an estimate from the magnitude of P_k as obtained in the weighted least squares estimate. In Section 4.4, it will be shown that, with proper selection of W, P_k is the variance-covariance matrix of the estimation error associated with $\hat{\mathbf{x}}_k$.

If an *a priori* value, $\bar{\mathbf{x}}_k$, is available for $\mathbf{x}_k$ and an associated weighting matrix, $\overline{W}_k$, is given, the weighted least squares estimate for $\mathbf{x}_k$ can be obtained by choosing for $\hat{\mathbf{x}}_k$ the value of $\mathbf{x}_k$, which minimizes the performance index

$$J(\mathbf{x}_k) = 1/2(\mathbf{y}-H\mathbf{x}_k)^T W(\mathbf{y}-H\mathbf{x}_k)+1/2(\bar{\mathbf{x}}_k-\mathbf{x}_k)^T \overline{W}_k(\bar{\mathbf{x}}_k-\mathbf{x}_k). \tag{4.3.24}$$

This results in

$$\hat{\mathbf{x}}_k = (H^T W H + \overline{W}_k)^{-1}(H^T W \mathbf{y} + \overline{W}_k \bar{\mathbf{x}}_k). \tag{4.3.25}$$

Here $\bar{\mathbf{x}}_k$ represents an *a priori* estimate of $\mathbf{x}_k$ and $\overline{W}_k$ represents a weighting matrix for the *a priori* estimate of $\mathbf{x}_k$. In Section 4.4 these terms will be introduced in terms of their statistical significance.

4.3.4 AN ALTERNATE LEAST SQUARES APPROACH

A somewhat classic approach to the problem of least squares that was introduced in Section 1.2.3 is described in this section. Assume that we have ℓ scalar observations. The following least squares performance index, or cost function, can be defined:

$$J = \sum_{j=1}^{\ell} (O_j - C_j)^2, \tag{4.3.26}$$

where O_j is the observation, such as range, provided by an instrument; C_j is the *computed observation* using the reference value of $\mathbf{X}_j^*$; and ℓ is the number of observations. C_j is computed from the appropriate observation-state model for the measurement, and is based on a reference set of coordinates for the instrument and for the satellite. In classic terminology, $(O_j - C_j)$ is referred to as *O minus C* and represents an *observation residual*. The coordinates of the satellite are available as the solution to the equations of motion; for example, Eq. (2.3.31) with a specified set of initial conditions, represented by the position and velocity of the satellite at time t_0. The state vector at t_0, $\mathbf{X}(t_0)$, represents the initial position and velocity, specifically

$$\mathbf{X}(t_0) = \begin{bmatrix} X_0 \\ Y_0 \\ Z_0 \\ \dot{X}_0 \\ \dot{Y}_0 \\ \dot{Z}_0 \end{bmatrix}. \tag{4.3.27}$$

As in the traditional least squares problem, the goal is to determine the initial conditions that minimize J in Eq. (4.3.26). Thus, the partial derivatives with respect to $\mathbf{X}(t_0)$ must be zero:

$$\frac{\partial J}{\partial X_0} = 0 = \sum_{j=1}^{\ell} 2(O_j - C_j)\left(-\frac{\partial C_j}{\partial X_0}\right)$$

$$\vdots \tag{4.3.28}$$

$$\frac{\partial J}{\partial \dot{Z}_0} = 0 = \sum_{j=1}^{\ell} 2(O_j - C_j)\left(-\frac{\partial C_j}{\partial \dot{Z}_0}\right).$$

These nonlinear algebraic equations can be written as

$$F_1 = \sum_{j=1}^{\ell} (O_j - C_j)\left(\frac{\partial C_j}{\partial X_0}\right) \tag{4.3.29}$$

$$\vdots$$

$$F_6 = \sum_{j=1}^{\ell} (O_j - C_j) \left(\frac{\partial C_j}{\partial \dot{Z}_0} \right).$$

From Eqs. (4.3.28) and (4.3.29), the following vector equation is defined:

$$\mathbf{F} = \begin{bmatrix} F_1 \\ F_2 \\ F_3 \\ F_4 \\ F_5 \\ F_6 \end{bmatrix} = 0. \tag{4.3.30}$$

The problem is to solve these equations for the value of $\mathbf{X}(t_0)$, which produces the condition that $\mathbf{F} = 0$. Eq. (4.3.30) represents a set of nonlinear algebraic equations in the n unknown components of $\mathbf{X}(t_0)$.

The solution to a set of nonlinear algebraic equations can be accomplished with the well-known Newton–Raphson method, which usually is derived for a single equation. As shown by Dahlquist and Björck (1974), a nonlinear set of equations can be solved in a recursive manner by

$$\mathbf{X}(t_0)_{n+1} = \mathbf{X}(t_0)_n - \left[\frac{\partial \mathbf{F}}{\partial \mathbf{X}} \right]_n^{-1} [\mathbf{F}]_n, \tag{4.3.31}$$

where

$$\frac{\partial \mathbf{F}}{\partial \mathbf{X}} = \begin{bmatrix} \dfrac{\partial F_1}{\partial X_0} & \dfrac{\partial F_1}{\partial Y_0} & \dfrac{\partial F_1}{\partial Z_0} & \dfrac{\partial F_1}{\partial \dot{X}_0} & \dfrac{\partial F_1}{\partial \dot{Y}_0} & \dfrac{\partial F_1}{\partial \dot{Z}_0} \\ \vdots & & & & & \vdots \\ \dfrac{\partial F_6}{\partial X_0} & \dfrac{\partial F_6}{\partial Y_0} & \dfrac{\partial F_6}{\partial Z_0} & \dfrac{\partial F_6}{\partial \dot{X}_0} & \dfrac{\partial F_6}{\partial \dot{Y}_0} & \dfrac{\partial F_6}{\partial \dot{Z}_0} \end{bmatrix} \tag{4.3.32}$$

and where n represents the iteration number. It is evident that the method requires an initial guess corresponding to $n = 0$. Consider, for example,

$$\frac{\partial F_1}{\partial X_0} = \sum_{j=1}^{\ell} \left[\left(-\frac{\partial C_j}{\partial X_0} \right) \left(\frac{\partial C_j}{\partial X_0} \right) + (O_j - C_j) \left(\frac{\partial^2 C_j}{\partial X_0^2} \right) \right]$$

and

$$\frac{\partial F_1}{\partial Y_0} = \sum_{j=1}^{\ell} \left[\left(-\frac{\partial C_j}{\partial Y_0} \right) \left(\frac{\partial C_j}{\partial X_0} \right) + (O_j - C_j) \left(\frac{\partial^2 C_j}{\partial X_0 \partial Y_0} \right) \right]. \tag{4.3.33}$$

In a converged solution, or nearly converged case, the $(O_j - C_j)$ can be expected to be small, thereby providing a reasonable justification for ignoring the terms involving the second partial derivative of C_j.

The solution to these nonlinear equations, $\mathbf{X}(t_0)$, can then be written as

$$\mathbf{X}(t_0)_{n+1} = \tag{4.3.34}$$

$$\mathbf{X}(t_0)_n + \begin{bmatrix} \sum_j \left(\frac{\partial C_j}{\partial X_0}\right)_n^2 & \cdots & \sum_j \left(\frac{\partial C_j}{\partial X_0}\right)_n \left(\frac{\partial C_j}{\partial \dot{Z}_0}\right)_n \\ \vdots & & \\ \sum_j \left(\frac{\partial C_j}{\partial X_0}\right)_n \left(\frac{\partial C_j}{\partial \dot{Z}_0}\right)_n & \cdots & \sum_j \left(\frac{\partial C_j}{\partial \dot{Z}_0}\right)_n^2 \end{bmatrix}^{-1}$$

$$\times \begin{bmatrix} \sum_j \left(\frac{\partial C_j}{\partial X_0}\right)_n (O_j - C_j)_n \\ \vdots \\ \sum_j \left(\frac{\partial C_j}{\partial \dot{Z}_0}\right)_n (O_j - C_j)_n \end{bmatrix}$$

where the terms involving the second partial derivatives have been ignored. Note that, for example,

$$\frac{\partial C_j}{\partial X_0} = \frac{\partial C_j}{\partial X}\frac{\partial X}{\partial X_0} + \frac{\partial C_j}{\partial Y}\frac{\partial Y}{\partial X_0} + \cdots + \frac{\partial C_j}{\partial \dot{Z}}\frac{\partial \dot{Z}}{\partial X_0}$$

and it can be readily shown from

$$H = \tilde{H}\Phi$$

that Eq. (4.3.31) is

$$\mathbf{X}(t_0)_{n+1} = \mathbf{X}(t_0)_n + (H_n^T H_n)^{-1} H_n^T \mathbf{y}_n \tag{4.3.35}$$

where $\mathbf{y}_n$ represents a vector of residuals $(\mathbf{O} - \mathbf{C})_n$ where $\mathbf{O}$ and $\mathbf{C}$ correspond to $\mathbf{Y}$ and $\mathbf{Y}^*$ in the notation used in Section 4.2. Furthermore, if

$$\hat{\mathbf{x}}_{n+1} = \mathbf{X}(t_0)_{n+1} - \mathbf{X}(t_0)_n, \tag{4.3.36}$$

it follows that

$$\hat{\mathbf{x}}_{n+1} = (H_n^T H_n)^{-1} H_n^T \mathbf{y}_n \tag{4.3.37}$$

corresponding to Eq. (4.3.6).

Some further conclusions can be drawn using this approach:

(1) For the orbit determination application, the least squares solution should be iterated, especially when the reference trajectory has significant deviations from the true trajectory.

(2) Ignoring the second partial derivatives in Eq. (4.3.33) may influence the convergence process if the deviations are large.

(3) Since the process is based on the Newton–Raphson method, it will exhibit *quadratic convergence* when near to the solution, namely

$$||\hat{\mathbf{x}}_n|| < ||\hat{\mathbf{x}}_{n-1}||^2.$$

(4) The least squares formulation allows *accumulation* in a sequential manner as illustrated by the summations in Eq. (4.3.33). That is, the method can be formulated to accumulate the measurements sequentially, followed by a matrix inversion or linear system solution.

(5) The iterative process for a specified set of observations can be repeated until, for example, $||\hat{\mathbf{x}}_{n+1}||$ is smaller than some convergence criteria.

Example 4.3.1

Section 3.6 illustrated the range residuals that would be obtained from two stations (Easter Island and Fortaleza, Brazil) for a specified error in the initial conditions of the two-body problem. The orbit error is illustrated in Fig. 2.2.9. For this case, the satellite is observed twice by Easter Island and once by Fortaleza in the time interval shown in Fig. 2.2.9, ignoring one pass from Fortaleza that rises less than one degree in elevation above the horizon. The true (and presumably unknown) initial conditions are given by Example 2.2.6.2 and the nominal or reference initial conditions (presumably known) are given by Example 2.2.4.1.

Using the algorithm described in this section for the three passes, assuming that error-free geometric ranges are collected every 20 seconds when the satellite is above the station's horizon, the results for the state corrections are shown in Table 4.3.1. To obtain these results, the Goodyear state transition matrix (Section 2.2.6) was used, but a numerically integrated state transition matrix, as given by Eqs. (4.2.1) and (4.2.10), could have been used. Note that the initial state being determined is expressed in the nonrotating system, so the partial derivatives comprising H must be found in that system. To form these partial derivatives, the station coordinates must be transformed from Earth-fixed to nonrotating. It is apparent from Table 4.3.1 that the method converges rapidly, since the error that exists after the first iteration is small (see Fig. 2.2.9).

Table 4.3.1:

Estimated Corrections[1]

	Iteration Number				Converged State[2]	
$\hat{\mathbf{x}}$	1	2	3		$(\mathbf{X}(t_0))$	
$\hat{x}_0$	0.808885	0.000566	0.000000	X_0	5492001.14945	m
$\hat{y}_0$	0.586653	0.000536	0.000000	Y_0	3984001.98719	m
$\hat{z}_0$	0.000015	0.000425	0.000000	Z_0	2955.81044	m
$\hat{\dot{x}}_0$	0.000000	0.000000	0.000000	$\dot{X}_0$	−3931.046491	m/sec
$\hat{\dot{y}}_0$	0.000000	0.000000	0.000000	$\dot{Y}_0$	5498.676921	m/sec
$\hat{\dot{z}}_0$	0.000000	0.000000	0.000000	$\dot{Z}_0$	3665.980697	m/sec

[1] Data: three passes described in Section 3.6
[2] Compare to Example 2.2.6.2

If only the first pass from Easter Island is used instead of three passes, a dramatically different result is obtained. With 33 observations, the correction on the first iteration is at the 100-meter level in position and the 0.1-m/sec level in velocity. The subsequent iteration produces corrections at a comparable level, but numerical instabilities are encountered on the third iteration. Close examination shows that $(H^T H)$ in Eq. (4.3.37) is ill conditioned. Explanation of this behavior is discussed in Section 4.12.

4.4 THE MINIMUM VARIANCE ESTIMATE

As noted, the least squares and weighted least squares methods do not include any information on the statistical characteristics of the measurement errors or the *a priori* errors in the values of the parameters to be estimated. The minimum variance approach is one method for removing this limitation. The minimum variance criterion is used widely in developing solutions to estimation problems because of the simplicity in its use. It has the advantage that the complete statistical description of the random errors in the problem is not required. Rather, only the first and second moments of the probability density function of the observation errors are required. This information is expressed in the mean and covariance matrix associated with the random error.

If it is assumed that the observation error ϵ_i is random with zero mean and specified covariance, the state estimation problem can be formulated as follows:

Given: The system of state-propagation equations and observation state equations

$$\mathbf{x}_i = \Phi(t_i, t_k)\mathbf{x}_k \tag{4.4.1}$$

$$\mathbf{y}_i = \tilde{H}_i\mathbf{x}_i + \epsilon_i \quad i = 1, \ldots, \ell. \tag{4.4.2}$$

Find: The linear, unbiased, minimum variance estimate, $\hat{\mathbf{x}}_k$, of the state $\mathbf{x}_k$.

The solution to this problem proceeds as follows. Using the state transition matrix and the definitions of Eq. (4.3.16), reduce Eq. (4.4.2) to the following form

$$\mathbf{y} = H\mathbf{x}_k + \epsilon \tag{4.4.3}$$

where

$$E[\epsilon] = \begin{bmatrix} E[\epsilon_1] \\ E[\epsilon_2] \\ \vdots \\ E[\epsilon_\ell] \end{bmatrix} = \begin{bmatrix} 0 \\ 0 \\ \vdots \\ 0 \end{bmatrix} \quad E[\epsilon\epsilon^T] = \begin{bmatrix} R_{11} R_{12} & \cdots & R_{1\ell} \\ R_{12}^T R_{22} & \cdots & R_{2\ell} \\ \vdots & \ddots & \vdots \\ R_{1\ell}^T & \cdots & R_{\ell\ell} \end{bmatrix} = R. \tag{4.4.4}$$

Generally, $R_{11} = R_{22} = \ldots = R_{\ell\ell}$ and $R_{ij} = 0$ $(i \neq j)$, but this is not a necessary restriction in the following argument. $R_{ij} \neq 0$ $(i \neq j)$ corresponds to the more general case of time-correlated observation errors.

From the problem statement, the estimate is to be the best linear, unbiased, minimum variance estimate. The consequences of each of these requirements are addressed in the following steps.

(1) *Linear:* The requirement of a linear estimate implies that the estimate is to be made up of a linear combination of the observations:

$$\hat{\mathbf{x}}_k = M\mathbf{y}. \tag{4.4.5}$$

The $(n \times m)$ matrix M is unspecified and is to be selected to obtain the best estimate.

(2) *Unbiased:* If the estimate is unbiased, then by definition

$$E[\hat{\mathbf{x}}] = \mathbf{x}. \tag{4.4.6}$$

Substituting Eqs. (4.4.5) and (4.4.3) into Eq. (4.4.6) leads to the following requirement:

$$E[\hat{\mathbf{x}}_k] = E[M\mathbf{y}] = E[MH\mathbf{x}_k + M\epsilon] = \mathbf{x}_k.$$

But, since $E[\epsilon] = 0$, this reduces to

$$MH\mathbf{x}_k = \mathbf{x}_k$$

from which the following constraint on M is obtained

$$MH = I. \tag{4.4.7}$$

That is, if the estimate is to be unbiased, the linear mapping matrix M must satisfy Eq. (4.4.7). This condition requires the rows of M to be orthogonal to the columns of H.

(3) *Minimum Variance:* If the estimate is unbiased, then the *estimation error covariance matrix* can be expressed as (see Appendix A)

$$P_k = E\left\{ [(\hat{\mathbf{x}}_k - \mathbf{x}_k) - E(\hat{\mathbf{x}}_k - \mathbf{x}_k)][(\hat{\mathbf{x}}_k - \mathbf{x}_k) - E(\hat{\mathbf{x}}_k - \mathbf{x}_k)]^T \right\}$$
$$= E[(\hat{\mathbf{x}}_k - \mathbf{x}_k)(\hat{\mathbf{x}}_k - \mathbf{x}_k)^T]. \tag{4.4.8}$$

Hence, the problem statement requires that $\hat{\mathbf{x}}_k$ be selected to minimize P_k while satisfying Eqs. (4.4.6) and (4.4.7). By minimizing P_k, we mean that $P_k^* - P_k$ is nonnegative definite for any P_k^* that results from an M that satisfies Eq. (4.4.7) (Deutsch, 1965). Substituting Eqs. (4.4.5) and (4.4.3) into Eq. (4.4.8) leads to the following result:

$$P_k = E[(M\mathbf{y} - \mathbf{x}_k)(M\mathbf{y} - \mathbf{x}_k)^T]$$
$$= E[\{M(H\mathbf{x}_k + \epsilon) - \mathbf{x}_k\}\{M(H\mathbf{x}_k + \epsilon) - \mathbf{x}_k\}^T]$$
$$= E[M\epsilon\epsilon^T M^T]$$

where we have used $MH = I$. It follows from Eq. (4.4.4) that the covariance matrix can be written as

$$P_k = MRM^T \tag{4.4.9}$$

where M is to be selected to satisfy Eq. (4.4.7). To involve the constraint imposed by Eq. (4.4.7) and to keep the constrained relation for P_k symmetric, Eq. (4.4.7) is adjoined to Eq. (4.4.9) in the following form

$$P_k = MRM^T + \Lambda^T(I - MH)^T + (I - MH)\Lambda \tag{4.4.10}$$

where Λ is a $n \times n$ matrix of unspecified Lagrange multipliers. The final term is added to ensure that P_k remains symmetric. For a minimum of P_k, it is necessary that its first variation with respect to M vanish, and that $I - MH = 0$. Accordingly,

$$\delta P_k = 0 = (MR - \Lambda^T H^T)\delta M^T + \delta M(RM^T - H\Lambda). \tag{4.4.11}$$

Now, if δP_k is to vanish for an arbitrary δM, one of the following conditions must be met:

1. $RM^T - H\Lambda = 0$.

2. δM and/or $RM^T - H\Lambda$ must not be of full rank.

We will impose condition 1 and show that this yields a minimum value of P. Hence, it is required that

$$MR - \Lambda^T H^T = 0, \quad I - MH = 0. \tag{4.4.12}$$

From the first of these conditions

$$M = \Lambda^T H^T R^{-1} \tag{4.4.13}$$

since R is assumed to be positive definite. Substituting Eq. (4.4.13) into the second of Eqs. (4.4.12) leads to the following result

$$\Lambda^T (H^T R^{-1} H) = I. \tag{4.4.14}$$

Now, if the matrix $H^T R^{-1} H$ is full rank, which requires that $m \geq n$, then the inverse matrix will exist and

$$\Lambda^T = (H^T R^{-1} H)^{-1}. \tag{4.4.15}$$

Then, in view of Eq. (4.4.13),

$$M = (H^T R^{-1} H)^{-1} H^T R^{-1}. \tag{4.4.16}$$

This is the value of M that satisfies the unbiased and minimum variance requirements. Substitution of Eq. (4.4.16) into Eq. (4.4.9) leads to the following expression for the covariance matrix:

$$P_k = (H^T R^{-1} H)^{-1}. \tag{4.4.17}$$

With Eqs. (4.4.16) and (4.4.5), the linear unbiased minimum variance estimate of $\mathbf{x}_k$ is given as

$$\hat{\mathbf{x}}_k = (H^T R^{-1} H)^{-1} H^T R^{-1} \mathbf{y}. \tag{4.4.18}$$

It is not obvious that requiring $RM^T - H\Lambda = 0$ yields the minimum estimation error covariance matrix. We now demonstrate that $P_k^* - P_k$ is nonnegative definite, where P_k^* is the covariance matrix associated with any other linear unbiased estimator, $\tilde{\mathbf{x}}$. Without loss of generality let

$$\tilde{\mathbf{x}} = \hat{\mathbf{x}} + B\mathbf{y}.$$

Then

$$E[\tilde{\mathbf{x}}] = E[\hat{\mathbf{x}}] + BE[\mathbf{y}]$$
$$= \mathbf{x} + BH\mathbf{x}.$$

Hence, $BH = 0$ in order for $\tilde{\mathbf{x}}$ to be unbiased. Since H is full rank, B cannot be full rank. We ignore the trivial solution $B = 0$.

Computing the estimation error covariance matrix associated with $\tilde{\mathbf{x}}$ yields

$$P_{\tilde{\mathbf{x}}} = E\left[\left(\tilde{\mathbf{x}} - E(\tilde{\mathbf{x}})\right)\left(\tilde{\mathbf{x}} - E(\tilde{\mathbf{x}})\right)^T\right]$$
$$= E\left[\left(\hat{\mathbf{x}} + B\mathbf{y} - \mathbf{x} - BH\mathbf{x}\right)\left(\hat{\mathbf{x}} + B\mathbf{y} - \mathbf{x} - BH\mathbf{x}\right)^T\right]$$
$$= E\left[\left((\hat{\mathbf{x}} - \mathbf{x}) + B\epsilon\right)\left((\hat{\mathbf{x}} - \mathbf{x}) + B\epsilon\right)^T\right]$$
$$= P + BRB^T + E\left[(\hat{\mathbf{x}} - \mathbf{x})\epsilon^T\right]B^T + BE\left[\epsilon(\hat{\mathbf{x}} - \mathbf{x})^T\right].$$

Also,

$$BE\left[\epsilon(\hat{\mathbf{x}} - \mathbf{x})^T\right] = BE\left[\epsilon(PH^T R^{-1}\mathbf{y} - \mathbf{x})^T\right]$$
$$= BE\left[\epsilon(PH^T R^{-1}(H\mathbf{x} + \epsilon) - \mathbf{x})^T\right]$$
$$= BE(\epsilon\epsilon^T)R^{-1}HP$$
$$= BHP = 0,$$

since $BH = 0$.

Hence,

$$P_{\tilde{\mathbf{x}}} - P = BRB^T$$

Because B is not of full rank, BRB^T must be positive semidefinite (i.e., its diagonal elements are each ≥ 0). Hence, the requirement that the difference of the two covariance matrices be nonnegative definite has been met. Consequently, all variances associated with $P_{\tilde{\mathbf{x}}}$ must be greater than or equal to those of P and the trace of $P_{\tilde{\mathbf{x}}}$ must be greater than that of P.

Note that computation of the estimate, $\hat{\mathbf{x}}_k$, requires inverting the $n \times n$ normal matrix, $H^T R^{-1} H$. For a large dimension system the computation of this inverse may be difficult. The solution given by Eq. (4.4.18) will agree with the weighted least squares solution if the weighting matrix, W, used in the least squares approach is equal to the inverse of the observation noise covariance matrix; that is, if $W = R^{-1}$.

4.4.1 PROPAGATION OF THE ESTIMATE AND COVARIANCE MATRIX

If the estimate at a time t_j is obtained by using Eq. (4.4.18), the estimate may be mapped to any later time by using Eq. (4.4.2):

$$\overline{\mathbf{x}}_k = \Phi(t_k, t_j)\hat{\mathbf{x}}_j. \tag{4.4.19}$$

The expression for propagating the covariance matrix can be obtained as follows:

$$\overline{P}_k \equiv E[(\overline{\mathbf{x}}_k - \mathbf{x}_k)(\overline{\mathbf{x}}_k - \mathbf{x}_k)^T]. \tag{4.4.20}$$

In view of Eq. (4.4.19), Eq. (4.4.20) becomes

$$\overline{P}_k = E[\Phi(t_k, t_j)(\hat{\mathbf{x}}_j - \mathbf{x}_j)(\hat{\mathbf{x}}_j - \mathbf{x}_j)^T \Phi^T(t_k, t_j)]. \tag{4.4.21}$$

Since the state transition matrix is deterministic, it follows from Eq. (4.4.8) that

$$\overline{P}_k = \Phi(t_k, t_j)P_j\Phi^T(t_k, t_j). \tag{4.4.22}$$

Equations (4.4.19) and (4.4.22) can be used to map the estimate of the state and its associated covariance matrix from t_j to t_k.

4.4.2 MINIMUM VARIANCE ESTIMATE WITH *A Priori* INFORMATION

If an estimate and the associated covariance matrix are obtained at a time t_j, and an additional observation or observation sequence is obtained at a time t_k, the estimate and the observation can be combined in a straightforward manner to obtain the new estimate $\hat{\mathbf{x}}_k$. The estimate, $\hat{\mathbf{x}}_j$, and associated covariance, P_j, are propagated forward to t_k using Eqs. (4.4.19) and (4.4.22) and are given by

$$\overline{\mathbf{x}}_k = \Phi(t_k, t_j)\hat{\mathbf{x}}_j, \ \ \overline{P}_k = \Phi(t_k, t_j)P_j\Phi^T(t_k, t_j). \tag{4.4.23}$$

The problem to be considered can be stated as follows:

Given: $\overline{\mathbf{x}}_k, \overline{P}_k$ and $\mathbf{y}_k = \widetilde{H}_k\mathbf{x}_k + \epsilon_k$, where $E[\epsilon_k] = 0$, $E[\epsilon_k\epsilon_j^T] = R_k\delta_{kj}$, and $E[(\overline{\mathbf{x}}_j - \mathbf{x}_j)\epsilon_k^T] = 0$, find the linear, minimum variance, unbiased estimate of $\mathbf{x}_k$.

The solution to the problem can be obtained by reducing it to the previously solved problem. To this end, note that if $\hat{\mathbf{x}}_j$ is unbiased, $\overline{\mathbf{x}}_k$ will be unbiased since

$E[\overline{\mathbf{x}}_k] = \Phi(t_k, t_j)E[\hat{\mathbf{x}}_j] = \mathbf{x}_k$. Hence, $\overline{\mathbf{x}}_k$ can be interpreted as an unbiased observation of $\mathbf{x}_k$ and we may treat it as an additional data equation at t_k,

$$\mathbf{y}_k = \widetilde{H}_k\mathbf{x}_k + \boldsymbol{\epsilon}_k$$
$$\overline{\mathbf{x}}_k = \mathbf{x}_k + \boldsymbol{\eta}_k \tag{4.4.24}$$

where

$$E[\boldsymbol{\epsilon}_k] = 0, \quad E[\boldsymbol{\epsilon}_k\boldsymbol{\epsilon}_k^T] = R_k, \quad E[\boldsymbol{\eta}_k] = 0,$$
$$E[\boldsymbol{\eta}_k\boldsymbol{\epsilon}_k^T] = 0, \text{ and } E[\boldsymbol{\eta}_k\boldsymbol{\eta}_k^T] = \overline{P}_k. \tag{4.4.25}$$

It is assumed that the errors in the observations, $\boldsymbol{\epsilon}_k$, are not correlated with the errors in the *a priori* estimate, $\boldsymbol{\eta}_k$. That is, $E[\boldsymbol{\eta}_k\boldsymbol{\epsilon}_k^T] = 0$. Now, if the following definitions are used

$$\mathbf{y} = \begin{bmatrix} \mathbf{y}_k \\ \dots \\ \overline{\mathbf{x}}_k \end{bmatrix}; H = \begin{bmatrix} \widetilde{H}_k \\ \dots \\ I \end{bmatrix};$$

$$\boldsymbol{\epsilon} = \begin{bmatrix} \boldsymbol{\epsilon}_k \\ \dots \\ \boldsymbol{\eta}_k \end{bmatrix}; R = \begin{bmatrix} R_k & 0 \\ & \dots & \\ 0 & \overline{P}_k \end{bmatrix}; \tag{4.4.26}$$

Eq. (4.4.24) can be expressed as $\mathbf{y} = H\mathbf{x}_k + \boldsymbol{\epsilon}$ as in Eq. (4.4.3), and the solution for $\hat{\mathbf{x}}_k$ is given by Eq. (4.4.18),

$$\hat{\mathbf{x}}_k = (H^T R^{-1} H)^{-1} H^T R^{-1} \mathbf{y}. \tag{4.4.27}$$

In view of the definitions in Eq. (4.4.26),

$$\hat{\mathbf{x}}_k = \left\{ [\widetilde{H}_k^T : I] \begin{bmatrix} R_k^{-1} & 0 \\ \dots & \dots \\ 0 & \overline{P}_k^{-1} \end{bmatrix} \begin{bmatrix} \widetilde{H}_k \\ \dots \\ I \end{bmatrix} \right\}^{-1}$$

$$\left\{ [\widetilde{H}_k^T : I] \begin{bmatrix} R_k^{-1} & 0 \\ \dots & \dots \\ 0 & \overline{P}_k^{-1} \end{bmatrix} \begin{bmatrix} \mathbf{y}_k \\ \dots \\ \overline{\mathbf{x}}_k \end{bmatrix} \right\} \tag{4.4.28}$$

or

$$\hat{\mathbf{x}}_k = (\widetilde{H}_k^T R_k^{-1} \widetilde{H}_k + \overline{P}_k^{-1})^{-1}(\widetilde{H}_k^T R_k^{-1} \mathbf{y}_k + \overline{P}_k^{-1}\overline{\mathbf{x}}_k). \tag{4.4.29}$$

Using Eq. (4.4.17) the covariance matrix associated with the estimation error in $\hat{\mathbf{x}}$ easily is shown to be

$$P_k = E[(\hat{\mathbf{x}}_k - \mathbf{x}_k)(\hat{\mathbf{x}}_k - \mathbf{x}_k)^T]$$
$$= (\widetilde{H}_k^T R_k^{-1} \widetilde{H}_k + \overline{P}_k^{-1})^{-1}. \tag{4.4.30}$$

The inverse of the covariance matrix is called the *information matrix*,

$$\Lambda_k = P_k^{-1}. \tag{4.4.31}$$

Equation (4.4.29) often is seen written in normal equation form as

$$\Lambda_k \hat{\mathbf{x}}_k = \widetilde{H}_k^T R_k^{-1} \mathbf{y}_k + \overline{P}_k^{-1} \overline{\mathbf{x}}_k. \tag{4.4.32}$$

The following remarks relate to Eq. (4.4.29):

1. The vector $\mathbf{y}_k$ may be only a single observation or it may include a batch of observations mapped to t_k.

2. The *a priori estimate*, $\overline{\mathbf{x}}_k$, may represent the estimate based on *a priori* initial conditions or the estimate based on the reduction of a previous batch of data.

3. The $n \times n$ normal matrix of Eq. (4.4.29) must be inverted and if the dimension n is large, this inversion can lead to computational problems. However, alternate solution techniques that avoid the accumulation and inversion of the normal matrix have been developed and are discussed in Chapter 5.

4. The algorithm for using Eq. (4.4.29) is referred to as the *batch processor*. The name derives from the fact that all data generally are accumulated prior to solving for $\hat{\mathbf{x}}_k$; that is, the data are processed in a single batch.

5. Note that Eq. (4.4.29) could also be implemented as a *sequential processor*; that is, after each observation the state estimate and covariance matrix could be mapped to the time of the next observation, where it would become the *a priori* information. This could then be combined with the observation at that time to yield the estimate for $\hat{\mathbf{x}}$ using Eq. (4.4.29).

4.5   MAXIMUM LIKELIHOOD AND BAYESIAN ESTIMATION

The method of *Maximum Likelihood Estimation* for determining the best estimate of a variable is due to Fisher (1912). The Maximum Likelihood Estimate (MLE) of a parameter Θ—given observations y_1, y_2, $\ldots y_k$ and the joint probability density function

$$f(y_1, y_2, \ldots y_k; \Theta) \tag{4.5.1}$$

is defined to be that value of Θ that maximizes the probability density function (Walpole and Myers, 1989). However, if Θ is a random variable and we have knowledge of its probability density function, the MLE of Θ is defined to be the value of Θ, which maximizes the probability density function of Θ conditioned on knowledge of the observations $y_1, y_2, \ldots y_k$:

$$f(\Theta/y_1, y_2, \ldots y_k). \tag{4.5.2}$$

The *Bayes estimate* for Θ is defined to be the mean of the conditional density function given by Eq. (4.5.2) (Walpole and Myers, 1989). The joint density function, Eq. (4.5.1), and the conditional density function, Eq. (4.5.2), are referred to as the *likelihood function*, L. The logic behind maximizing L is that of all the possible values of Θ we should choose the one that maximizes the probability of obtaining the observations that actually were observed. If Θ is a random variable, this corresponds to the mode, or peak, of the conditional density function. In the case of a symmetric, unimodal, density function such as a Gaussian function, this will correspond to the mean of the conditional density function. Hence, the MLE and the Bayes estimate for a Gaussian density function are identical.

Since the logarithm of the density function is a monotonically increasing function of the density function, it is often simpler to determine the value of Θ that maximizes $\ln(L)$.

For example, assume we are given the following joint probability density function of the independent random variables $y_i, i = 1, 2 \ldots k$ with common mean, α, and common standard deviation, σ, and we wish to determine the MLE of the parameter α:

$$f(y_1, , y_2 \cdots y_k; \alpha, \sigma) = f(y_1; \alpha, \sigma) f(y_2; \alpha, \sigma) \cdots f(y_k; \alpha, \sigma). \tag{4.5.3}$$

We are able to factor the joint density function into the product of the marginal density functions because the random variables y_i are independent. If the joint density function is Gaussian, we may write (see Eq. (A.19.1))

$$L = f(y_1, y_2 \cdots y_k; \alpha, \sigma) = \frac{1}{(2\pi)^{k/2} \sigma^k} \exp\left\{ \frac{-1}{2\sigma^2} \sum_{i=1}^{k} (y_i - \alpha)^2 \right\} \tag{4.5.4}$$

then

$$\ln L = -\frac{k}{2} \ln 2\pi - k \ln \sigma - \frac{1}{2\sigma^2} \sum_{i=1}^{k} (y_i - \alpha)^2 \tag{4.5.5}$$

and for a maximum

$$\frac{\partial \ln L}{\partial \alpha} = \frac{1}{\sigma^2} \sum_{i=1}^{k} (y_i - \alpha) = 0 \tag{4.5.6}$$

and

$$\frac{\partial^2 \ln L}{\partial \alpha^2} = -\frac{k}{\sigma^2} < 0.$$

Hence,

$$\sum_{i=1}^{k}(y_i - \hat{\alpha}) = 0 \qquad (4.5.7)$$

and the MLE of α is

$$\hat{\alpha} = \frac{1}{k}\sum_{i=1}^{k}y_i. \qquad (4.5.8)$$

In terms of the orbit determination problem, we are given observations y_1, $y_2 \ldots y_k$ and we wish to determine the MLE of the state, $\hat{x}$. Hence, we wish to find the value of the state vector, x_k, which maximizes the conditional density function

$$f(x_k/y_1, y_2 \ldots y_k). \qquad (4.5.9)$$

We will assume that all density functions for this derivation are Gaussian. Using the first of Eq. (A.21.1) of Appendix A, we may write

$$f(x_k/y_1, y_2 \ldots y_k) = \frac{f(x_k, y_1, y_2 \ldots y_k)}{f(y_1, y_2 \ldots y_k)}. \qquad (4.5.10)$$

Using the second of Eq. (A.21.1) yields

$$f(x_k, y_1, y_2 \ldots y_k) = f(y_k/x_k, y_1, \ldots y_{k-1})f(x_k, y_1, \ldots y_{k-1}). \quad (4.5.11)$$

Assuming independent observations results in

$$f(y_k/x_k, y_1, \ldots y_{k-1}) = f(y_k/x_k), \qquad (4.5.12)$$

and again using the first of Eq. (A.21.1) we have

$$f(x_k, y_1, \ldots y_{k-1}) = f(x_k/y_1, \ldots y_{k-1})f(y_1, \ldots y_{k-1}). \qquad (4.5.13)$$

Hence, Eq. (4.5.11) may be written as

$$f(x_k, y_1, y_2 \ldots y_k) = f(y_k/x_k)f(x_k/y_1, \ldots y_{k-1})f(y_1, \ldots y_{k-1}). \tag{4.5.14}$$

Substituting Eq. (4.5.14) into Eq. (4.5.10) yields

$$f(x_k/y_1, y_2 \ldots y_k) = \frac{f(y_k/x_k)f(x_k/y_1, \ldots y_{k-1})}{f(y_k)}, \qquad (4.5.15)$$

where we have used the fact that the observations, y_i, are independent so we can write

$$f(y_1, y_2 \ldots y_k) = f(y_1)f(y_2) \ldots f(y_k). \qquad (4.5.16)$$

For our system,

$$\mathbf{x}_k = \Phi(t_k, t_i)\mathbf{x}_i \tag{4.5.17}$$

$$\mathbf{y}_i = \widetilde{H}_i\,\mathbf{x}_i + \boldsymbol{\epsilon}_i, \quad i = 1 \ldots k \tag{4.5.18}$$

where $\boldsymbol{\epsilon}_i \sim N(0, R_i)$; that is, $\boldsymbol{\epsilon}_i$ has a normal distribution with zero mean and covariance R_i. We are assuming independent observations; hence,

$$E\left[\boldsymbol{\epsilon}_j\boldsymbol{\epsilon}_k^T\right] = \delta_{jk}\,R_k. \tag{4.5.19}$$

We seek the MLE estimate of $\mathbf{x}_k$, the value of $\mathbf{x}_k$ that maximizes the conditional density function of Eq. (4.5.15). Note that $f(\mathbf{y}_k)$, the marginal density function of $\mathbf{y}_k$, is by definition independent of $\mathbf{x}_k$. Hence, only the numerator of Eq. (4.5.15) is dependent on $\mathbf{x}_k$. $\mathbf{y}_k/\mathbf{x}_k$ has mean

$$E(\mathbf{y}_k/\mathbf{x}_k) = E\left[(\widetilde{H}_k\mathbf{x}_k + \boldsymbol{\epsilon}_k)/\mathbf{x}_k\right] \tag{4.5.20}$$

$$= \widetilde{H}_k\mathbf{x}_k$$

and covariance

$$E\left[(\mathbf{y}_k - \widetilde{H}_k\mathbf{x}_k)(\mathbf{y}_k - \widetilde{H}_k\mathbf{x}_k)^T/\mathbf{x}_k\right]$$

$$= E\left[\boldsymbol{\epsilon}_k\boldsymbol{\epsilon}_k^T\right] = R_k. \tag{4.5.21}$$

Hence,

$$\mathbf{y}_k/\mathbf{x}_k \sim N(\widetilde{H}_k\mathbf{x}_k, R_k). \tag{4.5.22}$$

Also,

$$E\left[\mathbf{x}_k/\mathbf{y}_1, \mathbf{y}_2 \ldots \mathbf{y}_{k-1}\right] \equiv \overline{\mathbf{x}}_k. \tag{4.5.23}$$

The associated covariance is

$$E[(\mathbf{x}_k - \overline{\mathbf{x}}_k)(\mathbf{x}_k - \overline{\mathbf{x}}_k)^T] \equiv \overline{P}_k. \tag{4.5.24}$$

Hence,

$$\mathbf{x}_k/\mathbf{y}_1, \ldots \mathbf{y}_{k-1} \sim N(\overline{\mathbf{x}}_k, \overline{P}_k). \tag{4.5.25}$$

The *likelihood function* defined by Eq. (4.5.15) is given by

$$L = f(\mathbf{x}_k/\mathbf{y}_1, \mathbf{y}_2 \ldots \mathbf{y}_k) = \frac{1}{(2\pi)^{p/2}|R_k|^{1/2}}$$

$$\times \exp{-1/2\left\{(\mathbf{y}_k - \widetilde{H}_k\mathbf{x}_k)^T R_k^{-1}(\mathbf{y}_k - \widetilde{H}_k\mathbf{x}_k)\right\}} \tag{4.5.26}$$

$$\times \frac{1}{(2\pi)^{n/2}|\overline{P}_k|^{1/2}} \exp{-1/2\left\{(\mathbf{x}_k - \overline{\mathbf{x}}_k)^T \overline{P}_k^{-1}(\mathbf{x}_k - \overline{\mathbf{x}}_k)\right\}} \frac{1}{f(\mathbf{y}_k)}.$$

Because this is a Gaussian density function, the mean, median (having equal probability weight on either side), and mode will be identical. Hence, any criterion that chooses one of these values for $\hat{\mathbf{x}}_k$ will yield the same estimator.

Accumulating the logarithms of all terms in $f(\mathbf{x}_k/\mathbf{y}_1, \mathbf{y}_2 \ldots \mathbf{y}_k)$ that are a function of $\mathbf{x}_k$ and calling this function $\ln L'$ yields

$$\ln L' = -1/2[(\mathbf{y}_k - \widetilde{H}_k\mathbf{x}_k)^T R_k^{-1}(\mathbf{y}_k - \widetilde{H}_k\mathbf{x}_k)$$
$$+(\mathbf{x}_k - \overline{\mathbf{x}}_k)^T \overline{P}_k^{-1}(\mathbf{x}_k - \overline{\mathbf{x}}_k)]. \tag{4.5.27}$$

Differentiating $\ln L'$ with respect to $\mathbf{x}_k$,

$$\frac{\partial \ln L'}{\partial \mathbf{x}_k} = -\left[-(\widetilde{H}_k^T R_k^{-1})(\mathbf{y}_k - \widetilde{H}_k\mathbf{x}_k) + \overline{P}_k^{-1}(\mathbf{x}_k - \overline{\mathbf{x}}_k)\right] = 0. \tag{4.5.28}$$

The value of $\mathbf{x}_k$ that satisfies Eq. (4.5.28) is $\hat{\mathbf{x}}_k$:

$$\left(\widetilde{H}_k^T R_k^{-1} \widetilde{H}_k + \overline{P}_k^{-1}\right) \hat{\mathbf{x}}_k = H_k^T R_k^{-1} \mathbf{y}_k + \overline{P}^{-1}\overline{\mathbf{x}}_k \tag{4.5.29}$$

$$\hat{\mathbf{x}}_k = \left(\widetilde{H}_k^T R_k^{-1} \widetilde{H}_k + \overline{P}_k^{-1}\right)^{-1} \left(\widetilde{H}_k^T R_k^{-1} \mathbf{y}_k + \overline{P}_k^{-1}\overline{\mathbf{x}}_k\right) \tag{4.5.30}$$

which is identical to the minimum variance estimate of $\hat{\mathbf{x}}_k$ as well as the weighted least squares estimate if $W_k = R_k^{-1}$.

Furthermore, from Eq. (4.5.28),

$$\frac{\partial^2 \ln L'}{\partial \mathbf{x}^2} = -\left[\widetilde{H}_k^T R_k^{-1} \widetilde{H}_k + \overline{P}_k^{-1}\right]. \tag{4.5.31}$$

Because $-\frac{\partial^2 \ln L'}{\partial \mathbf{x}^2}$ is positive definite, we have maximized the likelihood function (see Appendix B).

4.6   COMPUTATIONAL ALGORITHM FOR THE BATCH PROCESSOR

Assume that we wish to estimate the state deviation vector $\mathbf{x}_0$ at a reference time, t_0. Given a set of initial conditions $\mathbf{X}^*(t_0)$, an *a priori* estimate $\overline{\mathbf{x}}_0$ and the associated error covariance matrix, $\overline{P}_0$, the computational algorithm for the batch processor generally uses the normal equation form for $\hat{\mathbf{x}}_0$. Writing Eqs. (4.3.25) and (4.4.29) in normal equation form for a batch of observations and recognizing in Eq. (4.3.25) that $W = R^{-1}$ and $\overline{W} = \overline{P}_0^{-1}$ yields

$$(H^T R^{-1} H + \overline{P}_0^{-1})\hat{\mathbf{x}}_0 = H^T R^{-1} \mathbf{y} + \overline{P}_0^{-1}\overline{\mathbf{x}}_0. \tag{4.6.1}$$

Here t_0 is an arbitrary epoch and all quantities in Eq. (4.6.1) are assumed to have been mapped to this epoch using the appropriate state transition matrices as illustrated in Eqs. (4.3.15) and (4.3.16). Because we are dealing with a linearized system, Eq. (4.6.1) generally is iterated to convergence; that is, until $\hat{x}_0$ no longer changes. Note that the two matrices in Eq. (4.6.1) that must be accumulated are $H^T R^{-1} H$ and $H^T R^{-1} y$. If R is a block diagonal matrix—the observations are uncorrelated in time although correlations between the observations at any given time may exist—these matrices simply may be accumulated as follows:

$$H^T R^{-1} H = \sum_{i=1}^{\ell} \left[\tilde{H}_i \Phi(t_i, t_0) \right]^T R_i^{-1} \tilde{H}_i \Phi(t_i, t_0) \qquad (4.6.2)$$

$$H^T R^{-1} y = \sum_{i=1}^{\ell} [\tilde{H}_i \Phi(t_i, t_0)]^T R_i^{-1} y_i. \qquad (4.6.3)$$

In general $\mathbf{X}^*(t_0)$ would be chosen so that $\bar{x}_0 = 0$, and $\overline{P}_0$ would reflect the relative accuracy of the elements of the initial condition vector $\mathbf{X}^*(t_0)$. In theory $\bar{x}_0$ and $\bar{P}_0$ represent information and should be treated as data that are merged with the observation data, as indicated by Eq. (4.6.1). Consequently, the value of $\mathbf{X}_0^* + \bar{x}_0$ should be held constant for the beginning of each iteration. Since the initial condition vector $\mathbf{X}_0^*$ is augmented by the value of $\hat{x}_0$ after each iteration, that is, $(\mathbf{X}_0^*)_n = (\mathbf{X}_0^*)_{n-1} + (\hat{x}_0)_n$, holding $\mathbf{X}_0^* + \bar{x}_0$ constant results in the following expression for $(\bar{x}_0)_n$

$$\begin{aligned}
\mathbf{X}_0^* + \bar{x}_0 &= (\mathbf{X}_0^*)_{n-1} + (\bar{x}_0)_{n-1} \\
&= (\mathbf{X}_0^*)_n + (\bar{x}_0)_n \\
&= (\mathbf{X}_0^*)_{n-1} + (\hat{x}_0)_{n-1} + (\bar{x}_0)_n
\end{aligned}$$

or

$$(\bar{x}_0)_n = (\bar{x}_0)_{n-1} - (\hat{x}_0)_{n-1}. \qquad (4.6.4)$$

Recall from Section 4.2 that the state transition matrix is obtained by integrating

$$\dot{\Phi}(t, t_k) = A(t)\Phi(t, t_k)$$

subject to the initial conditions

$$\Phi(t_k, t_k) = I$$

along with the nonlinear equations, $\dot{\mathbf{X}}^* = F(\mathbf{X}^*, t)$, which define the nominal trajectory, $\mathbf{X}^*(t)$. The matrix $A(t)$ is evaluated on the reference trajectory,

$$A(t) = \frac{\partial F(\mathbf{X}^*, t)}{\partial \mathbf{X}} \qquad (4.6.5)$$

(A) Initialize for this iteration:

Set $i = 1$. Set $t_{i-1} = t_0$, $\mathbf{X}^*(t_{i-1}) = \mathbf{X}_0^*$, $\Phi(t_{i-1}, t_0) = \Phi(t_0, t_0) = I$.

If there is an *a priori* estimate, set $\Lambda = \overline{P}_0^{-1}$ and set $N = \overline{P}_0^{-1}\bar{\mathbf{x}}_0$.

If there is no *a priori* estimate, set $\Lambda = 0$ and set $N = 0$.

(B) Read the next observation: t_i, $\mathbf{Y}_i$, R_i.

Integrate reference trajectory and state transition matrix from t_{i-1} to t_i.

$\dot{\mathbf{X}}^* = F(\mathbf{X}^*(t), t)$ with initial conditions $\mathbf{X}^*(t_{i-1})$

$A(t) = [\partial F(\mathbf{X}, t)/\partial \mathbf{X}]^*$, where * means evaluated on the nominal trajectory.

$\dot{\Phi}(t, t_0) = A(t)\Phi(t, t_0)$ with initial conditions $\Phi(t_{i-1}, t_0)$.

This gives $\mathbf{X}^*(t_i)$, $\Phi(t_i, t_0)$.

Accumulate current observation

$$\tilde{H}_i = [\partial G(\mathbf{X}, t_i)/\partial \mathbf{X}]^*$$
$$\mathbf{y}_i = \mathbf{Y}_i - G(\mathbf{X}_i^*, t_i)$$
$$H_i = \tilde{H}_i \Phi(t_i, t_0)$$
$$\Lambda = \Lambda + H_i^T R_i^{-1} H_i$$
$$N = N + H_i^T R_i^{-1} \mathbf{y}_i$$

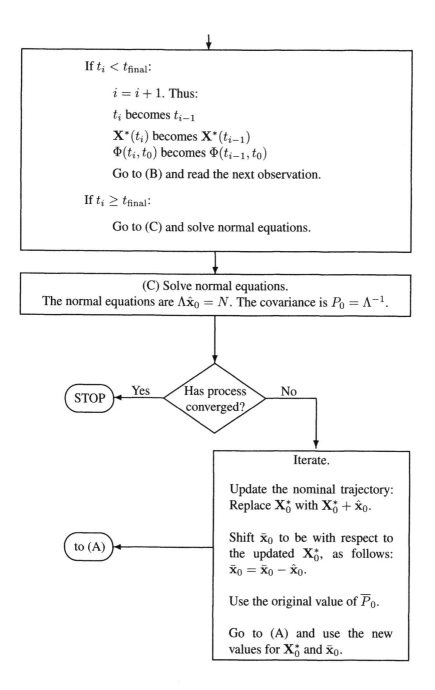

Figure 4.6.1: Batch processing algorithm flow chart.

where $F(\mathbf{X}^*, t)$ is the time derivative of the state vector in the differential equations governing the time evolution of the system. The observation-state mapping matrix is given by

$$\tilde{H}_i = \frac{\partial G(\mathbf{X}_i^*, t_i)}{\partial \mathbf{X}} \tag{4.6.6}$$

where $G(\mathbf{X}_i^*, t_i)$ are the observation-state relationships evaluated on the nominal or reference trajectory.

Notice that the solution for $\hat{\mathbf{x}}_0$ involved inversion of the information matrix, Λ_0, where

$$\Lambda_0 = H^T R^{-1} H + \overline{P}_0^{-1}. \tag{4.6.7}$$

Generally the normal equation would not be solved by a direct inversion of Λ_0 but rather would be solved by an indirect but more accurate technique, such as the Cholesky decomposition described in Chapter 5, Section 5.2. The sequence of operations required to implement the batch estimation process is outlined in Fig. 4.6.1. Note that the algorithm in Fig. 4.6.1 assumes that there are no observations at t_0. If observations exist at t_0, set $\Lambda = \overline{P}_0^{-1} + H_0^T R_0^{-1} H_0$ and $N = H_0^T R_0^{-1} \mathbf{y}_0$ in the initialization box, A. As previously stated, the entire sequence of computations are repeated until the estimation process has converged. If there are observations at t_0, the state transition matrix for processing these observations is the identity matrix.

This procedure yields a minimum value of the performance index (see Eq. (4.3.24))

$$J(x) = (\hat{\mathbf{x}}_0 - \bar{\mathbf{x}}_0)^T \overline{P}_0^{-1} (\hat{\mathbf{x}}_0 - \bar{\mathbf{x}}_0) + \sum_{i=1}^{\ell} \hat{\epsilon}_i^T R_i^{-1} \hat{\epsilon}_i \tag{4.6.8}$$

where

$$\hat{\epsilon}_i = \mathbf{y}_i - H_i \hat{\mathbf{x}}_0 \tag{4.6.9}$$

and $\hat{\epsilon}_i$ is the best estimate of the observation error.

In practice, $\overline{P}_0$ is generally not a realistic representation of the accuracy of $\bar{\mathbf{x}}_0$ and it is used only to better condition the estimation error covariance matrix, P. In this case, $\bar{\mathbf{x}}_0$ usually is set to zero for each iteration and $\overline{P}_0$ is chosen to be a diagonal matrix with large diagonal values. Hence, the first term in Eq. (4.6.8) will be very small and the tracking data residuals will determine the value of $J(x)$. The RMS of the observation residuals generally is computed and may be used as a measure of convergence; when the RMS no longer changes the solution is assumed to be converged. The RMS is computed from

$$\text{RMS} = \left\{ \frac{\sum_{i=1}^{\ell} \hat{\epsilon}_i^T R_i^{-1} \hat{\epsilon}_i}{m} \right\}^{1/2}. \tag{4.6.10}$$

$\hat{\boldsymbol{\epsilon}}_i$ is a p-vector and $m = \ell \times p$. Hence, m is the total number of observations. Eq. (4.6.10) is referred to as the weighted RMS. If the RMS is computed without including the weighting matrix, R_i^{-1}, it may be referred to as the unweighted RMS or just the RMS.

4.7 THE SEQUENTIAL ESTIMATION ALGORITHM

In this section, an alternate approach to the batch processor is discussed in which the observations are processed as soon as they are received. An advantage of this approach, referred to as the sequential processing algorithm, is that the matrix to be inverted will be of the same dimension as the observation vector. Hence, if the observations are processed individually, only scalar divisions will be required to obtain the estimate of $\mathbf{x}_k$. The algorithm was developed originally by Swerling (1959), but the treatment that received more popular acclaim is due to Kalman (1960). In fact, the sequential estimation algorithm discussed here often is referred to as the *Kalman filter*. A number of papers and textbooks have been written describing and providing variations of the Kalman filter. The collection of papers edited by Sorenson (1985) contains many of the pioneering papers in this area. Also, the treatments by Bierman (1977), Liebelt (1967), Tapley (1973), Gelb (1974), Maybeck (1979), Grewal and Andrews (1993), and Montenbruck and Gill (2000) are other references.

Recall that an estimate $\hat{\mathbf{x}}_j$ and a covariance matrix P_j can be propagated forward to a time t_k by the relations

$$\overline{\mathbf{x}}_k = \Phi(t_k, t_j)\hat{\mathbf{x}}_j \tag{4.7.1}$$
$$\overline{P}_k = \Phi(t_k, t_j)P_j\Phi^T(t_k, t_j).$$

Assume that we have an additional observation at t_k (see Eq. (4.2.6)),

$$\mathbf{y}_k = \widetilde{H}_k\mathbf{x}_k + \boldsymbol{\epsilon}_k \tag{4.7.2}$$

where $E[\boldsymbol{\epsilon}_k] = 0$ and $E[\boldsymbol{\epsilon}_k\boldsymbol{\epsilon}_j^T] = R_k\delta_{kj}$, where δ_{kj} is the *Kronicker delta*. We wish to process $\mathbf{y}_k$ in order to determine $\hat{\mathbf{x}}_k$. The best estimate of $\mathbf{x}_k$ is obtained in Eq. (4.4.29) as

$$\hat{\mathbf{x}}_k = (\widetilde{H}_k^T R_k^{-1} \widetilde{H}_k + \overline{P}_k^{-1})^{-1}(\widetilde{H}_k^T R_k^{-1}\mathbf{y}_k + \overline{P}_k^{-1}\overline{\mathbf{x}}_k). \tag{4.7.3}$$

The primary computational problems are associated with computing the $(n \times n)$ matrix inverse in Eq. (4.7.3). Recall that in the original derivation it was shown that the quantity to be inverted is the information matrix Λ_k, which yields the covariance matrix P_k associated with estimate $\hat{\mathbf{x}}_k$ (see Eq. (4.4.30)),

$$P_k = \Lambda_k^{-1} = (\widetilde{H}_k^T R_k^{-1} \widetilde{H}_k + \overline{P}_k^{-1})^{-1}. \tag{4.7.4}$$

From Eq. (4.7.4), it follows that

$$P_k^{-1} = \tilde{H}_k^T R_k^{-1} \tilde{H}_k + \overline{P}_k^{-1}. \tag{4.7.5}$$

Premultiplying each side of Eq. (4.7.5) by P_k and then postmultiplying by $\overline{P}_k$ leads to the following expressions:

$$\overline{P}_k = P_k \tilde{H}_k^T R_k^{-1} \tilde{H}_k \overline{P}_k + P_k \tag{4.7.6}$$

or

$$P_k = \overline{P}_k - P_k \tilde{H}_k^T R_k^{-1} \tilde{H}_k \overline{P}_k. \tag{4.7.7}$$

Now if Eq. (4.7.6) is postmultiplied by the quantity $H_k^T R_k^{-1}$, the following expression is obtained:

$$\begin{aligned}
\overline{P}_k \tilde{H}_k^T R_k^{-1} &= P_k \tilde{H}_k^T R_k^{-1} [\tilde{H}_k \overline{P}_k \tilde{H}_k^T R_k^{-1} + I] \\
&= P_k \tilde{H}_k^T R_k^{-1} [\tilde{H}_k \overline{P}_k \tilde{H}_k^T + R_k] R_k^{-1}. \tag{4.7.8}
\end{aligned}$$

Solving for the quantity $P_k \tilde{H}_k^T R_k^{-1}$ leads to

$$P_k \tilde{H}_k^T R_k^{-1} = \overline{P}_k \tilde{H}_k^T [\tilde{H}_k \overline{P}_k \tilde{H}_k^T + R_k]^{-1}. \tag{4.7.9}$$

This relates the *a priori* covariance matrix $\overline{P}_k$ to the *a posteriori* covariance matrix P_k. If Eq. (4.7.9) is used to eliminate $P_k \tilde{H}_k^T R_k^{-1}$ in Eq. (4.7.7), the following result is obtained:

$$P_k = \overline{P}_k - \overline{P}_k \tilde{H}_k^T [\tilde{H}_k \overline{P}_k \tilde{H}_k^T + R_k]^{-1} \tilde{H}_k \overline{P}_k. \tag{4.7.10}$$

Equation (4.7.10) also can be derived by using the *Schur identity* (Theorem 4 of Appendix B). Note that Eq. (4.7.10) is an alternate way of computing the inverse in Eq. (4.7.4). In Eq. (4.7.10), the matrix to be inverted is of dimension $p \times p$, the same dimensions as the observation error covariance matrix. If the observations are processed as scalars (i.e., one at a time), only a scalar division is required. If the weighting matrix, K_k (sometimes referred to as the *Kalman gain* or simply gain matrix), is defined as

$$K_k \equiv \overline{P}_k \tilde{H}_k^T [\tilde{H}_k \overline{P}_k \tilde{H}_k^T + R_k]^{-1}, \tag{4.7.11}$$

then Eq. (4.7.10) can be expressed in the compact form

$$P_k = [I - K_k \tilde{H}_k] \overline{P}_k. \tag{4.7.12}$$

If Eq. (4.7.4) is substituted into Eq. (4.7.3), the sequential form for computing $\hat{x}_k$ can be written as

$$\hat{x}_k = P_k [\tilde{H}_k^T R_k^{-1} \mathbf{y}_k + \overline{P}_k^{-1} \overline{\mathbf{x}}_k] \tag{4.7.13}$$

$$= P_k \widetilde{H}_k^T R_k^{-1} \mathbf{y}_k + P_k \overline{P}_k^{-1} \overline{\mathbf{x}}_k.$$

But from Eqs. (4.7.9) and (4.7.11)

$$K_k = P_k \widetilde{H}_k^T R_k^{-1}. \tag{4.7.14}$$

Using the preceding and Eq. (4.7.12), Eq. (4.7.13) becomes

$$\hat{\mathbf{x}}_k = K_k \mathbf{y}_k + [I - K_k \widetilde{H}_k] \overline{P}_k \overline{P}_k^{-1} \overline{\mathbf{x}}_k. \tag{4.7.15}$$

Collecting terms yields

$$\hat{\mathbf{x}}_k = \overline{\mathbf{x}}_k + K_k[\mathbf{y}_k - \widetilde{H}_k \overline{\mathbf{x}}_k]. \tag{4.7.16}$$

Equation (4.7.16), with Eqs. (4.7.1), (4.7.11), and (4.7.12), can be used in a recursive fashion to compute the estimate of $\hat{\mathbf{x}}_k$, incorporating the observation $\mathbf{y}_k$.

4.7.1 THE SEQUENTIAL COMPUTATIONAL ALGORITHM

The algorithm for computing the estimate sequentially is summarized as

Given: $\hat{\mathbf{x}}_{k-1}$, P_{k-1}, $\mathbf{X}_{k-1}^*$, and R_k, and the observation $\mathbf{Y}_k$, at t_k (at the initial time t_0, these would be $\mathbf{X}_0^*$, $\hat{\mathbf{x}}_0$, and P_0).

(1) Integrate from t_{k-1} to t_k,

$$\begin{aligned} \dot{\mathbf{X}}^* &= F(\mathbf{X}^*, t), & \mathbf{X}^*(t_{k-1}) &= \mathbf{X}_{k-1}^* \\ \dot{\Phi}(t, t_{k-1}) &= A(t)\Phi(t, t_{k-1}), & \Phi(t_{k-1}, t_{k-1}) &= I. \end{aligned} \tag{4.7.17}$$

(2) Compute

$$\overline{\mathbf{x}}_k = \Phi(t_k, t_{k-1})\hat{\mathbf{x}}_{k-1} \quad \overline{P}_k = \Phi(t_k, t_{k-1})P_{k-1}\Phi^T(t_k, t_{k-1}).$$

(3) Compute

$$\mathbf{y}_k = \mathbf{Y}_k - G(\mathbf{X}_k^*, t_k) \qquad \widetilde{H}_k = \frac{\partial G(\mathbf{X}_k^*, t_k)}{\partial \mathbf{X}}.$$

(4) Compute the measurement update

$$\begin{aligned} K_k &= \overline{P}_k \widetilde{H}_k^T [\widetilde{H}_k \overline{P}_k \widetilde{H}_k^T + R_k]^{-1} \\ \hat{\mathbf{x}}_k &= \overline{\mathbf{x}}_k + K_k[\mathbf{y}_k - \widetilde{H}_k \overline{\mathbf{x}}_k] \\ P_k &= [I - K_k \widetilde{H}_k]\overline{P}_k. \end{aligned} \tag{4.7.18}$$

(5) Replace k with $k + 1$ and return to (1).

Equation (4.7.1) is known as the *time update* and Eq. (4.7.18) is called the *measurement update* equation. The flow chart for the sequential computational algorithm is given in Fig. 4.7.1. If there is an observation at t_0, a time update is not performed but a measurement update is performed.

Note that we do not multiply $\widetilde{H}_k$ by the state transition matrix, since the observation at $\mathbf{y}_k$ is not accumulated at another time epoch, as is the case for the batch processor. Also, note that the differential equations for the state transition matrix are reinitialized at each observation epoch. Therefore, the state transition matrix is reinitialized at each observation epoch. If there is more than one observation at each epoch and we are processing them as scalars, we would set $\Phi(t_i, t_i) = I$ after processing the first observation at each epoch; P and $\hat{\mathbf{x}}$ are not time updated until we move to the next observation epoch. The estimate of the state of the nonlinear system at t_k is given by $\hat{\mathbf{X}}_k = \mathbf{X}_k^* + \hat{\mathbf{x}}_k$.

One disadvantage of both the batch and sequential algorithm lies in the fact that if the true state and the reference state are not close together then the linearization assumption leading to Eq. (4.2.6) may not be valid and the estimation process may diverge. The extended sequential filter algorithm (see Section 4.7.3) is often used to overcome problems with the linearity assumption.

A second unfavorable characteristic of the sequential estimation algorithm is that the state estimation error covariance matrix may approach zero as the number of observations becomes large. The sketch in Fig. 4.7.2 illustrates the behavior of the trace of the state estimation error covariance matrix as discrete observations are processed. As illustrated in this sketch, the trace grows between observations and is reduced by the amount, trace $(K\widetilde{H}\overline{P})$, after each observation. Hence, the magnitude of the covariance matrix elements will decrease depending on the density, information content, and accuracy of the observations.

Examination of the estimation algorithm shows that as $P_k \rightarrow 0$ the gain approaches zero, and the estimation procedure will become insensitive to the observations. Consequently, the estimate will diverge due to either errors introduced in the linearization procedure, computational errors, or errors due to an incomplete mathematical model. To overcome this problem, process noise often is added to the state propagation equations (see Section 4.9).

In addition to these two problems, the Kalman filter may diverge because of numerical difficulties associated with the covariance measurement update, given by Eq. (4.7.12). This problem is described in the next section.

Numerical Considerations

The conventional Kalman filter, which uses the covariance update, (4.7.12), can sometimes fail due to numerical problems with this equation. The covariance

Initialize at t_0

Set $i = 1$.   Set values of $t_{i-1} = t_0$ and $\mathbf{X}^*(t_{i-1}) = \mathbf{X}_0^*$.
$$\hat{\mathbf{x}}_{i-1} = \bar{\mathbf{x}}_0$$
$$P_{i-1} = \overline{P}_0$$

(A) Read the next observation: t_i, $\mathbf{Y}_i$, R_i.

Integrate reference trajectory and state transition matrix from t_{i-1} to t_i.

$\dot{\mathbf{X}}^* = F(\mathbf{X}^*(t), t)$ with initial conditions $\mathbf{X}^*(t_{i-1})$

$A(t) = [\partial F(\mathbf{X}, t)/\partial \mathbf{X}]^*$, where * means evaluated on the nominal trajectory.

$\dot{\Phi}(t, t_{i-1}) = A(t)\Phi(t, t_{i-1})$ with initial conditions $\Phi(t_{i-1}, t_{i-1}) = I$.

This gives $\mathbf{X}^*(t_i)$, $\Phi(t_i, t_{i-1})$.

Time Update

$$\bar{\mathbf{x}}_i = \Phi(t_i, t_{i-1})\hat{\mathbf{x}}_{i-1}$$
$$\overline{P}_i = \Phi(t_i, t_{i-1})P_{i-1}\Phi^T(t_i, t_{i-1})$$

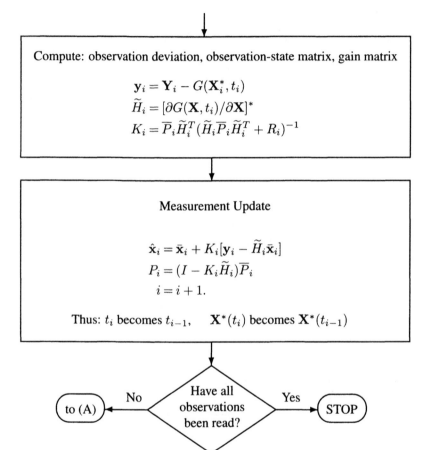

Figure 4.7.1: Sequential processing algorithm flow chart.

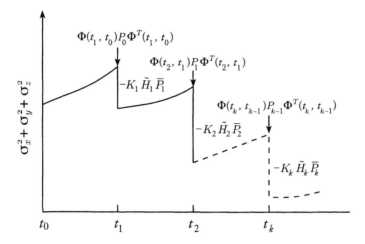

Figure 4.7.2: Illustration of the behavior of the trace of the state estimation error covariance matrix.

matrix may lose its properties of symmetry and become nonpositive definite when the computations are carried out with the finite digit arithmetic of the computer. In particular, this equation can fail to yield a symmetric positive definite result when a large *a priori* covariance is reduced by the incorporation of very accurate observation data.

In order to prevent these problems, several alternate algorithms have been suggested. A common alternative is to use the form of the equation given by Bucy and Joseph (1968):

$$P_k = (I - K_k \widetilde{H}_k)\overline{P}_k(I - K_k \widetilde{H}_k)^T + K_k R_k K_k^T. \qquad (4.7.19)$$

Note that this formulation will always yield a symmetric result for P_k, although it may lose its positive definite quality for a poorly observed system.

The most common solution to numerical problems with the covariance update is to use a square root formulation to update the covariance matrix. The square root algorithms are discussed in Chapter 5.

An example of a poorly conditioned system taken from Bierman (1977, p. 96, Example V.2) illustrates the numerical instability problem. Consider the problem of estimating x_1 and x_2 from scalar measurements z_1 and z_2, where

$$z_1 = x_1 + \epsilon x_2 + v_1$$
$$z_2 = x_1 + x_2 + v_2.$$

We assume that the observation noise v_1 and v_2 are uncorrelated, zero mean random variables with unit variance. If v_1 and v_2 do not have unit variance or are correlated, we can perform a simple whitening transformation so that their covariance matrix is the identity matrix (Bierman, 1977, or see Section 5.7.1). In matrix form,

$$\begin{bmatrix} z_1 \\ z_2 \end{bmatrix} = \begin{bmatrix} 1 & \epsilon \\ 1 & 1 \end{bmatrix} \begin{bmatrix} x_1 \\ x_2 \end{bmatrix} + \begin{bmatrix} v_1 \\ v_2 \end{bmatrix}. \tag{4.7.20}$$

Assume that the *a priori* covariance associated with our knowledge of $[x_1 \ x_2]$ is given by

$$\overline{P}_1 = \sigma^2 I$$

where $\sigma = 1/\epsilon$ and $0 < \epsilon \ll 1$. The quantity ϵ is assumed to be small enough such that computer round-off produces

$$\begin{aligned} 1 + \epsilon^2 &= 1, \\ \text{and} & \\ 1 + \epsilon &\neq 1; \end{aligned} \tag{4.7.21}$$

that is, $1 + \epsilon^2$ rounds to 1 and $1 + \epsilon$ does not. Note that $\sigma = 1/\epsilon$ is large; hence, this is an illustration of the problem where accurate data is being combined with a large *a priori* covariance.

This estimation problem is well posed. The observation z_1 provides an estimate of x_1 which, when combined with z_2, should determine x_2. However, the conventional Kalman update to P given by Eq. (4.7.12) will result in failure of the filter.

The observations will be processed one at a time. Let the gain and estimation error covariance associated with z_1 be denoted as K_1 and P_1, respectively. Similar definitions apply for z_2.

The exact solution yields

$$K_1 = \frac{1}{1 + 2\epsilon^2} \begin{bmatrix} 1 \\ \epsilon \end{bmatrix}. \tag{4.7.22}$$

The estimation error covariance associated with processing z_1 is

$$P_1 = \frac{1}{1 + 2\epsilon^2} \begin{bmatrix} 2 & -\sigma \\ -\sigma & \sigma^2 + 1 \end{bmatrix}. \tag{4.7.23}$$

After processing z_2 it can be shown that the exact solution for the estimation error covariance is

$$P_2 = \frac{1}{\beta} \begin{bmatrix} 1 + 2\epsilon^2 & -(1+\epsilon) \\ -(1+\epsilon) & 2+\epsilon^2 \end{bmatrix}, \tag{4.7.24}$$

where

$$\beta = 1 - 2\epsilon + 2\epsilon^2(2 + \epsilon^2)$$

and we see that both P_1 and P_2 are symmetric and positive definite.

The conventional Kalman filter, subject to Eq. (4.7.21), yields

$$K_1 = \begin{bmatrix} 1 \\ \epsilon \end{bmatrix}$$

and

$$P_1 = \begin{bmatrix} 0 & -\sigma \\ -\sigma & \sigma^2 \end{bmatrix}. \qquad (4.7.25)$$

Note that P_1 is no longer positive definite and the filter will not produce correct results. If z_2 is processed, the conventional Kalman filter yields

$$P_2 = \frac{1}{1 - 2\epsilon} \begin{bmatrix} -1 & 1 \\ 1 & -1 \end{bmatrix}. \qquad (4.7.26)$$

Note that P_2 does not have positive terms on the diagonal and $|P_2| = 0$. Hence, the conventional Kalman filter has failed.

By way of comparison, the Joseph formulation yields

$$P_1 = \begin{bmatrix} 2 & -\sigma \\ -\sigma & \sigma^2 \end{bmatrix} \qquad (4.7.27)$$

$$P_2 = \begin{bmatrix} 1 + 2\epsilon & -(1 + 3\epsilon) \\ -(1 + 3\epsilon) & (2 + \epsilon) \end{bmatrix}$$

which are symmetric and positive definite and agree with the exact solution to $O(\epsilon)$.

It should be pointed out that the batch processor for this problem, under the same assumption that $1 + \epsilon^2 = 1$, yields

$$P_2 = \begin{bmatrix} 1 + 2\epsilon & -(1 + 3\epsilon) \\ -(1 + 3\epsilon) & 2(1 + 2\epsilon) \end{bmatrix}.$$

This agrees to $O(\epsilon)$ with the exact solution for P_2. The batch solution is not as sensitive as the conventional Kalman filter to computer round-off errors. An intuitive understanding of this can be had by examining the equations for P.

For the batch processor

$$P = (\overline{P}_0^{-1} + \sum_{i=1}^{\ell} H_i^T R^{-1} H_i)^{-1},$$

and for the sequential processor

$$P = (I - \overline{P}\widetilde{H}^T (\widetilde{H}\overline{P}\widetilde{H}^T + R)^{-1}\widetilde{H})\overline{P}.$$

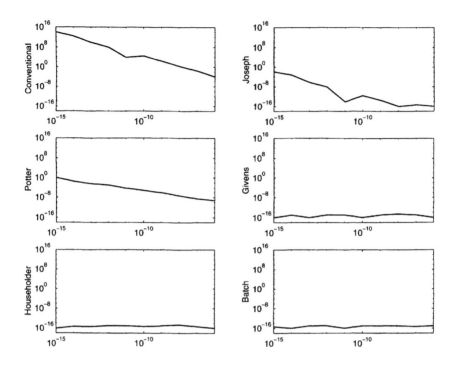

Figure 4.7.3: Difference of the exact trace of P_2 and that for various algorithms versus the value of ϵ on the abscissa.

Hence, if very accurate tracking data (small values for the elements of R) are combined with a large *a priori* (large values for the elements of $\overline{P}$), the sequential algorithm will ignore the data contribution, R, when it adds $\widetilde{H}\overline{P}\widetilde{H}^T + R$; whereas the batch processor will tend to ignore $\overline{P}_0^{-1}$ and base its solution on the strength of the data when it computes $(\Sigma H^T R^{-1} H + \overline{P}_0^{-1})$. It is emphasized that for sequential processing of the data the preferred solution to this problem is to use one of the square root formulations for propagating the covariance or information matrix discussed in Chapter 5.

Figure 4.7.3 shows the difference between the exact value of the trace of P_2 and the trace computed using various algorithms on a 16-digit machine. Notice that the conventional Kalman update yields an error of O(1) for $\epsilon \simeq 10^{-8}$, whereas the Joseph and Potter algorithms approach this value for $\epsilon = 10^{-15}$. Also, the Givens and Householder algorithms, which are introduced in Chapter 5, and the batch algorithm are all highly accurate.

4.7.2 THE EXTENDED SEQUENTIAL ESTIMATION ALGORITHM

To minimize the effects of errors due to the neglect of higher order terms in the linearization procedure leading to Eq. (4.2.6), the extended form of the sequential estimation algorithm is sometimes used. This algorithm is often referred to as the *Extended Kalman Filter* (EKF). The primary difference between the sequential and the extended sequential algorithm is that the reference trajectory for the extended sequential algorithm is updated after each observation to reflect the best estimate of the true trajectory. For example, after processing the kth observation, the best estimate of the state vector at t_k is used to provide new initial conditions for the reference trajectory,

$$(\mathbf{X}_k^*)_{\text{new}} = \hat{\mathbf{X}}_k = \mathbf{X}_k^* + \hat{\mathbf{x}}_k. \qquad (4.7.28)$$

Using $\hat{\mathbf{X}}_k$ for the reference trajectory leads to $\hat{\mathbf{x}}_k = 0$, which will result in $\bar{\mathbf{x}}_{k+1} = 0$. The integration for the reference trajectory and the state transition matrix is reinitialized at each observation epoch, and the equations are integrated forward from t_k to t_{k+1}. The estimate for $\hat{\mathbf{x}}_{k+1}$ is then computed from

$$\hat{\mathbf{x}}_{k+1} = K_{k+1}\mathbf{y}_{k+1} \qquad (4.7.29)$$

where K_{k+1} and $\mathbf{y}_{k+1}$ are computed based on the new reference orbit. Then, the reference orbit is updated at time t_{k+1} by incorporating $\hat{\mathbf{x}}_{k+1}$ and the process proceeds to t_{k+2}. The process of incorporating the estimate at each observation point into the reference trajectory for propagating to the next observation epoch leads to the reference trajectory being the prediction of the estimate of the nonlinear state; for example, $\mathbf{X}^*(t) = \hat{\mathbf{X}}(t)$.

In actual practice, it is not a good idea to update the reference trajectory using the first observations. This is particularly true if the observations contain significant noise. After a few observations have been processed, the estimates of $\hat{\mathbf{x}}$ will stabilize, and the trajectory update can be initiated.

The advantage of the extended sequential algorithm is that convergence to the best estimate will be more rapid because errors introduced in the linearization process are reduced. In addition, because the state estimate deviation, $\hat{\mathbf{x}}(t)$, need not be mapped between observations, it is not necessary to compute the state transition matrix. The estimation error covariance matrix, $P(t)$, can be mapped by integrating the matrix differential equation (4.9.35) discussed in Section 4.9.

The major disadvantage of the extended sequential algorithm is that the differential equations for the reference trajectory must be reinitialized after each observation is processed.

4.7.3 THE EXTENDED SEQUENTIAL COMPUTATIONAL ALGORITHM

The algorithm for computing the extended sequential estimate can be summarized as follows:

Given: P_{k-1}, $\hat{\mathbf{X}}_{k-1}$ and $\mathbf{Y}_k$, R_k.

(1) Integrate from t_{k-1} to t_k,

$$
\begin{aligned}
\dot{\mathbf{X}}^* &= F(\mathbf{X}^*, t), & \mathbf{X}^*(t_{k-1}) &= \hat{\mathbf{X}}_{k-1} \\
\dot{\Phi}(t, t_{k-1}) &= A(t)\Phi(t, t_{k-1}), & \Phi(t_{k-1}, t_{k-1}) &= I.
\end{aligned}
\tag{4.7.30}
$$

(2) Compute

$$
\begin{aligned}
\overline{P}_k &= \Phi(t_k, t_{k-1}) P_{k-1} \Phi^T(t_k, t_{k-1}) \\
\mathbf{y}_k &= \mathbf{Y}_k - G(\mathbf{X}_k^*, t_k) \\
\widetilde{H}_k &= \partial G(\mathbf{X}_k^*, t_k)/\partial \mathbf{X}_k.
\end{aligned}
\tag{4.7.31}
$$

(3) Compute

$$
\begin{aligned}
K_k &= \overline{P}_k \widetilde{H}_k^T [\widetilde{H}_k \overline{P}_k \widetilde{H}_k^T + R_k]^{-1} \\
\hat{\mathbf{X}}_k &= \mathbf{X}_k^* + K_k \mathbf{y}_k \\
P_k &= [I - K_k \widetilde{H}_k] \overline{P}_k.
\end{aligned}
\tag{4.7.32}
$$

(4) Replace k with $k+1$ and return to (1).

The flow chart for the extended sequential computational algorithm is given in Fig. 4.7.4.

4.7.4 THE PREDICTION RESIDUAL

It is of interest to examine the variance of the predicted residuals, which are sometimes referred to as the *innovation*, or new information, which comes from each measurement. The *predicted residual*, or innovation, is the observation residual based on the *a priori* or predicted state, $\bar{\mathbf{x}}_k$, at the observation time, t_k, and is defined as

$$
\beta_k = \mathbf{y}_k - \widetilde{H}_k \bar{\mathbf{x}}_k.
\tag{4.7.33}
$$

As noted previously,

$$
\begin{aligned}
\bar{\mathbf{x}}_k &= \mathbf{x}_k + \boldsymbol{\eta}_k \\
\mathbf{y}_k &= \widetilde{H}_k \mathbf{x}_k + \boldsymbol{\epsilon}_k
\end{aligned}
$$

where $\mathbf{x}_k$ is the true value of the state deviation vector and $\boldsymbol{\eta}_k$ is the error in $\bar{\mathbf{x}}_k$. Also,

$$E[\boldsymbol{\eta}_k] = 0, \; E[\boldsymbol{\eta}_k \boldsymbol{\eta}_k^T] = \overline{P}_k$$

and

$$E[\epsilon_k] = 0, \; E[\epsilon_k \epsilon_k^T] = R_k$$
$$E[\epsilon_k \boldsymbol{\eta}_k^T] = 0.$$

From these conditions it follows that $\boldsymbol{\beta}_k$ has mean

$$\begin{aligned} E[\boldsymbol{\beta}_k] \equiv \overline{\boldsymbol{\beta}}_k &= E(\widetilde{H}_k \mathbf{x}_k + \epsilon_k - \widetilde{H}_k \bar{\mathbf{x}}_k) \\ &= E(\epsilon_k - \widetilde{H} \boldsymbol{\eta}_k) = 0 \end{aligned}$$

and variance-covariance

$$\begin{aligned} P_{\beta_k} = E[(\boldsymbol{\beta}_k - \overline{\boldsymbol{\beta}}_k)(\boldsymbol{\beta}_k - \overline{\boldsymbol{\beta}}_k)^T] &= E[\boldsymbol{\beta}_k \boldsymbol{\beta}_k^T] \\ &= E[(\mathbf{y}_k - \widetilde{H}_k \bar{\mathbf{x}}_k)(\mathbf{y}_k - \widetilde{H}_k \bar{\mathbf{x}}_k)^T] \\ &= E[(\epsilon_k - \widetilde{H}_k \boldsymbol{\eta}_k)(\epsilon_k - \widetilde{H}_k \boldsymbol{\eta}_k)^T] \\ P_{\beta_k} &= R_k + \widetilde{H}_k \overline{P}_k \widetilde{H}_k^T. \end{aligned} \quad (4.7.34)$$

Hence, for a large prediction residual variance-covariance, the Kalman gain

$$K_k = \overline{P}_k \widetilde{H}_k^T P_{\beta_k}^{-1} \quad (4.7.35)$$

will be small, and the observation will have little influence on the estimate of the state. Also, large values of the prediction residual relative to the prediction residual standard deviation may be an indication of bad tracking data and hence may be used to edit data from the solution.

4.8 EXAMPLE PROBLEMS

This section provides several examples that involve processing observations with the batch and sequential processors.

4.8.1 LINEAR SYSTEM

Given: A system described by

$$\mathbf{x}(t_{i+1}) = \begin{bmatrix} x_1(t_{i+1}) \\ x_2(t_{i+1}) \end{bmatrix} = \begin{bmatrix} 1 & 1 \\ 0 & 1 \end{bmatrix} \begin{bmatrix} x_1(t_i) \\ x_2(t_i) \end{bmatrix}$$

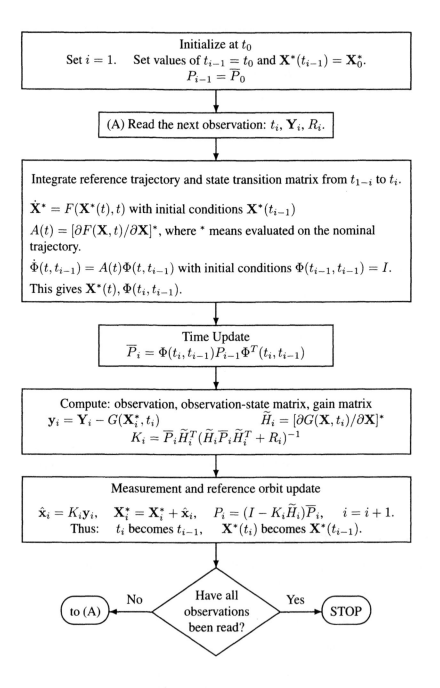

Figure 4.8.1: Extended sequential processing algorithm flow chart.

and observation-state relationship

$$\begin{bmatrix} y_1(t_i) \\ y_2(t_i) \end{bmatrix} = \begin{bmatrix} 0 & 1 \\ 1/2 & 1/2 \end{bmatrix} \begin{bmatrix} x_1(t_i) \\ x_2(t_i) \end{bmatrix} + \begin{bmatrix} \epsilon_1 \\ \epsilon_2 \end{bmatrix}$$

with *a priori* information

$$\overline{\mathbf{x}}_0 = \overline{\mathbf{x}}(t_0) = \begin{bmatrix} \overline{x}_1(t_0) \\ \overline{x}_2(t_0) \end{bmatrix} = \begin{bmatrix} 3 \\ 2 \end{bmatrix}, \overline{P}_0 = \overline{P}(t_0) = \begin{bmatrix} 1 & 0 \\ 0 & 1 \end{bmatrix}$$

$$W^{-1} = R = E \left\{ \begin{bmatrix} \epsilon_1 \\ \epsilon_2 \end{bmatrix} [\epsilon_1 \epsilon_2] \right\} = \begin{bmatrix} 2 & 0 \\ 0 & 3/4 \end{bmatrix}$$

and observations at t_1 given by

$$\begin{bmatrix} y_1(t_1) \\ y_2(t_1) \end{bmatrix} = \begin{bmatrix} 6 \\ 4 \end{bmatrix}.$$

(1) Using the batch processor algorithm, determine the best estimate of $\mathbf{x}(t_0)$ based on observations $y_1(t_1)$ and $y_2(t_1)$, $\overline{\mathbf{x}}(t_0)$, $\overline{P}(t_0)$, and R.

(2) Using the sequential algorithm and the information given, determine $\hat{\mathbf{x}}(t_1)$, the best estimate of $\mathbf{x}(t_1)$.

(3) Show that $\hat{\mathbf{x}}(t_0)$ obtained in (1), when mapped forward to t_1 is equal to $\hat{\mathbf{x}}(t_1)$ obtained in (2). Show that P_0 from the batch processor mapped forward to t_1 agrees with P_1 from the sequential processor.

(1) *Batch Processor*

$$\Phi(t_{i+1}, t_i) = \begin{bmatrix} 1 & 1 \\ 0 & 1 \end{bmatrix}$$

$$\tilde{H} = \begin{bmatrix} 0 & 1 \\ 1/2 & 1/2 \end{bmatrix}, \quad \overline{P}_o = \begin{bmatrix} 1 & 0 \\ 0 & 1 \end{bmatrix}, \quad \overline{\mathbf{x}}_o = \begin{bmatrix} 3 \\ 2 \end{bmatrix}$$

$$W^{-1} = R = \begin{bmatrix} 2 & 0 \\ 0 & 3/4 \end{bmatrix}$$

$$\mathbf{y} = \begin{bmatrix} y_1(t_1) \\ y_2(t_1) \end{bmatrix} = \begin{bmatrix} 6 \\ 4 \end{bmatrix}.$$

The computational algorithm for the batch processor is given in Section 4.6.

$$\hat{\mathbf{x}}_0 = (\overline{P}_0^{-1} + H_1^T W H_1)^{-1} (H_1^T W \mathbf{y} + \overline{P}_0^{-1} \overline{\mathbf{x}}_0)$$

$$H_1 = \tilde{H}_1 \Phi(t_1, t_0) = \begin{bmatrix} 0 & 1 \\ 1/2 & 1/2 \end{bmatrix} \begin{bmatrix} 1 & 1 \\ 0 & 1 \end{bmatrix} = \begin{bmatrix} 0 & 1 \\ 1/2 & 1 \end{bmatrix}$$

$$\hat{\mathbf{x}}_0 = \left\{ \begin{bmatrix} 1 & 0 \\ 0 & 1 \end{bmatrix} + \begin{bmatrix} 0 & 1/2 \\ 1 & 1 \end{bmatrix} \begin{bmatrix} 1/2 & 0 \\ 0 & 4/3 \end{bmatrix} \begin{bmatrix} 0 & 1 \\ 1/2 & 1 \end{bmatrix} \right\}^{-1}$$

$$\left\{ \begin{bmatrix} 0 & 1/2 \\ 1 & 1 \end{bmatrix} \begin{bmatrix} 1/2 & 0 \\ 0 & 4/3 \end{bmatrix} \begin{bmatrix} 6 \\ 4 \end{bmatrix} + \begin{bmatrix} 3 \\ 2 \end{bmatrix} \right\}$$

$$\hat{\mathbf{x}}_o = \begin{bmatrix} 4/3 & 2/3 \\ 2/3 & 17/6 \end{bmatrix}^{-1} \begin{bmatrix} 17/3 \\ 31/3 \end{bmatrix}$$

$$\hat{\mathbf{x}}_o = \begin{bmatrix} 17/20 & -1/5 \\ -1/5 & 2/5 \end{bmatrix} \begin{bmatrix} 17/3 \\ 31/3 \end{bmatrix}$$

$$\hat{\mathbf{x}}_0 = \begin{bmatrix} 2\frac{3}{4} \\ 3 \end{bmatrix}$$

Note that

$$P_0 = \begin{bmatrix} 17/20 & -1/5 \\ -1/5 & 2/5 \end{bmatrix} .$$

The correlation matrix, computed from P_0, has standard deviations on the diagonal and correlation coefficients for the lower off-diagonal terms,

$$\begin{bmatrix} \sigma_{x_1} \\ \rho_{x_1 x_2} \sigma_{x_2} \end{bmatrix} = \begin{bmatrix} .922 \\ -.343 & .632 \end{bmatrix} .$$

(2) *Sequential Processor*

$$\hat{\mathbf{x}}_1 = \bar{\mathbf{x}}_1 + \bar{P}_1 \tilde{H}_1^T (W_1^{-1} + \tilde{H}_1 \bar{P}_1 \tilde{H}_1^T)^{-1} (\mathbf{y}_1 - \tilde{H}_1 \bar{\mathbf{x}}_1)$$
$$= \bar{\mathbf{x}}_1 + K_1 (\mathbf{y}_1 - \tilde{H}_1 \bar{\mathbf{x}}_1)$$

From the problem definition

$$\bar{\mathbf{x}}_1 = \Phi(t_1, t_0) \bar{\mathbf{x}}_o$$
$$= \begin{bmatrix} 1 & 1 \\ 0 & 1 \end{bmatrix} \begin{bmatrix} 3 \\ 2 \end{bmatrix} = \begin{bmatrix} 5 \\ 2 \end{bmatrix}$$

$$\bar{P}_1 = \Phi(t_1, t_0) \bar{P}_o \Phi^T(t_1, t_0)$$
$$= \begin{bmatrix} 1 & 1 \\ 0 & 1 \end{bmatrix} \begin{bmatrix} 1 & 0 \\ 1 & 1 \end{bmatrix} = \begin{bmatrix} 2 & 1 \\ 1 & 1 \end{bmatrix}$$

$$\bar{P}_1 \tilde{H}_1^T = \begin{bmatrix} 2 & 1 \\ 1 & 1 \end{bmatrix} \begin{bmatrix} 0 & 1/2 \\ 1 & 1/2 \end{bmatrix} = \begin{bmatrix} 1 & 3/2 \\ 1 & 1 \end{bmatrix}$$

$$\tilde{H}_1 \overline{P}_1 \tilde{H}_1^T = \begin{bmatrix} 0 & 1 \\ 1/2 & 1/2 \end{bmatrix} \begin{bmatrix} 1 & 3/2 \\ 1 & 1 \end{bmatrix} = \begin{bmatrix} 1 & 1 \\ 1 & 5/4 \end{bmatrix}$$

$$(W_1^{-1} + \tilde{H}_1 \overline{P}_1 \tilde{H}_1^T)^{-1} = \left[\begin{pmatrix} 2 & 0 \\ 0 & 3/4 \end{pmatrix} + \begin{pmatrix} 1 & 1 \\ 1 & 5/4 \end{pmatrix} \right]^{-1}$$

$$= \begin{bmatrix} 2/5 & -1/5 \\ -1/5 & 3/5 \end{bmatrix}$$

$$K_1 = \overline{P}_1 \tilde{H}_1^T (W_1^{-1} + \tilde{H}_1 \overline{P}_1 \tilde{H}_1^T)^{-1}$$

$$= \begin{bmatrix} 1 & 3/2 \\ 1 & 1 \end{bmatrix} \begin{bmatrix} 2/5 & -1/5 \\ -1/5 & 3/5 \end{bmatrix} = \begin{bmatrix} 1/10 & 7/10 \\ 1/5 & 2/5 \end{bmatrix}.$$

With these results,

$$\hat{\mathbf{x}}_1 = \begin{bmatrix} 5 \\ 2 \end{bmatrix} + \begin{bmatrix} 1/10 & 7/10 \\ 1/5 & 2/5 \end{bmatrix} \left\{ \begin{bmatrix} 6 \\ 4 \end{bmatrix} - \begin{bmatrix} 0 & 1 \\ 1/2 & 1/2 \end{bmatrix} \begin{bmatrix} 5 \\ 2 \end{bmatrix} \right\}$$

$$= \begin{bmatrix} 5 \\ 2 \end{bmatrix} + \begin{bmatrix} 1/10 & 7/10 \\ 1/5 & 2/5 \end{bmatrix} \begin{bmatrix} 4 \\ 1/2 \end{bmatrix}$$

$$\hat{\mathbf{x}}_1 = \begin{bmatrix} 5\frac{3}{4} \\ 3 \end{bmatrix}$$

$$P_1 = (I - K_1 \tilde{H}_1) \overline{P}_1$$

$$= \left\{ \begin{bmatrix} 1 & 0 \\ 0 & 1 \end{bmatrix} - \begin{bmatrix} 7/20 & 9/20 \\ 1/5 & 2/5 \end{bmatrix} \right\} \begin{bmatrix} 2 & 1 \\ 1 & 1 \end{bmatrix}$$

$$P_1 = \begin{bmatrix} 17/20 & 1/5 \\ 1/5 & 2/5 \end{bmatrix}.$$

(3) Map $\hat{\mathbf{x}}_o$ obtained with the batch processor to t_1.

$$\hat{\mathbf{x}}_1 = \Phi(t_1, t_0) \hat{\mathbf{x}}_0$$

$$= \begin{bmatrix} 1 & 1 \\ 0 & 1 \end{bmatrix} \begin{bmatrix} 2\frac{3}{4} \\ 3 \end{bmatrix}$$

$$= \begin{bmatrix} 5\frac{3}{4} \\ 3 \end{bmatrix}$$

Hence, the batch and sequential estimates and covariances at t_1 agree.

Note that P_0 from the batch processor may be mapped to t_1 by

$$P_1 = \Phi(t_1, t_0) P_0 \Phi^T(t_1, t_0)$$

$$= \begin{bmatrix} 1 & 1 \\ 0 & 1 \end{bmatrix} \begin{bmatrix} 17/20 & -1/5 \\ -1/5 & 2/5 \end{bmatrix} \begin{bmatrix} 1 & 0 \\ 1 & 1 \end{bmatrix}$$

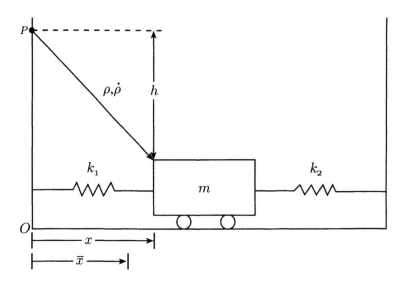

Figure 4.8.2: Spring-mass system.

$$P_1 = \begin{bmatrix} 17/20 & 1/5 \\ 1/5 & 2/5 \end{bmatrix},$$

which agrees with the sequential processor results.

4.8.2 SPRING-MASS PROBLEM

A block of mass m is attached to two parallel vertical walls by two springs as shown in Fig. 4.8.2. k_1 and k_2 are the spring constants. h is the height of the position P on one of the walls, from which the distance, ρ, and the rate of change of distance of the block from P, $\dot{\rho}$, can be observed.

Let the horizontal distances be measured with respect to the point O where the line OP, the lengths of the springs, and the center of mass of the block are all assumed to be in the same vertical plane. Then, if $\bar{x}$ denotes the position of the block at static equilibrium, the equation of motion of the block is given by

$$\ddot{x} \equiv \dot{v} = -(k_1 + k_2)(x - \bar{x})/m. \tag{4.8.1}$$

Let

$$\omega^2 = (k_1 + k_2)/m.$$

Since $\bar{x}$ is an arbitrary constant we have set it to zero so that Eq. (4.8.1) can be written as

$$\dot{v} + \omega^2 x = 0. \tag{4.8.2}$$

Consider the problem of estimating the position and the velocity of the block with respect to the reference point O, by using the range and range-rate measurements of the block from the point, P. In order to formulate this problem mathematically, the estimation state vector is taken as $\mathbf{X}^T = [x \ v]$. Then the system dynamics are represented by

$$\dot{\mathbf{X}} = F(\mathbf{X}, t) = \begin{bmatrix} \dot{x} \\ \dot{v} \end{bmatrix} = \begin{bmatrix} v \\ -\omega^2 x \end{bmatrix} = \begin{bmatrix} 0 & 1 \\ -\omega^2 & 0 \end{bmatrix} \begin{bmatrix} x \\ v \end{bmatrix}. \tag{4.8.3}$$

The observation vector is

$$\mathbf{Y}(t) = \begin{bmatrix} \rho \\ \dot{\rho} \end{bmatrix}.$$

Also,

$$G(\mathbf{X}^*, t) = \begin{bmatrix} \rho \\ \dot{\rho} \end{bmatrix}^* = \begin{bmatrix} \sqrt{x^2 + h^2} \\ xv/\rho \end{bmatrix}^*. \tag{4.8.4}$$

The system parameters are m, k_1, k_2 (from dynamics), and h (from geometry).

Note that the state propagation equation, Eq. (4.8.3), is linear in the state variables x and v. However, the observation-state equation, Eq. (4.8.4), is nonlinear. Hence, the system must be linearized about a reference solution. Accordingly, let

$$\delta\mathbf{X}(t) = \begin{bmatrix} x(t) - x^*(t) \\ v(t) - v^*(t) \end{bmatrix}$$

and we will solve for $\delta\hat{\mathbf{X}}(t_0) \equiv \delta\hat{\mathbf{X}}_0$ and

$$\delta\dot{\mathbf{X}}(t) = A(t)\delta\mathbf{X}(t)$$
$$\mathbf{y}(t) = \tilde{H}\delta\mathbf{X}(t)$$

where

$$\mathbf{y}(t) = \mathbf{Y}(t) - G(\mathbf{X}^*, t)$$

$$A(t) = \begin{bmatrix} 0 & 1 \\ -\omega^2 & 0 \end{bmatrix}$$

$$\tilde{H} = \begin{bmatrix} \frac{x}{\rho} & 0 \\ \left(\frac{v}{\rho} - \frac{x^2 v}{\rho^3}\right) & \frac{x}{\rho} \end{bmatrix}.$$

The solution to Eq. (4.8.3) is given by (assuming $t_0 = 0$)

$$x(t) = x_0 \cos \omega t + \frac{v_0}{\omega} \sin \omega t$$
$$v(t) = v_0 \cos \omega t - x_0 \omega \sin \omega t. \tag{4.8.5}$$

Hence, we could determine the state transition matrix by differentiating Eq. (4.8.5),

$$\Phi(t,0) = \frac{\partial \mathbf{X}(t)}{\partial \mathbf{X}(t_0)}.$$

Alternatively, the solution to the variational equation

$$\dot{\Phi}(t,0) = A(t)\Phi(t,0), \quad \Phi(0,0) = I$$

is given by

$$\Phi(t,0) = \begin{bmatrix} \cos \omega t & \frac{1}{\omega}\sin \omega t \\ -\omega \sin \omega t & \cos \omega t \end{bmatrix}$$

hence,

$$\delta \mathbf{X}(t) = \Phi(t,0)\delta \mathbf{X}(t_0).$$

Also,

$$H_i = \widetilde{H}_i \Phi(t_i, 0).$$

Perfect observations were simulated for a period of 10 seconds at one-second intervals (see Table 4.8.1), assuming the following values for the system parameters and the initial condition for the state parameters:

Table 4.8.1:

Observation Data

Time (s)	$\rho(m)$	$\dot{\rho}\ (m/s)$
0.00	6.1773780845922	0
1.00	5.56327661282686	1.31285863495514
2.00	5.69420161397342	−1.54488114381612
3.00	6.15294262127432	0.534923988815733
4.00	5.46251322092491	0.884698415328368
5.00	5.83638064328625	−1.56123248918054
6.00	6.08236452736002	1.00979943157547
7.00	5.40737619817037	0.31705117039215
8.00	5.97065615746125	−1.37453070975606
9.00	5.97369258835895	1.36768169443236
10.00	5.40669060248179	−0.302111588503166

$$k_1 = 2.5 \text{ N/m}$$
$$k_2 = 3.7 \text{ N/m}$$
$$m = 1.5 \text{ kg} \quad\quad (4.8.6)$$
$$h = 5.4 \text{ m}$$
$$x_0 = 3.0 \text{ m}$$
$$v_0 = 0.0 \text{ m/s}.$$

The 11 perfect observations were processed using the batch processor with the following *a priori* values:

$$\mathbf{X}_0^* = \begin{bmatrix} 4.0 \\ 0.2 \end{bmatrix}, \overline{P}_0 = \begin{bmatrix} 1000 & 0 \\ 0 & 100 \end{bmatrix}, \text{ and } \delta \overline{\mathbf{X}}_0 = \begin{bmatrix} 0 \\ 0 \end{bmatrix}.$$

The computed observations were calculated with the exact values of k_1, k_2, m, and h from Eq. (4.8.6); $\mathbf{X}_0^*$ was used for the values of x_0^* and v_0^*.

The least squares solution for $\delta \mathbf{X}_0$ is given by (assuming that $W_i^{-1} = R_i = I$)

$$\delta \hat{\mathbf{X}}_0 = \left(\sum_{i=0}^{10} H_i^T H_i + \overline{P}_0^{-1} \right)^{-1} \left(\sum_{i=0}^{10} H_i^T y_i + \overline{P}_0^{-1} \delta \overline{\mathbf{X}}_0 \right)$$

and the best estimate of the initial state is

$$\hat{\mathbf{X}}_0 = \mathbf{X}_0^* + \delta \hat{\mathbf{X}}_0.$$

After each iteration shift the *a priori* state deviation vector so that

$$(\delta \overline{\mathbf{X}}_0)_n = (\delta \overline{\mathbf{X}}_0)_{n-1} - (\delta \hat{\mathbf{X}}_0)_{n-1}$$

following Eq. (4.6.4) and Fig. 4.6.1.

After four iterations the estimate for the state at $t = 0$ is

$$\begin{bmatrix} \hat{\mathbf{X}}_0 \end{bmatrix} = \begin{bmatrix} \hat{x}_0 \\ \hat{v}_0 \end{bmatrix} = \begin{bmatrix} 3.00019 \text{ m} \\ 1.18181 \times 10^{-3} \text{ m/s} \end{bmatrix}.$$

The computed standard deviations and correlation coefficient are

$$\sigma_{x_0} = 0.411 \text{ m}, \ \sigma_{v_0} = 0.765 \text{ m/s}, \text{ and } \rho_{x_0 v_0} = 0.0406.$$

The mean of the residuals is $\rho_{\text{mean}} = -4.30 \times 10^{-5}$m, $\dot{\rho}_{\text{mean}} = -1.76 \times 10^{-6}$ m/s. The RMS of residuals is $\rho_{RMS} = 1.16 \times 10^{-4}$m, $\dot{\rho}_{\text{RMS}} = 4.66 \times 10^{-4}$ m/s.

4.9   STATE NOISE COMPENSATION ALGORITHM

In addition to the effects of the nonlinearities, the effects of errors in the dynamical model can lead to divergence in the estimate. See, for example, the discussion in Schlee *et al.*, (1967) . As pointed out previously, for a sufficiently large number of observations the elements of the covariance matrix P_k will asymptotically approach zero and the estimation algorithm will be insensitive to any further observations. This condition can lead to filter divergence. One approach to preventing this divergence is to recognize that the linearized equations for propagating the estimate of the state are in error and to compensate for this by assuming that the error in the linearized dynamics can be approximated by process noise.

The state dynamics of a linear system under the influence of process noise are described by

$$\dot{\mathbf{x}}(t) = A(t)\mathbf{x}(t) + B(t)\mathbf{u}(t) \tag{4.9.1}$$

where $A(t)$ and $B(t)$ are known functions of time. The vector $\mathbf{u}(t)$ is dimension $m \times 1$, and $B(t)$ is an $n \times m$ matrix. The functional form of $\mathbf{u}(t)$ can include a number of processes, including constant, piecewise constant, correlated, or white noise. In the following discussion, the function $\mathbf{u}(t)$ (called *state or process noise*) is assumed to be a white noise process with

$$\begin{aligned} E[\mathbf{u}(t)] &= 0 \\ E[\mathbf{u}(t)\mathbf{u}^T(\tau)] &= Q(t)\delta(t - \tau) \end{aligned} \tag{4.9.2}$$

where $\delta(t - \tau)$ is the Dirac Delta, and Q is called the *process noise covariance matrix*.

The algorithm that results from the assumption that $\mathbf{u}(t)$ is white noise with known covariance is known as *State Noise Compensation (SNC)*. The use of more sophisticated models such as the process to compensate for state and/or measurement model errors generally is referred to as *Dynamic Model Compensation* (DMC). In the case of DMC, process noise parameters are often included in the state vector to be estimated. Appendix F provides more details on SNC and DMC.

The solution of Eq. (4.9.1) can be obtained by the method of variation of parameters. The homogeneous equation is given by

$$\dot{\mathbf{x}}(t) = A(t)\mathbf{x}(t) \tag{4.9.3}$$

which, as previously noted, has a solution of the form

$$\mathbf{x}(t) = \Phi(t, t_0)\mathbf{C}_0. \tag{4.9.4}$$

The method of variation of parameters selects $\mathbf{C}_0$ as a function of time so that

$$\mathbf{x}(t) = \Phi(t, t_0)\mathbf{C}(t). \tag{4.9.5}$$

It follows then that

$$\dot{\mathbf{x}}(t) = \dot{\Phi}(t, t_0)\mathbf{C}(t) + \Phi(t, t_0)\dot{\mathbf{C}}(t). \tag{4.9.6}$$

Substituting Eq. (4.9.6) into Eq. (4.9.1) yields

$$\dot{\Phi}(t, t_0)\mathbf{C}(t) + \Phi(t, t_0)\dot{\mathbf{C}}(t) = A(t)\mathbf{x}(t) + B(t)\mathbf{u}(t). \tag{4.9.7}$$

Recall that

$$\dot{\Phi}(t, t_0) = A(t)\Phi(t, t_0). \tag{4.9.8}$$

Substituting Eq. (4.9.8) into Eq. (4.9.7) and using Eq. (4.9.5) reduces Eq. (4.9.7) to

$$\Phi(t, t_0)\dot{\mathbf{C}}(t) = B(t)\mathbf{u}(t). \tag{4.9.9}$$

Hence,

$$\dot{\mathbf{C}}(t) = \Phi^{-1}(t, t_0)B(t)\mathbf{u}(t). \tag{4.9.10}$$

Integrating Eq. (4.9.10) yields

$$\mathbf{C}(t) = \mathbf{C}_0 + \int_{t_0}^{t} \Phi^{-1}(\tau, t_0)B(\tau)\mathbf{u}(\tau)d\tau. \tag{4.9.11}$$

Substituting Eq. (4.9.11) into Eq. (4.9.5) results in

$$\mathbf{x}(t) = \Phi(t, t_0)\mathbf{C}_0 + \int_{t_0}^{t} \Phi(t, t_0)\Phi^{-1}(\tau, t_0)B(\tau)\mathbf{u}(\tau)d\tau. \tag{4.9.12}$$

Using the properties of the transition matrix, we may write

$$\Phi(t, t_0)\Phi^{-1}(\tau, t_0) = \Phi(t, t_0)\Phi(t_0, \tau) = \Phi(t, \tau). \tag{4.9.13}$$

At $t = t_0$, $\mathbf{x}(t_0) = \mathbf{x}_0$; hence, Eq. (4.9.12) can be used to determine that $\mathbf{C}_0 = \mathbf{x}_0$. With these results, Eq. (4.9.12) can be written as

$$\mathbf{x}(t) = \Phi(t, t_0)\mathbf{x}_0 + \int_{t_0}^{t} \Phi(t, \tau)B(\tau)\mathbf{u}(\tau)d\tau. \tag{4.9.14}$$

Equation (4.9.14) is the general solution for the inhomogeneous Eq. (4.9.1) and indicates how the true state propagates under the influence of process noise.

The equations for propagating the state estimate $\hat{\mathbf{x}}(t_{k-1})$ to the next observation time, t_k, are obtained by recalling that

$$\bar{\mathbf{x}}(t) = E[\mathbf{x}(t)|\mathbf{y}_{k-1}] \quad \text{for} \quad t \geq t_{k-1}; \tag{4.9.15}$$

that is, the expected value of $\mathbf{x}(t)$ given observations through t_{k-1}. Also, because we have observations through t_{k-1}

$$\overline{\mathbf{x}}(t_{k-1}) = \hat{\mathbf{x}}(t_{k-1}). \tag{4.9.16}$$

Differentiating Eq. (4.9.15) and using Eq. (4.9.1) gives

$$\begin{aligned}
\dot{\overline{\mathbf{x}}}(t) &= E\left[\dot{\mathbf{x}}(t)|\mathbf{y}_{k-1}\right] \\
&= E\left[\{A(t)\mathbf{x}(t) + B(t)\mathbf{u}(t)\}|\mathbf{y}_{k-1}\right] \\
&= A(t)\,E\left[\mathbf{x}(t)|\mathbf{y}_{k-1}\right].
\end{aligned}$$

Since $E[\mathbf{u}(t)] = 0$, it follows then that

$$\dot{\overline{\mathbf{x}}}(t) = A(t)\overline{\mathbf{x}}(t) \tag{4.9.17}$$

with initial conditions $\overline{\mathbf{x}}(t_{k-1}) = \hat{\mathbf{x}}(t_{k-1})$. In Eq. (4.9.17), the assumption has been made that the state noise $\mathbf{u}(t)$ has zero mean and is independent of the observations,

$$E[\mathbf{u}(t)|\mathbf{y}_{k-1}] = E[\mathbf{u}(t)] = 0. \tag{4.9.18}$$

Hence, if the mean of the process noise is zero, $(E[\mathbf{u}(t)] = 0)$, then the equation for propagating the state estimate is the same as without process noise,

$$\overline{\mathbf{x}}(t) = \Phi(t, t_{k-1})\hat{\mathbf{x}}_{k-1}. \tag{4.9.19}$$

One could derive a solution for the case where the mean is nonzero. In the case where $E[\mathbf{u}(t)] = \overline{\mathbf{u}}$, the solution would be obtained by applying the expectation operator to Eq. (4.9.14) to yield

$$\overline{\mathbf{x}}(t) = \Phi(t, t_{k-1})\hat{\mathbf{x}}_{k-1} + \Gamma(t_k, t_{k-1})\overline{\mathbf{u}} \tag{4.9.20}$$

where $\Gamma(t_k, t_{k-1})$ is defined by Eq. (4.9.47).

The equation for propagation of the estimation error covariance matrix is obtained by using the definition for $\overline{P}(t)$, given by

$$\overline{P}(t) = E[(\overline{\mathbf{x}}(t) - \mathbf{x}(t))(\overline{\mathbf{x}}(t) - \mathbf{x}(t))^T|\mathbf{y}_{k-1}]\ t \geq t_{k-1}. \tag{4.9.21}$$

Let

$$\Delta\mathbf{x}(t) \equiv \overline{\mathbf{x}}(t) - \mathbf{x}(t).$$

Then

$$\overline{P}(t) = E[\Delta\mathbf{x}\Delta\mathbf{x}^T|\mathbf{y}_{k-1}]$$

and differentiating $\overline{P}(t)$ yields

$$\dot{\overline{P}}(t) = E\left[\frac{d}{dt}\,\Delta\mathbf{x}\Delta\mathbf{x}^T|\mathbf{y}_{k-1}\right]$$
$$= E\left[\Delta\dot{\mathbf{x}}\Delta\mathbf{x}^T + \Delta\mathbf{x}\Delta\dot{\mathbf{x}}^T|\mathbf{y}_{k-1}\right]. \tag{4.9.22}$$

From Eqs. (4.9.1) and (4.9.17)

$$\Delta\dot{\mathbf{x}}(t) = \dot{\overline{\mathbf{x}}}(t) - \dot{\mathbf{x}}(t)$$
$$= A(t)\overline{\mathbf{x}}(t) - A(t)\mathbf{x}(t) - B(t)\mathbf{u}(t).$$

Hence,

$$\Delta\dot{\mathbf{x}}(t) = A(t)\Delta\mathbf{x}(t) - B(t)\mathbf{u}(t). \tag{4.9.23}$$

Substituting Eq. (4.9.23) into Eq. (4.9.22) yields

$$\dot{\overline{P}}(t) = E\left[\{(A(t)\Delta\mathbf{x} - B(t)\mathbf{u}(t))\Delta\mathbf{x}^T + \Delta\mathbf{x}(A(t)\Delta\mathbf{x} - B(t)\mathbf{u}(t))^T\}\,|\mathbf{y}_{k-1}\right]$$
$$= A(t)E\left[\Delta\mathbf{x}\Delta\mathbf{x}^T|\mathbf{y}_{k-1}\right] - B(t)E\left[\mathbf{u}(t)\Delta\mathbf{x}^T|\mathbf{y}_{k-1}\right]$$
$$+E\left[\Delta\mathbf{x}\Delta\mathbf{x}^T|\mathbf{y}_{k-1}\right]A^T(t) - E\left[\Delta\mathbf{x}\mathbf{u}^T(t)|\mathbf{y}_{k-1}\right]B^T(t)$$

and using Eq. (4.9.21)

$$\dot{\overline{P}}(t) = A(t)\overline{P}(t) + \overline{P}(t)A^T(t) - B(t)E\left[\mathbf{u}(t)\Delta\mathbf{x}^T|\mathbf{y}_{k-1}\right]$$
$$-E\left[\Delta\mathbf{x}\mathbf{u}^T(t)|\mathbf{y}_{k-1}\right]B^T(t). \tag{4.9.24}$$

The solution for $\Delta\mathbf{x}(t)$ now is needed to substitute into Eq. (4.9.24). Note that the solution for Eq. (4.9.23) is identical in form to the solution of Eq. (4.9.1) given by Eq. (4.9.14). Substituting Eq. (4.9.14) with appropriate subscripts into the last term of Eq. (4.9.24) yields

$$E\left[\Delta\mathbf{x}(t)\mathbf{u}^T(t)|\mathbf{y}_{k-1}\right] = E\left[\Phi(t, t_{k-1})\Delta\mathbf{x}_{k-1}\mathbf{u}^T(t)|\mathbf{y}_{k-1}\right] \tag{4.9.25}$$
$$- E\left[\int_{t_{k-1}}^{t}\Phi(t, \tau)B(\tau)\mathbf{u}(\tau)\mathbf{u}^T(t)d\tau|\mathbf{y}_{k-1}\right]$$

where $t \geq t_{k-1}$. However,

$$E\left[\Delta\mathbf{x}_{k-1}\mathbf{u}^T(t)|\mathbf{y}_{k-1}\right] = 0$$

since the forcing function $\mathbf{u}(t)$ cannot influence the state at a time t_{k-1} for $t \geq$

t_{k-1}. Hence,

$$E\left[\Delta\mathbf{x}(t)\mathbf{u}^T(t)|\mathbf{y}_{k-1}\right] = -\int_{t_{k-1}}^{t} \Phi(t,\tau)B(\tau)E\left[\mathbf{u}(\tau)\mathbf{u}^T(t)d\tau|\mathbf{y}_{k-1}\right].$$

(4.9.26)

From Eq. (4.9.2)

$$E\left[\mathbf{u}(\tau)\mathbf{u}^T(t)\right] = Q(t)\delta(t-\tau)$$ (4.9.27)

where $\delta(t-\tau)$ is the *Dirac delta function*, which is defined to be zero everywhere except at $\tau = t$, where it is infinite in such a way that the integral of $\delta(t-\tau)$ across the singularity is unity; the Dirac delta function is defined by

$$\delta(t) = \lim_{\epsilon\to 0} \delta_\epsilon(t)$$

where $\delta_\epsilon(t) = \begin{cases} 0 & |t| > \frac{\epsilon}{2} \\ \frac{1}{\epsilon} & |t| < \frac{\epsilon}{2} \end{cases}$.

This definition implies that

$$\int_{t_1}^{t_2} \delta(t-\tau)d\tau = \int_{t-\epsilon/2}^{t+\epsilon/2} \delta(t-\tau)d\tau = 1.$$

Substituting Eq. (4.9.27) into Eq. (4.9.26) gives

$$E\left[\Delta\mathbf{x}\mathbf{u}^T(t)|\mathbf{y}_{k-1}\right] = -\int_{t_{k-1}}^{t} \Phi(t,\tau)B(\tau)Q(t)\delta(t-\tau)d\tau.$$ (4.9.28)

Now

$$\int_{-\infty}^{\infty} \delta(t-\tau)d\tau = \lim_{\epsilon\to 0}\left[\int_{-\infty}^{t-\epsilon/2} 0\,d\tau + \int_{t-\epsilon/2}^{t+\epsilon/2} \frac{1}{\epsilon}d\tau + \int_{t+\epsilon/2}^{\infty} 0\,d\tau\right].$$

Hence, Eq. (4.9.28) becomes

$$E\left[\Delta\mathbf{x}(t)\mathbf{u}^T(t)|\mathbf{y}_{k-1}\right] =$$

$$-\lim_{\epsilon\to 0}\left[\int_{t_{k-1}}^{t-\epsilon/2} 0\,d\tau + \int_{t-\epsilon/2}^{t} \Phi(t,\tau)B(\tau)Q(t)(\frac{1}{\epsilon})d\tau\right].$$ (4.9.29)

Since ϵ is small, $\Phi(t,\tau)$ and $B(\tau)$ can be expanded in a Taylor series about $\Phi(t,t)$ and $B(t)$, respectively. To this end, let

$$\Phi(t,\tau) = \Phi(t,t) - \dot{\Phi}(t,t)(t-\tau) + O(t-\tau)^2. \tag{4.9.30}$$

The second term of this equation is negative, since $t \geq \tau$. Using the fact that $\Phi(t,t) = I$ and $\dot{\Phi}(t,t) = A(t)\Phi(t,t)$, we can write

$$\Phi(t,\tau) = I - A(t)(t-\tau) + O(t-\tau)^2. \tag{4.9.31}$$

Similarly we can expand $B(\tau)$:

$$B(\tau) = B(t) - \frac{d}{dt}B(t)(t-\tau) + O(t-\tau)^2. \tag{4.9.32}$$

Substituting Eqs. (4.9.31) and (4.9.32) into Eq. (4.9.29) yields

$$
\begin{aligned}
E\left[\Delta \mathbf{x} \mathbf{u}^T(t) | \mathbf{y}_{k-1}\right] = & -\lim_{\epsilon \to 0} \frac{1}{\epsilon} \int_{t-\epsilon/2}^{t} \left[[I - A(t)(t-\tau)]\, [B(t)Q(t) \right.\\
& \left. -\frac{d}{dt}B(t)Q(t)(t-\tau)] + O(t-\tau)^2 \right] d\tau \\
= & -\lim_{\epsilon \to 0} \int_{t-\epsilon/2}^{t} \left[B(t)Q(t) - \frac{d}{dt}B(t)Q(t)(t-\tau) \right. \\
& -A(t)B(t)Q(t)(t-\tau) \\
& +A(t)\frac{d}{dt}B(t)Q(t)(t-\tau)^2 \\
& \left. +O(t-\tau)^2 \right] \frac{d\tau}{\epsilon}.
\end{aligned} \tag{4.9.33}
$$

Ignoring higher order terms $(O(t-\tau)^2)$ in Eq. (4.9.33) results in

$$
\begin{aligned}
E\left[\Delta \mathbf{x}(t)\mathbf{u}^T(t)|\mathbf{y}_{k-1}\right] = & -\lim_{\epsilon \to 0} \frac{1}{\epsilon} \int_{t-\epsilon/2}^{t} \left\{ B(t)Q(t) - \left[A(t)B(t)Q(t) \right. \right. \\
& \left. \left. +\frac{d}{dt}B(t)Q(t) \right](t-\tau) \right\} d\tau.
\end{aligned}
$$

Defining $K(t) = A(t)B(t)Q(t) + \frac{d}{dt}B(t)Q(t)$ and carrying out the integration

$$
\begin{aligned}
E\left[\Delta \mathbf{x} \mathbf{u}^T(t)|\mathbf{y}_{k-1}\right] = & -\lim_{\epsilon \to 0} \frac{1}{\epsilon}\left[B(t)Q(t)\tau - K(t)(t\tau - \frac{\tau^2}{2}) \right]_{t-\frac{\epsilon}{2}}^{t} \\
= & -\lim_{\epsilon \to 0} \frac{1}{\epsilon}\left[B(t)Q(t)\frac{\epsilon}{2} - K(t)\left(\frac{\epsilon^2}{8}\right) \right]
\end{aligned}
$$

$$= -\lim_{\epsilon \to 0} \left[\frac{B(t)Q(t)}{2} - \frac{K(t)}{8}\epsilon \right]$$
$$= -\frac{B(t)Q(t)}{2}. \tag{4.9.34}$$

Substituting (4.9.34) and its transpose into (4.9.24) leads to

$$\dot{\overline{P}}(t) = A(t)\overline{P}(t) + \overline{P}(t)A^T(t) + B(t)Q(t)B^T(t). \tag{4.9.35}$$

Equation (4.9.35) is an $n \times n$ matrix differential equation whose solution may be obtained by integrating with the initial conditions $\overline{P}(t_k) = P_k$; that is, the measurement update of the estimation error covariance matrix at t_k.

Equation (4.9.35) also can be expressed in integral form by using the method of variation of parameters. The homogeneous differential equation is given by

$$\dot{\overline{P}}(t) = A(t)\overline{P}(t) + \overline{P}(t)A^T(t), \tag{4.9.36}$$

which has the solution

$$\overline{P}(t) = \Phi(t, t_0)P_0\Phi^T(t, t_0), \tag{4.9.37}$$

where for convenience, t_{k-1} has been replaced by t_0. Letting P_0 become a function of time, Eq. (4.9.37) becomes

$$\dot{\overline{P}}(t) = \dot{\Phi}(t, t_0)P_0\Phi^T(t, t_0)$$
$$+ \Phi(t, t_0)P_0\dot{\Phi}^T(t, t_0) + \Phi(t, t_0)\dot{P}_0\Phi^T(t, t_0). \tag{4.9.38}$$

Equating Eqs. (4.9.35) and (4.9.38)

$$A(t)\overline{P}(t) + \overline{P}(t)A^T(t) + B(t)Q(t)B^T(t) = \dot{\Phi}(t, t_0)P_0\Phi^T(t, t_0)$$
$$+ \Phi(t, t_0)P_0\dot{\Phi}^T(t, t_0) + \Phi(t, t_0)\dot{P}_0\Phi^T(t, t_0). \tag{4.9.39}$$

However, from Eqs. (4.9.36) and (4.9.37)

$$A(t)\overline{P}(t) + \overline{P}(t)A^T(t) = \dot{\Phi}(t, t_0)P_0\Phi^T(t, t_0)$$
$$+ \Phi(t, t_0)P_0\dot{\Phi}^T(t, t_0). \tag{4.9.40}$$

Using Eq. (4.9.40), Eq. (4.9.39) reduces to

$$\Phi(t, t_0)\dot{P}_0\Phi^T(t, t_0) = B(t)Q(t)B^T(t),$$

or

$$\dot{P}_o = \Phi^{-1}(t, t_0)B(t)Q(t)B^T(t)\Phi^{-T}(t, t_0). \tag{4.9.41}$$

Integrating Eq. (4.9.41) results in

$$P_0(t) - P_0 = \int_{t_0}^t \Phi(t_0, \tau) B(\tau) Q(\tau) B^T(\tau) \Phi^T(t_0, \tau) d\tau. \tag{4.9.42}$$

Substituting Eq. (4.9.42) into (4.9.37)

$$\overline{P}(t) = \Phi(t, t_0)[P_0 + \int_{t_0}^t \Phi(t_0\ \tau) B(\tau) Q(\tau) B^T(\tau) \Phi^T(t_0, \tau) d\tau] \Phi^T(t, t_0)$$

$$= \Phi(t, t_0) P_0 \Phi^T(t, t_0) + \int_{t_0}^t \Phi(t, t_0) \Phi(t_0, \tau) B(\tau) Q(\tau) B^T(\tau)$$

$$\times \Phi^T(t_0, \tau) \Phi^T(t, t_0) d\tau. \tag{4.9.43}$$

If we use

$$\Phi(t, t_0) \Phi(t_0\ \tau) = \Phi(t, \tau),$$

and

$$\Phi^T(t_0, \tau) \Phi^T(t, t_0) = [\Phi(t, t_0) \Phi(t_0, \tau)]^T = \Phi^T(t, \tau),$$

then after letting $t_0 = t_{k-1}$, Eq. (4.9.43) becomes

$$\overline{P}(t) = \Phi(t, t_{k-1}) P_{k-1} \Phi^T(t, t_{k-1})$$

$$+ \int_{t_{k-1}}^t \Phi(t, \tau) B(\tau) Q(\tau) B^T(\tau) \Phi^T(t, \tau) d\tau. \tag{4.9.44}$$

Equations (4.9.19) and (4.9.44) are the equations for propagating the estimate of the state and the covariance for a *continuous system*. Since the orbit determination problem generally consists of a continuous system (the trajectory) subjected to discrete observations, it is convenient to use Eq. (4.9.19) to propagate the state estimate and to discretize Eq. (4.9.44). This can be accomplished by replacing t with t_{k+1} and assuming that $\mathbf{u}(\tau)$ is a white *random sequence* rather than a process. Thus, as indicated in Fig. 4.9.1, $\mathbf{u}(t)$ is considered to be a piecewise constant function with covariance

$$E[\mathbf{u}(t_i) \mathbf{u}^T(t_j)] = Q_i \delta_{ij} \quad \delta_{ij} = \begin{cases} 1\ i = j \\ 0\ i \neq j \end{cases} \tag{4.9.45}$$

where the Dirac delta function has been replaced by its analog for the discrete case, the Kroneker delta function. In the discrete case, Eq. (4.9.14) becomes

$$\mathbf{x}(t_{k+1}) \equiv \mathbf{x}_{k+1} = \Phi(t_{k+1}, t_k) \mathbf{x}_k + \Gamma(t_{k+1}, t_k) \mathbf{u}_k \tag{4.9.46}$$

where

$$\Gamma(t_{k+1}, t_k) = \int_{t_k}^{t_{k+1}} \Phi(t_{k+1}, \tau) B(\tau) d\tau. \tag{4.9.47}$$

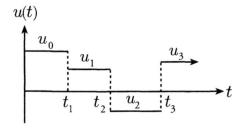

Figure 4.9.1: A white random sequence.

$\Gamma(t_{k+1}, t_k)$ is referred to as the *process noise transition matrix*, and Eq. (4.9.47) is an $n \times m$ quadrature since $\Phi(t_{k+1}, \tau)$ and $B(\tau)$ are known functions. Using the definition of the estimation error covariance matrix

$$\overline{P}_{k+1} = E[(\overline{\mathbf{x}}_{k+1} - \mathbf{x}_{k+1})(\overline{\mathbf{x}}_{k+1} - \mathbf{x}_{k+1})^T] \qquad (4.9.48)$$

and substituting Eqs. (4.9.46) and (4.9.19) into (4.9.48) leads to

$$\overline{P}_{k+1} = E\Big\{ \big[\Phi(t_{k+1}, t_k)\hat{\mathbf{x}}_k - \Phi(t_{k+1}, t_k)\mathbf{x}_k - \Gamma(t_{k+1}, t_k)\mathbf{u}_k\big]$$
$$\times \big[\Phi(t_{k+1}, t_k)\hat{\mathbf{x}}_k - \Phi(t_{k+1}, t_k)\mathbf{x}_k - \Gamma(t_{k+1}, t_k)\mathbf{u}_k\big]^T \Big\}. \quad (4.9.49)$$

Note that $E[(\hat{\mathbf{x}}_k - \mathbf{x}_k)\mathbf{u}_k^T] = 0$; that is, $\mathbf{u}_k$ cannot affect the estimation error at t_k since a finite time must evolve for $\mathbf{u}_k$ to affect the state. Finally, carrying out the expectation operation in Eq. (4.9.49) yields

$$\overline{P}_{k+1} = \Phi(t_{k+1}, t_k)P_k\Phi^T(t_{k+1}, t_k) + \Gamma(t_{k+1}, t_k)Q_k\Gamma^T(t_{k+1}, t_k). \quad (4.9.50)$$

The estimation error covariance matrix $\overline{P}_{k+1}$ can be obtained by integrating the differential equation for $\overline{P}$ given by Eq. (4.9.35), or $\overline{P}_{k+1}$ may be obtained by using the state and process noise transition matrices as indicated in Eq. (4.9.50). A comparison of Eq. (4.9.35) and Eq. (4.9.50) indicates the following:

1. Since $P(t)$ is symmetric, only $n(n+1)/2$ of the $n \times n$ system of equations represented by Eq. (4.9.35) must be integrated. However, the $n(n+1)/2$ equations are coupled and must be integrated as a single first-order system of dimension $n(n+1)/2$.

2. The $n \times n$ system represented by Eq. (4.9.50) can be separated into an $n \times n$ system of differential equations for Φ and an $n \times m$ quadrature for

Γ. Furthermore, the $n \times n$ system of equations represented by the solution for $\Phi(t_{k+1}, t_k)$ can be integrated as a sequence of $n \times 1$ column vectors.

The comparison between the two methods indicates that integration of fewer equations is required to obtain the solution for $\overline{P}(t)$ with Eq. (4.9.35). However, the integration of these equations may be more difficult than the integration associated with the larger system represented by Eq. (4.9.50) since they are coupled.

The equations for determining $\hat{x}$ using the sequential processing algorithm are unchanged whenever a zero-mean process noise is included. However, as has been shown, the equations that propagate the estimation error covariance matrix do change; that is, the first of Eq. (4.7.31) is replaced by Eq. (4.9.50). Generally, the batch processing algorithm is not used with process noise because mapping of the process noise effects from the observation times to the epoch time is cumbersome. It further results in a normal matrix with an observation weighting matrix that will be nondiagonal and whose dimension is equal to $m \times m$, where m is the total number of observations. Computation of the inverse of the normal matrix will be so cumbersome that the normal equation solution involving process noise in the data accumulation interval is impractical. For example, a one-day tracking period for the TOPEX/Poseidon satellite by the French tracking system, DORIS, typically yields 5000–7000 doppler observations.

The advantage of using the process noise compensated sequential estimation algorithm lies in the fact that the asymptotic value of $\overline{P}(t)$ will approach a nonzero value determined by the magnitude of $Q(t)$. That is, for certain values of $Q(t)$, the increase in the state error covariance matrix $\overline{P}(t)$ during the interval between observations will balance the decrease in the covariance matrix that occurs at the observation point. In this situation, the estimation procedure will always be sensitive to new observations.

The question of how to choose the process noise covariance matrix, $Q(t)$, is complex. In practice, it is often chosen as a simple diagonal matrix and its elements are determined by trial and error. Although this method can be effective for a particular estimation scenario, such a process noise matrix is not generally applicable to other scenarios. The dynamic evolution of the true states of parameters estimated in a filter typically is affected by stochastic processes that are not modeled in the filter's deterministic dynamic model. The process noise model is a characterization of these stochastic processes, and the process noise covariance matrix should be determined by this process noise model. Development of the process noise model will not be presented in depth here; however, extensive discussions are given by Cruickshank (1998), Ingram (1970), and Lichten (1990).

The Gauss-Markov process is used as a process noise model and will be introduced here. It is computationally well suited for describing unmodeled forces since it obeys Gaussian probability laws and is exponentially correlated in time.

4.9.1   THE GAUSS-MARKOV PROCESS

A first-order Gauss-Markov process is often used for dynamic model compensation in orbit determination problems to account for unmodeled or inaccurately modeled accelerations acting on a spacecraft. A Gauss-Markov process is one that obeys a Gaussian probability law and displays the Markov property. The Markov property means that the probability density function at t_n given its past history at $t_{n-1}, t_{n-2}, \ldots$ is equal to its probability density function at t_n given its value at t_{n-1}.

A Gauss-Markov process obeys a differential equation (often referred to as a Langevin equation) of the form

$$\dot{\eta}(t) = -\beta\eta(t) + u(t) \tag{4.9.51}$$

where $u(t)$ is white Gaussian noise with

$$\begin{aligned} E(u) &= 0 \\ E\left[u(t)u(\tau)\right] &= \sigma^2\delta(t-\tau) \end{aligned} \tag{4.9.52}$$

and

$$\beta = \frac{1}{\tau},$$

where τ is the time constant or correlation time (not the same as τ in Eq. (4.9.52)).

Equation (4.9.51) can be solved by the method of variation of parameters to yield

$$\eta(t) = \eta(t_0)e^{-\beta(t-t_0)} + \int_{t_0}^{t} e^{-\beta(t-\tau)}u(\tau)d\tau. \tag{4.9.53}$$

Hence, $\eta(t)$ consists of a deterministic part and a random part. The autocorrelation function is (Maybeck, 1979)

$$\begin{aligned} E\left[\eta(t_j)\eta(t_i)\right] &= e^{-\beta(t_j-t_i)}E\left[\eta(t_i)\eta(t_i)\right] \\ &\quad + E\left[\left(\int_{t_i}^{t_j} e^{-\beta(t_j-\tau)}u(\tau)d\tau\right)\eta(t_i)\right] \\ &= e^{-\beta(t_j-t_i)}E\left[\eta(t_i)\eta(t_i)\right] \end{aligned} \tag{4.9.54}$$

since the stochastic process represented by the integral consists of independent increments. The remaining expectation is the autocorrelation of $\eta(t)$ at t_i:

$$\begin{aligned} E\left[\eta(t_i)\eta(t_i)\right] &\equiv \Psi(t_i, t_i) \\ &= \eta^2(t_0)e^{-2\beta(t_i-t_0)} + \frac{\sigma^2}{2\beta}\left(1 - e^{-2\beta(t_i-t_0)}\right) \end{aligned} \tag{4.9.55}$$

where σ^2 is the variance (strength) of the driving noise in Eq. (4.9.51). Using Eq. (4.9.55), equation (4.9.54) can then be written as

$$E\left[\eta(t_j)\eta(t_i)\right] = \Psi(t_i, t_i)e^{-\beta(t_j - t_i)}. \tag{4.9.56}$$

This is important because it points out one of the salient characteristics of the first-order Gauss-Markov process; that is, its autocorrelation fades exponentially with the rate of the fade governed by the time constant, $\tau = 1/\beta$.

Equation (4.9.53) contains a stochastic integral that cannot, in general, be evaluated except in a statistical sense. The mean of the stochastic integral is zero since $E[\mathbf{u}(t)] = 0$. Its variance can be shown to be (Myers, 1973)

$$\frac{\sigma^2}{2\beta}(1 - e^{-2\beta(t_j - t_i)}). \tag{4.9.57}$$

Because the stochastic integral is a Gaussian process it is uniquely defined by its mean and variance. Hence, if a function can be found with the same mean and variance it will be an equivalent process. Such a discrete process is given by

$$L_k \equiv u_k \sqrt{\frac{\sigma^2}{2\beta}(1 - e^{-2\beta(t_j - t_i)})} \tag{4.9.58}$$

where u_k is a discrete, Gaussian random sequence with mean and variance

$$E[u_k] = 0, \quad E[u_{k_i} u_{k_j}] = \delta_{ij}. \tag{4.9.59}$$

It is evident that L_k has the same mean and variance as the stochastic integral in Eq. (4.9.53); hence, the solution for $\eta(t)$ is given by

$$\eta(t_j) = e^{-\beta(t_j - t_i)}\eta(t_i)$$
$$+ u_k(t_i)\sqrt{\frac{\sigma^2}{2\beta}(1 - e^{-2\beta(t_j - t_i)})} \tag{4.9.60}$$

where $u_k(t_i)$ is a random number chosen by sampling from a Gaussian density function with a mean of zero and variance of 1.

The degree of correlation of the random process $\eta(t)$ is determined by the choice of σ and β. For a finite value of β and $\sigma^2 \simeq 0$, Eq. (4.9.60) yields

$$\eta(t_j) = e^{-\beta(t_j - t_i)}\eta(t_i) \tag{4.9.61}$$

and as β becomes small, $\eta(t_j) \rightarrow \eta(t_i)$, a constant. For a finite value of σ and $\beta \simeq 0$, Eq. (4.9.60) yields

$$\eta(t_j) = \eta(t_i) + u_k(t_i)\sigma\sqrt{t_j - t_i}; \tag{4.9.62}$$

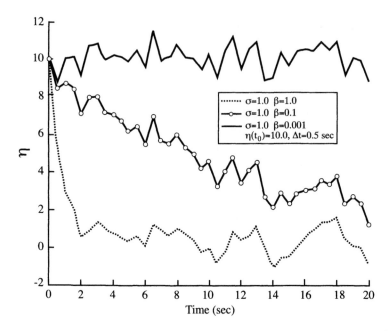

Figure 4.9.2: Gauss-Markov process for three time constants.

that is, a random walk process.

The term $\frac{\sigma^2}{2\beta}$ represents the steady-state variance of $\eta(t)$, the variance after a large enough time so that any transients in $\eta(t)$ have died out and it is in steady state.

With the proper choice of σ and β any number of time-correlated random functions can be generated. Figure 4.9.2 illustrates three histories of $\eta(t)$ for $\beta = 0.001, 0.1, 1.0,$ and $\sigma = 1.0$. Note that for $\beta = 10^{-3}$, $\eta(t)$ is essentially constant. The shape of this curve can be varied by changing σ or β or both.

To use the Gauss-Markov process to account for unmodeled accelerations in the orbit determination procedure we may formulate the problem as follows. The equations of motion are

$$\dot{\mathbf{r}} = \mathbf{v}$$
$$\dot{\mathbf{v}} = -\frac{\mu\mathbf{r}}{r^3} + \mathbf{f}(t) + \boldsymbol{\eta}(t) \qquad (4.9.63)$$
$$\dot{\boldsymbol{\eta}}(t) = -\beta\boldsymbol{\eta}(t) + \mathbf{u}(t),$$

where $-\frac{\mu\mathbf{r}}{r^3} + \mathbf{f}(t)$ represents the known accelerations and $\boldsymbol{\eta}(t)$ is a 3×1 vector of unknown accelerations. The procedure is to estimate the deterministic portion of $\eta(t)$ and perhaps the associated time constants as part of the state vector. The random portion of $\eta(t)$ contributes to the process noise covariance matrix Q. For details on implementing this algorithm see Tapley and Ingram (1973), Ingram and Tapley (1974), Cruickshank (1998), Lichten (1990), or Goldstein *et al.* (2001).

An example of the use of process noise for SNC and DMC for a simple problem is given in Appendix F.

4.10 INFORMATION FILTER

A sequential estimation algorithm can be developed by propagating the *information matrix*, $\Lambda \equiv P^{-1}$ (Maybeck, 1979). This form of the filter offers some numerical properties with better characteristics than the covariance filter. Writing Eq. (4.7.3) in terms of the information matrix gives

$$\left\{ \overline{\Lambda}_k + \widetilde{H}_k^T R_k^{-1} \widetilde{H}_k \right\} \hat{\mathbf{x}}_k = \overline{\Lambda}_k \overline{\mathbf{x}}_k + \widetilde{H}_k^T R_k^{-1} \mathbf{y}_k \tag{4.10.1}$$

or

$$\Lambda_k \hat{\mathbf{x}}_k = \overline{\Lambda}_k \overline{\mathbf{x}}_k + \widetilde{H}_k^T R_k^{-1} \mathbf{y}_k.$$

Recall that $\overline{P}_{k+1}$ is expressed in terms of P_k by Eq. (4.9.50) and $\overline{\mathbf{x}}_k$ is obtained by propagating $\hat{\mathbf{x}}_{k-1}$ according to Eq. (4.9.19). The Schur identity is given by (see Appendix B)

$$(A + BC)^{-1} = A^{-1} - A^{-1}B(I + CA^{-1}B)^{-1}CA^{-1}. \tag{4.10.2}$$

In Eq. (4.9.50) let

$$\begin{aligned} A &= \Phi\left(t_{k+1}, t_k\right) P_k \Phi^T\left(t_{k+1}, t_k\right) \\ B &= \Gamma(t_{k+1}, t_k) Q_k \\ C &= \Gamma^T(t_{k+1}, t_k). \end{aligned} \tag{4.10.3}$$

Then, for Q_k nonsingular, the Schur identity yields

$$\begin{aligned} \overline{\Lambda}_{k+1} = \overline{P}_{k+1}^{-1} &= M(t_{k+1}) - M(t_{k+1}) \Gamma(t_{k+1}, t_k) \\ &\times [\Gamma^T(t_{k+1}, t_k) M(t_{k+1}) \Gamma(t_{k+1}, t_k) + Q_k^{-1}]^{-1} \\ &\times \Gamma^T(t_{k+1}, t_k) M(t_{k+1}) \end{aligned} \tag{4.10.4}$$

where

$$M(t_{k+1}) = A^{-1} = \Phi^T(t_k, t_{k+1}) P_k^{-1} \Phi(t_k, t_{k+1}). \tag{4.10.5}$$

If there is no process noise, Eq. (4.10.4) reduces to

$$\overline{\Lambda}_{k+1} = \overline{P}_{k+1}^{-1} = M(t_{k+1}).$$

Define

$$L_{k+1} \equiv M(t_{k+1})\Gamma(t_{k+1}, t_k)$$
$$\times \left[\Gamma^T(t_{k+1}, t_k)M(t_{k+1})\Gamma(t_{k+1}, t_k) + Q_k^{-1}\right]^{-1}. \quad (4.10.6)$$

In terms of L_{k+1}, Eq. (4.10.4) becomes

$$\bar{\Lambda}_{k+1} = M(t_{k+1}) - L_{k+1}\Gamma^T(t_{k+1}, t_k)M(t_{k+1}). \quad (4.10.7)$$

The measurement update for Λ_{k+1} is obtained by inverting Eq. (4.7.4); that is,

$$\Lambda_{k+1} = \bar{\Lambda}_{k+1} + \tilde{H}_{k+1}^T R_{k+1}^{-1} \tilde{H}_{k+1}. \quad (4.10.8)$$

Define the following quantities:

$$\bar{D}_k \equiv \bar{\Lambda}_k \bar{x}_k \quad (4.10.9)$$
$$\hat{D}_k \equiv \Lambda_k \hat{x}_k. \quad (4.10.10)$$

The recursion relations for D are given by

$$\bar{D}_{k+1} = \bar{\Lambda}_{k+1} \bar{x}_{k+1}. \quad (4.10.11)$$

Using Eq. (4.10.7),

$$\bar{D}_{k+1} = \left\{I - L_{k+1}\Gamma^T(t_{k+1}, t_k)\right\}$$
$$\times \Phi^T(t_k, t_{k+1}) P_k^{-1} \Phi(t_k, t_{k+1})\bar{x}_{k+1} \quad (4.10.12)$$

but $\hat{x}_k = \Phi(t_k, t_{k+1})\bar{x}_{k+1}$ and $\hat{D}_k = P_k^{-1}\hat{x}_k$. Hence,

$$\bar{D}_{k+1} = \left\{I - L_{k+1}\Gamma^T(t_{k+1}, t_k)\right\} \Phi^T(t_k, t_{k+1})\hat{D}_k. \quad (4.10.13)$$

Also, from Eq. (4.10.9) and the second of Eq. (4.10.1),

$$\hat{D}_{k+1} = \Lambda_{k+1}\hat{x}_{k+1} = \bar{D}_{k+1} + \tilde{H}_{k+1}^T R_{k+1}^{-1} y_{k+1}. \quad (4.10.14)$$

The procedure starts with *a priori* values $\bar{P}_0$ and $\bar{x}_0$. From these compute

$$\bar{\Lambda}_0 = \bar{P}_0^{-1}$$
$$\bar{D}_0 = \bar{\Lambda}_0\bar{x}_0. \quad (4.10.15)$$

Next compute M_1, L_1, $\bar{\Lambda}_1$, $\bar{D}_1$, $\hat{D}_1$, and Λ_1 from Eqs. (4.10.5), (4.10.6), (4.10.7), (4.10.13), (4.10.14), and (4.10.8). Once Λ_k becomes nonsingular, its inverse can be computed to obtain P_k and the optimal estimate of $\mathbf{x}$ can be computed from Eq. (4.10.10); that is,

$$\hat{\mathbf{x}}_k = P_k \hat{\mathbf{D}}_k. \tag{4.10.16}$$

Time updates are given by Eqs. (4.10.5), (4.10.6), (4.10.7), and (4.10.13), with initial conditions given by Eq. (4.10.15). The measurement update is accomplished using Eqs. (4.10.8) and (4.10.14).

In summary, the covariance filter and the information filter time and measurement updates are given by

Time Update—Covariance Filter

$$\overline{\mathbf{x}}_{k+1} = \Phi(t_{k+1}, t_k)\hat{\mathbf{x}}_k \tag{4.10.17}$$
$$\overline{P}_{k+1} = \Phi(t_{k+1}, t_k)P_k\Phi^T(t_{k+1}, t_k)$$
$$+\Gamma(t_{k+1}, t_k)Q_k\Gamma^T(t_{k+1}, t_k). \tag{4.10.18}$$

Measurement Update—Covariance Filter

$$K_{k+1} = \overline{P}_{k+1}\widetilde{H}_{k+1}^T(R_{k+1} + \widetilde{H}_{k+1}\overline{P}_{k+1}\widetilde{H}_{k+1}^T)^{-1} \tag{4.10.19}$$
$$\hat{\mathbf{x}}_{k+1} = \overline{\mathbf{x}}_{k+1} + K_{k+1}(\mathbf{y}_{k+1} - \widetilde{H}_{k+1}\overline{\mathbf{x}}_{k+1}) \tag{4.10.20}$$
$$P_{k+1} = (I - K_{k+1}\widetilde{H}_{k+1})\overline{P}_{k+1}. \tag{4.10.21}$$

Time Update—Information Filter

$$M(t_{k+1}) = \Phi^T(t_k, t_{k+1})\Lambda_k\Phi(t_k, t_{k+1}) \tag{4.10.22}$$
$$L_{k+1} = M(t_{k+1})\Gamma(t_{k+1}, t_k)$$
$$\times [\Gamma^T(t_{k+1}, t_k)M(t_{k+1})\Gamma(t_{k+1}, t_k)$$
$$+ Q_k^{-1}]^{-1} \tag{4.10.23}$$
$$\overline{\mathbf{D}}_{k+1} = \{I - L_{k+1}\Gamma^T(t_{k+1}, t_k)\}$$
$$\times \Phi^T(t_k, t_{k+1})\hat{\mathbf{D}}_k \tag{4.10.24}$$
$$\overline{\Lambda}_{k+1} = \overline{P}_{k+1}^{-1} = [I - L_{k+1}\Gamma^T(t_{k+1}, t_k)]$$
$$\times M(t_{k+1}). \tag{4.10.25}$$

Measurement Update—Information Filter

$$\hat{\mathbf{D}}_{k+1} = \overline{\mathbf{D}}_{k+1} + \widetilde{H}_{k+1}^T R_{k+1}^{-1}\mathbf{y}_{k+1} \tag{4.10.26}$$
$$P_{k+1}^{-1} = \Lambda_{k+1} = \overline{\Lambda}_{k+1} + \widetilde{H}_{k+1}^T R_{k+1}^{-1}\widetilde{H}_{k+1} \tag{4.10.27}$$
$$\hat{\mathbf{x}}_{k+1} = \Lambda_{k+1}^{-1}\hat{\mathbf{D}}_{k+1}. \tag{4.10.28}$$

We can initialize the information filter with $P_0 = \infty$ or with P_0 singular, and obtain valid results for $\hat{x}$ once P_k becomes nonsingular. The Cholesky Algorithm of Chapter 5 may be used to solve for $\hat{x}$. However, the solution for $\hat{x}$ is not required by the algorithm and needs to be computed only when desired. The conventional sequential estimation algorithm fails in these cases. Also, as indicated in Section 4.7.1, the conventional sequential estimator can fail in the case where very accurate measurements are processed, which rapidly reduce the estimation error covariance matrix. This can be mitigated with the information filter.

Consider the example discussed in Section 4.7.1,

$$\overline{P}_0 = \sigma^2 I, \quad \sigma = 1/\epsilon, \quad \tilde{H}_1 = \begin{bmatrix} 1 & \vdots & \epsilon \end{bmatrix}, \quad R = I$$

where $\epsilon \ll 1$ and we assume that our computer word length is such that

$$1 + \epsilon \neq 1$$
$$1 + \epsilon^2 = 1.$$

The objective is to find P_1; that is, the estimation error covariance matrix after processing one observation using the information filter. The information filter yields

$$M_1 = \epsilon^2 I$$
$$\overline{\Lambda}_1 = M_1$$
$$\Lambda_1 = \overline{\Lambda}_1 + \tilde{H}_1^T R^{-1} \tilde{H}_1$$
$$= \epsilon^2 I + \begin{bmatrix} 1 & \epsilon \\ \epsilon & \epsilon^2 \end{bmatrix}$$
$$\Lambda_1 = P_1^{-1} = \begin{bmatrix} 1 & \epsilon \\ \epsilon & 2\epsilon^2 \end{bmatrix}$$

and

$$P_1 = \begin{bmatrix} 2 & -\sigma \\ -\sigma & \sigma^2 \end{bmatrix}.$$

This is the same symmetric, positive definite result obtained from the batch processor and agrees with the exact solution to $O(\epsilon)$. Because we are accumulating the information matrix at each stage, the accuracy of the information filter should be comparable to that of the batch processor. Hence, the conventional covariance filter fails for this example, but the information filter does not.

4.11 BATCH AND SEQUENTIAL ESTIMATION

As described in previous sections, two general categories of estimators are used, the batch processor and the sequential processor, both with distinct advantages and disadvantages. The batch formulation provides an estimate of the state at some chosen epoch using an entire batch or set of data. This estimate and its associated covariance matrix can then be mapped to other times. The sequential processor, on the other hand, provides an estimate of the state at each observation time based on observations up to that time. This solution and its associated covariance also can be mapped to another time.

In the sequential formulation without process noise, a mathematical equivalence can be shown between the batch and sequential algorithms; given the same data set, both algorithms produce the same estimates when the estimates are mapped to the same times. In the extended form of the sequential algorithm, where the reference orbit is updated at each observation time, the algorithms are not equivalent, but numerical experiments have shown a very close agreement.

Normally, the batch and sequential algorithm will need to be iterated to convergence, whereas the extended sequential will obtain near convergence in a single iteration. The sequential algorithm, unlike the batch, requires restarting a numerical integrator at each observation time. In general, the sequential processor is used in real-time applications supporting control or other decision functions and it is appropriate to incorporate some representation of the state noise to ensure that divergence does not occur. This implementation provides a means of compensating for various error sources in the processing of ground-based or onboard data. As indicated previously, inclusion of process noise in the batch algorithm substantially complicates the solution of the normal equations by increasing the dimensions of the normal matrix from n (the number of state parameters) to m (the number of observations).

4.12 OBSERVABILITY

The property of *observability* refers to the ability to apply an estimator to a particular system and obtain a unique estimate for all components of the state vector. As applied to orbital mechanics, the observability property refers to the ability to obtain a unique estimate of the spacecraft state as opposed to whether the satellite can be physically observed. Cases of unobservability rarely occur for properly formulated problems. However, these cases illustrate clearly the caution that must be exercised in applying an estimation algorithm to a particular problem. An unobservable parameter can best be illustrated by the following example.

Consider a satellite moving in an orbit about a perfectly spherical planet with a homogeneous mass distribution. Assuming no forces other than the gravitational attraction of the planet, which can be represented by a point mass, the orbit

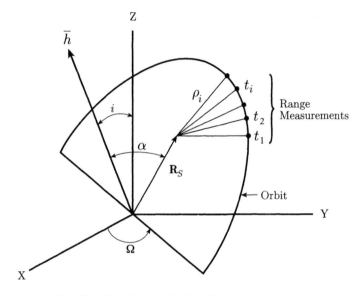

$\mathbf{R}_S$ = Tracking Station Position Vector

Figure 4.11.1: Single-station range geometry.

will be a conic section. Suppose a laser or radar range system exists on the surface of the planet capable of measuring to infinite precision the distance between the instrument and the satellite. Assuming the planet does not rotate and only a single range measuring system exists, the previously described estimators can be applied to the problem of estimating the state of the spacecraft at some appropriate time. Consider a single pass of the satellite over the station as shown in Fig. 4.11.1, where $\mathbf{R_S}$ is the station position vector, assumed known. Given a sequence of range measurements, ρ_i, as shown, an orbit could be determined from which the computed ranges match the measured values. However, a subtle problem exists because there is no change in the location of the tracking station relative to the orbit plane; hence, multiple solutions are possible—the same range history occurs for each orbit as we rotate the orbit's angular momentum vector about $\mathbf{R_S}$ keeping α constant. This results in an infinite number of solutions that could have the same range measurement sequence. This circumstance leads to an ambiguity in the inclination, i, and longitude of the ascending node, Ω. The difficulty is not alleviated by using multiple passes, since each pass would have the same geometry relative to the station. The possible multiple solutions will be manifested by a singular normal matrix $(H^T H)$ in the ordinary least squares formulation, regardless of the number of observations used.

A unique solution can be obtained if (1) i or Ω is known and not estimated, resulting in five elements being estimated, or (2) an *a priori* covariance matrix is assigned to the state indicative of knowledge of i or Ω. The problem can be avoided by other techniques as well; for example, using angular data such as azimuth and elevation or by using a second ranging station. However, this difficulty is not altered by varying the accuracy of the range measurements, or if range-rate is used instead of, or in addition to, range.

In practice, in a single pass the earth rotates slightly, resulting in a change in the station-orbit geometry. However, the resulting least squares system is frequently ill conditioned; that is, nearly singular. Numerous other examples of nonobservability exist, particularly in the recovery of spherical harmonic coefficients of the earth's gravity field from inadequate amounts and distributions of tracking data.

The mathematical conditions for parameter observability can be derived from the observability criterion for a linear dynamical system.

Theorem: For the linear system and linear observation set,

$$\mathbf{x}_i = \Phi(t_i, t_k)\mathbf{x}_k$$
$$\mathbf{y}_i = \widetilde{H}_i\mathbf{x}_i + \boldsymbol{\epsilon}_i; \qquad i = 1, \ldots, \ell \qquad (4.12.1)$$

complete observability of the $n \times 1$ state vector, at the general time, t_k, requires satisfying the condition

$$\delta\mathbf{x}_k^T \Lambda \delta\mathbf{x}_k > 0 \qquad (4.12.2)$$

for all arbitrary real vectors, $\delta\mathbf{x}_k$, where the $n \times n$ information matrix, Λ, is defined as

$$\Lambda = \sum_{i=1}^{\ell} \Phi^T(t_i, t_k)\widetilde{H}_i^T R_i^{-1} \widetilde{H}_i \Phi(t_i, t_k) = H^T R^{-1} H \qquad (4.12.3)$$

and H is defined by Eq. (4.2.38). Hence, complete observability of the state with the data in the batch accumulated from the observation set $\mathbf{y}_i; i = 1, \ldots, \ell$ requires that the symmetric information matrix, Λ, be positive definite. Note that in order for Λ to be positive definite, H must be full rank.

In addition to the constraints placed on the data as a necessary criterion for observability, care in the mathematical formulation of the problem must be exercised to ensure that only a minimal set of parameters is estimated. If spurious or unnecessary parameters are included in the problem formulation, the solution will be nonobservable regardless of the data characteristics. As an example, the expression for the acceleration on a satellite due to effects of atmospheric drag can be written as

$$\mathbf{D} = -\frac{1}{2}\frac{C_D A}{m}\rho(h)|\mathbf{V}_{rel}|\mathbf{V}_{rel} \qquad (4.12.4)$$

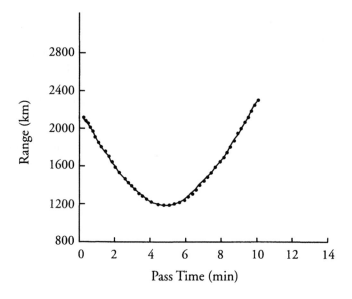

Figure 4.12.1: Range vs time from ground-based station.

where $\mathbf{V}_{rel} = \mathbf{V} - \boldsymbol{\omega}_e \times \mathbf{r}$, $\mathbf{V}$ is the inertial satellite velocity, $\boldsymbol{\omega}_e$ is the rotational velocity of the atmosphere (which is assumed to be the same as the earth's rotational velocity), C_D is the satellite drag coefficient, A is the satellite projected cross-sectional area, m is the satellite mass, and $\rho(h)$ is the atmospheric density at the satellite altitude, h. In many applications, C_D, A, and m will be sufficiently uncertain that the errors in their *a priori* values may degrade the estimate of the satellite state. However, if any two or all three of these parameters are added to the state vector, a singular or nonobservable state vector will result, no matter how dense or how complete the observation set is. In this case, the H matrix will have two rows that are identical except for a scalar mutiplication factor. This condition will cause H to be non-full-rank and $H^T H$ to be nonpositive definite. The single scalar parameter,

$$\beta = \frac{C_D A}{m} \qquad (4.12.5)$$

can be included as a state variable, and a well-posed estimation problem will occur. Two constants, say C_1 and C_2, that do not appear separately in other force functions cannot be observed. Rather the product $C_3 = C_1 C_2$ may be observed. Any state variables that are linearly related to one another cannot be uniquely determined (see Exercise 2 of Chapter 1). In general, the observability criterion tests not only the completeness of the data set in both the spatial and temporal sense, but it also tests the validity of the problem formulation.

4.13 ERROR SOURCES

In the application of an estimation procedure to a satellite or trajectory problem, measurements are obtained by various ground-based or on-board instruments. For example, a ground-based ranging system may make the measurements shown in Fig. 4.13.1 with time measured in minutes since the first measurement. Based on a mathematical model of the dynamical system and the measurement system, a predicted or computed measurement could be generated and compared with the actual measurement. If, in fact, the models are quite accurate, the difference (or residual) between the actual and predicted (or computed) measurements (O-C) will exhibit the random component in the measurement system as in Fig. 4.13.3. On the other hand, as is usually the case, the model has some inaccuracies associated with it, and the residual pattern will exhibit the character shown in Fig. 4.13.2. By using a polynomial function of time, the systematic component in Fig. 4.13.2 can be removed to determine the noise-only components. The data editing functions to eliminate the spurious or erroneous measurements are applied to these residuals. Finally, the edited data in Fig. 4.13.2 are used by the estimators to improve the state and the force and measurement models.

In the ideal case, the nonzero difference between the actual measurement and the predicted value should be due to the noise and biasing that occur in making the measurement. In practice, however, the mathematical models that describe the satellite force environment and those that describe the instrument performing some measurement are not completely accurate, or certain approximations are made for the benefit of computer storage and/or computer execution time, which introduce some discrepancy or error in the data processing. It is frequently necessary to ascribe the source of an error to a phenomenon in the physical world or to an approximation made in the model of the real world. Knowledge of various parameters in the mathematical models, such as the mass of the Earth or the coefficients that describe the inhomogeneous mass distribution within the Earth, have been obtained through various experiments or through use of many measurements and are only approximately known.

Typical error sources are as follows.

- **SATELLITE FORCE MODEL:**

  Gravitation parameters
  - Mass of the earth (GM)
  - Geopotential coefficients, (C_{lm} and S_{lm})
  - Solid earth and ocean tide perturbations
  - Mass and position of the moon and planets
  - General relativistic perturbation

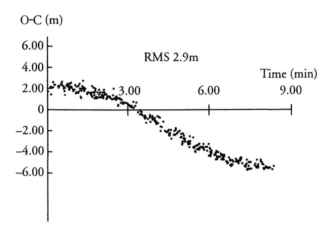

Figure 4.13.1: O-C, random and systematic component.

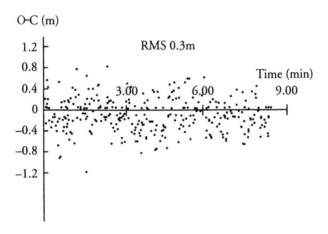

Figure 4.13.2: Random component.

Nongravitational parameters

– Drag (C_D, atmospheric density)
– Solar and earth radiation pressure
– Thrust (including venting and momentum dumping)
– Other (magnetic origin, etc.)

- **MEASUREMENT MODEL:**

Inertial and terrestrial coordinate systems

– Precession and nutation
– Polar motion

Ground-based measurements

– Coordinates of tracking station
– Atmospheric effects (tropospheric and
  ionospheric refraction)
– Instrument modeling
– Clock accuracy
– Tectonic plate motion

The error sources are dependent on the satellite under consideration, that is, satellite configuration, the orbit altitude and inclination, and measurement systems. Some of these error sources have distinct signatures in the data, and others may be very similar, thus producing aliasing between these components and making it difficult or impossible to separate their effects into individual components.

As a consequence, one constantly attempts to improve the model of the satellite environment. However, this improvement is normally done outside the operations aspect in which state estimates may be needed for an ongoing mission. To improve the gravitational model, for example, observations of various types such as range, range-rate, and angular data from different satellites would be used in a large parameter least squares solution. These large parameter solutions, in which the number of parameters may be 5000 or more, consume large amounts of computer time and, consequently, can be performed only infrequently. A family of such solutions has been generated since 1972 at the Goddard Space Flight Center and the Center for Space Research, and in several international centers. The models start with the Goddard Earth Model, GEM-1, and progress through the recent JGM-3 model (Tapley *et al.*, 1996) developed for the TOPEX/Poseidon mission. In the case of JGM-3, a full degree-70 and order-70 solution was produced. Other solutions have been obtained at the Smithsonian Astrophysical Observatory (SAO), the Department of Defense, and by foreign institutions, including GFZ in Germany and GRGS in France.

Implicit in all of these solutions is the time standard used in solving Newton's Equations, as well as time tagging the observations by various instruments. In the case of Newton's Laws, a uniform time scale is inherent, and we attempt to provide such a system in time tagging the measurements. Uniform time scales are provided by atomic clocks, aside from small relativistic effects, and an operational problem exists to ensure that all clocks used in making measurements are adequately synchronized. For ground-based systems, station clocks are usually synchronized to a national standard. Clearly, in the event that a station is out of synchronization with the other stations, the measurements made at that station will produce residuals that are somewhat larger than the others.

4.14 ORBIT ACCURACY

In general, the orbit accuracy is dependent on the following factors:

1. Truncation error in the application of an approximating algorithm to a mathematical process;

2. Round-off error in the application of a finite computational precision;

3. Mathematical model simplifications;

4. Errors in the parameters used in the mathematical model of the physical system or the instrument model;

5. Amount, type, and accuracy of tracking data.

For missions with high orbit-accuracy requirements, the limitation on the accuracy usually is imposed by the accuracy of the dynamical model, particularly the geopotential. But even the geopotential errors are dependent on the particular orbit; for example, an orbit at an altitude of one earth radius will be less affected by those errors than a satellite at an altitude of 800 km. The accuracy of the models is reflected in the current ability to determine an orbit for TOPEX/Poseidon (1334 km altitude) of ten days duration, which yields 1 cm root mean square (RMS) of the laser range residuals (Tapley *et al.*, 1994). This RMS reflects an overall orbit accuracy of about 8 cm.

Determining the true accuracy of an orbit determination solution based on actual tracking data is difficult because the true trajectory is never known. Furthermore, the estimation error covariance matrix for a given solution is generally overly optimistic depending on the weights assigned to the tracking data. If realistic process noise has been included, the estimation error covariance matrix may be a good indicator of orbit accuracy but it is difficult to determine the level of process noise needed to properly scale the covariance matrix. This is best done through simulation studies, but these require quantitative knowledge of the error sources. Covariance analysis, described in Chapter 6, may also aid in accuracy assessment assuming statistical knowledge of the error sources is available.

A few quantitative indicators of accuracy can be examined depending on what tracking information is available. If solutions derived from different independent tracking systems are available they can be compared. The locations of most laser tracking sites are known to the centimeter level and the measurements themselves are accurate to this level. Therefore, if the satellite is tracked by lasers as it flies at high elevation over a site, the laser residuals will be a good indication of radial orbit accuracy. Furthermore, the estimation error covariance matrix can be mapped to these times, rotated to radial, along-track and cross-track directions and the radial standard deviation compared with the RMS of laser residuals. This comparison will provide a calibration of the radial component of the covariance matrix. The RMS of tracking data residuals is also an indicator of accuracy. However, small tracking residuals do not guarantee an accurate orbit because there may be a component of the satellite's position that is insensitive to the particular tracking data type (see Section 4.12).

Another measure of accuracy is the RMS of orbit solution overlaps. For example, five days of contiguous tracking data may be fit as two arcs of three days length with one day of overlap. The RMS of the three components of overlapping position is an indicator of orbit accuracy. However, any error common to a given coordinate during the overlap period will not be evident in the RMS. Finally, orbit solutions for satellites that make any kind of metric measurements (radar altimeters, laser altimeters, SAR, etc.) can be evaluated by examining the quality of parameters based on these measurements themselves. For example, for laser or radar altimeter satellites the RMS of crossover residuals discussed in Chapter 3 are an indicator of orbit accuracy.

The accuracy just described relates to the estimation accuracy. Another important accuracy consideration occurs in the problem of prediction. Given some estimate of the satellite state, how well can the state of the spacecraft be predicted at some future time? Such prediction accuracy is important for (1) predicting and scheduling events at ground-based stations, including antenna or telescope pointing, and (2) scheduling events for future orbital maneuvers. The prediction accuracy is influenced by the same effects that influence the estimation accuracy; however, it is also dependent on the estimation accuracy itself. If, for instance, a perfect physical model was known but the state estimate used to initiate the prediction was in error, this error would grow during the prediction interval. As an example of prediction accuracy, the position of Lageos can be predicted to about 200 meters after two months. For the TOPEX/Poseidon Mission, the orbit can be predicted to about 0.5 km after a week based on tracking with laser ranging or the Global Positioning System.

4.15 SMOOTHING

It is often desirable to perform a smoothing operation when using a sequential filter. In this case, we are searching for the best estimate of the state at some time t_k based on all observations through time t_ℓ where $\ell > k$. For the case where there is no random component to the dynamical equation of state—for example, the no-process noise case—the batch estimation algorithm along with the prediction equation, Eqs. (4.4.19) and (4.4.22), will give the smoothed solution. However, as noted, the batch estimation approach has difficulty including the effects of process noise. The smoothing algorithms have been developed to overcome this difficulty. Following Jazwinski (1970), the smoothing algorithm can be derived using a Bayesian approach of maximizing the density function of the state conditioned on knowledge of the observations through time, t_ℓ. Our system is described in Section 4.9 (see Eq. (4.9.46)).

$$\mathbf{x}_{k+1} = \Phi(t_{k+1}, t_k)\mathbf{x}_k + \Gamma(t_{k+1}, t_k)\mathbf{u}_k$$
$$\mathbf{y}_k = \tilde{H}_k\mathbf{x}_k + \boldsymbol{\epsilon}_k.$$

We will use the notation $\hat{\mathbf{x}}_k^\ell$ to indicate the best estimate of $\mathbf{x}$ at t_k based on observations through t_ℓ, where in general $\ell > k$. Following the Maximum Likelihood philosophy, we wish to find a recursive expression for $\hat{\mathbf{x}}_k^\ell$ in terms of $\hat{\mathbf{x}}_{k+1}^\ell$, which maximizes the conditional density function

$$p(\mathbf{x}_k,\ \mathbf{x}_{k+1}/\mathbf{Y}_\ell), \quad \text{where} \quad \mathbf{Y}_\ell = \mathbf{y}_1, \mathbf{y}_2 \cdots \mathbf{y}_k \cdots \mathbf{y}_\ell. \tag{4.15.1}$$

From Bayes Rule

$$
\begin{aligned}
p(\mathbf{x}_k, \mathbf{x}_{k+1}/\mathbf{Y}_\ell) &= \frac{p(\mathbf{x}_k, \mathbf{x}_{k+1}, \mathbf{Y}_\ell)}{p(\mathbf{Y}_\ell)} = \frac{p(\mathbf{x}_k, \mathbf{x}_{k+1}, \mathbf{Y}_k, \mathbf{y}_{k+1} \cdots \mathbf{y}_\ell)}{p(\mathbf{Y}_\ell)} \\
&= \frac{p(\mathbf{Y}_k)}{p(\mathbf{Y}_\ell)} p(\mathbf{x}_k, \mathbf{x}_{k+1}, \mathbf{y}_{k+1} \cdots \mathbf{y}_\ell / \mathbf{Y}_k) \tag{4.15.2} \\
&= \frac{p(\mathbf{Y}_k)}{p(\mathbf{Y}_\ell)} p(\mathbf{y}_{k+1} \cdots \mathbf{y}_\ell / \mathbf{x}_k, \mathbf{x}_{k+1}, \mathbf{Y}_k) \\
&\quad \times p(\mathbf{x}_k, \mathbf{x}_{k+1} / \mathbf{Y}_k).
\end{aligned}
$$

Notice that

$$p(\mathbf{y}_{k+1} \cdots \mathbf{y}_\ell / \mathbf{x}_k, \mathbf{x}_{k+1}, \mathbf{Y}_k) = p(\mathbf{y}_{k+1} \cdots \mathbf{y}_\ell / \mathbf{x}_{k+1}), \tag{4.15.3}$$

and

$$
\begin{aligned}
p(\mathbf{x}_k, \mathbf{x}_{k+1} / \mathbf{Y}_k) &= p(\mathbf{x}_{k+1} / \mathbf{x}_k, \mathbf{Y}_k)\ p(\mathbf{x}_k / \mathbf{Y}_k) \\
&= p(\mathbf{x}_{k+1} / \mathbf{x}_k)\ p(\mathbf{x}_k / \mathbf{Y}_k), \tag{4.15.4}
\end{aligned}
$$

since knowledge of $\mathbf{x}_k$ and $\mathbf{Y}_k$ is redundant. Using Eqs. (4.15.3) and (4.15.4), Eq. (4.15.2) may be written

$$p\left(\mathbf{x}_k, \mathbf{x}_{k+1} / \mathbf{Y}_\ell\right) = \frac{p(\mathbf{Y}_k)}{p(\mathbf{Y}_\ell)} p\left(\mathbf{y}_{k+1} \cdots \mathbf{y}_\ell / \mathbf{x}_{k+1}\right)$$
$$\times p\left(\mathbf{x}_{k+1} / \mathbf{x}_k\right) p\left(\mathbf{x}_k / \mathbf{Y}_k\right). \tag{4.15.5}$$

The first three density functions on the right-hand side of Eq. (4.15.5) are independent of $\mathbf{x}_k$; hence, we need to be concerned only with $p(\mathbf{x}_{k+1}/\mathbf{x}_k)$ and $p(\mathbf{x}_k/\mathbf{Y}_k)$. Assuming that these are Gaussian and that the process noise is zero mean, it is easily shown that

$$p(\mathbf{x}_{k+1}/\mathbf{x}_k) \sim N\left(\Phi(t_{k+1}, t_k)\mathbf{x}_k, \Gamma(t_{k+1}, t_k)Q_k\Gamma^T(t_{k+1}, t_k)\right)$$
$$p(\mathbf{x}_k/\mathbf{Y}_k) \sim N(\hat{\mathbf{x}}_k, P_k). \tag{4.15.6}$$

It may seem like the covariance of $p(\mathbf{x}_{k+1}/\mathbf{x}_k)$ should be

$$\Phi(t_{k+1}, t_k)P_k\Phi^T(t_{k+1}, t_k) + \Gamma(t_{k+1}, t_k)Q_k\Gamma^T(t_{k+1}, t_k).$$

However, notice that this is the density function of $\mathbf{x}_{k+1}$ conditioned on knowledge of $\mathbf{x}_k$. Since P_k describes the error in $\hat{\mathbf{x}}_k$ and $\mathbf{x}_k$ has occurred and is known, P_k must be a null matrix.

In order to maximize the conditional probability density function given in Eq. (4.15.5), we may maximize the logarithm of $p\left(\mathbf{x}_{k+1} / \mathbf{x}_k\right) p\left(\mathbf{x}_k / \mathbf{Y}_k\right)$,

$$\ln L = -\frac{1}{2}[\mathbf{x}_{k+1} - \Phi\left(t_{k+1}, t_k\right)\mathbf{x}_k]^T[\Gamma\left(t_{k+1}, t_k\right)Q_k\Gamma^T(t_{k+1}, t_k)]^{-1}$$
$$\times [\mathbf{x}_{k+1} - \Phi\left(t_{k+1}, t_k\right)\mathbf{x}_k] - \frac{1}{2}[\mathbf{x}_k - \hat{\mathbf{x}}_k]^T P_k^{-1}[\mathbf{x}_k - \hat{\mathbf{x}}_k]. \tag{4.15.7}$$

Suppose that $\hat{\mathbf{x}}_{k+1}^\ell$, the maximizing value of $\mathbf{x}_{k+1}$ based on observations through t_ℓ, is available. We wish to find the value of $\hat{\mathbf{x}}_k^\ell$ that maximizes $\ln L$. Differentiating $\ln L$ with respect to $\mathbf{x}_k$ and setting this to zero yields (for simplicity we have dropped time identifiers on $\Phi(t_{k+1}, t_k)$ and $\Gamma(t_{k+1}, t_k)$)

$$\hat{\mathbf{x}}_k^\ell = [(P_k^k)^{-1} + \Phi^T(\Gamma Q \Gamma^T)^{-1}\Phi]^{-1}$$
$$\times [(P_k^k)^{-1}\hat{\mathbf{x}}_k^k + \Phi^T(\Gamma Q \Gamma^T)^{-1}\hat{\mathbf{x}}_{k+1}^\ell], \tag{4.15.8}$$

using our current notation,

$$P_k^k \equiv P_k$$
$$\hat{\mathbf{x}}_k^k \equiv \hat{\mathbf{x}}_k.$$

Applying the Schur identity we can write this in a more conventional form

$$\hat{\mathbf{x}}_k^\ell = \hat{\mathbf{x}}_k^k + S_k \left(\hat{\mathbf{x}}_{k+1}^\ell - \Phi(t_{k+1}, t_k) \hat{\mathbf{x}}_k^k \right) \tag{4.15.9}$$

where

$$\begin{aligned} S_k &= P_k^k \Phi^T(t_{k+1}, t_k) [\Phi(t_{k+1}, t_k) P_k^k \Phi^T(t_{k+1}, t_k) \\ &\quad + \Gamma(t_{k+1}, t_k) Q_k \Gamma^T(t_{k+1}, t_k)]^{-1} \\ &= P_k^k \Phi^T(t_{k+1}, t_k)(P_{k+1}^k)^{-1}. \end{aligned} \tag{4.15.10}$$

Eq. (4.15.9) is the smoothing algorithm. Computation goes backward in index k, with $\hat{\mathbf{x}}_\ell^\ell$, the filter solution, as initial conditions. Note that the filter solutions for $\hat{\mathbf{x}}_k^k$, P_k^k, $\Phi(t_{k+1}, t_k)$, and $\Gamma(t_{k+1}, t_k)$ are required and should be stored in the filtering process. The time update of the covariance matrix, P_{k+1}^k, may be stored or recomputed.

The equation for propagating the smoothed covariance is derived next (Jazwinski, 1970; Rausch et al., 1965) . It can easily be shown from Eq. (4.15.9) that $\hat{\mathbf{x}}_k^\ell$ is unbiased; hence, the smoothed covariance is defined by

$$P_k^\ell = E \left[(\hat{\mathbf{x}}_k^\ell - \mathbf{x}_k)(\hat{\mathbf{x}}_k^\ell - \mathbf{x}_k)^T \right]. \tag{4.15.11}$$

Subtracting $\mathbf{x}_k$ from both sides of Eq. (4.15.9) and moving all terms involving smoothed quantities to the LHS yields:

$$\tilde{\mathbf{x}}_k^\ell - S_k \hat{\mathbf{x}}_{k+1}^\ell = \tilde{\mathbf{x}}_k^k - S_k \Phi(t_{k+1}, t_k) \hat{\mathbf{x}}_k^k \tag{4.15.12}$$

where

$$\tilde{\mathbf{x}}_k^\ell \equiv \hat{\mathbf{x}}_k^\ell - \mathbf{x}_k, \quad \tilde{\mathbf{x}}_k^k \equiv \hat{\mathbf{x}}_k^k - \mathbf{x}_k.$$

Multiplying both sides of Eq. (4.15.12) by their respective transpose and taking the expected value yields

$$E \left(\tilde{\mathbf{x}}_k^\ell \tilde{\mathbf{x}}_k^{\ell T} \right) + S_k E \left(\hat{\mathbf{x}}_{k+1}^\ell \hat{\mathbf{x}}_{k+1}^{\ell T} \right) S_k^T = \tag{4.15.13}$$
$$E \left(\tilde{\mathbf{x}}_k^k \tilde{\mathbf{x}}_k^{k T} \right) + S_k \Phi(t_{k+1}, t_k) E \left(\hat{\mathbf{x}}_k^k \hat{\mathbf{x}}_k^{k T} \right) \Phi^T(t_{k+1}, t_k) S_k^T.$$

By definition

$$P_k^\ell \equiv E \left[\tilde{\mathbf{x}}_k^\ell \tilde{\mathbf{x}}_k^{\ell T} \right] \tag{4.15.14}$$

and

$$P_k^k \equiv E \left[\tilde{\mathbf{x}}_k^k \tilde{\mathbf{x}}_k^{k T} \right]. \tag{4.15.15}$$

The cross product terms that have been dropped in Eq. (4.15.13) can be shown to be null matrices,

$$
\begin{aligned}
E\left(\tilde{\mathbf{x}}_k^\ell \hat{\mathbf{x}}_{k+1}^{\ell T}\right) &= E\left(\hat{\mathbf{x}}_k^\ell - \mathbf{x}_k\right)\hat{\mathbf{x}}_{k+1}^{\ell T} \\
&= E\left(\hat{\mathbf{x}}_k^\ell \hat{\mathbf{x}}_{k+1}^{\ell T}\right) - E\left(\mathbf{x}_k \hat{\mathbf{x}}_{k+1}^{\ell T}\right) \\
&= \hat{\mathbf{x}}_k^\ell \hat{\mathbf{x}}_{k+1}^{\ell T} - E\left(\mathbf{x}_k/\mathbf{y}_\ell\right)\hat{\mathbf{x}}_{k+1}^{\ell T} = 0.
\end{aligned}
\tag{4.15.16}
$$

Here $\hat{\mathbf{x}}$ is the conditional mean of the appropriate conditional density function and is not a random variable; hence,

$$
E\left[\mathbf{x}_k/\mathbf{y}_\ell\right] \equiv \hat{\mathbf{x}}_k^\ell.
\tag{4.15.17}
$$

Likewise, for the filter results,

$$
\begin{aligned}
E\left(\tilde{\mathbf{x}}_k^k \hat{\mathbf{x}}_k^{k T}\right) &= E\left(\hat{\mathbf{x}}_k^k - \mathbf{x}_k\right)\hat{\mathbf{x}}_k^{k T} \\
&= \hat{\mathbf{x}}_k^k \hat{\mathbf{x}}_k^{k T} - E\left(\mathbf{x}_k/\mathbf{y}_k\right)\hat{\mathbf{x}}_k^{k T} = 0.
\end{aligned}
\tag{4.15.18}
$$

Hence, Eq. (4.15.13) becomes

$$
P_k^\ell + S_k \hat{\mathbf{x}}_{k+1}^\ell \hat{\mathbf{x}}_{k+1}^{\ell T} S_k^T = P_k^k + S_k \Phi(t_{k+1}, t_k)\hat{\mathbf{x}}_k^k \hat{\mathbf{x}}_k^{k T}\Phi^T(t_{k+1}, t_k)S_k^T.
\tag{4.15.19}
$$

Now,

$$
\begin{aligned}
P_{k+1}^\ell &\equiv E\left\{\left(\hat{\mathbf{x}}_{k+1}^\ell - \mathbf{x}_{k+1}\right)\left(\hat{\mathbf{x}}_{k+1}^\ell - \mathbf{x}_{k+1}\right)^T/\mathbf{y}_\ell\right\} \\
&= -\hat{\mathbf{x}}_{k+1}^\ell \hat{\mathbf{x}}_{k+1}^{\ell T} + E\left(\mathbf{x}_{k+1}\mathbf{x}_{k+1}^T\right)
\end{aligned}
$$

or

$$
\hat{\mathbf{x}}_{k+1}^\ell \hat{\mathbf{x}}_{k+1}^{\ell T} = E(\mathbf{x}_{k+1}\mathbf{x}_{k+1}^T) - P_{k+1}^\ell.
\tag{4.15.20}
$$

Also, in terms of $\mathbf{x}_k$ (let $G = \Phi(t_{k+1}, t_k)\mathbf{x}_k + \Gamma(t_{k+1}, t_k)\mathbf{u}_k$)

$$
\begin{aligned}
E(\mathbf{x}_{k+1}\mathbf{x}_{k+1}^T) &= E\left[GG^T\right] \\
&= \Phi(t_{k+1}, t_k)E(x_k, x_k^T)\Phi^T(t_{k+1}, t_k) \\
&\quad + \Gamma(t_{k+1}, t_k)Q_k \Gamma^T(t_{k+1}, t_k)
\end{aligned}
\tag{4.15.21}
$$

and

$$
\begin{aligned}
P_k^k &= E\left\{\left(\hat{\mathbf{x}}_k^k - \mathbf{x}_k\right)\left(\hat{\mathbf{x}}_k^k - \mathbf{x}_k\right)^T/\mathbf{y}_k\right\} \\
&= -\hat{\mathbf{x}}_k^k \hat{\mathbf{x}}_k^{k T} + E(\mathbf{x}_k\mathbf{x}_k^T)
\end{aligned}
$$

or

$$\hat{\mathbf{x}}_k^k \hat{\mathbf{x}}_k^{k^T} = E\left(\mathbf{x}_k \mathbf{x}_k^T\right) - P_k^k. \tag{4.15.22}$$

Substituting Eqs. (4.15.20), (4.15.21), and (4.15.22) into (4.15.19) yields

$$P_k^\ell + S_k \left[-P_{k+1}^\ell + \Gamma(t_{k+1}, t_k) Q_k \Gamma^T(t_{k+1}, t_k)\right] S_k^T$$
$$= P_k^k - S_k \Phi(t_{k+1}, t_k) P_k^k \Phi^T(t_{k+1}, t_k) S_k^T. \tag{4.15.23}$$

Finally,

$$P_k^\ell = P_k^k + S_k \left(P_{k+1}^\ell - P_{k+1}^k\right) S_k^T. \tag{4.15.24}$$

Note that neither the smoothed covariance nor the observation data appear explicitly in the smoothing algorithm. The algorithm derived previously is identical to the Rauch, Tung, and Striebel smoother (1965).

Suppose there is no process noise (i.e., $Q = 0$), then the smoothing algorithm reduces to

$$S_k = \Phi^{-1}(t_{k+1}, t_k), \tag{4.15.25}$$

and

$$\begin{aligned}
\hat{\mathbf{x}}_k^\ell &= \Phi^{-1}(t_{k+1}, t_k)\, \hat{\mathbf{x}}_{k+1}^\ell \\
&= \Phi(t_k, t_{k+1})\, \hat{\mathbf{x}}_{k+1}^\ell \\
&= \Phi(t_k, t_\ell)\hat{\mathbf{x}}_\ell^\ell.
\end{aligned} \tag{4.15.26}$$

Also,

$$\begin{aligned}
P_k^\ell &= \Phi^{-1}(t_{k+1}, t_k) P_{k+1}^\ell \Phi^{-T}(t_{k+1}, t_k) \\
&= \Phi(t_k, t_{k+1}) P_{k+1}^\ell \Phi^T(t_k, t_{k+1}) \\
&= \Phi(t_k, t_\ell) P_\ell^\ell \Phi^T(t_k, t_\ell).
\end{aligned} \tag{4.15.27} \tag{4.15.28}$$

Hence, with no process noise the smoothing algorithm simply maps the final filter state estimate and covariance matrix to earlier epochs.

4.15.1 COMPUTATIONAL ALGORITHM FOR SMOOTHER

Given (from the filtering algorithm)

$$\hat{\mathbf{x}}_\ell^\ell,\ \hat{\mathbf{x}}_{\ell-1}^{\ell-1},\ P_\ell^{\ell-1},\ P_{\ell-1}^{\ell-1},\ \Phi(t_\ell, t_{\ell-1});$$

set $k = \ell - 1$

$$\begin{aligned}
S_{\ell-1} &= P_{\ell-1}^{\ell-1}\, \Phi^T(t_\ell, t_{\ell-1}) \left(P_\ell^{\ell-1}\right)^{-1} \\
\hat{\mathbf{x}}_{\ell-1}^\ell &= \hat{\mathbf{x}}_{\ell-1}^{\ell-1} + S_{\ell-1}(\hat{\mathbf{x}}_\ell^\ell - \Phi(t_\ell, t_{\ell-1})\, \hat{\mathbf{x}}_{\ell-1}^{\ell-1}).
\end{aligned} \tag{4.15.29}$$

Given (from the filtering algorithm and the previous step of the smoothing algorithm)

$$\hat{\mathbf{x}}_{\ell-2}^{\ell-2}, \quad P_{\ell-1}^{\ell-2}, \quad P_{\ell-2}^{\ell-2}, \quad \hat{\mathbf{x}}_{\ell-1}^{\ell}, \quad \Phi(t_{\ell-1}, t_{\ell-2});$$

set $k = \ell - 2$, and compute

$$S_{\ell-2} = P_{\ell-2}^{\ell-2} \Phi^T (t_{\ell-1}, t_{\ell-2}) (P_{\ell-1}^{\ell-2})^{-1} \qquad (4.15.30)$$

$$\hat{\mathbf{x}}_{\ell-2}^{\ell} = \hat{\mathbf{x}}_{\ell-2}^{\ell-2} + S_{\ell-2} (\hat{\mathbf{x}}_{\ell-1}^{\ell} - \Phi (t_{\ell-1}, t_{\ell-2}) \hat{\mathbf{x}}_{\ell-2}^{\ell-2})$$

$$\vdots$$

and so on.

4.16 THE PROBABILITY ELLIPSOID

Given a normally distributed random vector, $\mathbf{x}$, with mean $\overline{\mathbf{x}}$, and variance-covariance P, the function

$$(\mathbf{x} - \overline{\mathbf{x}})^T P^{-1} (\mathbf{x} - \overline{\mathbf{x}}) = \ell^2 \qquad (4.16.1)$$

is a positive definite quadratic form representing a family of hyperellipsoids of constant probability density (Mikhail, 1976; Bryson and Ho, 1975). The 3D case is important because we often are interested in the 3D ellipsoids associated with the position uncertainty of a satellite. For example, in the case of interplanetary missions, we are interested in the probability ellipsoid of the spacecraft as it impacts the plane that contains the center of gravity of the target planet and that is normal to the asymptote of the spacecraft trajectory relative to the planet. This plane, shown in Fig. 4.16.1 and referred to as the B-plane (Dorroh and Thornton, 1970), is the reference plane used in interplanetary navigation applications. The associated coordinate system has two orthogonal axes in this plane and one normal to it. The axes of the ellipsoid in the B-plane give information on the uncertainty of the nearest approach distance to the planet that is needed to give assurance that the spacecraft will not impact the planet. The out-of-plane axis relates to the accuracy of the encounter time. If the spacecraft is to be placed in orbit about the target planet, information from the covariance matrix used to generate the probability ellipsoid is used to design the capture maneuver, and to compute the *a priori* uncertainty of the resulting orbit parameters. Construction of the probability ellipsoid is most easily accomplished relative to the principal axes. To this end, we introduce the following theorem (Kreyszig, 1993).

Theorem: If $\mathbf{u}_1, \mathbf{u}_2, \dots, \mathbf{u}_n$ is an orthonormal system of eigenvectors associated, respectively, with the eigenvalues $\lambda_1, \lambda_2, \dots, \lambda_n$ of an $n \times n$ symmetric positive definite matrix, P, and if

$$U = [\mathbf{u}_1, \mathbf{u}_2, \dots, \mathbf{u}_n]_{n \times n},$$

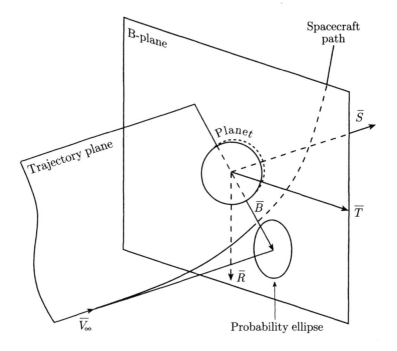

Figure 4.16.1: Probability ellipse on the B-plane.

then

$$U^T P U = \begin{bmatrix} \lambda_1 & 0 & \cdots & 0 \\ 0 & \lambda_2 & \cdots & 0 \\ \vdots & & \ddots & \vdots \\ 0 & 0 & \cdots & \lambda_n \end{bmatrix} = D\,[\lambda_1,\,\lambda_2,\dots,\,\lambda_n]; \qquad (4.16.2)$$

that is, $U^T P U$ is a diagonal matrix containing the eigenvalues of P. The normalized vectors $\mathbf{u}_1, \mathbf{u}_2, \dots, \mathbf{u}_n$ are called principal axes of P, and the transformation matrix U^T used to diagonalize P is called a *principal axes transformation*. The matrix U^T is the matrix of normalized eigenvectors and is orthogonal. For the random vector $\mathbf{x}$ with mean $\bar{\mathbf{x}}$ and variance-covariance, P, the principal axes, $\mathbf{x}'$, are given by

$$\mathbf{x}' = U^T \mathbf{x}. \qquad (4.16.3)$$

The variance-covariance matrix, P', associated with the principal axes is given by

$$\begin{aligned} P' &\equiv E[(\mathbf{x}' - \bar{\mathbf{x}}')(\mathbf{x}' - \bar{\mathbf{x}}')^T] \\ &= U^T E[(\mathbf{x} - \bar{\mathbf{x}})(\mathbf{x} - \bar{\mathbf{x}})^T]\,U \\ &= U^T P U = D\,[\lambda_1 \dots \lambda_n]. \end{aligned} \qquad (4.16.4)$$

In our case, $\Delta\mathbf{x}$ represents the estimation error vector defined by

$$\Delta\mathbf{x} \equiv \hat{\mathbf{x}} - \mathbf{x} \equiv [\tilde{x}\ \tilde{y}\ \tilde{z}]^T,$$

with zero mean and variance-covariance given by

$$P = E[\Delta\mathbf{x}\Delta\mathbf{x}^T]. \tag{4.16.5}$$

Although this restriction is unnecessary, we will simplify matters and deal only with the three position coordinates of $\Delta\mathbf{x}$ and the associated 3×3 portion of the estimation error covariance matrix.

In this case, the equation for the probability ellipsoid is

$$[\tilde{x}\ \tilde{y}\ \tilde{z}]P^{-1}\begin{bmatrix}\tilde{x}\\\tilde{y}\\\tilde{z}\end{bmatrix} = \ell^2. \tag{4.16.6}$$

The ellipsoids for $\ell = 1, 2,$ and 3 are called the 1σ, 2σ, and 3σ error ellipsoids. The probability of the state estimate error falling inside these ellipsoids, assuming a trivariate Gaussian density function, is 0.200, 0.739, and 0.971, respectively.

To obtain the principal axes, we use the theorem just introduced and determine the matrix of normalized eigenvectors, U, and the eigenvalues λ_i, $i = 1, 2, 3$ of P. The principal axes are given by

$$\begin{bmatrix}\tilde{x}'\\\tilde{y}'\\\tilde{z}'\end{bmatrix} = U^T\begin{bmatrix}\tilde{x}\\\tilde{y}\\\tilde{z}\end{bmatrix}, \tag{4.16.7}$$

and the associated covariance matrix is

$$P' = U^T P U.$$

The probability ellipsoids are given by

$$[\tilde{x}'\ \tilde{y}'\ \tilde{z}']\begin{bmatrix}1/\lambda_1 & & \\ & 1/\lambda_2 & \\ & & 1/\lambda_3\end{bmatrix}\begin{bmatrix}\tilde{x}'\\\tilde{y}'\\\tilde{z}'\end{bmatrix} = \ell^2, \tag{4.16.8}$$

or

$$\frac{\tilde{x}'^2}{\lambda_1} + \frac{\tilde{y}'^2}{\lambda_2} + \frac{\tilde{z}'^2}{\lambda_3} = \ell^2. \tag{4.16.9}$$

It is convenient to arrange the eigenvectors so that $\lambda_1 > \lambda_2 > \lambda_3$ (i.e., in order of descending values of λ_i). The axes of the 1σ ellipsoid are given by solving Eq.

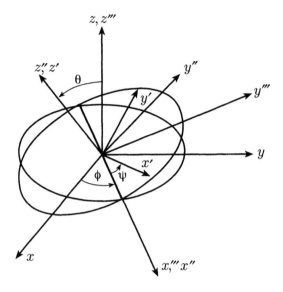

Figure 4.16.2: Euler angles defined.

(4.16.9) for $\ell = 1$ and sequentially setting two of the coordinate values to zero; that is, to obtain the semimajor axis, a, set $\tilde{x}' = a$ and $\tilde{y}' = \tilde{z}' = 0$. This yields

$$a^2 = \lambda_1, b^2 = \lambda_2, c^2 = \lambda_3. \tag{4.16.10}$$

The orientation of the ellipse relative to the original axis system is obtained by solving for the three Euler angles given by the transformation matrix, U. It is convenient to choose the sign of the normalized eigenvectors so that U defines a right-hand triad. This can be accomplished by requiring that $\mathbf{u}_1 \times \mathbf{u}_2 = \mathbf{u}_3$.

The Euler angles are defined in Fig. 4.16.2. The first rotation is about the z axis through the angle ϕ and takes the unprimed frame into the ( )$'''$ frame. The second rotation is about the x''' axis through the angle θ to the ( )$''$ frame, and the final rotation is about the z'' axis through the angle ψ to the ( )$'$ or principal axes frame (Reddy and Rasmussen, 1990). This transformation is analogous to that of transforming Earth-centered, Earth-fixed coordinates into orbital plane coordinates with the x' axis through perigee, the z' axis along the angular momentum vector, and the y' axis completing the right-hand triad. The transpose of the transformation matrix that accomplishes this is the matrix we have identified as U; that is, $\mathbf{x}' = U^T \mathbf{x}$ where

$$U = \begin{bmatrix} C_\phi C_\psi - C_\theta S_\phi S_\psi & -C_\phi S_\psi - C_\theta S_\phi C_\psi & S_\theta S_\phi \\ S_\phi C_\psi + C_\theta C_\phi S_\psi & -S_\phi S_\psi + C_\theta C_\phi C_\psi & -S_\theta C_\phi \\ S_\theta S_\psi & S_\theta C_\psi & C_\theta \end{bmatrix}, \quad (4.16.11)$$

and where C and S represent cosine and sine, respectively. The Euler angles are given by

$$\phi = \text{atan2} \left[\frac{U_{13}}{-U_{23}} \right], \quad 0 \le \phi \le 2\pi \qquad (4.16.12)$$

$$\theta = \text{acos} \left[U_{33} \right], \quad 0 \le \theta \le \pi \qquad (4.16.13)$$

$$\psi = \text{atan2} \left[\frac{U_{31}}{U_{32}} \right], \quad 0 \le \psi \le 2\pi. \qquad (4.16.14)$$

Example: Consider a normally distributed 2D random vector, $\mathbf{x}$, where

$$\mathbf{x} \sim N(0, P),$$

and

$$P = \begin{bmatrix} 4 & 2 \\ 2 & 2 \end{bmatrix}.$$

Sketch the 1-, 2-, and 3-σ probability ellipses.

The eigenvalues are given by the polynomial

$$|P - \lambda I| = 0,$$

or

$$\begin{vmatrix} 4 - \lambda & 2 \\ 2 & 2 - \lambda \end{vmatrix} = 0;$$

hence,

$$\lambda^2 - 6\lambda + 4 = 0,$$

and

$$\lambda_1 = 5.236, \quad \lambda_2 = 0.764.$$

The corresponding eigenvectors are given by

$$[P - \lambda_i I] \begin{bmatrix} x_1 \\ x_2 \end{bmatrix} = 0, \quad i = 1, 2.$$

The normalized eigenvectors are given by

$$U = \begin{bmatrix} .851 & -.526 \\ .526 & .851 \end{bmatrix}.$$

The angle between the principal and original axes system is obtained by recognizing that the coordinate transformation matrix is given by (this can be determined by setting $\theta = \psi = 0$ in Eq. (4.16.11))

$$U = \begin{bmatrix} \cos\phi & -\sin\phi \\ \sin\phi & \cos\phi \end{bmatrix}.$$

Hence,

$$\phi = \tan^{-1}\frac{\sin\phi}{\cos\phi} = 31.7°$$

where ϕ is the angle between the x_1 and x_1' axes.

The semimajor axes a_i and minor axes b_i are given by

$$a_i = \sqrt{\ell_i^2\lambda\,(\text{Max})}$$

$$i = 1, 2, 3;\quad \ell_1 = 1,\ \ell_2 = 2,\ \ell_3 = 3$$

$$b_i = \sqrt{\ell_i^2\lambda(\text{Min})}$$

with numerical values,

$$\begin{aligned} a_1 &= 2.29, & a_2 &= 4.58, & a_3 &= 6.86 \\ b_1 &= 0.87, & b_2 &= 1.75, & b_3 &= 2.62 \end{aligned}.$$

The error ellipses can now be constructed:

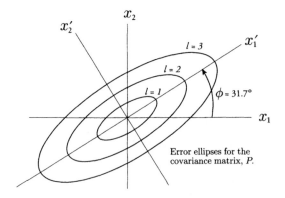

Error ellipses for the covariance matrix, P.

4.16.1 TRANSFORMATION OF THE COVARIANCE MATRIX BETWEEN COORDINATE SYSTEMS

Sometimes it is desirable to transform the state vector and the estimation error covariance into alternate coordinate systems. For example, it may be of interest to view these quantities in a radial, transverse, and normal (RTN) system. Here the transverse direction is normal to the radius vector and the normal direction lies along the instantaneous angular momentum vector. RTN forms a right-hand triad.

The general transformation between any two coordinate frames (say prime to unprime) for a position vector is given by

$$\mathbf{r} = \gamma \mathbf{r}' + \mathbf{a}, \tag{4.16.15}$$

where γ is an orthogonal transformation matrix, $\mathbf{r}$ is the vector in the unprimed frame, and $\mathbf{a}$ is the vector offset of the origin of the two systems expressed in the unprimed frame. Generally $\mathbf{a}$ will be zero unless, for example, we are transforming from a geocentric to a topocentric frame.

The velocity transforms according to

$$\dot{\mathbf{r}} = \gamma \dot{\mathbf{r}}' + \dot{\gamma} \mathbf{r}'. \tag{4.16.16}$$

Generally $\dot{\gamma}$ will be zero unless we are transforming from a rotating to a nonrotating frame or vice versa; for example, Earth-centered-Earth-fixed to Earth-centered inertial (ECEF to ECI). Let the unprimed system be the inertial, nonrotating frame. It can be shown (Wiesel, 1997) that $\dot{\gamma} \mathbf{r}' = \boldsymbol{\omega} \times \mathbf{r}$, where $\boldsymbol{\omega}$ is the angular velocity vector of the rotating frame expressed in the nonrotating frame coordinate system.

The transformation we want is ECI to RTN. We assume that the RTN frame is fixed to the osculating orbit at each point in time; hence, $\dot{\gamma} = 0$ and

$$\begin{bmatrix} \mathbf{r} \\ \mathbf{v} \end{bmatrix}_{\mathrm{RTN}} = \begin{bmatrix} \gamma & 0 \\ 0 & \gamma \end{bmatrix} \begin{bmatrix} \mathbf{r} \\ \mathbf{v} \end{bmatrix}_{\mathrm{ECI}}. \tag{4.16.17}$$

The covariance of the estimation error is transformed as follows:

$$[\hat{\mathbf{x}} - \mathbf{x}]_{\mathrm{RTN}} = \psi \, [\hat{\mathbf{x}} - \mathbf{x}]_{\mathrm{ECI}},$$

where

$$\psi = \begin{bmatrix} \gamma & 0 & 0 \\ 0 & \gamma & 0 \\ 0 & 0 & I \end{bmatrix} \quad \text{and} \quad [\hat{\mathbf{x}} - \mathbf{x}] = \begin{bmatrix} \hat{\mathbf{r}} - \mathbf{r} \\ \hat{\mathbf{v}} - \mathbf{v} \\ \hat{\boldsymbol{\beta}} - \boldsymbol{\beta} \end{bmatrix}, \tag{4.16.18}$$

and $\mathbf{r}$, $\mathbf{v}$, and β represent the true values of the position, velocity, and all other quantities in the state vector, respectively. It is assumed that none of the elements of β are affected by the coordinate transformation. The desired covariance is given by

$$P_{\mathrm{RTN}} = E\left[(\hat{\mathbf{x}} - \mathbf{x})(\hat{\mathbf{x}} - \mathbf{x})^T\right]_{\mathrm{RTN}}$$
$$= \psi E\left[(\hat{\mathbf{x}} - \mathbf{x})(\hat{\mathbf{x}} - \mathbf{x})^T\right]_{\mathrm{ECI}} \psi^T. \qquad (4.16.19)$$

The elements of γ for the ECI to RTN transformation are given by

$$\mathbf{u}_R = \frac{\mathbf{r}^*}{|\mathbf{r}^*|} = \epsilon_X \mathbf{i} + \epsilon_Y \mathbf{j} + \epsilon_Z \mathbf{k}$$
$$\mathbf{u}_T = \mathbf{u}_N \times \mathbf{u}_R = \delta_X \mathbf{i} + \delta_Y \mathbf{j} + \delta_Z \mathbf{k} \qquad (4.16.20)$$
$$\mathbf{u}_N = \frac{\mathbf{r}^* \times \mathbf{v}^*}{|\mathbf{r}^* \times \mathbf{v}^*|} = \alpha_X \mathbf{i} + \alpha_Y \mathbf{j} + \alpha_Z \mathbf{k},$$

where $\mathbf{u}_R$, $\mathbf{u}_T$, $\mathbf{u}_N$ are unit vectors in the RTN frame, $\mathbf{i}$, $\mathbf{j}$, and $\mathbf{k}$ are unit vectors in the ECI frame, and $\mathbf{r}^*$ and $\mathbf{v}^*$ are the position and velocity vectors of the reference orbit.

Equation (4.16.20) may be written

$$\begin{bmatrix} \mathbf{u}_R \\ \mathbf{u}_T \\ \mathbf{u}_N \end{bmatrix} = \begin{bmatrix} \epsilon_X & \epsilon_Y & \epsilon_Z \\ \delta_X & \delta_Y & \delta_Z \\ \alpha_X & \alpha_Y & \alpha_Z \end{bmatrix} \begin{bmatrix} \mathbf{i} \\ \mathbf{j} \\ \mathbf{k} \end{bmatrix}. \qquad (4.16.21)$$

Hence, the transformation matrix relating the RTN and ECI frame is

$$\gamma = \begin{bmatrix} \epsilon_X & \epsilon_Y & \epsilon_Z \\ \delta_X & \delta_Y & \delta_Z \\ \alpha_X & \alpha_Y & \alpha_Z \end{bmatrix}. \qquad (4.16.22)$$

4.17 COMBINING ESTIMATES

Assume we are given two unbiased and uncorrelated estimates $\hat{\mathbf{x}}_1$ and $\hat{\mathbf{x}}_2$ for the n-vector $\mathbf{x}$. Assume that the associated estimation errors η_1 and η_2 are Gaussian with covariance matrices P_1 and P_2. Our objective is to establish a performance index and determine the combination of $\hat{\mathbf{x}}_1$ and $\hat{\mathbf{x}}_2$ that is an optimal, unbiased $(E[\hat{\mathbf{x}}] = \mathbf{x})$ estimate of $\mathbf{x}$.

Using the fact that the errors in $\hat{\mathbf{x}}_1$ and $\hat{\mathbf{x}}_2$ are zero mean and uncorrelated (hence, being Gaussian, they are independent), the joint density function for the estimation errors, $\boldsymbol{\eta}_1$ and $\boldsymbol{\eta}_2$, where

$$\boldsymbol{\eta}_1 = \hat{\mathbf{x}}_1 - \mathbf{x} \qquad (4.17.1)$$
$$\boldsymbol{\eta}_2 = \hat{\mathbf{x}}_2 - \mathbf{x},$$

is given by

$$f(\boldsymbol{\eta}_1, \boldsymbol{\eta}_2) = f(\boldsymbol{\eta}_1)f(\boldsymbol{\eta}_2)$$

$$= \frac{1}{(2\pi)^{n/2}} \frac{1}{|P_1|^{\frac{1}{2}}} e^{-\frac{1}{2}(\boldsymbol{\eta}_1^T P_1^{-1} \boldsymbol{\eta}_1)}$$

$$\times \frac{1}{(2\pi)^{n/2}|P_2|^{\frac{1}{2}}} e^{-\frac{1}{2}(\boldsymbol{\eta}_2^T P_2^{-1} \boldsymbol{\eta}_2)}. \qquad (4.17.2)$$

The method of Maximum Likelihood selects the value of $\mathbf{x}$ that maximizes the likelihood function $L = f(\boldsymbol{\eta}_1, \boldsymbol{\eta}_2)$. Because the density functions involved are Gaussian, all classical estimation techniques, Bayesian, Minimum Variance, or Maximum Likelihood, will yield the same results.

Maximizing the logarithm of L is equivalent to maximizing L. Hence, we wish to maximize

$$\ln L = K - 1/2 \left[\boldsymbol{\eta}_1^T P_1^{-1} \boldsymbol{\eta}_1 + \boldsymbol{\eta}_2^T P_2^{-1} \boldsymbol{\eta} \right]$$
$$= K - 1/2 \left[(\hat{\mathbf{x}}_1 - \mathbf{x})^T P_1^{-1} (\hat{\mathbf{x}}_1 - \mathbf{x}) \right.$$
$$\left. + (\hat{\mathbf{x}}_2 - \mathbf{x})^T P_2^{-1} (\hat{\mathbf{x}}_2 - \mathbf{x}) \right] \qquad (4.17.3)$$

where

$$K = \ln \left(\frac{1}{(2\pi)^{n/2}} \frac{1}{|P_1|^{1/2}} \right) + \ln \left(\frac{1}{(2\pi)^{n/2}} \frac{1}{|P_2|^{1/2}} \right).$$

Using Eq. (B.7.3) of Appendix B, for a maximum of $\ln L$, it is necessary that

$$\frac{d \ln L}{d\mathbf{x}} = -\frac{1}{2} \left[-2P_1^{-1}(\hat{\mathbf{x}}_1 - \mathbf{x}) - 2P_2^{-1}(\hat{\mathbf{x}}_2 - \mathbf{x}) \right] = 0.$$

Hence,

$$P_1^{-1}(\hat{\mathbf{x}}_1 - \hat{\mathbf{x}}) + P_2^{-1}(\hat{\mathbf{x}}_2 - \hat{\mathbf{x}}) = 0$$

and

$$\hat{\mathbf{x}} = (P_1^{-1} + P_2^{-1})^{-1} [P_1^{-1} \hat{\mathbf{x}}_1 + P_2^{-1} \hat{\mathbf{x}}_2]. \qquad (4.17.4)$$

Also,

$$\frac{d^2 \ln L}{d\mathbf{x}^2} = -(P_1^{-1} + P_2^{-1}) \qquad (4.17.5)$$

which is negative definite; therefore, this is a maximum of $\ln L$.

It is easily shown that $\hat{\mathbf{x}}$ is unbiased,

$$E[\hat{\mathbf{x}}] = (P_1 + P_2)^{-1}[P_1^{-1} E(\hat{\mathbf{x}}_1) + P_2^{-1} E(\hat{\mathbf{x}}_2)]$$
$$= \mathbf{x},$$

since $E(\hat{\mathbf{x}}_1) = E(\hat{\mathbf{x}}_2) = \mathbf{x}$.

The estimation error covariance, P, associated with $\hat{\mathbf{x}}$ is given by

$$P = E[(\hat{\mathbf{x}} - \mathbf{x})(\hat{\mathbf{x}} - \mathbf{x})^T], \tag{4.17.6}$$

where

$$\hat{\mathbf{x}} = \gamma[P_1^{-1}\hat{\mathbf{x}}_1 + P_2^{-1}\hat{\mathbf{x}}_2],$$

and

$$\gamma \equiv (P_1^{-1} + P_2^{-1})^{-1}.$$

Using Eqs. (4.17.1) and the fact that

$$E[\boldsymbol{\eta}_1] = E[\boldsymbol{\eta}_2] = 0,$$

$$E[\boldsymbol{\eta}_1 \boldsymbol{\eta}_1^T] = P_1, \; E[\boldsymbol{\eta}_2 \boldsymbol{\eta}_2^T] = P_2, \; E[\boldsymbol{\eta}_1 \boldsymbol{\eta}_2^T] = 0, \tag{4.17.7}$$

we can write

$$\hat{\mathbf{x}} = \gamma[P_1^{-1}(\mathbf{x} + \boldsymbol{\eta}_1) + P_2^{-1}(\mathbf{x} + \boldsymbol{\eta}_2)].$$

Hence,

$$\hat{\mathbf{x}} - \mathbf{x} = \gamma[P_1^{-1}\boldsymbol{\eta}_1 + P_2^{-1}\boldsymbol{\eta}_2], \tag{4.17.8}$$

therefore,

$$\begin{aligned}
P &= \gamma E\left\{[P_1^{-1}\boldsymbol{\eta}_1 + P_2^{-1}\boldsymbol{\eta}_2][P_1^{-1}\boldsymbol{\eta}_1 + P_2^{-1}\boldsymbol{\eta}_2]^T\right\}\gamma \\
&= \gamma[P_1^{-1}P_1 P_1^{-1} + P_2^{-1}P_2 P_2^{-1}]\gamma \\
&= \gamma \\
&= (P_1^{-1} + P_2^{-1})^{-1}. \tag{4.17.9}
\end{aligned}$$

It is not necessary to assume that the errors in $\hat{\mathbf{x}}_1$ and $\hat{\mathbf{x}}_2$ are Gaussian. Knowing that the two estimates $\hat{\mathbf{x}}_1$ and $\hat{\mathbf{x}}_2$ are unbiased and uncorrelated, we could have simply chosen to minimize a performance index that yields a value of $\mathbf{x}$ that minimizes the weighted sum of squares of the estimation errors for $\hat{\mathbf{x}}_1$ and $\hat{\mathbf{x}}_2$,

$$Q = \frac{1}{2}[(\hat{\mathbf{x}}_1 - \mathbf{x})^T P_1^{-1}(\hat{\mathbf{x}}_1 - \mathbf{x}) + (\hat{\mathbf{x}}_2 - \mathbf{x})^T P_2^{-1}(\hat{\mathbf{x}}_2 - \mathbf{x})]. \tag{4.17.10}$$

Choosing $\mathbf{x}$ to minimize Q will yield the result given by Eq. (4.17.4).

If there are n independent solutions to combine, it is easily shown that

$$\hat{\mathbf{x}} = \left(\sum_{i=1}^{n} P_i^{-1} \right)^{-1} \sum_{i=1}^{n} P_i^{-1} \hat{\mathbf{x}}_i. \qquad (4.17.11)$$

4.18 REFERENCES

Battin, R., *An Introduction to the Mathematics and Methods of Astrodynamics*, American Institute of Aeronautics and Astronautics, Reston, VA, 1999.

Bertiger, W., Y. Bar-Sever, E. Christensen, E. Davis, J. Guinn, B. Haines, R. Ibanez-Meier, J. Jee, S. Lichten, W. Melbourne, R. Muellerschoen, T. Munson, Y. Vigue, S. Wu, T. Yunck, B. Schutz, P. Abusali, H. Rim, W. Watkins, and P. Willis , "GPS precise tracking of TOPEX/Poseidon: Results and implications," *J. Geophys. Res.*, Vol. 99, No. C12, pp. 24,449–24,462, December 15, 1995.

Bierman, G. J., *Factorization Methods for Discrete Sequential Estimation*, Academic Press, New York, 1977.

Björck, A., *Numerical Methods for Least Squares Problems*, SIAM, Philadelphia, PA, 1996.

Bryson, A. E., and Y. C. Ho, *Applied Optimal Control*, Hemisphere Publishing Corp., Washington, DC, 1975.

Bucy, R., and P. Joseph, *Filtering for Stochastic Processes*, John Wiley & Sons, Inc., New York, 1968.

Cruickshank, D. R., *Genetic Model Compensation: Theory and Applications*, Ph. D. Dissertation, The University of Colorado at Boulder, 1998.

Dahlquist, G., and A. Björck, *Numerical Methods*, Prentice-Hall, Englewood Cliffs, NJ, 1974 (translated to English: N. Anderson).

Desai, S. D., and B. J. Haines, "Near real-time GPS-based orbit determination and sea surface height observations from the Jason-1 mission," *Marine Geodesy*, Vol. 26, No. 3–4, pp. 187–199, 2003.

Deutsch, R., *Estimation Theory*, Prentice-Hall, Inc., Englewood Cliffs, NJ, 1965.

Dorroh, W. E., and T. H. Thornton, "Strategies and systems for navigation corrections," *Astronautics and Aeronautics*, Vol. 8, No. 5, pp. 50–55, May 1970.

Fisher, R. A., "On an absolute criteria for fitting frequency curves," *Mess. Math.*, Vol. 41, pp. 155–160, 1912.

Gauss, K. F., *Theoria Motus Corporum Coelestium*, 1809 (Translated into English: Davis, C. H., *Theory of the Motion of the Heavenly Bodies Moving about the Sun in Conic Sections*, Dover, New York, 1963).

Gelb, A. (ed.), *Applied Optimal Estimation*, MIT Press, Cambridge, MA, 1974.

Goldstein, D. B., G. H. Born, and P. Axelrad, "Real-time, autonomous, precise orbit determination using GPS," *J. ION*, Vol. 48, No. 3, pp. 169–179, Fall 2001.

Grewal, M. S., and A. P. Andrews, *Kalman Filtering: Theory and Practice*, Prentice Hall, 1993.

Ingram, D. S., *Orbit determination in the presence of unmodeled accelerations*, Ph.D. Dissertation, The University of Texas at Austin, 1970.

Ingram, D. S., and B. D. Tapley, "Lunar orbit determination in the presence of unmodeled accelerations," *Celest. Mech.*, Vol. 9, No. 2, pp. 191–211, 1974.

Jazwinski, A. H., *Stochastic Process and Filtering Theory*, Academic Press, New York, 1970.

Kalman, R. E., "A New Approach to Linear Filtering and Prediction Theory," *J. Basic Eng.*, Vol. 82, Series E, No. 1, pp. 35–45, March, 1960.

Kreyszig, E., *Advanced Engineering Mathematics*, John Wiley & Sons, Inc., New York, 1993.

Lawson, C. L., and R. J. Hanson, *Solving Least Squares Problems*, Prentice-Hall, Inc. Englewood Cliffs, NJ, 1974 (republished by SIAM, Philadelphia, PA, 1995).

Lichten, S. M., "Estimation and filtering for high precision GPS positioning applications," *Manuscripta Geodaetica*, Vol. 15, pp. 159–176, 1990.

Liebelt, P. B., *An Introduction to Optimal Estimation*, Addison-Wesley, Reading, MA, 1967.

Lutchke, S. B., N. P. Zelenski, D. D. Rowlands, F. G. Lemoine, and T. A. Williams, "The 1-centimeter orbit: Jason-1 precision orbit determination using GPS, SLR, DORIS and altimeter data," *Marine Geodesy*, Vol. 26, No. 3-4, pp. 399–421, 2003.

Marshall, J. A., N. P. Zelensky, S. M. Klosko, D. S. Chinn, S. B. Luthcke, K. E. Rachlin, and R. G. Williamson, "The temporal and spatial characteristics of TOPEX/Poseidon radial orbit error," *J. Geophys. Res.*, Vol. 99, No. C12, pp. 25,331–25,352, December 15, 1995.

Maybeck, P. S., *Stochastic Models, Estimation and Control*, Vol. 1, Academic Press, 1979.

Mikhail, E. M., *Observations and Least Squares*, University Press of America, Lanham, MD, 1976.

Moler, C., and C. Van Loan, "Nineteen dubious ways to compute the exponential of a matrix," *SIAM Review*, Vol. 20, No. 4, pp. 801–836, October 1978.

Montenbruck, O., and E. Gill, *Satellite Orbits: Models, Methods and Applications*, Springer, 2001.

Myers, K. A., *Filtering theory methods and applications to the orbit determination problem for near-Earth satellites*, Ph.D. Dissertation, The University of Texas at Austin, November 1973.1973.

Rausch, H. E., F. Tung, and C. T. Striebel, "Maximum likelihood estimates of linear dynamic systems," *AIAA J.*, Vol. 3, No. 7, pp. 1445–1450, August 1965.

Reddy, J. N., and M. L. Rasmussen, *Advanced Engineering Analysis*, Robert E. Krieger Publishing Co., Malabar, Florida, 1990.

Schlee, F. H., C. J. Standish, and N. F. Toda, "Divergence in the Kalman filter," *AIAA J.*, Vol. 5, No. 6, pp. 1114–1120, June 1967.

Schutz, B., B. D. Tapley, P. Abusali, H. Rim, "Dynamic orbit determination using GPS measurements from TOPEX/Poseidon," *Geophys. Res. Ltrs.*, Vol. 21, No. 19, pp. 2179–2182, 1994.

Sorenson, H. W. (ed.), *Kalman Filtering: Theory and Applications*, IEEE Press, 1985.

Swerling, P., "First order error propagation in a stagewise differential smoothing procedure for satellite observations," *J. Astronaut. Sci.*, Vol. 6, pp. 46–52, 1959.

Tapley, B. D., "Statistical orbit determination theory," in *Recent Advances in Dynamical Astronomy*, B. D. Tapley and V. Szebehely (eds.), D. Reidel, pp. 396–425, 1973.

Tapley, B. D.," Fundamentals of orbit determination", in *Theory of Satellite Geodesy and Gravity Field Determination*, Vol. 25, pp. 235-260, Springer-Verlag, 1989.

Tapley, B. D., and D. S. Ingram, "Orbit determination in the presence of unmodeled accelerations", *Trans. Auto. Cont.*, Vol. AC-18, No. 4, pp. 369–373, August, 1973.

Tapley, B. D., J. Ries, G. Davis, R. Eanes, B. Schutz, C. Shum, M. Watkins, J. Marshall, R. Nerem, B. Putney, S. Klosko, S. Luthcke, D. Pavlis, R. Williamson, and N. P. Zelensky, "Precision orbit determination for TOPEX/Poseidon," *J. Geophys. Res.*, Vol. 99, No. C12, pp. 24,383–24,404, December 15, 1994.

Tapley, B. D., M. Watkins, J. Ries, G. Davis, R. Eanes, S. Poole, H. Rim, B. Schutz, C. Shum, R. Nerem, F. Lerch, J. Marshall, S. Klosko, N. Pavlis, and R. Williamson, "The Joint Gravity Model 3", *J. Geophys. Res.*, Vol. 101, No. B12, pp. 28,029–28,049, December 10, 1996.

Walpole, R. E., R. H. Myers, S. L. Myers, and Y. Keying, *Probability and Statistics for Engineers and Scientists*, Prentice Hall, Englewood Cliffs, NJ, 2002.

Wiesel, W. E., *Spaceflight Dynamics*, McGraw-Hill, 1997.

4.19 EXERCISES

(1) A dynamical system is described by

$$\dot{\mathbf{X}}(t) = A\mathbf{X}(t).$$

Given that the state transition matrix for this system is

$$\Phi = \begin{bmatrix} e^{-2at} & 0 \\ 0 & e^{bt} \end{bmatrix},$$

determine the matrix A.

(2) Given the solution to the differential equation

$$\dot{\mathbf{x}}(t_i) = A(t_i)\mathbf{x}(t_i)$$

is

$$\mathbf{x}(t_i) = \Phi(t_i, t_k)\mathbf{x}(t_k)$$

where

$$\Phi(t_k, t_k) = I,$$

show that

(a) $\dot{\Phi}(t_i, t_k) = A(t_i)\Phi(t_i, t_k)$

(b) $\Phi(t_i, t_j) = \Phi(t_i, t_k)\Phi(t_k, t_j)$

(c) $\Phi^{-1}(t_i, t_k) = \Phi(t_k, t_i)$

(d) $\dot{\Phi}^{-1}(t_i, t_k) = -\Phi^{-1}(t_i, t_k)A(t_i).$

(3) Given a vector of observations, $\mathbf{y} = H\mathbf{x} + \boldsymbol{\epsilon}$ with weighting matrix W, and *a priori* information, $(\bar{\mathbf{x}}, \overline{W})$, determine the least squares estimate for $\hat{\mathbf{x}}$. (Note that W corresponds to R^{-1} and $\overline{W}$ to $\bar{P}_0^{-1}$ in the case where we have statistical information on $\boldsymbol{\epsilon}$ and $\bar{\mathbf{x}}$.) Let the performance index be

$$J(x) = 1/2\ \boldsymbol{\epsilon}^T W \boldsymbol{\epsilon} + 1/2\ \overline{\boldsymbol{\eta}}^T \overline{W} \overline{\boldsymbol{\eta}}$$

where $\overline{\boldsymbol{\eta}}$ is the error in the *a priori* estimate $\bar{\mathbf{x}}$,

$$\overline{\boldsymbol{\eta}} = \bar{\mathbf{x}} - \mathbf{x}.$$

Answer: $\hat{\mathbf{x}} = (H^T W H + \overline{W})^{-1}(H^T W \mathbf{y} + \overline{W}\bar{\mathbf{x}})$

(4) Determine the state transition matrix associated with the matrix

$$A = \begin{bmatrix} a & 0 \\ b & g \end{bmatrix}, \quad a \neq g, \text{ and } \dot{\Phi} = A\Phi, \ \Phi(t_0, t_0) = I.$$

(5) Express the linear system of equations

$$\ddot{x} = -abx$$

in the matrix form

$$\dot{\mathbf{x}} = A\mathbf{x},$$

where

$$\mathbf{x} = \begin{bmatrix} x \\ \dot{x} \end{bmatrix}.$$

Find the state transition matrix for this system.

(6) Show that the matrix

$$\Phi\,(t, t_0) = \begin{bmatrix} 3e^{at} & 0 \\ 0 & 2e^{-bt} \end{bmatrix}$$

satisfies the relation

$$\dot{\Phi} = A\Phi$$

but that $\Phi(t, t_0)$ is not a transition matrix. (Assume $t_0 = 0$.)

(7) Show that whenever $\Phi(t, t_k)$ satisfies Eq. (4.2.22), it is symplectic; that is, $\Phi(t, t_k)^T J\phi(t, t_k) = J$.

(8) The displacement of a particle, under the influence of a constant acceleration $\ddot{x}_0$, can be expressed as

$$x(t) = x_0 + \dot{x}_0 t + \frac{1}{2}\ddot{x}_0 t^2,$$

where x_0 is the initial displacement, $\dot{x}_0$ is the initial velocity, and $\ddot{x}_0$ is the acceleration at the initial time, $t_0 = 0$.

(a) By successive differentiation of this expression, show that the linear system

$$\dot{\mathbf{x}} = A\mathbf{x}$$

describes the motion, where

$$\mathbf{x} = \begin{bmatrix} x \\ \dot{x} \\ \ddot{x} \end{bmatrix} ; A = \begin{bmatrix} 0 & 1 & 0 \\ 0 & 0 & 1 \\ 0 & 0 & 0 \end{bmatrix}.$$

(b) Prove that the transition matrix $\Phi(t, t_0)$ of the system in (a) is

$$\Phi(t, t_0) = \begin{bmatrix} 1 & (t - t_0) & (t - t_0)^2/2 \\ 0 & 1 & (t - t_0) \\ 0 & 0 & 1 \end{bmatrix}$$

first by differentiating Φ and showing that $\dot{\Phi}(t, t_0) = A\Phi(t, t_0)$ where $\Phi(t_0, t_0) = I$, and then by integrating this system of differential equations.

(c) Show that $\Phi(t, t_0)$ can be represented as

$$\Phi(t, t_0) = e^{A(t-t_0)}$$

$$= I + \sum_{n=1}^{\infty} \frac{1}{n!} A^n (t - t_0)^n.$$

(d) Calculate $\Phi(t_0, t)$ by direct inversion.

(e) Let $\Phi^{-1}(t, t_0) = \Theta(t, t_0)$ and show that

$$\dot{\Theta}(t, t_0) = -\Theta(t, t_0) A, \; \Theta(t_0, t_0) = I$$

by integration and comparison with the results of d.

(f) Calculate $\Phi(t_2, t_1)$ by finding the product $\Phi(t_2, t_0) \Phi(t_0, t_1)$.

(g) Compare this result with the result obtained by integrating the equation

$$\dot{\Phi}(t, t_1) = A \Phi(t, t_1),$$

with the condition $\Phi(t_1, t_1) = I$.

(h) Show that

$$\frac{\partial \mathbf{x}}{\partial \mathbf{x}_0} = \Phi(t, t_0),$$

where $\mathbf{x}^T = (x \; \dot{x} \; \ddot{x})$ and $\mathbf{x}_0^T = (x_0 \; \dot{x}_0 \; \ddot{x}_0)$.

(9) The equations of motion for a satellite moving in the vicinity of a body with a homogeneous mass distribution can be expressed as

$$\ddot{\mathbf{r}} = -\frac{\mu \mathbf{r}}{r^3}$$

where $\mathbf{r}$ is the position vector, $\ddot{\mathbf{r}}$ is the acceleration vector, and where $r = |\mathbf{r}|$. Let $\mathbf{v} = \dot{\mathbf{r}}$ denote the velocity vector, and express the equations in first-order form as

$$\begin{bmatrix} \dot{\mathbf{r}} \\ \dot{\mathbf{v}} \end{bmatrix} = \begin{bmatrix} \mathbf{v} \\ -\dfrac{\mu}{r^3}\mathbf{r} \end{bmatrix}.$$

(a) The relations that define the deviations from a given reference orbit due to deviations in $\mathbf{r}$ and $\mathbf{v}$ and μ at a given time t_0 can be used to analyze the trajectory sensitivity. Show that

$$\frac{d}{dt} \begin{bmatrix} \delta\mathbf{r} \\ \delta\mathbf{v} \end{bmatrix} = \begin{bmatrix} 0 & I \\ \dfrac{\partial \mathbf{f}}{\partial \mathbf{r}} & 0 \end{bmatrix} \begin{bmatrix} \delta\mathbf{r} \\ \delta\mathbf{v} \end{bmatrix} - \begin{bmatrix} 0 \\ \dfrac{\mathbf{r}}{r^3} \end{bmatrix} \delta\mu,$$

where

$$\mathbf{f} = -\frac{\mu\,\mathbf{r}}{r^3}$$

$$-\frac{\partial\mathbf{f}}{\partial\mathbf{r}} = \begin{bmatrix} \dfrac{\mu}{r^3} - 3\mu\dfrac{x_1^2}{r^5} & -3\mu\dfrac{x_1x_2}{r^5} & -3\mu\dfrac{x_1x_3}{r^5} \\[2ex] -3\mu\dfrac{x_1x_2}{r^5} & \dfrac{\mu}{r^3} - 3\mu\dfrac{x_2^2}{r^5} & -3\mu\dfrac{x_2x_3}{r^5} \\[2ex] -3\mu\dfrac{x_1x_3}{r^5} & -3\mu\dfrac{x_2x_3}{r^5} & \dfrac{\mu}{r^3} - 3\mu\dfrac{x_3^2}{r^5} \end{bmatrix}.$$

Note that $\partial\mathbf{f}/\partial\mathbf{r}$ is symmetric.

(b) The constant μ is a physical constant and $\delta\mu$ represents the error in the knowledge of μ. Show that the error in $\mathbf{r}(t)$ and $\mathbf{v}(t)$ can be related to the error in μ by the following expression (Hint: Use the solution for $x(t)$ given by Eq. (4.9.14).):

$$\begin{bmatrix} \delta\mathbf{r}(t) \\ \delta\mathbf{v}(t) \end{bmatrix} = \Phi(t, t_0)\begin{bmatrix} \delta\mathbf{r}_0 \\ \delta\mathbf{v}_0 \end{bmatrix} - \delta\mu \int_{t_0}^{t} \Phi(t, \tau)\begin{bmatrix} 0 \\ \frac{\mathbf{r}}{r^3} \end{bmatrix} d\tau.$$

(10) Assume an orbit plane coordinate system for Exercise 9 with $\mu = 1$.

(a) Generate a "true" solution by numerically integrating the resulting differential equations

$$\ddot{x} = -\frac{x}{r^3}$$
$$\ddot{y} = -\frac{y}{r^3}$$
$$r^2 = x^2 + y^2$$

for the initial conditions

$$\mathbf{X}(t_0) = \begin{pmatrix} x \\ y \\ \dot{x} \\ \dot{y} \end{pmatrix}_{t=t_0} = \begin{pmatrix} 1.0 \\ 0.0 \\ 0.0 \\ 1.0 \end{pmatrix}.$$

Save the values of the state vector $\mathbf{X}(t_i)$ for $t_i = i*10.; i = 0, \ldots, 10$.

(b) Perturb the previous set of initial conditions by an amount

$$\mathbf{X}^*(t_0) = \mathbf{X}(t_0) - \delta\mathbf{X}(t_0)$$

where

$$\delta \mathbf{X}(t_0) = \begin{pmatrix} 1 \times 10^{-6} \\ -1 \times 10^{-6} \\ 1 \times 10^{-6} \\ 1 \times 10^{-6} \end{pmatrix}$$

and numerically integrate this "nominal" trajectory along with the associated state transition matrix to find $\mathbf{X}^*(t_i)$ and $\Phi(t_i, t_0)$ at $t_i = i * 10.; i = 0, \ldots, 10$.

(c) For this problem, $\Phi(t_i, t_0)$ is symplectic. Demonstrate this for $\Phi(t_{100}, t_0)$ by multiplying it by $\Phi^{-1}(t_{100}, t_0)$, given by Eq. (4.2.22), and showing that the result is the identity matrix.

(d) Calculate the perturbation vector, $\delta \mathbf{X}(t_i)$, by the following two methods:

(1) $\delta \mathbf{X}(t_i) = \mathbf{X}(t_i) - \mathbf{X}^*(t_i)$

(2) $\delta \mathbf{X}(t_i) = \Phi(t_i, t_0) \delta \mathbf{X}(t_0)$

and compare the results of (1) and (2). A program such as Matlab works well for this problem.

(11) Show that if the acceleration on a spacecraft is derivable from a potential function,

$$\ddot{\mathbf{r}} = \nabla U$$

then the state transition matrix is symplectic; that is, Eq. (4.2.13) is true under the assumption that the state vector is

$$\mathbf{x} = \begin{bmatrix} \mathbf{r} \\ \mathbf{v} \end{bmatrix}.$$

(12) Given the spring-mass system, as shown,

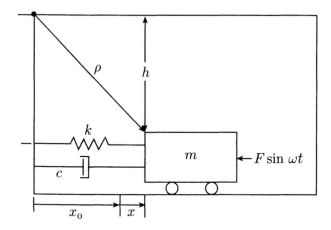

$$m\ddot{x} + c\dot{x} + kx + F\sin\omega t = 0.$$

Assume that we wish to estimate x, $\dot{x}$, F, c, and h using observations ρ. Derive the A and $\widetilde{H}$ matrix for this state vector.

(13) Given a point mass falling under the acceleration of gravity and being observed by range observations, ρ, as shown. Determine the state transition matrix, and the $\widetilde{H}$ and H matrices. Assume the state vector is to be estimated at $t_0 = 0$.

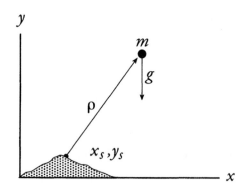

$$
\text{let } \mathbf{X} =
\begin{bmatrix}
x \\
y \\
\dot{x} \\
\dot{y} \\
g \\
x_s
\end{bmatrix},
\qquad
\begin{array}{l}
@\ t_0 \\
x = x_0 \\
\dot{x} = \dot{x}_0 \\
y = y_0 \\
\dot{y} = \dot{y}_0
\end{array}
$$

(14) Given a simple pendulum with range observations $|\rho|$, from a fixed point, as shown in the figure.

    (a) Write the equations of motion and form the observation-state matrix ($\widetilde{H}$) and the state propagation matrix (A). Assume the state vector is

$$
\mathbf{X} =
\begin{bmatrix}
\Theta \\
\dot{\Theta} \\
x_0
\end{bmatrix}.
$$

    (b) Assume small oscillations; that is, $\sin\Theta \approx \Theta$, $\cos\Theta \approx 1$. Solve the equations of motion and derive expressions for the state transition matrix.

    (c) How does the assumption that Θ is small differ from a linearized formulation of this problem?

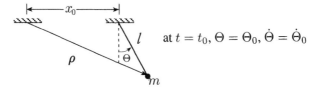

at $t = t_0$, $\Theta = \Theta_0$, $\dot{\Theta} = \dot{\Theta}_0$

    (d) Write a computer program to process range observations of the pendulum mass using both a batch processor and a sequential filter.

    (e) Generate a set of observations and process these observations using both the batch and sequential estimation algorithms. Use both perfect observations and observations with random noise.

(15) Write a computer program to process range and range-rate data for the spring-mass system of Section 4.8.2. Use the following data set (or create your own), which has Gaussian noise with mean zero and $\sigma_\rho = 0.25$ m

and $\sigma_{\dot\rho} = 0.10$ m/s added. Use the same initial conditions given in Example 4.8.2. Use a weighting matrix, W, that reflects the noise on the data,

$$E[\epsilon\epsilon^T] = R = \begin{bmatrix} \sigma_\rho^2 & 0 \\ 0 & \sigma_{\dot\rho}^2 \end{bmatrix} = \begin{bmatrix} .0625 & 0 \\ 0 & .01 \end{bmatrix} = W^{-1}.$$

Observation Data

Time	$\rho(m)$	$\dot\rho(m/s)$
0.00	6.37687486186586	−0.00317546143535849
1.00	5.50318198665912	1.17587430814596
2.00	5.94513302809067	−1.47058865193489
3.00	6.30210798411686	0.489030779000695
4.00	5.19084347133671	0.993054430595876
5.00	6.31368240334678	−1.40470245576321
6.00	5.80399842220377	0.939807575607138
7.00	5.45115048359871	0.425908088320457
8.00	5.91089305965839	−1.47604467619908
9.00	5.6769731201352	1.42173765213734
10.00	5.25263404969825	−0.12082311844776

Answer after three iterations:

$$x_0 = 2.9571m, \quad v_0 = -0.1260m/s$$
$$\rho_{\text{RMS}} = 0.247m, \ \dot\rho_{\text{RMS}} = 0.0875m/s$$
$$\sigma_{x_0} = 0.0450m, \ \sigma_{v_0} = 0.0794m/s, \quad \rho_{x_0v_0} = 0.0426$$

(16) Repeat Exercise 15, except program the sequential processor algorithm. Use the same initial conditions. Solve for the state at each observation time. Map the state estimate and covariance matrix from the final time to the initial time and show that they agree with the batch results.

(17) Given: The equation of motion of a particle moving in a uniform gravity field influenced by a resistive drag force; for example,

$$m\ddot{\mathbf{r}} = -mg\mathbf{j} - m\mathbf{D}$$

where

$$\mathbf{D} = \frac{1}{2}\rho\beta\,\dot{r}\,\dot{\mathbf{r}},$$

and the sequence of observations

$$\rho_i = \sqrt{(x - x_s)_i^2 + (y - y_s)_i^2}\ i = 1, \ldots, m.$$

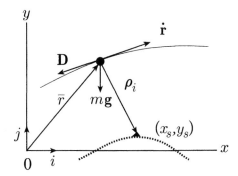

Set up the necessary equations to estimate the initial position and velocity of the particle assuming that β and x_s, y_s are unknown but constant force-model and measurement-model parameters to be estimated. Discuss the conditions necessary to ensure complete parameter observability.

(18) Given the following dynamic system

$$\ddot{x} = -kx - \beta x^2 + c \sin \omega t$$
$$\rho_i = \sqrt{d^2 + (\ell + x_i)^2} + \epsilon_i; i = 1, \ldots, m.$$

$$E[\epsilon_i] = 0, \qquad E[\epsilon_i \epsilon_j] = \sigma_i^2 \delta_{ij}$$

Set up all necessary equations for sequentially estimating the position and velocity of the system from observations (ρ_i). Assume that $[d, \ell, k, \beta, c]$ are unknowns to be estimated. State clearly all information that must be given as input information beyond what is given in the problem statement.

(19) Given the observation-state relation

$$y(t) = \sum_{i=1}^{3} (t)^i x_i,$$

and the observation sequence at $t = 1$, $y(1) = 2$, and at $t = 2$, $y(2) = 1$.

Find the "best" estimate of x_i, $i = 1$, 2, 3. (Hint: There are fewer observations than unknowns, so use the minimum norm solution.)

(20) Given the observation state relation $y = H x + \epsilon$, where x is a scalar and given that

$$y = \begin{bmatrix} 1 \\ 2 \\ 1 \end{bmatrix}, \quad \text{and } E[\epsilon] = 0,$$

$$R = E\left[\epsilon\epsilon^T\right] = \begin{bmatrix} \frac{1}{2} & 0 & 0 \\ 0 & 1 & 0 \\ 0 & 0 & 1 \end{bmatrix}, H = \begin{bmatrix} 1 \\ 1 \\ 1 \end{bmatrix}$$

with *a priori* information, $\bar{x} = 2$ and $\sigma^2(\bar{x}) = \frac{1}{2}$.

(a) Using R^{-1} as the weighting matrix, find $\hat{x}$ using the batch processing algorithm.

(b) What is the standard deviation of the estimation error associated with $\hat{x}$?

(c) What is the best estimate of the observation error, $\hat{\epsilon}$?

(21) Given the system

$$x(t) = x_0 + \dot{x}_0 t + \frac{1}{2}\ddot{x}_0 t^2; \quad t_0 = 0$$
$$\dot{x}(t) = \dot{x}_0 + \ddot{x}_0 t$$
$$\ddot{x}(t) = \ddot{x}_0$$

with

$$\mathbf{X}(t) = \begin{bmatrix} x \\ \dot{x} \\ \ddot{x} \end{bmatrix}, \quad \overline{P}_0 = \begin{bmatrix} 4 & 0 & 0 \\ 0 & 2 & 0 \\ 0 & 0 & 1 \end{bmatrix}, \quad \overline{\mathbf{X}}(t_0) = \begin{bmatrix} 1 \\ 1 \\ 1 \end{bmatrix}.$$

At $t_1 = 1$, an observation of $x(t)$ is made; that is, $Y(t_1) = x(t_1) = 2$. The variance of the error in $Y(t_1)$ is $\sigma^2(\epsilon_Y) = 1$.

(a) Compute $\hat{\mathbf{X}}(t_0)$ using the batch processing algorithm.
Answer:

$$\hat{\mathbf{X}}(t_0) = \begin{bmatrix} 21/29 \\ 25/29 \\ 28/29 \end{bmatrix}$$

(b) Using the sequential processing algorithm, determine $\hat{\mathbf{X}}(t_0)$. Remember you are to solve for $\hat{\mathbf{X}}(t_0)$, not $\hat{\mathbf{X}}(t_1)$. You should get identical results to part a.

(c) Compute $\hat{\mathbf{X}}(t_1)$ using the sequential algorithm, then map $\hat{\mathbf{X}}(t_1)$ to t_0 using the state transition matrix. You should get the same value of $\hat{\mathbf{X}}(t_0)$ as you computed in parts a and b.

Answer:

$$\hat{\mathbf{X}}(t_1) = \begin{bmatrix} 60/29 \\ 53/29 \\ 28/29 \end{bmatrix}$$

(22) Assume a vehicle is moving along a rectilinear path with a constant acceleration. The generic equation of motion is

$$x(t) = x_0 + v_0 t + \frac{1}{2} a t^2.$$

At discrete constant intervals of time $t = t_i$, $i = 0, 1, 2, 3 \ldots 10$ the position and velocity of the vehicle are observed.

(a) Set up a flow chart for the batch and sequential estimation algorithm to estimate a state vector consisting of the position, velocity, and acceleration. For the batch algorithm estimate the state at t_0, and for the sequential estimate the state at t_i.

(b) Write a program (using Matlab or a similar software package) to solve for the state vector using the batch processor. Include an *a priori* covariance matrix and a measurement weighting matrix and the capability to map the state estimate and the estimation error covariance matrix from the epoch to any desired time. Generate a sequence of simulated observations to which random noise is added. Solve for the state vector and associated covariance matrix at t_0. Map these quantities to the final time for comparison with results from the sequential processor in part c.

(c) Repeat part b for the sequential processor and estimate the state at t_i, $i = 1 \ldots 10$. Compare results at the final measurement time with the mapped results of the batch solution of part b.

(d) Experiment with your program by changing the *a priori* covariance and weighting matrix. Write a brief report describing your results.

(23) For the problem posed in Section 4.7.1:

(a) Show that the exact value for P_2 is given by Eq. (4.7.24).

(b) Under the assumption that $1 + \epsilon^2 = 1$, show that the conventional Kalman filter yields

$$P_2 = \frac{1}{1 - 2\epsilon} \begin{bmatrix} -1 & 1 \\ 1 & -1 \end{bmatrix}.$$

(c) Under the assumption that $1 + \epsilon^2 = 1$, show that the batch processor yields

$$P_2 = \frac{1}{1 - 2\epsilon} \begin{bmatrix} 1 & -(1 + \epsilon) \\ -(1 + \epsilon) & 2 \end{bmatrix} = \begin{bmatrix} 1 + 2\epsilon & -(1 + 3\epsilon) \\ -(1 + 3\epsilon) & 2(1 + 2\epsilon) \end{bmatrix},$$

which agrees with the exact result to $O(\epsilon)$, thus illustrating the improved numerical behavior of the batch over the sequential processor.

(24) Two observations are taken of the parameter x,

$$y_1 = x + \epsilon_1, y_2 = x + \epsilon_2,$$

where ϵ_1 and ϵ_2 are zero mean random noise. It is known that the variance of ϵ_2 is twice the variance of ϵ_1. Determine the weighted least squares estimate of x.

(25) Derive the following two equations for the estimation error covariance matrix:

(a) Beginning with Eq. (4.4.29), derive the expression for the estimation error covariance matrix given by Eq. (4.4.30). Assume that

$$E\left[(\bar{x} - x)\epsilon^T\right] = 0.$$

(b) Beginning with the equation for $\hat{x}$ (Eq. 4.7.16), derive the Joseph formulation for the measurement update of the estimation error covariance matrix given by Eq. (4.7.19).

(c) Show that the Joseph formulation for the measurement update of P is equivalent to the conventional Kalman update given by Eq. (4.7.18),

$$P = (I - K\tilde{H})\bar{P}.$$

(26) For the joint density function given by Eq. (4.5.4),

(a) Determine the MLE of σ.

(b) Show that the MLE of α given by Eq. (4.5.8) is unbiased.

(27) Let $\mathbf{X} = [x_1\ x_2\ \dots\ x_n]^T$ have a multivariate Gaussian distribution with each component of $\mathbf{X}$ having the same (unknown) mean α,

$$f\,(\mathbf{X};\,\alpha) = \frac{1}{(2\pi)^{n/2}\,|P|^{\frac{1}{2}}}\,\exp[-\frac{1}{2}(\mathbf{X} - A\alpha)^T P^{-1}(\mathbf{X} - A\alpha)]$$

with

$$A = \begin{pmatrix} 1 \\ 1 \\ \vdots \\ 1 \end{pmatrix}.$$

(a) Identify the likelihood function and show that the expression for the maximum likelihood estimate of α is $\hat{\alpha} = (A^T P^{-1} A)^{-1} A^T P^{-1} \mathbf{X}$.

(b) Show that $\hat{\alpha}$ is an unbiased estimator of α.

(28) Given the estimator $\hat{\mathbf{x}} = \overline{\mathbf{x}} + K\,(\mathbf{y} - H\overline{\mathbf{x}})$, state all necessary conditions on $\overline{\mathbf{x}}$ and $\mathbf{y}$ for $\hat{\mathbf{x}}$ to be an unbiased estimate of $\mathbf{x}$.

(29) Given: Random variables X and Y and

$$Y = a \pm bX$$

where a and b are constants. Show that the correlation coefficient $\rho\,(X,\,Y)$ has the value

$$\rho\,(X,\,Y) = \pm 1.$$

(30) Given that $\mathbf{x}$ is an $n \times 1$ vector of random variables with mean $\overline{\mathbf{x}}$ and covariance P. Let $\mathbf{y}$ be an $m \times 1$ vector of random variables related to $\mathbf{x}$ by

$$\mathbf{y} = H\mathbf{x} + \boldsymbol{\epsilon}$$

where $\boldsymbol{\epsilon}$ is zero mean with covariance R and is independent of $\mathbf{x}$.

(a) Find the mean of $\mathbf{y}$.

(b) Show that the variance-covariance of $\mathbf{y}$ is given by

$$P_y = E\left[(\mathbf{y} - \overline{\mathbf{y}})(\mathbf{y} - \overline{\mathbf{y}})^T\right]$$
$$= H P H^T + R.$$

(c) Show that the covariance of $\mathbf{x}$ and $\mathbf{y}$ is

$$P_{\mathbf{xy}} \equiv E[(\mathbf{x} - \bar{\mathbf{x}})(\mathbf{y} - \bar{\mathbf{y}})^T] = PH^T.$$

(31) An estimate, $\hat{\mathbf{x}}$ is made of $\mathbf{x}$ (an $n \times 1$ vector), based on m observations $\mathbf{y}$ $(m > n)$, where

$$\mathbf{y} = H\mathbf{x} + \epsilon$$

and

$$\epsilon \text{ is } N\,(\bar{\epsilon}, R).$$

An *a priori* value of $\mathbf{x}$ is given:

$$\bar{\mathbf{x}} = \mathbf{x} + \mathbf{e}$$

where $\mathbf{e}$ is $N\,(0,\,\bar{P})$. Assume $E\,[\mathbf{e}\,\epsilon^T] = 0$.

The estimate is

$$\hat{\mathbf{x}} = \bar{\mathbf{x}} + K\,(\mathbf{y} - H\bar{\mathbf{x}}).$$

(a) What is the bias for this estimator?

(b) What is the variance-covariance associated with the estimation error for $\hat{\mathbf{x}}$? Note that $\hat{\mathbf{x}}$ is biased so use the definition of P given in the answer.

(c) Show that by redefining the noise vector to be $\epsilon = \bar{\epsilon} + \epsilon'$, and by including $\bar{\epsilon}$ in the state vector, an unbiased estimator may be formed. Assume that an *a priori* estimate of $\bar{\epsilon}$ with covariance $P_{\bar{\epsilon}}$ is available.

Answer:

(a) $E[\hat{\mathbf{x}}] = \mathbf{x} + K\bar{\epsilon}$, bias $= K\bar{\epsilon}$

(b) $P = (I - KH)\bar{P}(I - KH)^T + KRK^T$ where

$$R = E\left[(\epsilon - \bar{\epsilon})(\epsilon - \bar{\epsilon})^T\right]$$

and

$$P \equiv E\left[(\tilde{\mathbf{e}} - E(\tilde{\mathbf{e}}))(\tilde{\mathbf{e}} - E(\tilde{\mathbf{e}}))^T\right]$$
$$\tilde{\mathbf{e}} \equiv \hat{\mathbf{x}} - \mathbf{x}, \quad E(\tilde{\mathbf{e}}) = K\bar{\epsilon}$$

(32) Assuming we are given no *a priori* information on $\hat{\mathbf{x}}$,

(a) show that there is no correlation between the residual observation error, $\hat{\mathbf{y}} - \mathbf{y} = \hat{\boldsymbol{\epsilon}}$, and the estimation error, $\hat{\mathbf{x}} - \mathbf{x}$, that

$$E\left[\hat{\boldsymbol{\epsilon}}(\hat{\mathbf{x}} - \mathbf{x})^T\right] = 0.$$

If this were not zero, it would mean that we had not extracted all of the available information from the observations.

Hint: Use

$$\hat{\boldsymbol{\epsilon}} = \mathbf{y} - H\hat{\mathbf{x}}$$
$$\hat{\mathbf{x}} = (H^T R^{-1} H)^{-1} H^T R^{-1} \mathbf{y}$$
$$\mathbf{y} = H\mathbf{x} + \boldsymbol{\epsilon}, \quad E[\boldsymbol{\epsilon}] = 0, \quad E[\boldsymbol{\epsilon}\boldsymbol{\epsilon}^T] = R.$$

(b) Assuming that *a priori* information is given; that is, $\bar{\mathbf{x}} = \mathbf{x} + \boldsymbol{\eta}$ where $E[\boldsymbol{\eta}] = 0$, $E[\boldsymbol{\eta}\boldsymbol{\eta}^T] = \bar{P}$ and $E[\boldsymbol{\epsilon}\boldsymbol{\eta}^T] = 0$. Show that $E[\hat{\boldsymbol{\epsilon}}(\hat{\mathbf{x}} - \mathbf{x})^T] = 0$ is still true.

(33) Occasionally *a priori* information, $\bar{\mathbf{x}}$ and $\bar{P}$, used to initiate a batch filter is obtained from processing an earlier batch of data so that the errors in $\bar{\mathbf{x}}$ and the observation error, $\boldsymbol{\epsilon}$, are correlated. Derive the expression for the covariance matrix of the estimation error, P, assuming

$$P = E\left[(\hat{\mathbf{x}} - \mathbf{x})(\hat{\mathbf{x}} - \mathbf{x})^T\right]$$
$$\hat{\mathbf{x}} = (\bar{P}^{-1} + H^T R^{-1} H)^{-1}(\bar{P}^{-1}\bar{\mathbf{x}} + H^T R^{-1}\mathbf{y}),$$

$$\bar{\mathbf{x}} = \mathbf{x} + \boldsymbol{\epsilon}_{\bar{\mathbf{x}}}, \quad E(\boldsymbol{\epsilon}_{\bar{\mathbf{x}}}) = 0, \quad E(\boldsymbol{\epsilon}_{\bar{\mathbf{x}}}\boldsymbol{\epsilon}_{\bar{\mathbf{x}}}^T) = \bar{P},$$
$$\mathbf{y} = H\mathbf{x} + \boldsymbol{\epsilon}, \quad E[\boldsymbol{\epsilon}] = 0, \quad E[\boldsymbol{\epsilon}\boldsymbol{\epsilon}^T] = R,$$
$$E\left[\boldsymbol{\epsilon}\,\boldsymbol{\epsilon}_{\bar{\mathbf{x}}}^T\right] = M.$$

Answer:

$$P = (\bar{P}^{-1} + H^T R^{-1} H)^{-1} + (\bar{P}^{-1} + H^T R^{-1} H)^{-1}$$
$$\times\, [H^T R^{-1} M \bar{P}^{-1} + \bar{P}^{-1} M^T R^{-1} H]\,(\bar{P}^{-1} + H^T R^{-1} H)^{-1}.$$

(34) Given that

$$\hat{\mathbf{x}}_k = \mathbf{x}_k + \boldsymbol{\eta}_k$$
$$\hat{\mathbf{x}}_{k+1} = \mathbf{x}_{k+1} + \boldsymbol{\eta}_{k+1}$$

and

$$\mathbf{x}_{k+1} = \Phi(t_{k+1}, t_k)\mathbf{x}_k + \Gamma(t_{k+1}, t_k)\mathbf{u}_k,$$

where

$$E[\eta_k] = E[\eta_{k+1}] = 0$$

and x_k and x_{k+1} are the true values, show that

$$P_{\eta_k \eta_{k+1}} = E[\eta_k \eta_{k+1}^T] = P_k \Phi^T(t_{k+1}, t_k)(I - K_{k+1}H_{k+1})^T.$$

(35) Consider the linear system defined in Excrcisc 21. Assume $t_0 = 0$; otherwise replace t with $(t - t_0)$.

$$\Phi(t, t_0) = \frac{\partial X(t)}{\partial X(t_0)} = \begin{bmatrix} 0 & t & \frac{t^2}{2} \\ 0 & 1 & t \\ 0 & 0 & 1 \end{bmatrix}$$

and

$$\overline{P}_0 = \begin{bmatrix} 4 & 0 & 0 \\ 0 & 2 & 0 \\ 0 & 0 & 1 \end{bmatrix}.$$

(a) Show that the time update for P at the first measurement time, t_1, obtained by integrating the differential equation for $\dot{\overline{P}}$ (Eq. 4.9.36) with initial conditions $\overline{P}_0$ is given by

$$\overline{P}(t_1) = \begin{bmatrix} 4 + 2t_1 + \frac{t_1^4}{4} & 2t_1 + \frac{t_1^3}{3} & \frac{t_1^2}{2} \\ 2t_1 + \frac{t_1^3}{3} & 2 + t_1^2 & t_1 \\ \frac{t_1^2}{2} & t_1 & 1 \end{bmatrix}.$$

(b) Show that this agrees with the conventional time update given by

$$\overline{P}(t_1) = \Phi(t_1, t_0)\overline{P}_0 \Phi^T(t_1, t_0).$$

(36) Given the information matrix

$$\Lambda = \begin{bmatrix} 13 & 2 & 6 \\ 2 & 2 & 6 \\ 6 & 6 & 18 \end{bmatrix}.$$

Are all three elements of the state vector observable? Why or why not?

(37) In Example 4.2.1 the state vector includes X, Y, X_s, and Y_s. Are all of these quantities observable? Why or why not?

(38) Given a 2D state vector, $\mathbf{X}$, with

$$E[\mathbf{X}] = 0, \quad E[\mathbf{X}\mathbf{X}^T] = P = \begin{bmatrix} 4 & 1 \\ 1 & 2 \end{bmatrix}.$$

Sketch the 2σ probability ellipse.

(39) For the transformation of coordinates

$$\begin{bmatrix} x' \\ y' \end{bmatrix} = \begin{bmatrix} \cos\theta & \sin\theta \\ -\sin\theta & \cos\theta \end{bmatrix} \begin{bmatrix} x+2 \\ y+2 \end{bmatrix},$$

determine $\bar{x}'$, $\bar{y}'$, and P' where $\bar{x}' = E(x')$, $\bar{y}' = E(y')$ and

$$P' = \begin{bmatrix} \sigma^2(x') & \mu_{11}(x'y') \\ \mu_{11}(x'y') & \sigma^2(y') \end{bmatrix}.$$

(40) The differential equation for the Ornstein-Uhlenbeck process, a first-order Gauss-Markov process, is

$$\dot{\eta}(t) + \beta\eta(t) = u(t)$$

with initial conditions $\eta = \eta(t_0)$, at $t = t_0$ where $u(t)$ is a Gaussian white noise process with mean zero. Show that the solution is given by

$$\eta(t) = \eta(t_0)e^{-\beta(t-t_0)} + \int_{t_0}^{t} e^{-\beta(t-\tau)}u(\tau)d\tau.$$

(41) Generate 1000 equally spaced observations of one cycle of a sine wave with amplitude 1 and period 10. Add Gaussian random noise with zero mean and variance $= 0.25$. Set up a sequential estimation procedure to estimate the amplitude of the sine wave as a function of time using the noisy raw data. Model the sine wave as a Gauss-Markov process as given by Eq. (4.9.60),

$$\eta_{i+1} = m_{i+1}\eta_i + \Gamma_{i+1}u_i$$

where

$$u_i = N(0, 1)$$

$$m_{i+1} = e^{-\beta(t_{i+1}-t_i)}$$

$$\Gamma_{i+1} = \sqrt{\frac{\sigma^2}{2\beta}(1 - m_{i+1}^2)}$$

$$\beta = \frac{1}{\tau}$$

and τ is the time constant. The sequential algorithm is given by

1. $\bar{\eta}_i = \Phi(t_i, t_{i-1})\hat{\eta}_{i-1}$ $(i = 1, 2 \ldots 1000)$

   $\Phi(t_i, t_{i-1}) = m_i = e^{-\beta(t_i - t_{i-1})}$

   $\overline{P}_i = \Phi(t_i, t_{i-1})P_{i-1}\Phi^T(t_i, t_{i-1}) + \Gamma_i Q_{i-1} \Gamma_i^T$

   Note that P, Φ, Q, and Γ are scalars

   $\qquad Y_i = \eta_i$, thus $\tilde{H}_i = 1$, assume $R_i = 1$, $Q_i = 1$, $\bar{\eta}_0 = 0$, $\overline{P}_0 = 1$

   $\qquad K_i = \overline{P}_i\tilde{H}_i^T(\tilde{H}_i\overline{P}_i\tilde{H}_i^T + R_i)^{-1} = \dfrac{\overline{P}_i}{\overline{P}_i + 1}$

   $\qquad \hat{\eta}_i = \bar{\eta}_i + K_i(Y_i - \tilde{H}_i\bar{\eta}_i) = \bar{\eta}_i + K_i(Y_i - \bar{\eta}_i)$, ($Y_i$ is the observation
   data)

   $\qquad P_i = (I - K_i\tilde{H}_i)\overline{P}_i = K_i$

   Next i

Plot your observations, the truth data, and $\hat{\eta}$ versus time. You will need to guess initial values for σ and β. Try several values to see which gives you the best results (i.e., the smallest RMS of estimation errors).

$$\text{RMS} = \left\{ \sum_{i=1}^N \frac{(T_i - \hat{\eta}_i)^2}{N} \right\}^{1/2},$$

where T_i is the true amplitude of the sine wave and $N = 1000$. Fig. 4.19.1 illustrates the truth, the raw data, and one example of the filtered results (not necessarily the optimal) for the indicated values of σ and β. The truth is the smooth curve. You may also plot a histogram of the post-fit residuals to see if they conform to a Gaussian distribution.

(42) Using the optimum values of σ and τ determined in Exercise 41, solve for the smoothed history of $\hat{\eta}(t)$ using the algorithm of Section 4.15.1. Plot the true values of η, the filter values determined in Exercise 41, and the smoothed values. Compute the RMS of the smoothed fit. You should find the RMS from the smoothed solution somewhat smaller than the filter result. Fig. 4.19.2 is an example of the truth, the filter, and the smoothed results for the case illustrated in Fig. 4.19.1. Is the histogram of residuals

Process Noise / Sine Wave Recovery

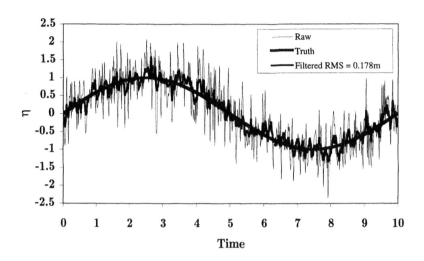

Figure 4.19.1: Process noise/sine wave recovery showing truth, raw data (truth plus noise) and the filtered solution. $\bar{\eta}_0 = 0$, $\sigma = 2.49$, $\beta = .045$.

for the smoothed solution more nearly Gaussian than those from the filter in Exercise 41?

(43) Given a system described by

$$\mathbf{X}(t_{i+1}) = \begin{bmatrix} x_1(t_{i+1}) \\ x_2(t_{i+1}) \end{bmatrix} = \begin{bmatrix} 1 & 1 \\ 0 & 1 \end{bmatrix} \begin{bmatrix} x_1(t_i) \\ x_2(t_i) \end{bmatrix} + \begin{bmatrix} 1 & 1 \\ 1 & 0 \end{bmatrix} \begin{bmatrix} u_1(t_i) \\ u_2(t_i) \end{bmatrix}$$

$$\begin{bmatrix} y_1(t_i) \\ y_2(t_i) \end{bmatrix} = \begin{bmatrix} 0 & 1/2 \\ 1 & 1/2 \end{bmatrix} \begin{bmatrix} x_1(t_i) \\ x_2(t_i) \end{bmatrix} + \begin{bmatrix} \epsilon_1 \\ \epsilon_2 \end{bmatrix}$$

$$\overline{\mathbf{X}}(t_0) = \begin{bmatrix} \bar{x}_1(t_0) \\ \bar{x}_2(t_0) \end{bmatrix} = \begin{bmatrix} 2 \\ 3 \end{bmatrix}, \overline{P}(t_0) = \begin{bmatrix} 1 & 0 \\ 0 & 1 \end{bmatrix}$$

$$R = E\left\{ \begin{bmatrix} \epsilon_1 \\ \epsilon_2 \end{bmatrix} [\epsilon_1 \epsilon_2] \right\} = \begin{bmatrix} 1 & 0 \\ 0 & 2 \end{bmatrix},$$

$$E[u_1(t_i)] = E[u_2(t_i)] = 0$$

Process Noise / Sine Wave Recovery

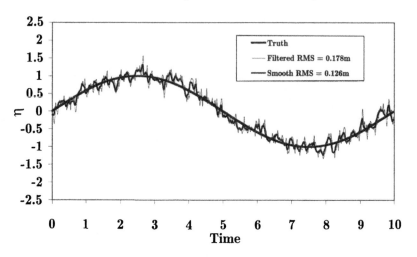

Figure 4.19.2: Process noise/sine wave recovery showing the truth, the filtered, and the smoothed solution. $\bar{\eta}_0 = 0$, $\sigma = 2.49$, $\beta = .045$.

$$E\left\{ \begin{bmatrix} u_1(t_i) \\ u_2(t_i) \end{bmatrix} [u_1(t_j)u_2(t_j)] \right\} = I\delta_{ij}.$$

Observations at t_1 are given by

$$\begin{bmatrix} y_1(t_1) \\ y_2(t_1) \end{bmatrix} = \begin{bmatrix} 3 \\ 2 \end{bmatrix}.$$

(a) Using the sequential estimation algorithm, determine the optimal estimate of $\mathbf{X}(t_1)$ and $P(t_1)$.

(b) What is the smoothed estimate of $\mathbf{X}(t_0)$?

(c) What is the optimal estimate of $\mathbf{X}(t_1)$ if no process noise were present?

(44) Use the information filter to determine the optimal estimate of $\mathbf{X}(t_1)$ and $P(t_1)$ for Exercise 43a.

Chapter 5

Square Root Solution Methods for the Orbit Determination Problem

5.1 INTRODUCTION

In the previous chapter, the solution to the least squares estimation problem including *a priori* information, $\overline{P}$ and $\overline{\mathbf{x}}$, is represented in the normal equation form as

$$(H^T W H + \overline{P}^{-1})\hat{\mathbf{x}} = H^T W y + \overline{P}^{-1}\overline{\mathbf{x}} \qquad (5.1.1)$$

which can be expressed as

$$M\hat{\mathbf{x}} = N. \qquad (5.1.2)$$

The solution for $\hat{\mathbf{x}}$ is obtained by computing the inverse of M. In practice, computational problems are encountered in forming and inverting the matrix $M = H^T W H + \overline{P}^{-1}$. An orthogonal transformation approach can be used to write Eq. (5.1.2) in the form

$$R\hat{\mathbf{x}} = b \qquad (5.1.3)$$

where R is an $n \times n$ upper triangular matrix (R is not the observation error covariance matrix). Then $\hat{\mathbf{x}}$ can be obtained by backward substitution. This approach has the advantage that increased numerical accuracy is achieved over conventional matrix inversion methods for solving Eq. (5.1.2). Using the orthogonal transformation approach, accuracy can be achieved with a single-precision computation that is equal to the accuracy obtained by inverting M with the normal equation approach using double-precision computations.

The normal equation approach has several operational and conceptual advantages that have led to the widespread adoption of this technique for many operational orbit determination systems. Both the normal equation and orthogonal transformation approaches are described in the subsequent sections.

The solution of the linear system

$$M\hat{\mathbf{x}} = \mathbf{N} \tag{5.1.4}$$

is expressed as

$$\hat{\mathbf{x}} = M^{-1}\mathbf{N} \tag{5.1.5}$$

where the operation M^{-1} implies that the inverse of the $(n \times n)$ matrix M is computed and then postmultiplied by the column vector $\mathbf{N}$. A solution based on the Cholesky decomposition of M is more efficient and, in most cases, more accurate. The Cholesky decomposition is applicable only if M is symmetric and positive definite, a condition satisfied for the case considered here. The following discussion first outlines the Cholesky decomposition, and describes this approach to solving the normal equations. Then a discussion of solutions based on orthogonal transformations is presented in subsequent sections.

5.2 CHOLESKY DECOMPOSITION

Let M be a symmetric positive definite matrix, and let R be an upper triangular matrix computed such that

$$R^T R = M. \tag{5.2.1}$$

(Note that R is not the observation error covariance matrix.) Then Eq. (5.1.4) can be expressed as

$$R^T R\hat{\mathbf{x}} = \mathbf{N}. \tag{5.2.2}$$

If the definition

$$\mathbf{z} = R\hat{\mathbf{x}} \tag{5.2.3}$$

is used, Eq. (5.2.2) can be written as

$$R^T \mathbf{z} = \mathbf{N} \tag{5.2.4}$$

where R^T is lower triangular. The components of $\mathbf{z}$ can be determined using a forward recursion relation. Then, Eq. (5.2.3) can be solved using a backward recursion to obtain the elements of $\hat{\mathbf{x}}$.

The elements of the error covariance matrix, $P = (H^T W H + \overline{P}^{-1})^{-1} = M^{-1}$, can be obtained from the condition

$$P = M^{-1} = (R^T R)^{-1} = R^{-1} R^{-T} = SS^T \tag{5.2.5}$$

where S, the inverse of the upper triangular matrix, R, can be computed by an efficient backward recursion.

Equation (5.2.1) represents a set of $(n^2 + n)/2$ equations for the $(n^2 + n)/2$ unknown elements of the upper triangular matrix, R. The expression for the Cholesky decomposition of M is obtained by expanding Eq. (5.2.1) and solving term by term for the elements of R (e.g., R_{ij} is determined in terms of the elements of M).

Given the elements of the $n \times n$ positive definite matrix M and the $n \times 1$ column vector $\mathbf{N}$, the following Cholesky algorithm will yield a solution for the elements of R, $\mathbf{z}$, $\hat{\mathbf{x}}$ and the upper triangular matrix, S. Step 1 of the following algorithm determines the elements of R and the vector $\mathbf{z}$. Steps 2 and 3 perform a backward recursion to form $\hat{\mathbf{x}}$ and the elements of the matrix $S = R^{-1}$.

5.2.1 THE CHOLESKY ALGORITHM

The Cholesky algorithm for R is derived easily by equating the elements of $R^T R$ to the elements of M that are known. For example, from expanding $R^T R$ it is shown that $r_{11} = \sqrt{M_{11}}$.

1) For $i = 1, 2, \ldots, n$

$$r_{ii} = \left(M_{ii} - \sum_{k=1}^{i-1} r_{ki}^2 \right)^{1/2}$$

$$r_{ij} = \left(M_{ij} - \sum_{k=1}^{i-1} r_{ki} r_{kj} \right) / r_{ii}; \quad j = i+1, \ldots, n. \quad (5.2.6)$$

The elements of $\mathbf{z}$ are obtained from an expansion of Eq. (5.2.4):

$$z_i = \left(N_i - \sum_{j=1}^{i-1} r_{ji} z_j \right) / r_{ii} \quad (5.2.7)$$

2) For $i = n, n-1, \ldots, 1$

$$\hat{x}_i = \left(z_i - \sum_{j=i+1}^{n} r_{ij} \hat{x}_j \right) / r_{ii}. \quad (5.2.8)$$

The elements of S are obtained from an expansion of $SR = I$.

3) For $i = 1, 2 \ldots, n$

$$s_{ii} = \frac{1}{r_{ii}}$$

$$s_{ij} = -s_{jj} \left[\sum_{k=i}^{j-1} r_{kj} s_{ik} \right]; \quad j = i+1, \ldots, n. \qquad (5.2.9)$$

Examples of the application of this algorithm are given in Sections 5.6.1 and 5.6.5.

This Cholesky algorithm is a nonblock algorithm. That is, it does not use matrix multiplication. Because matrix multiplication is much faster in terms of floating point operations per second than matrix-vector operations on modern computers, a block Cholesky algorithm often will be faster than a nonblock version. In fact, the increase in speed may be a factor of three or more. See Golub and Van Loan (1996) for examples of a block Cholesky algorithm.

5.2.2 THE SQUARE ROOT FREE CHOLESKY ALGORITHM

Notice that calculation of the diagonal elements of R requires that n square roots be taken. Computationally the square root operation is expensive compared to multiplication, division, addition, or subtraction; hence, it is desirable to avoid square roots if possible. A square root free Cholesky algorithm may be developed by defining

$$M = UDU^T \qquad (5.2.10)$$

where U is unit upper triangular (i.e., has ones on the diagonal), and D is a diagonal matrix. As in the previous section, Eq. (5.2.10) represents the $(n^2 + n)/2$ equation in the $(n^2 + n)/2$ unknown elements of U and D. The algorithm for the elements of U and D is obtained by expanding Eq. (5.2.10) and is given by

$$D_j U_{ij} = M_{ij} - \sum_{k=j+1}^{n} M'_{jk} U_{ik} \equiv M'_{ij} \ j = n, \ldots, 1 \qquad (5.2.11)$$

$$i = 1, \ldots, j-1$$

$$D_j = M_{jj} - \sum_{k=j+1}^{n} M'_{jk} U_{jk}. \qquad (5.2.12)$$

The procedure is to set $j = n$ and cycle through the algorithm for $i = 1 \ldots n-1$, solving for D_n and the elements of U_{in} (i.e., the last column of U). Then set $j = n-1$ and cycle through for $i = 1 \ldots n-2$, solving for D_{n-1} and the $n-1$ column of U. Repeat this procedure for the remaining values of j.

Knowing U and D, it is possible to solve for $\hat{\mathbf{x}}$ from Eq. (5.1.2) and Eq. (5.2.10) as follows. Note that

$$U D U^T \hat{\mathbf{x}} = \mathbf{N} \tag{5.2.13}$$

and let

$$\mathbf{z} = D U^T \hat{\mathbf{x}}. \tag{5.2.14}$$

Then

$$U \mathbf{z} = \mathbf{N}. \tag{5.2.15}$$

The intermediate vector $\mathbf{z}$ can be determined from Eq. (5.2.15). The solution is a backward substitution (i.e., we solve for $z_n, z_{n-1} \dots z_1$):

$$z_n = N_n \tag{5.2.16}$$

$$z_i = N_i - \sum_{j=i+1}^{n} U_{ij} z_j \quad i = n-1, \dots, 1. \tag{5.2.17}$$

Then the elements of $\hat{\mathbf{x}}$ can be determined from Eq. (5.2.14) by using a forward substitution, as

$$\hat{x}_i = \frac{z_i}{D_{ii}} - \sum_{j=1}^{i-1} U_{ji} \hat{x}_j \quad i = 1, 2, \dots n. \tag{5.2.18}$$

The associated estimation error covariance is obtained from

$$\begin{aligned} P &= (U D U^T)^{-1} \\ &= U^{-T} D^{-1} U^{-1} \end{aligned} \tag{5.2.19}$$

where

$$D_{ii}^{-1} = 1/D_{ii} \qquad i = 1, \dots n$$
$$j = i+1, \dots n$$

$$U_{ij}^{-1} = -\sum_{k=i}^{j-1} U_{ik}^{-1} U_{kj}$$

$$\tag{5.2.20}$$

$$U_{ij}^{-1} = 0 \qquad i > j$$

$$D_{ij}^{-1} = 0 \qquad i \neq j.$$

Note that none of the algorithms required to compute $\hat{\mathbf{x}}$ involve the calculation of a square root.

5.3   LEAST SQUARES SOLUTION VIA ORTHOGONAL TRANSFORMATION

An alternate approach that avoids some of the numerical problems encountered in the normal equation approach is described in the following discussion. The method obtains the solution by applying successive orthogonal transformations to the information array, $(H, \mathbf{y})$. Enhanced numerical accuracy is obtained by this approach. Consider the quadratic performance index, $J(\mathbf{x})$, which minimizes the weighted sum of squares of the observation errors, $\epsilon = \mathbf{y} - H\mathbf{x}$ (for the moment we will assume no *a priori* information; i.e., $\overline{P}^{-1} = 0, \overline{\mathbf{x}} = 0$):

$$J(\mathbf{x}) = \epsilon^T W \epsilon = \left\| W^{\frac{1}{2}}(H\mathbf{x} - \mathbf{y}) \right\|^2 = (H\mathbf{x} - \mathbf{y})^T W (H\mathbf{x} - \mathbf{y}). \qquad (5.3.1)$$

If W is not diagonal, $W^{1/2}$ can be computed by the Cholesky decomposition. Or the prewhitening transformation described at the end of Section 5.7.1 can be applied so that $W = I$. For notational convenience we are using $-\epsilon$ in Eq. (5.3.1).

The solution to the least squares estimation problem (as well as the minimum variance and the maximum likelihood estimation problem, under certain restrictions) is obtained by finding the value $\hat{\mathbf{x}}$ that minimizes the performance index $J(\mathbf{x})$. To achieve the minimum value of $J(\mathbf{x})$, we introduce the $m \times m$ orthogonal matrix, Q. An orthogonal matrix has the following properties:

1. $QQ^T = I.$ $\qquad\qquad\qquad\qquad\qquad\qquad\qquad\qquad\qquad\qquad\qquad (5.3.2)$

2. $Q^{-1} = Q^T$ hence $Q^T Q = I.$

3. If Q_1 and Q_2 are orthogonal matrices, then so is $Q_1 Q_2$.

4. For any vector $\mathbf{x}$,

$$\|Q\mathbf{x}\| = \|\mathbf{x}\| = (\mathbf{x}^T \mathbf{x})^{\frac{1}{2}}. \qquad (5.3.3)$$

Multiplying by Q does not change the Euclidean norm of a vector.

5. If ϵ is an m vector of random variables with $\epsilon \sim (O, I)$ (i.e., $E(\epsilon) = 0$ and $E(\epsilon\epsilon^T) = I$), then $\overline{\epsilon} = Q\epsilon$ has the same properties,

$$E(\overline{\epsilon}) = QE(\epsilon) = 0, \;\; E(\overline{\epsilon}\,\overline{\epsilon}^T) = QE(\epsilon\epsilon^T)Q^T = I. \qquad (5.3.4)$$

It follows then that (5.3.1) can be expressed as

$$\begin{aligned}
J(\mathbf{x}) &= \left\| QW^{\frac{1}{2}}(H\mathbf{x} - \mathbf{y}) \right\|^2 \\
&= \left(H\mathbf{x} - \mathbf{y} \right)^T W^{\frac{1}{2}} Q^T Q W^{\frac{1}{2}} \left(H\mathbf{x} - \mathbf{y} \right).
\end{aligned} \qquad (5.3.5)$$

Select Q such that

$$QW^{\frac{1}{2}}H = \begin{bmatrix} R \\ O \end{bmatrix} \text{ and define } QW^{\frac{1}{2}}y \equiv \begin{bmatrix} \mathbf{b} \\ \mathbf{e} \end{bmatrix} \qquad (5.3.6)$$

where

R is a $n \times n$ upper-triangular matrix of rank n

O is a $(m - n) \times n$ null matrix

$\mathbf{b}$ is a $n \times 1$ column vector

$\mathbf{e}$ is a $(m - n) \times 1$ column vector.

The results given by Eq. (5.3.6) assume that $m > n$ and H is of rank n. Using Eq. (5.3.6), Eq. (5.3.5) can be written as

$$J(\mathbf{x}) = \left\| \begin{bmatrix} R \\ O \end{bmatrix} \mathbf{x} - \begin{bmatrix} \mathbf{b} \\ \mathbf{e} \end{bmatrix} \right\|^2 . \qquad (5.3.7)$$

Expanding leads to

$$J(\mathbf{x}) = \left\| R\mathbf{x} - \mathbf{b} \right\|^2 + \left\| \mathbf{e} \right\|^2 . \qquad (5.3.8)$$

Only the first term in Eq. (5.3.8) is a function of $\mathbf{x}$, so the value of $\mathbf{x}$ that minimizes $J(\mathbf{x})$ is obtained by requiring that

$$R\hat{\mathbf{x}} = \mathbf{b} \qquad (5.3.9)$$

and the minimum value of the performance index becomes (equating $J(\hat{x})$ in Eq. (5.3.1) and Eq. (5.3.8))

$$J(\hat{\mathbf{x}}) = \left\| \mathbf{e} \right\|^2 = \left\| W^{\frac{1}{2}}(H\hat{\mathbf{x}} - \mathbf{y}) \right\|^2 . \qquad (5.3.10)$$

That is, $\|\mathbf{e}\|$ is the norm of the observation residual vector, which for a linear system will be equal to the weighted sum of the squares of observation residuals determined by using $\hat{\mathbf{x}}$ in Eq. (5.3.10).

5.4 GIVENS TRANSFORMATIONS

The procedure described in the previous section is direct and for implementation requires only that a convenient procedure for computing Q be obtained. One such procedure can be developed based on the Givens plane rotation (Givens, 1958). Let $\mathbf{x}$ be a 2×1 vector having components $\mathbf{x}^T = [x_1 \; x_2]$ and let G be a 2×2 orthogonal matrix associated with the plane rotation through the angle θ. Then select G such that

$$G\mathbf{x} = \mathbf{x}' = \begin{pmatrix} x_1' \\ 0 \end{pmatrix}. \tag{5.4.1}$$

To this end, consider the transformation

$$\begin{bmatrix} x_1' \\ x_2' \end{bmatrix} = \begin{bmatrix} \cos\theta & \sin\theta \\ -\sin\theta & \cos\theta \end{bmatrix} \begin{bmatrix} x_1 \\ x_2 \end{bmatrix} \tag{5.4.2}$$

or

$$\begin{aligned} x_1' &= \cos\theta x_1 + \sin\theta x_2 \\ x_2' &= -\sin\theta x_1 + \cos\theta x_2. \end{aligned} \tag{5.4.3}$$

Equations (5.4.3) represent a system of two equations in three unknowns; that is, x_1', x_2', and θ. The Givens rotation is defined by selecting the rotation θ such that $x_2' = 0$. That is, let

$$x_1' = \cos\theta x_1 + \sin\theta x_2 \tag{5.4.4}$$
$$0 = -\sin\theta x_1 + \cos\theta x_2. \tag{5.4.5}$$

From Eq. (5.4.5), it follows that

$$\tan\theta = \frac{x_2}{x_1}, \; \sin\theta = \frac{x_2}{\sqrt{x_1^2 + x_2^2}}, \; \cos\theta = \frac{x_1}{\sqrt{x_1^2 + x_2^2}}. \tag{5.4.6}$$

The positive value associated with the square root operation is selected for the following discussion. Substituting the expression for $\sin\theta$ and $\cos\theta$ into Eq. (5.4.4) leads to

$$x_1' = \frac{x_1^2}{\sqrt{x_1^2 + x_2^2}} + \frac{x_2^2}{\sqrt{x_1^2 + x_2^2}} = \sqrt{x_1^2 + x_2^2}. \tag{5.4.7}$$

Consider the application of the transformation

$$G(\theta) = \begin{bmatrix} \cos\theta & \sin\theta \\ -\sin\theta & \cos\theta \end{bmatrix} \tag{5.4.8}$$

to two general row vectors $\mathbf{h}_i$ and $\mathbf{h}_k$; for example,

$$G \begin{bmatrix} h_{ii} & h_{ii+1} & \ldots & h_{in} \\ h_{ki} & h_{ki+1} & \ldots & h_{kn} \end{bmatrix} = \begin{bmatrix} h_{ii}' & h_{ii+1}' & \ldots & h_{in}' \\ 0 & h_{ki+1}' & \ldots & h_{kn}' \end{bmatrix}. \tag{5.4.9}$$

That is, for any two general row vectors, $\mathbf{h}_i$ and $\mathbf{h}_k$, the transformation is applied to the first column so as to null h_{ki}. The transformation that accomplishes this is applied to each remaining column to obtain the transformed matrix. Hence,

$$\begin{bmatrix} \cos\theta & \sin\theta \\ -\sin\theta & \cos\theta \end{bmatrix} \begin{bmatrix} h_{ii} \\ h_{ki} \end{bmatrix} = \begin{bmatrix} h'_{ii} \\ 0 \end{bmatrix} \tag{5.4.10}$$

or

$$\sin\theta = h_{ki}/\sqrt{h_{ii}^2 + h_{ki}^2} = h_{ki}/h'_{ii}$$
$$\cos\theta = h_{ii}/\sqrt{h_{ii}^2 + h_{ki}^2} = h_{ii}/h'_{ii} \tag{5.4.11}$$
$$h'_{ii} = \sqrt{h_{ii}^2 + h_{ki}^2}.$$

Then for all other columns,

$$h'_{ij} = h_{ij}\cos\theta + h_{kj}\sin\theta$$
$$\qquad\qquad\qquad\qquad j = i+1,\ldots,n \tag{5.4.12}$$
$$h'_{kj} = -h_{ij}\sin\theta + h_{kj}\cos\theta.$$

By using this transformation repetitively as k goes from $i+1$ to m, the remaining elements of the ith column can be nulled. Then by moving down the diagonal and applying the transformation to successive columns whose first element lies on the diagonal, a rank n matrix can be reduced to an upper triangular $n \times n$ matrix with a lower $(m-n) \times n$ null matrix. If the element to be nulled already has a value of zero, the transformation matrix will be the identity matrix and the corresponding transformation may be skipped.

As an example, the transformation to null the fourth element in the third column is shown as follows:

$$\begin{bmatrix} 1 & & & & & & & & \\ & 1 & & & & & & & \\ & & C^{4,3} & S^{4,3} & & & & & \\ & & -S^{4,3} & C^{4,3} & & & & & \\ & & & & 1 & & & & \\ & & & & & 1 & & & \\ & & & & & & 1 & & \\ & & & & & & & \ddots & \\ & & & & & & & & 1 \end{bmatrix} \begin{bmatrix} h_{11} & h_{12} & h_{13} & \cdots & h_{1n} & y_1 \\ 0 & h_{22} & h_{23} & \cdots & h_{2n} & y_2 \\ 0 & 0 & h_{33} & \cdots & h_{3n} & y_3 \\ 0 & 0 & h_{43} & \cdots & h_{4n} & y_4 \\ 0 & 0 & h_{53} & \cdots & h_{5n} & y_5 \\ 0 & 0 & h_{63} & \cdots & h_{6n} & y_6 \\ \vdots & \vdots & \vdots & & \vdots & \vdots \\ 0 & 0 & h_{m3} & \cdots & h_{mn} & y_m \end{bmatrix}$$

$$
=
\begin{bmatrix}
h_{11} & h_{12} & h_{13} & \cdots & h_{1n} & y_1 \\
0 & h_{22} & h_{23} & \cdots & h_{2n} & y_2 \\
0 & 0 & h'_{33} & \cdots & h'_{3n} & y'_3 \\
0 & 0 & 0 & \cdots & h'_{4n} & y'_4 \\
0 & 0 & h_{53} & \cdots & h_{5n} & y_5 \\
0 & 0 & h_{63} & \cdots & h_{6n} & y_6 \\
\vdots & \vdots & \vdots & & \vdots & \vdots \\
0 & 0 & h_{m3} & \cdots & h_{mn} & y_m
\end{bmatrix}
. \tag{5.4.13}
$$

The prime superscript identifies the two rows that are affected by this transformation (rows three and four). Notice that the location of $-\sin\theta$ in the Givens transformation matrix corresponds to the location of the element to be nulled in the H matrix. For example, in the previous example both are element (4,3).

By using the transformation

$$
Q^{5,3} =
\begin{bmatrix}
1 & & & & & & & & & \\
& 1 & & & & & & & & \\
& & C^{5,3} & 0 & S^{5,3} & & & & & \\
& & 0 & 1 & 0 & & & & & \\
& & -S^{5,3} & 0 & C^{5,3} & & & & & \\
& & & & & 1 & & & & \\
& & & & & & 1 & & & \\
& & & & & & & 1 & & \\
& & & & & & & & \ddots & \\
& & & & & & & & & 1
\end{bmatrix}
, \tag{5.4.14}
$$

the third and fifth rows will be transformed so that the term h_{53} will be zero.

5.4.1 *A Priori* INFORMATION AND INITIALIZATION

The formulation given earlier does not specifically address the question of *a priori* information. Assume *a priori* information, $\overline{x}$ and $\overline{P}$, are available. The procedure is initialized by writing the *a priori* information in the form of a data equation; that is, in the form of $y = Hx + \epsilon$. This is accomplished by writing

$$
\overline{x} = x + \eta \tag{5.4.15}
$$

where $\mathbf{x}$ is the true value and η is the error in $\overline{\mathbf{x}}$. We assume that

$$E[\eta] = 0, \ E[\eta \eta^T] = \overline{P}. \tag{5.4.16}$$

Compute $\overline{S}$, the upper triangular square root of $\overline{P}$,

$$\overline{P} = \overline{S}\,\overline{S}^T. \tag{5.4.17}$$

If $\overline{P}$ is not diagonal, the Cholesky decomposition may be used to accomplish this. Next compute $\overline{R}$, the square root of the *a priori* information matrix, $\overline{\Lambda}$,

$$\overline{\Lambda} = \overline{P}^{-1} = \overline{S}^{-T}\overline{S}^{-1} = \overline{R}^T\overline{R} \tag{5.4.18}$$

hence,

$$\overline{R} = \overline{S}^{-1}. \tag{5.4.19}$$

Multiplying Eq. (5.4.15) by $\overline{R}$ yields

$$\overline{R}\overline{\mathbf{x}} = \overline{R}\mathbf{x} + \overline{R}\eta. \tag{5.4.20}$$

Define

$$\overline{\mathbf{b}} \equiv \overline{R}\overline{\mathbf{x}}, \ \overline{\eta} \equiv \overline{R}\eta, \tag{5.4.21}$$

then

$$\overline{\mathbf{b}} = \overline{R}\mathbf{x} + \overline{\eta} \tag{5.4.22}$$

where $\overline{\eta} \sim (O, I)$. Note that Eq. (5.4.22) expresses the *a priori* information in the form of the data equation, $\mathbf{y} = H\mathbf{x} + \epsilon$. Hence, the equations we wish to solve for $\hat{\mathbf{x}}$, using orthogonal transformations, are

$$\begin{aligned} \overline{\mathbf{b}} &= \overline{R}\mathbf{x} + \overline{\eta} \\ \mathbf{y} &= H\mathbf{x} + \epsilon \end{aligned} \tag{5.4.23}$$

where $\mathbf{x}$ is an n vector and $\mathbf{y}$ is an m vector.

The least squares solution for $\mathbf{x}$ in Eq. (5.4.23) is found by minimizing the performance index (we assume that ϵ has been prewhitened so that $\epsilon \sim (O, I)$; if not, replace ϵ with $W^{1/2}\epsilon$ in J)

$$\begin{aligned} J &= \|\overline{\eta}\|^2 + \|\epsilon\|^2 \\ &= \|\overline{R}\mathbf{x} - \overline{\mathbf{b}}\|^2 + \|H\mathbf{x} - \mathbf{y}\|^2 \\ &= \left\| \begin{bmatrix} \overline{R} \\ H \end{bmatrix} \mathbf{x} - \begin{bmatrix} \overline{\mathbf{b}} \\ \mathbf{y} \end{bmatrix} \right\|^2. \end{aligned} \tag{5.4.24}$$

After multiplying by an orthogonal transformation, Q, Eq. (5.4.24) may be written as

$$J = \left\{ \begin{bmatrix} \overline{R} \\ H \end{bmatrix} \mathbf{x} - \begin{bmatrix} \overline{\mathbf{b}} \\ \mathbf{y} \end{bmatrix} \right\}^T Q^T Q \left\{ \begin{bmatrix} \overline{R} \\ H \end{bmatrix} \mathbf{x} - \begin{bmatrix} \overline{\mathbf{b}} \\ \mathbf{y} \end{bmatrix} \right\}. \tag{5.4.25}$$

Choose Q so that

$$Q \begin{bmatrix} \overline{R} \\ H \end{bmatrix} = \begin{bmatrix} R \\ O \end{bmatrix}, \text{ and define } Q \begin{bmatrix} \overline{\mathbf{b}} \\ \mathbf{y} \end{bmatrix} \equiv \begin{bmatrix} \mathbf{b} \\ \mathbf{e} \end{bmatrix} \tag{5.4.26}$$

where R is upper triangular. Eq. (5.4.24) can now be written as

$$J = \left\| \begin{bmatrix} R \\ O \end{bmatrix} \mathbf{x} - \begin{bmatrix} \mathbf{b} \\ \mathbf{e} \end{bmatrix} \right\|^2 \tag{5.4.27}$$

or

$$J = \left\| R\mathbf{x} - \mathbf{b} \right\|^2 + \left\| \mathbf{e} \right\|^2 \tag{5.4.28}$$

as noted before. The minimum value of J is found by choosing $\hat{\mathbf{x}}$ so that

$$R\hat{\mathbf{x}} - \mathbf{b} = 0. \tag{5.4.29}$$

The vector $\hat{\mathbf{x}}$ is obtained by the backward substitution described by Eq. (5.2.8), where z and r are replaced by b and R, respectively. Observe that $\hat{\mathbf{x}}$ usually would be determined after processing all observations. However, intermediate values of $\hat{\mathbf{x}}$ could be determined at any point in the process.

The minimum value of J is given by substituting $\hat{\mathbf{x}}$ into Eq. (5.4.24):

$$J = \left\| \mathbf{e} \right\|^2 = \sum_{i=1}^{m} e_i^2 = \left\| \overline{R}\hat{\mathbf{x}} - \overline{\mathbf{b}} \right\|^2 + \sum_{i=1}^{m} (H_i\hat{\mathbf{x}} - y_i)^2. \tag{5.4.30}$$

Note that the first term on the right-hand side of Eq. (5.4.30) corresponds to the norm of the error in the *a priori* value for $\mathbf{x}$ multiplied by the square root of the inverse of the *a priori* covariance matrix,

$$\left\| \overline{R}\hat{\mathbf{x}} - \overline{\mathbf{b}} \right\|^2 = \left\| \overline{R}(\hat{\mathbf{x}} - \overline{\mathbf{x}}) \right\|^2 \tag{5.4.31}$$

which also can be expressed as

$$\left\| \overline{R}(\hat{\mathbf{x}} - \overline{\mathbf{x}}) \right\|^2 = (\hat{\mathbf{x}} - \overline{\mathbf{x}})^T \overline{P}^{-1} (\hat{\mathbf{x}} - \overline{\mathbf{x}}). \tag{5.4.32}$$

From Eqs. (5.4.30) and (5.4.31) it is seen that

$$\sum_{i=1}^{m} e_i^2 = \left\| \overline{R}(\hat{\mathbf{x}} - \overline{\mathbf{x}}) \right\|^2 + \sum_{i=1}^{m} (H_i\hat{\mathbf{x}} - y_i)^2$$

or

$$\sum_{i=1}^{m} e_i^2 = \|\overline{R}\hat{\boldsymbol{\eta}}\|^2 + \sum_{i=1}^{m} \hat{\epsilon}_i^2 \tag{5.4.33}$$

$$= \hat{\boldsymbol{\eta}}^T \overline{R}^T \overline{R}\hat{\boldsymbol{\eta}} + \sum_{i=1}^{m} \hat{\epsilon}_i^2,$$

where

$$\hat{\boldsymbol{\eta}} = \hat{\mathbf{x}} - \overline{\mathbf{x}}, \quad \hat{\epsilon}_i = y_i - H_i\hat{\mathbf{x}}. \tag{5.4.34}$$

Consequently, $e_i \neq \hat{\epsilon}_i$; that is, the elements of the error vector $[\mathbf{e}]_{m \times 1}$ contain a contribution from errors in the *a priori* information as well as the observation residuals. The RMS of the observation residuals, $\hat{\epsilon}_i$, is given by

$$\text{RMS} = \sqrt{\frac{1}{m}\sum_{i=1}^{m} \hat{\epsilon}_i^2} \tag{5.4.35}$$

and from Eq. (5.4.33)

$$\sum_{i=1}^{m} \hat{\epsilon}_i^2 = \sum_{i=1}^{m} e_i^2 - \hat{\boldsymbol{\eta}}^T \overline{R}^T \overline{R}\hat{\boldsymbol{\eta}}. \tag{5.4.36}$$

If the procedure is initialized with n observations in place of *a priori* information, $\overline{\mathbf{x}}$ and $\overline{P}$, Eq. (5.4.33) becomes

$$\sum_{i=1}^{m-n} e_i^2 = \sum_{i=1}^{m} \hat{\epsilon}_i^2 \tag{5.4.37}$$

and again $e_i \neq \hat{\epsilon}_i$ because the first n observations serve the same function as *a priori* values of $\overline{\mathbf{x}}$ and $\overline{P}$. Here we have assumed that the weighting matrix is the identity matrix. If not, $(H_i\hat{\mathbf{x}} - y_i)^2$ in Eq. (5.4.30) and subsequent equations should be replaced by $W_i(H_i\hat{\mathbf{x}} - y_i)^2$.

An advantage of the orthogonal transformation approach, in addition to improved accuracy, is that the sum of squares of the residuals, e_i, based on the final value of $\hat{\mathbf{x}}$ is computed automatically as part of the solution procedure. To obtain this result with the conventional batch processor, one would have to go through the additional computation of evaluating Eq. (5.4.30) after solving the normal equations for $\hat{\mathbf{x}}$.

This procedure of solving the least squares problem using orthogonal transformations can be described as a square root information batch processor. If the state vector is independent of time, it could be thought of as a filter because the best estimate of $\mathbf{x}$ could be generated after processing each observation. If the state vector is a function of time, the H matrix must be combined with the state transition matrix to map the state to an epoch time as described in Chapter 4. As formulated here, it could not be implemented as a filter without mapping the square root information matrix, R, and the state estimate, $\hat{\mathbf{x}}$, to the appropriate time for each observation. The square root information filter with time-dependent effects is described in Section 5.10.1.

In summary, given *a priori* information $\overline{R}$ and $\overline{\mathbf{b}}$ and observations

$$y_i = H_i\mathbf{x} + \epsilon_i, \quad i = 1,\ldots,m,$$

the matrix we wish to reduce to upper triangular form is

$$
\overbrace{}^{n} \quad \overbrace{}^{1}
$$

$$
\left.
\begin{bmatrix}
\overline{R} & \overline{\mathbf{b}} \\
H_1 & y_1 \\
H_2 & y_2 \\
\vdots & \vdots \\
H_m & y_m
\end{bmatrix}
\begin{array}{l}
\left.\rule{0pt}{8pt}\right\} n \\[18pt]
\left.\rule{0pt}{24pt}\right\} m
\end{array}
\right.
\tag{5.4.38}
$$

where $\overline{R}$ is upper triangular.

5.4.2   GIVENS COMPUTATIONAL ALGORITHM

For purposes of the computational algorithm we will write Eq. (5.4.38) as

$$
\overbrace{}^{n}\;\overbrace{}^{1}\qquad\quad \overbrace{}^{n+1}
$$

$$
\begin{bmatrix}
\overline{R} & \overline{\mathbf{b}} \\
H & \mathbf{y}
\end{bmatrix}
\begin{array}{l} \}n \\ \}m \end{array}
=
\begin{bmatrix}
\tilde{R} \\
\tilde{H}
\end{bmatrix}
\begin{array}{l} \}n \\ \}m. \end{array}
\tag{5.4.39}
$$

Lowercase r and h in the following algorithm represent the elements of $\tilde{R}$ and $\tilde{H}$, respectively, in Eq. (5.4.39).

The algorithm using the Givens rotation for reducing the $(m+n) \times (n+1)$ matrix of Eq. (5.4.39) to upper triangular form can be expressed as follows:

Sum = 0.

1. Do $k = 1, \ldots, m$

2. Do $i = 1, \ldots, n$

   If $(h_{ki} = 0)$ Go to 2

   $$r'_{ii} = \sqrt{r_{ii}^2 + h_{ki}^2}$$
   $$S_{ik} = h_{ki}/r'_{ii}$$
   $$C_{ik} = r_{ii}/r'_{ii}$$
   $$h_{ki} = 0$$
   $$r_{ii} = r'_{ii}$$

3. Do $j = i + 1, \ldots, n + 1$

   $$r'_{ij} = C_{ik}r_{ij} + S_{ik}h_{kj}$$
   $$h_{kj} = -S_{ik}r_{ij} + C_{ik}h_{kj} \qquad (5.4.40)$$
   $$r_{ij} = r'_{ij}$$

   Next j
   Next i
   $e_k = h_{kj}$
   Sum = Sum $+e_k^2$
   Next k

After application of this algorithm, the $(n + m) \times (n + 1)$ matrix will appear as

$$Q \begin{bmatrix} \overline{R} \vdots \overline{b} \\ H \vdots y \end{bmatrix} = \begin{bmatrix} R & b \\ O & e \end{bmatrix} \begin{matrix} \}n \\ \}m \end{matrix} \qquad (5.4.41)$$

which is the required form for solution of the least squares estimation problem as given by Eq. (5.3.6). Note that $r_{i,n+1}(i = 1, \ldots, n)$ and $h_{k,n+1}(k = 1, \ldots, m)$ given by the algorithm represent b and e, respectively, in Eq. (5.4.41). Also, Sum $= \sum_{k=1}^{m} e_k^2$.

Once the array has been reduced to the form given by Eq. (5.4.41), subsequent observations can be included by considering the following array:

$$\begin{bmatrix} R & b \\ H_{m+1} & y_{m+1} \\ 0 & e^2 \end{bmatrix} = \qquad (5.4.42)$$

$$
\begin{bmatrix}
R_{11} & R_{12} & \cdots & R_{1n} & b_1 \\
0 & R_{22} & \cdots & R_{2n} & b_2 \\
0 & 0 & \cdots & R_{3n} & b_3 \\
\vdots & & & & \\
0 & 0 & \cdots & R_{nn} & b_n \\
H_{m+1,1} & H_{m+1,2} & \cdots & H_{m+1,n} & y_{m+1} \\
0 & 0 & \cdots & 0 & e^2
\end{bmatrix}
$$

where

$$
e^2 = \sum_{k=1}^{m} e_k^2 = \text{Sum.}
$$

Then by application of a Givens rotation to rows 1 and $n+1$, $H_{m+1,1}$ can be nulled. Successive applications moving down the main diagonal can be used to null the remaining $n-1$ elements of the $n+1^{\text{st}}$ row and reduce the array to upper triangular form:

$$
\begin{bmatrix}
R' & b' \\
0 & e_{m+1} \\
0 & e^2
\end{bmatrix}.
$$

Next e^2 is replaced by $e^2 + e_{m+1}^2$ and the procedure is repeated with the next observation, and so on. It is also obvious that a group of m' observations could be included by replacing the array (H_{m+1}, y_{m+1}) with an array in which H_{m+1} has dimension $(m' \times n)$ and y_{m+1} has dimension m'. The Givens rotation would be used as before to reduce the augmented array to upper triangular form. Note that Sum is set to zero only before processing the first observation or batch of observations. Also note that if there are different observation types, e.g., range and range rate, the values of e for each observation type should be stored in separate arrays.

The Givens algorithm operates on an individual row (i.e., observation) until it is in the proper form and then moves to the next row of the matrix. The algorithm can also be applied so that it operates column by column simply by interchanging the k and i loops. The same procedure just described for processing a new observation or batch of observations still applies. The following Givens algorithm operates on the successive columns.

    Sum = 0.

1. Do $i = 1, \ldots, n$

2. Do $k = 1, \ldots, m$

If $(h_{ki} = 0)$ Go to 2

$$r'_{ii} = \sqrt{r_{ii}^2 + h_{ki}^2}$$
$$S_{ik} = h_{ki}/r'_{ii}$$
$$C_{ik} = r_{ii}/r'_{ii}$$
$$h_{ki} = 0$$
$$r_{ii} = r'_{ii}$$

3. Do $j = i + 1, \ldots, n + 1$

$$r'_{ij} = C_{ik}r_{ij} + S_{ik}h_{kj}$$
$$h_{kj} = -S_{ik}r_{ij} + C_{ik}h_{kj}$$
$$r_{ij} = r'_{ij}$$

Next j

Next k

Next i

Do $\ell = 1, \ldots, m$

Sum = Sum $+h_{\ell,n+1}^2$

Next ℓ

5.4.3 SQUARE ROOT FREE GIVENS TRANSFORMATION

The original Givens transformation requires the formation of square roots, which are more complex to compute than divisions or multiplications. The following procedure leads to a square root free algorithm.

Rather than seek the orthogonal transformation Q, which leads to $QW^{1/2}H = [R^T \vdots 0]^T$, we seek a factorization of R of the form

$$R = D^{\frac{1}{2}}U$$

where D is an $n \times n$ diagonal matrix and U is $n \times n$ unit upper triangular matrix, i.e.

$$D^{1/2} = \begin{bmatrix} d_1^{1/2} & & 0 \\ & d_2^{1/2} & \\ & & \cdot \\ 0 & & d_n^{1/2} \end{bmatrix}, U = \begin{bmatrix} 1 & U_{12} & \cdots & U_{1n} \\ 0 & 1 & U_{23} & \cdots & U_{2n} \\ \vdots & & & & \vdots \\ 0 & \cdots & \cdots & \cdots & 1 \end{bmatrix}. \qquad (5.4.43)$$

Following Eq. (5.3.6) select the orthogonal matrix Q, such that

$$QW^{\frac{1}{2}}H = \begin{bmatrix} D^{\frac{1}{2}}U \\ 0 \end{bmatrix} = \begin{bmatrix} D^{\frac{1}{2}} \vdots 0 \\ \cdots \quad \cdots \\ 0 \vdots 0 \end{bmatrix} \begin{bmatrix} U \\ \cdots \\ 0 \end{bmatrix} \qquad (5.4.44)$$

and

$$QW^{\frac{1}{2}}\mathbf{y} = \begin{bmatrix} \mathbf{b} \\ \mathbf{e} \end{bmatrix}. \qquad (5.4.45)$$

Let

$$\mathbf{b} = D^{\frac{1}{2}}\widetilde{\mathbf{b}}. \qquad (5.4.46)$$

It follows from Eq. (5.3.7) that the least squares performance index can be expressed as

$$J(\mathbf{x}) = \left\| \begin{bmatrix} D^{\frac{1}{2}}U \\ 0 \end{bmatrix} \mathbf{x} - \begin{bmatrix} D^{\frac{1}{2}}\widetilde{\mathbf{b}} \\ \mathbf{e} \end{bmatrix} \right\|^2 \qquad (5.4.47)$$

which for a minimum requires

$$D^{\frac{1}{2}}U\hat{\mathbf{x}} = D^{\frac{1}{2}}\widetilde{\mathbf{b}}. \qquad (5.4.48)$$

Since $D^{1/2}$ is common to both sides of Eq. (5.4.48), the solution is

$$U\hat{\mathbf{x}} = \widetilde{\mathbf{b}} \qquad (5.4.49)$$

and $\hat{\mathbf{x}}$ is obtained by backward recursion. Because the diagonals of U are unitary, division by the n diagonal elements is eliminated.

Consider now the use of the square root free Givens transformation to obtain the orthogonal decomposition (Gentleman, 1973). The product $W^{1/2}H$ can be expressed in component form as

$$W^{\frac{1}{2}}H = \begin{bmatrix} \sigma_1 h_{11} & \sigma_1 h_{12} & \cdots & \sigma_1 h_{1n} \\ \sigma_2 h_{21} & \sigma_2 h_{22} & \cdots & \sigma_2 h_{2n} \\ & \vdots & & \\ \sigma_m h_{m1} & \sigma_m h_{m2} & \cdots & \sigma_m h_{mn} \end{bmatrix} \tag{5.4.50}$$

where

$$\sigma_i = W_{ii}^{\frac{1}{2}}, W_{ij}^{\frac{1}{2}} = 0, i \neq j.$$

Now consider the application of the Givens rotation, where any two rows of Eq. (5.4.19) are expressed in the form

$$G \begin{bmatrix} \sqrt{d_i} & \sqrt{d_i}\, l_{i,i+1} & \cdots & \sqrt{d_i}\, l_{in} \\ \sqrt{\delta_k}\, h_{ki} & \sqrt{\delta_k}\, h_{k,i+1} & \cdots & \sqrt{\delta_k}\, h_{kn} \end{bmatrix}$$

$$= \begin{bmatrix} \sqrt{d_i'} & \sqrt{d_i'}\, l_{i,i+1}' & \cdots & \sqrt{d_i'}\, l_{in}' \\ 0 & \sqrt{\delta_k'}\, h_{k,i+1}' & \cdots & \sqrt{\delta_k'}\, h_{kn}' \end{bmatrix}. \tag{5.4.51}$$

From Eq. (5.4.50), it follows that

$$\sqrt{d_i} = \sigma_i h_{ii}$$

$$l_{ij} = \sigma_i h_{ij}/\sigma_i h_{ii}, j = i+1, \ldots, , n$$

$$\sqrt{\delta_k} = \sigma_k. \tag{5.4.52}$$

Then from Eq. (5.4.51), the Givens transformation, Eq. (5.4.2), applied to the first column leads to

$$\sqrt{d'} = \cos\theta\sqrt{d_i} + \sin\theta\sqrt{\delta_k}h_{ki}$$
$$0 = -\sin\theta\sqrt{d_i} + \cos\theta\sqrt{\delta_k}h_{ki}. \tag{5.4.53}$$

The second of Eq. (5.4.53) can be used to obtain

$$\tan\theta = \frac{\sin\theta}{\cos\theta} = \frac{\sqrt{\delta_k}h_{ki}}{\sqrt{d_i}}$$

and hence

$$\sin\theta = \sqrt{\delta_k}h_{ki} \Big/ \sqrt{d_i + \delta_k h_{ki}^2}$$

$$\cos \theta = \sqrt{d_i} \bigg/ \sqrt{d_i + \delta_k h_{ki}^2}. \tag{5.4.54}$$

Substituting (5.4.54) into (5.4.53) leads to

$$\sqrt{d_i'} = \sqrt{d_i + \delta_k h_{ki}^2}. \tag{5.4.55}$$

Hence, Eq. (5.4.54) becomes

$$\sin \theta = \sqrt{\delta_k}\, h_{ki}/\sqrt{d_i'}; \cos \theta = \sqrt{d_i}/\sqrt{d_i'}. \tag{5.4.56}$$

Then for the general transformation, the ij^{th} element of the first row is

$$\sqrt{d_i'}\ell_{ij}' = \cos \theta \sqrt{d_i}\ell_{ij} + \sin \theta \sqrt{\delta_k}\, h_{kj}$$
$$= \frac{\sqrt{d_i}\sqrt{d_i}}{\sqrt{d_i'}}\ell_{ij} + \frac{\sqrt{\delta_k}h_{ki}\sqrt{\delta_k}}{\sqrt{d_i'}}h_{kj}.$$

Now, dividing by $\sqrt{d_i'}$, yields

$$\ell_{ij}' = \frac{d_i}{d_i'}\ell_{ij} + \frac{\delta_k}{d_i'}h_{ki}h_{kj}. \tag{5.4.57}$$

Using the definitions

$$\overline{C}_i = d_i/d_i',\ \overline{S}_i = \delta_k h_{ki}/d_i'$$

Eq. (5.4.57) becomes

$$\ell_{ij}' = \overline{C}_i\ell_{ij} + \overline{S}_i h_{kj}. \tag{5.4.58}$$

Similarly, the kj^{th} element of the kth row is

$$\sqrt{\delta_k'}h_{kj}' = -\sin \theta \sqrt{d_i}\ell_{ij} + \cos \theta \sqrt{\delta_k}h_{kj}.$$

Then, using Eq. (5.4.56) results in

$$\sqrt{\delta_k'}h_{kj}' = \frac{-\sqrt{\delta_k}h_{ki}}{\sqrt{d_i'}}\sqrt{d_i}\, \ell_{ij} + \frac{\sqrt{d_i}}{\sqrt{d_i'}}\sqrt{\delta_k}h_{kj}. \tag{5.4.59}$$

Dividing by $\sqrt{\delta_k'}$ leads to

$$h_{kj}' = \frac{\sqrt{d_i}}{\sqrt{d_i'}}\frac{\sqrt{\delta_k}}{\sqrt{\delta_k'}}h_{kj} - \frac{\sqrt{\delta_k}\sqrt{d_i}}{\sqrt{d_i'}\sqrt{\delta_k'}}h_{ki}\ell_{ij}. \tag{5.4.60}$$

Because $\sqrt{\delta_k'}$ is an arbitrary scaling factor for each element of the kth row, it is convenient to let

$$\sqrt{\delta_k'} = \sqrt{\delta_k} \times \frac{\sqrt{d_i}}{\sqrt{d_i'}}. \tag{5.4.61}$$

It then follows that Eq. (5.4.60) can be expressed as

$$h_{kj}' = h_{kj} - h_{ki}l_{ij}. \tag{5.4.62}$$

The final values of l_{ij}' become the elements of U_{ij} and $\tilde{b}_i$, and h_{kj}' are the interim elements of the H matrix. The final value of h_{kj}' in each row is the observation error. This will become clear from the examples of Section 5.6.

5.4.4 SQUARE ROOT FREE GIVENS COMPUTATIONAL ALGORITHM

If *a priori* information, $\overline{x}$ and $\overline{P}$, are available the algorithm is initialized by computing $\overline{S}$ and $\overline{R}$ where

$$\overline{P} = \overline{S}\,\overline{S}^T \tag{5.4.63}$$

and

$$\overline{R} = \overline{S}^{\,-1}. \tag{5.4.64}$$

The *a priori* information we need is $\overline{D}, \overline{U},$ and $\tilde{b}$, where

$$\overline{R} = \overline{D}^{\frac{1}{2}}\overline{U}. \tag{5.4.65}$$

Hence,

$$\overline{d}_i = \overline{R}_{ii}^2 \ \ i = 1 \ldots n \tag{5.4.66}$$

and

$$\overline{U} = \begin{bmatrix} 1 & \overline{U}_{12} & \cdots & \cdots & \overline{U}_{1n} \\ 0 & 1 & \overline{U}_{23} & \cdots & \overline{U}_{2n} \\ 0 & 0 & 1 & & \vdots \\ \vdots & & & \ddots & \vdots \\ 0 & \cdots & \cdots & \cdots & 1 \end{bmatrix}, \tag{5.4.67}$$

where $\overline{d}_i$ are the square of the diagonal elements of $\overline{R}$; for example, $d_i = R_{ii}^2$, $i = 1, \ldots, n$, and $\overline{U}_{ij} = \overline{R}_{ij}/\overline{R}_{ii}$; $i = 1, \ldots n$; $j = i + 1, n$. Also, recall from Eq. (5.4.21) that

$$\overline{b} = \overline{R}\,\overline{x}$$

and the *a priori* value of $\tilde{\mathbf{b}} \equiv \overline{\overline{\mathbf{b}}}$ is

$$\overline{\overline{\mathbf{b}}} = \overline{D}^{-\frac{1}{2}} \, \overline{\mathbf{b}}. \qquad (5.4.68)$$

Given observations y_i and $i = 1 \ldots m$, we now have the necessary information to execute the algorithm; that is, we wish to apply a series of orthogonal transformations, G, so that

$$
G
\begin{bmatrix}
\overline{U} & \overline{\overline{\mathbf{b}}} \\
H_1 & y_1 \\
\vdots & \\
H_m & y_m
\end{bmatrix}
\left.\begin{array}{l} \\ \\ \\ \\ \end{array}\right\}
\begin{array}{l} n \\ \\ m \end{array}
=
\begin{bmatrix}
U & \tilde{\mathbf{b}} \\
& e_1 \\
0 & \vdots \\
& e_m
\end{bmatrix}
\left.\begin{array}{l} \\ \\ \\ \\ \end{array}\right\}
\begin{array}{l} n \\ \\ m \end{array}
. \qquad (5.4.69)
$$

The computational algorithm for operating on the data one row, or one observation, at a time is as follows:

Sum $= 0$

$U_{ii} = 1 \qquad i = 1, \ldots, n$

1. Do $k = 1, \ldots, m$

$$\delta_k = 1$$

2. Do $i = 1, \ldots, n$

If $(h_{ki} = 0)$ Go to 2

$$
\begin{aligned}
d_i' &= d_i + \delta_k h_{ki}^2 \\
\overline{C} &= d_i / d_i' \\
\overline{S} &= \delta_k h_{ki} / d_i' \\
y_k' &= y_k - \tilde{b}_i h_{ki} \\
\tilde{b}_i &= \tilde{b}_i \overline{C} + y_k \overline{S} \\
y_k &= y_k' \\
\delta_k &= \delta_k \overline{C} \\
d_i &= d_i'
\end{aligned}
\qquad (5.4.70)
$$

3. Do $j = i + 1, \ldots, n$

$$
\begin{aligned}
h_{kj}' &= h_{kj} - U_{ij} h_{ki} \\
U_{ij} &= U_{ij} \overline{C} + h_{kj} \overline{S} \\
h_{kj} &= h_{kj}'
\end{aligned}
$$

$$\text{Next } j$$
$$\text{Next } i$$
$$e_k = \sqrt{\delta_k}\, y_k$$
$$\text{Sum} = \text{Sum} + e_k^2$$

$$\text{Next } k$$

The diagonal elements of D are given by $d_i (i = 1, \ldots, n)$, the upper triangular elements of U are given by $U_{ij} (i = 1, \ldots, n, \; j = i+1, \ldots, n+1)$, the elements of $\tilde{\mathbf{b}}$ are given by $\tilde{b}_i \; (i = 1, \ldots, n)$ and the elements of $\mathbf{e}$ are given by $e_k (k = 1, \ldots, m)$. The same procedure described for the Givens algorithm at the end of Section 5.4.2 can be used to handle multiple batches of observation data. Note that Sum and δ_k are set to zero and one, respectively, only for the first batch of observations.

The vector $\hat{\mathbf{x}}$ is obtained from U and $\tilde{\mathbf{b}}$ by performing a back substitution using Eq. (5.4.49). Note that D is not needed to compute $\hat{\mathbf{x}}$ but is needed to compute the estimation error covariance matrix, $P = U^{-1}D^{-1}U^{-T}$ (see Eq. (5.4.80)).

5.4.5 A SIMPLIFIED SQUARE ROOT FREE GIVENS TRANSFORMATION

The square root free Givens algorithm can be simplified further (Gentleman, 1973) by noting that we may write

$$y_k' = y_k - \tilde{b}_i h_{ki} \tag{5.4.71}$$
$$\tilde{b}_i' = \tilde{b}_i \overline{C} + y_k \overline{S}$$

and

$$h_{kj}' = h_{kj} - U_{ij} h_{ki} \tag{5.4.72}$$
$$U_{ij}' = U_{ij}\overline{C} + h_{kj}\overline{S}$$

as

$$y_k' = y_k - \tilde{b}_i h_{ki} \tag{5.4.73}$$
$$\tilde{b}_i' = \tilde{b}_i + y_k'\overline{S}$$

and

$$h_{kj}' = h_{kj} - U_{ij}\, h_{ki} \tag{5.4.74}$$
$$U_{ij}' = U_{ij} + h_{kj}'\, \overline{S}$$

by noting that

$$\overline{C} = 1 - \overline{S}\, h_{ki}. \tag{5.4.75}$$

Hence, the algorithm may be simplified by eliminating the need to compute $\overline{C}$ explicitly. The computational algorithm becomes

Sum = 0.

1. Do $k = 1, \ldots, m$

$$\delta_k = 1$$

2. Do $i = 1, \ldots, n$

If ($h_{ki} = 0$) go to 2

$$
\begin{aligned}
d_i' &= d_i + \delta_k\, h_{ki}^2 \\
\overline{S} &= \delta_k\, h_{ki}/d_i' \\
y_k' &= y_k - \tilde{b}_i\, h_{ki} \\
\tilde{b}_i &= \tilde{b}_i + y_k'\, \overline{S} \\
y_k &= y_k' \\
\delta_k &= \delta_k\, d_i/d_i' \\
d_i &= d_i'
\end{aligned}
\tag{5.4.76}
$$

3. Do $j = i + 1, \ldots, n$

$$
\begin{aligned}
h_{kj}' &= h_{kj} - U_{ij}\, h_{ki} \\
U_{ij} &= U_{ij} + h_{kj}'\, \overline{S} \\
h_{kj} &= h_{kj}'
\end{aligned}
$$

Next j

Next i

$$e_k = \sqrt{\delta_k}\; y_k$$

Sum = Sum + e_k^2

Next k

This version of the square root free Givens algorithm is preferred because it involves fewer operations than that of Eq. (5.4.70).

5.4.6 IMPLEMENTATION CONSIDERATIONS

The following observations on the square root free algorithm are given:

1. Note that no square roots are required.

2. The algorithm assumes that *a priori* information $\overline{D}$, $\overline{U}$, and $\overline{\widetilde{b}}$ computed from $\overline{P}$ and $\overline{x}$ is available. These are computed using Eqs. (5.4.63) through (5.4.68):

$$
\begin{aligned}
\overline{P} &= \overline{S}\,\overline{S}^T \\
\overline{R} &= \overline{S}^{-1} \\
\bar{d}_i &= \overline{R}_{ii}^2, \quad i = 1\ldots n \\
\overline{U} &= \overline{D}^{-1/2}\,\overline{R} \\
\overline{\widetilde{b}} &= \overline{D}^{-1/2}\,\overline{R}\,\overline{x} = \overline{U}\overline{x}.
\end{aligned}
\tag{5.4.77}
$$

3. If no *a priori* information is given the algorithm may be initialized using

$$
\begin{aligned}
\overline{D} &= 10^{-16\ddagger} \\
\overline{U} &= I \\
\overline{\widetilde{b}} &= 0.
\end{aligned}
\tag{5.4.78}
$$

4. When an unprimed variable appears on both sides of an equation, this is taken to represent a replacement.

5. Remember that an orthogonal transformation does not change the norm or Euclidian length of each column of the original matrix.

6. Each of the Givens algorithms (Eqs. (5.4.40), (5.4.70), and (5.4.76)) operates on the subject matrix row by row. They can be modified to operate column by column by interchanging the k and i loops.

The value of $\hat{x}$ is computed from Eq. (5.4.49),

$$
U\hat{x} = \widetilde{b}
$$

---

‡$\overline{D}$ is an $n \times n$ diagonal matrix with 10^{-16} on the diagonal, using $\overline{D} = 0$ causes the algorithm to fail when computing $\widetilde{b}$; hence, a small number should be used. This is equivalent to $\overline{P}$ being diagonal with values of 10^{16} and $\overline{x} = 0$.

by using the back substitution algorithm obtained by a slight modification of Eq. (5.2.8):

$$i = n, \ n - 1, \ldots, 1$$

$$\hat{x}_i = \tilde{b}_i - \sum_{j=i+1}^{n} U_{ij} \hat{x}_j. \tag{5.4.79}$$

The estimation error covariance is computed from

$$\begin{aligned} P &= R^{-1} R^{-T} \\ &= (D^{\frac{1}{2}} U)^{-1} (U^T D^{\frac{1}{2}})^{-1} \\ &= U^{-1} D^{-\frac{1}{2}} D^{-\frac{1}{2}} U^{-T} \\ &= U^{-1} D^{-1} U^{-T} \end{aligned} \tag{5.4.80}$$

where U^{-1} may be computed by using Eq. (5.2.20).

5.5   THE HOUSEHOLDER TRANSFORMATION

An alternate computational approach can be developed by using orthogonal reflections rather than the planar rotations used in the previous discussions. Such transformations, which are referred to as Householder transformations (Householder, 1958), have the advantage of nulling a complete column in a single operation. Consider the following matrix

$$T = I - 2\hat{u}\hat{u}^T$$

where $\hat{u}^T \hat{u} = 1$. The matrix T satisfies the following conditions:

1) T is symmetric.

2) T is idempotent: $T^2 = I$. $\qquad\qquad$ (5.5.1)

3) T is orthogonal: $TT^T = I$.

Proof: The first condition follows from the definition. Then, since $\hat{u}\hat{u}^T$ is symmetric and $\hat{u}^T \hat{u} = 1$,

$$T^2 = (I - 2\hat{u}\hat{u}^T)(I - 2\hat{u}\hat{u}^T) = I - 4\hat{u}\hat{u}^T + 4\hat{u}(\hat{u}^T \hat{u})\hat{u}^T \tag{5.5.2}$$
$$= I - 4\hat{u}\hat{u}^T + 4\hat{u}\hat{u}^T = I.$$

From Properties 1 and 2, it follows that

$$T^2 = TT^T = I. \tag{5.5.3}$$

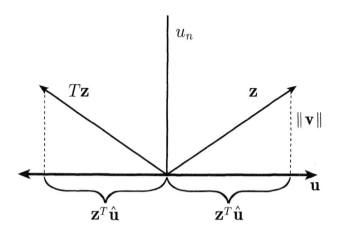

Figure 5.5.1: An elementary reflection.

Matrices of the form T are called elementary Hermitian matrices, elementary reflectors, or Householder transformations. The geometric notion of a reflection is shown in Fig. 5.5.1 and can be developed as follows. Let u_n be the plane perpendicular to $\mathbf{u}$. Let

$$\hat{\mathbf{u}} = \mathbf{u}\big/\|\mathbf{u}\|$$

and let $T\mathbf{z}$ be the reflection of $\mathbf{z}$ in the plane u_n—the mirror image of $\mathbf{z}$ where the plane u_n represents the mirror. That is,

$$\mathbf{z} = (\mathbf{z}^T\hat{\mathbf{u}})\hat{\mathbf{u}} + \mathbf{v} \qquad (5.5.4)$$

and

$$T\mathbf{z} = -(\mathbf{z}^T\hat{\mathbf{u}})\hat{\mathbf{u}} + \mathbf{v}. \qquad (5.5.5)$$

Eliminating $\mathbf{v}$ leads to

$$-T\mathbf{z} + \mathbf{z} = 2(\mathbf{z}^T\hat{\mathbf{u}})\hat{\mathbf{u}}.$$

Therefore,

$$\begin{aligned}
T\mathbf{z} &= \mathbf{z} - 2(\mathbf{z}^T\hat{\mathbf{u}})\hat{\mathbf{u}} \\
&= \mathbf{z} - 2\hat{\mathbf{u}}(\hat{\mathbf{u}}^T\mathbf{z}) \\
&= \mathbf{z} - 2\hat{\mathbf{u}}\hat{\mathbf{u}}^T\mathbf{z}
\end{aligned}$$

where we have used the fact that $\mathbf{z}^T\hat{\mathbf{u}} = \hat{\mathbf{u}}^T\mathbf{z}$ because this quantity is a scalar. Finally,

$$T\mathbf{z} = [I - 2\hat{\mathbf{u}}\hat{\mathbf{u}}^T]\mathbf{z}. \qquad (5.5.6)$$

Hence,

$$T = I - 2\frac{\mathbf{u}\mathbf{u}^T}{\mathbf{u}^T\mathbf{u}}$$
$$= I - \beta\mathbf{u}\mathbf{u}^T \tag{5.5.7}$$

where

$$\beta = 2/\mathbf{u}^T\mathbf{u} = 2\big/\|\mathbf{u}\|^2. \tag{5.5.8}$$

The following additional properties are of use:

4) If u_j, the jth component of $\mathbf{u}$, is zero, then $(T\mathbf{z})_j = z_j$. That is, if the jth component of $\mathbf{u}$ is zero, then the transformation T leaves the jth component of $\mathbf{z}$ unchanged.

Proof: This can be verified by writing down the product $\mathbf{u}\mathbf{u}^T$ and noting that the jth row and j^{th} column vanish if u_j is zero.

5) If $\mathbf{u}$ is perpendicular to $\mathbf{z}$, then $T\mathbf{z} = \mathbf{z}$.

Proof: Since $T\mathbf{z} = [I - \beta\mathbf{u}\mathbf{u}^T]\mathbf{z} = \mathbf{z} - \beta\mathbf{u}(\mathbf{u}^T\mathbf{z})$, it follows that $\mathbf{u}^T\mathbf{z} = 0$ if $\mathbf{u}$ is perpendicular to $\mathbf{z}$ and hence that $T\mathbf{z} = \mathbf{z}$.

6) $T\mathbf{z} = \mathbf{z} - \gamma\mathbf{u}$, where $\gamma = 2\mathbf{z}^T\mathbf{u}/\mathbf{u}^T\mathbf{u}$. $\tag{5.5.9}$

Proof:

$$T\mathbf{z} = \left[I - \frac{2\mathbf{u}\mathbf{u}^T}{\mathbf{u}^T\mathbf{u}}\right]\mathbf{z} = \mathbf{z} - 2\mathbf{u}\mathbf{u}^T\mathbf{z}/\mathbf{u}^T\mathbf{u}$$
$$= \mathbf{z} - \frac{2(\mathbf{z}^T\mathbf{u})\mathbf{u}}{\mathbf{u}^T\mathbf{u}} = \mathbf{z} - 2\left(\frac{\mathbf{z}^T\mathbf{u}}{\mathbf{u}^T\mathbf{u}}\right)\mathbf{u} \tag{5.5.10}$$
$$= \mathbf{z} - \gamma\mathbf{u}.$$

Here we use the fact that $\mathbf{u}^T\mathbf{z}$ is a scalar; hence, $\mathbf{u}^T\mathbf{z} = \mathbf{z}^T\mathbf{u}$. Note that the computation of T requires significantly more computation than the computation of $T\mathbf{z}$, if property 6 is used. Furthermore, using property 6, it is not necessary that T be stored; only the elements of $T\mathbf{z}$ need to be stored.

The feature of the Householder transformation, T, that is of crucial interest in this discussion is that T can be used to introduce zeros into any column vector, $\mathbf{z}$. One can always find an orthogonal matrix, T, such that $T\mathbf{z} = -\sigma\mathbf{e}_1$ where σ is a scalar and $\mathbf{e}_1$ is a unit vector in the direction of z_1, the first component of the column vector $\mathbf{z}$. The constant σ can be determined up to a sign by the fact that T is orthogonal. That is,

$$\|T\mathbf{z}\|^2 = \| -\sigma\mathbf{e}_1\|^2 = \sigma^2 = \|\mathbf{z}\|^2 = \mathbf{z}^T\mathbf{z}.$$

Therefore,

$$\sigma = \pm \|\mathbf{z}\|. \tag{5.5.11}$$

To obtain the specific transformation, T, which accomplishes this reduction, the vector, $\mathbf{u}$, must be determined. If we define $\mathbf{u}$ as

$$\mathbf{u} = \mathbf{z} + \sigma\mathbf{e} = \begin{bmatrix} z_1 \\ z_2 \\ \vdots \\ z_n \end{bmatrix} + \begin{bmatrix} \sigma \\ 0 \\ \vdots \\ 0_n \end{bmatrix} \tag{5.5.12}$$

where

$$\mathbf{e} = [1\ \ 0\ \ 0\ \cdots\ 0]^T \quad \text{is an } n \text{ vector}$$

$$\sigma = \text{sign}\ (z_1)(\mathbf{z}^T\mathbf{z})^{\frac{1}{2}} \tag{5.5.13}$$

then the orthogonal transformation $T = I - \beta\mathbf{u}\mathbf{u}^T$ is an elementary reflector and

$$T\mathbf{z} = -\sigma\mathbf{e}$$

where

$$\beta = \frac{2}{\mathbf{u}^T\mathbf{u}}. \tag{5.5.14}$$

Proof: Writing the expression for $T\mathbf{z}$ yields

$$\begin{aligned}
T\mathbf{z} &= \mathbf{z} - \beta\mathbf{u}\,\mathbf{u}^T\,\mathbf{z} \\
&= \mathbf{z} - \beta(\mathbf{u}^T\,\mathbf{z})\mathbf{u} \\
&= \mathbf{z} - \beta(\mathbf{z}^T\mathbf{z} + \sigma\mathbf{e}^T\mathbf{z})\mathbf{u} \\
&= \mathbf{z} - \beta(\sigma^2 + \sigma\mathbf{e}^T\mathbf{z})\mathbf{u}.
\end{aligned} \tag{5.5.15}$$

However,

$$\begin{aligned}
\beta &= \frac{2}{\mathbf{u}^T\mathbf{u}} = 2/(\mathbf{z}^T\mathbf{z} + 2\sigma\mathbf{e}^T\mathbf{z} + \sigma^2) \\
&= 1/(\sigma^2 + \sigma\mathbf{e}^T\mathbf{z})
\end{aligned} \tag{5.5.16}$$

or

$$\beta = \frac{1}{\sigma u_1} \tag{5.5.17}$$

where u_1 is the first element of $\mathbf{u}$.

Substituting Eq. (5.5.16) into (5.5.15) yields

$$T\mathbf{z} = \mathbf{z} - \mathbf{u}$$
$$= -\sigma \mathbf{e}. \tag{5.5.18}$$

From Eq. (5.5.14) and (5.5.17) note that

$$\mathbf{u}^T\mathbf{u} = 2\sigma u_1. \tag{5.5.19}$$

Also, from Eq. (5.5.9) we may write

$$\gamma = \beta \mathbf{z}^T \mathbf{u}$$

or

$$\gamma = \beta \mathbf{u}^T \mathbf{z}. \tag{5.5.20}$$

For each application of T to a given matrix, we zero out all elements below the main diagonal of the first column. In this operation all elements will be multiplied by β. Hence, we want to choose the sign of σ to maximize $\|u_1\|$. Because $u_1 = z_1 + \sigma$, we choose the sign of σ to be the same as that of z_1, the first element of $\mathbf{z}$.

In summary, the Householder equations to achieve the desired transformation are

$$\begin{aligned}
\sigma &= \text{sign } (z_1)(\mathbf{z}^T\mathbf{z})^{1/2} \\
z_i' &= -\sigma \\
u_1 &= z_1 + \sigma \\
u_i &= z_i, \quad z_i' = 0, \quad i = 2, \ldots \text{number of rows} \\
\beta &= 1/(\sigma u_1).
\end{aligned} \tag{5.5.21}$$

The remaining columns also must have T applied. From Eq. (5.5.20)

$$\gamma_j = \beta \, \mathbf{u}^T \mathbf{z}_j, \quad j = 2, \ldots \text{number of columns} \tag{5.5.22}$$

where γ_j is a scalar and $\mathbf{z}_j$ is the jth column of the matrix being transformed. The transformed columns, $\mathbf{z}_j'$, are computed using Eq. (5.5.9):

$$\mathbf{z}_j' = \mathbf{z}_j - \gamma_j\mathbf{u}. \tag{5.5.23}$$

A more detailed algorithm is given in Section 5.5.2.

5.5.1 APPLICATION TO THE SOLUTION OF THE LEAST SQUARES PROBLEM

Our goal is to upper triangularize an $(n + m) \times (n + 1)$ matrix using the Householder transformation:

$$T_n T_{n-1} \ldots T_1 \begin{bmatrix} \overline{R} & \overline{\mathbf{b}} \\ H & \mathbf{y} \end{bmatrix} \begin{matrix} \} n \\ \} m \end{matrix} = \begin{bmatrix} R & \mathbf{b} \\ 0 & \mathbf{e} \end{bmatrix} \begin{matrix} \} n \\ \} m. \end{matrix} \tag{5.5.24}$$

The procedure for accomplishing this is as follows. Let

$$A \equiv \begin{bmatrix} \overline{R} & \overline{\mathbf{b}} \\ H & \mathbf{y} \end{bmatrix}. \tag{5.5.25}$$

The first transformation $T_1 A$ will zero all elements of the first column except for the first element, which will have the opposite sign of $\overline{R}_{11}$ and magnitude equal to the Euclidian norm of the first column,

$$T_1 A = \left. \begin{array}{c} 1\{ \\ \\ m+n-1 \end{array} \right\{ \left[\begin{array}{c|ccccc} -\sigma_1 & \tilde{a}_{11} & \tilde{a}_{12} & \cdots & \tilde{a}_{1n} \\ \hline 0 & & & & \\ 0 & & & \tilde{A}_1 & \\ \vdots & & & & \\ 0 & & & & \end{array} \right] \begin{array}{c} \}1 \\ \\ \} m+n-1. \end{array} \tag{5.5.26}$$

The next transformation operates on $\tilde{A}_1$ and does not change the first row or column.

$$T_2 \tilde{A}_1 = \left. \begin{array}{c} 1\{ \\ \\ m+n-2 \end{array} \right\{ \left[\begin{array}{c|ccccc} -\sigma_2 & \tilde{a}_{21} & \tilde{a}_{22} & \cdots & \tilde{a}_{2,n-1} \\ \hline 0 & & & & \\ 0 & & & \tilde{A}_2 & \\ \vdots & & & & \\ 0 & & & & \end{array} \right] \begin{array}{c} \}1 \\ \\ \} m+n-2 \end{array} \tag{5.5.27}$$

This procedure is continued, zeroing elements below the diagonal until the upper $n \times n$ portion of A is upper triangular.

5.5.2   HOUSEHOLDER COMPUTATIONAL ALGORITHM

The computational algorithm for applying the Householder transformation is given by Bierman (1977). A few steps have been added here to aid in implementing the algorithm.

Given:

$$
\overbrace{\qquad}^{n}\ \overbrace{\quad}^{1}
$$

$$
\left.\begin{bmatrix} \overline{R} & \overline{\mathbf{b}} \\ H & \mathbf{y} \end{bmatrix}\begin{array}{l} \} \, n \\ \} \, m \end{array}\right. \equiv [A]_{(m+n)\times(n+1)} \tag{5.5.28}
$$

Do $k = 1, \, n$

$$
\sigma = \text{sign}(A_{kk})\left(\sum_{i=k}^{m+n}[A_{ik}]^2\right)^{1/2}
$$

$$
u_k = A_{kk} + \sigma
$$

$$
A_{kk} = -\sigma \tag{5.5.29}
$$

$$
u_i = A_{ik} \quad i = k+1, \ldots, m+n
$$

$$
\beta = \frac{1}{\sigma u_k}
$$

Do $j = k+1, \ldots, n+1$

$$
\gamma = \beta \sum_{i=k}^{m+n} u_i A_{ij}
$$

$$
A_{ij} = A_{ij} - \gamma u_i \quad i = k, \ldots, m+n
$$

Next j
$$
A_{ik} = 0 \quad i = k+1, \ldots, m+n
$$

Next k
$$
\text{Sum} = \sum_{i=1}^{m}[A_{n+i,n+1}]^2
$$

Upon completion, the elements of A represent R, b, and e, respectively:

$$A = \begin{bmatrix} \overset{\overbrace{n}}{R} & \overset{\overbrace{1}}{b} \\ 0 & e \end{bmatrix} \begin{matrix} \}n \\ \}m \end{matrix} \tag{5.5.30}$$

and

$$\text{Sum} = \sum_{i=1}^{m} e_i^2.$$

The algorithm can be initialized with *a priori* $\overline{R}$ and $\overline{b}$. If no *a priori* is available it can be initialized with $\overline{R} = 0$, $\overline{b} = 0$. Alternatively we may define A as

$$A_{m \times (n+1)} = m\{ \begin{bmatrix} \overset{\overbrace{n}}{H} & \vdots & \overset{\overbrace{1}}{y} \end{bmatrix} \tag{5.5.31}$$

and the operations with index i range to m instead of $m + n$. In this case the final result is

$$A = \begin{bmatrix} \overset{\overbrace{n}}{R} & \overset{\overbrace{1}}{b} \\ 0 & e \end{bmatrix} \begin{matrix} \}n \\ \}m - n \end{matrix}. \tag{5.5.32}$$

Recall from the discussion on Givens transformation in Section 5.4.1 that

$$\sum_{i=1}^{m} e_i^2 = \|\overline{R}(\hat{\mathbf{x}} - \overline{\mathbf{x}})\|^2 + \sum_{i=1}^{m} \hat{e}_i^2 \tag{5.5.33}$$

where

$$\hat{\epsilon}_i = y_i - H_i \hat{\mathbf{x}}. \tag{5.5.34}$$

If the initialization procedure of Eq. (5.5.31) is used, Eq. (5.5.33) becomes

$$\sum_{i=1}^{m-n} e_i^2 = \sum_{i=1}^{m} \hat{\epsilon}_i^2. \tag{5.5.35}$$

As with the Givens transformation, the value of $\hat{\mathbf{x}}$ is obtained by a backward substitution using Eq. (5.2.8); that is, $R\hat{\mathbf{x}} = b$ and

$$\hat{x}_i = \left(b_i - \sum_{j=i+1}^{n} R_{ij} \hat{x}_j \right) \Big/ R_{ii} \quad i = n, \ldots, 1 \tag{5.5.36}$$

and the estimation error covariance is obtained from

$$
\begin{aligned}
P = \Lambda^{-1} = (R^T \, R)^{-1} \\
= R^{-1} \, R^{-T} \\
= S \, S^T
\end{aligned}
\tag{5.5.37}
$$

where S is obtained from Eq. (5.2.9):

$$
i = 1, \ldots, n
$$

$$
S_{ii} = \frac{1}{R_{ii}}
\tag{5.5.38}
$$

$$
S_{ij} = -S_{jj} \left[\sum_{k=i}^{j-1} R_{kj} S_{ik} \right] ; \quad j = i+1, \ldots, n \, .
$$

5.6   NUMERICAL EXAMPLES

First consider a case with no *a priori* information. Assume we are given the observation-state matrix, H, and the observation vector, **y**, as follows:

$$
\mathbf{y} = H\mathbf{x} + \boldsymbol{\epsilon}
$$

where

$$
\mathbf{y} = \begin{bmatrix} -1 \\ 1 \\ 2 \end{bmatrix}, \, H = \begin{bmatrix} 1 & -2 \\ 2 & -1 \\ 1 & 1 \end{bmatrix}, \, \mathbf{x} = \begin{bmatrix} x_1 \\ x_2 \end{bmatrix}.
\tag{5.6.1}
$$

Assume that the prewhitening transformation described in Section 5.7.1 has been applied so that $\boldsymbol{\epsilon} \sim (O, I)$; that is, $W = I$. Use the Cholesky decomposition algorithm and the Givens and Householder transformations to solve for $\hat{\mathbf{x}}$.

5.6.1   CHOLESKY DECOMPOSITION

The basic equations are

$$
H^T H \mathbf{x} = H^T \mathbf{y}, \qquad M\mathbf{x} = \mathbf{N}
$$

$$
M = H^T H = \begin{bmatrix} 1 & 2 & 1 \\ -2 & -1 & 1 \end{bmatrix} \begin{bmatrix} 1 & -2 \\ 2 & -1 \\ 1 & 1 \end{bmatrix} = \begin{bmatrix} 6 & -3 \\ -3 & 6 \end{bmatrix}
\tag{5.6.2}
$$

$$\mathbf{N} = H^T \mathbf{y} = \begin{bmatrix} 1 & 2 & 1 \\ -2 & -1 & 1 \end{bmatrix} \begin{bmatrix} -1 \\ 1 \\ 2 \end{bmatrix} = \begin{bmatrix} 3 \\ 3 \end{bmatrix}. \tag{5.6.3}$$

Find R such that

$$R^T R = M, \text{ where } R \text{ is upper triangular (UT).}$$

Using the Cholesky algorithm given by Eq. (5.2.6),

$$i = 1, \ldots, n, \, j = i + 1, \ldots, n$$

$$r_{ii} = \left(M_{ii} - \sum_{k=1}^{i-1} r_{ki}^2 \right)^{1/2} \tag{5.6.4}$$

$$r_{ij} = \left(M_{ij} - \sum_{k=1}^{i-1} r_{ki} r_{kj} \right) / r_{ii},$$

the following results are obtained:

$$i = 1, 2, \text{ and } j = 2 \text{ (for this example)}$$

$$r_{11} = \sqrt{6}, r_{12} = -3/\sqrt{6}, r_{22} = 3/\sqrt{2}$$

$$R = \begin{bmatrix} \sqrt{6} & -3/\sqrt{6} \\ 0 & 3/\sqrt{2} \end{bmatrix}$$

$$R^T \mathbf{z} = H^T \mathbf{y}$$

$$\begin{bmatrix} \sqrt{6} & 0 \\ -3/\sqrt{6} & 3/\sqrt{2} \end{bmatrix} \begin{bmatrix} z_1 \\ z_2 \end{bmatrix} = \begin{bmatrix} 3 \\ 3 \end{bmatrix}$$

$$\mathbf{z} = \begin{bmatrix} 3/\sqrt{6} \\ 3/\sqrt{2} \end{bmatrix}$$

$$R\hat{\mathbf{x}} = \mathbf{z}$$

$$\begin{bmatrix} \sqrt{6} & -3/\sqrt{6} \\ 0 & 3/\sqrt{2} \end{bmatrix} \begin{bmatrix} x_1 \\ x_2 \end{bmatrix} = \begin{bmatrix} 3/\sqrt{6} \\ 3/\sqrt{2} \end{bmatrix}$$

$$\hat{\mathbf{x}} = \begin{bmatrix} 1 \\ 1 \end{bmatrix}.$$

If this were a more complex problem, we would use Eqs. (5.2.7) and (5.2.8) to compute $\mathbf{z}$ and $\hat{\mathbf{x}}$, respectively.

5.6.2   GIVENS TRANSFORMATION

For this example,

$$G[H \vdots \mathbf{y}] = \begin{bmatrix} R \vdots \mathbf{b} \\ \phi \vdots e \end{bmatrix},$$

where $R \equiv 2 \times 2$ UT,
$\phi \equiv 1 \times 2$ null matrix
it follows then that $b \equiv 2 \times 1$
and $e \equiv 1 \times 1$.

(5.6.5)

G is computed directly here using Eq. (5.4.8).

Operating on rows 1 and 2 ($i = 1, k = 2$) yields

$$\begin{bmatrix} \cos\theta & \sin\theta & 0 \\ -\sin\theta & \cos\theta & 0 \\ 0 & 0 & 1 \end{bmatrix} \begin{bmatrix} 1 & -2 & \vdots & -1 \\ 2 & -1 & \vdots & 1 \\ 1 & 1 & \vdots & 2 \end{bmatrix}$$

$$h'_{ii} = \sqrt{h^2_{ii} + h^2_{ki}}, \qquad h'_{11} = \sqrt{1 + 4} = \sqrt{5}$$

$$\sin\theta = S = h_{ki} / h'_{ii} = 2 / \sqrt{5}$$
$$\cos\theta = C = h_{ii} / h'_{ii} = 1 / \sqrt{5}$$

$$\begin{bmatrix} 1/\sqrt{5} & 2/\sqrt{5} & 0 \\ -2/\sqrt{5} & 1/\sqrt{5} & 0 \\ 0 & 0 & 1 \end{bmatrix} \begin{bmatrix} 1 & -2 & -1 \\ 2 & -1 & 1 \\ 1 & 1 & 2 \end{bmatrix} =$$

$$\begin{bmatrix} \sqrt{5} & -4/\sqrt{5} & 1/\sqrt{5} \\ 0 & 3/\sqrt{5} & 3/\sqrt{5} \\ 1 & 1 & 2 \end{bmatrix}.$$

Operating on rows 1 and 3 ($i = 1, k = 3$) yields

$$h'_{11} = \sqrt{6}$$

$$S = h_{31} / h'_{11} = 1 / \sqrt{6}, C = h_{11} / h'_{11} = \sqrt{5} / \sqrt{6}$$

$$
\begin{bmatrix}
\sqrt{5}/\sqrt{6} & 0 & 1/\sqrt{6} \\
0 & 1 & 0 \\
-1/\sqrt{6} & 0 & \sqrt{5}/\sqrt{6}
\end{bmatrix}
\begin{bmatrix}
\sqrt{5} & -4/\sqrt{5} & 1/\sqrt{5} \\
0 & 3/\sqrt{5} & 3/\sqrt{5} \\
1 & 1 & 2
\end{bmatrix}
$$

$$
=
\begin{bmatrix}
\sqrt{6} & -3/\sqrt{6} & 3/\sqrt{6} \\
0 & 3/\sqrt{5} & 3/\sqrt{5} \\
0 & 9/\sqrt{30} & 9/\sqrt{30}
\end{bmatrix}.
$$

The final operation is on rows 2 and 3 ($i = 2$, $k = 3$), which results in

$$
h'_{22} = \sqrt{9/5 + 81/30} = \sqrt{135/30}
$$
$$
S = h_{32}/h'_{22} = 9/\sqrt{135}
$$
$$
C = h_{22}/h'_{22} = 3\sqrt{6}/\sqrt{135}
$$

$$
\begin{bmatrix}
1 & 0 & 0 \\
0 & 3\sqrt{6}/\sqrt{135} & 9/\sqrt{135} \\
0 & -9/\sqrt{135} & 3\sqrt{6}/\sqrt{135}
\end{bmatrix}
\begin{bmatrix}
\sqrt{6} & -3/\sqrt{6} & 3/\sqrt{6} \\
0 & 3/\sqrt{5} & 3/\sqrt{5} \\
0 & 9/\sqrt{30} & 9/\sqrt{30}
\end{bmatrix}
$$

$$
=
\begin{bmatrix}
\sqrt{6} & -3/\sqrt{6} & \vdots & 3/\sqrt{6} \\
0 & 3/\sqrt{2} & \vdots & 3/\sqrt{2} \\
0 & 0 & \vdots & 0
\end{bmatrix}.
$$

We now have the necessary information to solve for $\hat{\mathbf{x}}$. Using

$$
R\hat{\mathbf{x}} = \mathbf{b}
$$

yields

$$
\begin{bmatrix}
\sqrt{6} & -3/\sqrt{6} \\
0 & 3/\sqrt{2}
\end{bmatrix}
\begin{bmatrix}
\hat{x}_1 \\
\hat{x}_2
\end{bmatrix}
=
\begin{bmatrix}
3/\sqrt{6} \\
3/\sqrt{2}
\end{bmatrix}.
$$

Solving for $\hat{\mathbf{x}}$ by a backward recursion yields

$$
\hat{\mathbf{x}} =
\begin{bmatrix}
1 \\
1
\end{bmatrix}.
$$

The associated covariance matrix is

$$
P = (H^T H)^{-1} = (H^T G^T G H)^{-1} = R^{-1} R^{-T}
$$

$$P = \begin{bmatrix} 2/9 & 1/9 \\ 1/9 & 2/9 \end{bmatrix}.$$

Note that in this example the observation residual, e, has a value of zero. This is because the three observations were perfect, from Eq. (5.5.34):

$$\hat{\epsilon} = \mathbf{y} - H\hat{\mathbf{x}} = 0.$$

5.6.3 HOUSEHOLDER TRANSFORMATION

This algorithm is given by Eq. (5.5.29). Because there is no *a priori* information, we use the initialization procedure of Eq. (5.5.31).

For this example,

$$[H \vdots \mathbf{y}] = A = \begin{bmatrix} 1 & -2 & -1 \\ 2 & -1 & 1 \\ 1 & 1 & 2 \end{bmatrix}.$$

Each transformation, T, results in

$$T A = \begin{bmatrix} -\sigma & \cdots \\ \vdots & \tilde{A} \\ 0 & \end{bmatrix}.$$

Note that the algorithm given by Eq. (5.5.29) does not use the array, $\tilde{A}$. It simply redefines the elements of A, thereby requiring less computer storage. Beginning with

$$k = 1$$

$$\sigma = +\sqrt{1 + 2^2 + 1} = \sqrt{6}$$
$$A_{11} = -\sigma = -\sqrt{6}$$
$$u_1 = A_{11} + \sigma = 1 + \sqrt{6} = 3.4495$$
$$u_2 = A_{21} = 2$$
$$u_3 = A_{31} = 1$$
$$\beta = \frac{1}{\sigma\, u_1} = \frac{1}{\sqrt{6} + 6} = 0.11835$$

$$j = 2$$

$$\gamma = \beta \sum_{i=1}^{3} u_i A_{ij}$$
$$= 0.11835 \left[3.4495 \left(-2 \right) + 2 \left(-1 \right) + 1 \right]$$
$$= -0.9348$$

$$A_{12} = A_{12} - \gamma u_1$$
$$= -2 + 0.9348 \left(3.4495 \right)$$
$$= 1.2247$$
$$A_{22} = A_{22} - \gamma u_2 = -1 + 0.9348 \left(2 \right)$$
$$= 0.8696$$
$$A_{32} = 1 + 0.9348 = 1.9348$$

$$j = 3$$

$$\gamma = 0.11835 \left[3.4495(-1) + 2 + 2 \right]$$
$$= 0.06515$$

$$A_{13} = -1 - .06515 \left(3.4495 \right)$$
$$= -1.2247$$

$$A_{23} = 1 - 0.06515 \left(2 \right)$$
$$= 0.8697$$

$$A_{33} = A_{33} + \gamma u_3$$
$$= 2 - 0.06515$$
$$= 1.9348.$$

Hence,

$$\begin{bmatrix} \sigma \cdots \\ \vdots \ \widetilde{A} \\ 0 \end{bmatrix} = \begin{bmatrix} -2.4495 & \vdots & 1.2247 & -1.2247 \\ \cdots\cdots & \vdots & \cdots\cdots & \cdots\cdots \\ 0 & \vdots & 0.8697 & 0.8697 \\ 0 & \vdots & 1.9348 & 1.9348 \end{bmatrix} \qquad (5.6.6)$$

$$k = 2$$

$$\sigma = \sqrt{(0.8697)^2 + (1.9348)^2}$$
$$= 2.1212$$
$$\tilde{A}_{11} = -2.1212$$
$$u_2 = A_{22} + \sigma = 0.8697 + 2.1212$$
$$= 2.991$$
$$u_3 = A_{32} = 1.9348$$

$$\beta = \frac{1}{\sigma\, u_2} = \frac{1}{(2.1212)(2.991)} = 0.1576$$

$$j = 3$$

$$\gamma = \beta [u_2 u_3] \begin{bmatrix} A_{22} \\ A_{32} \end{bmatrix}$$

$$= 0.1576\, [2.991 \quad 1.9348] \begin{bmatrix} 0.8697 \\ 1.9348 \end{bmatrix}$$

$$\gamma = 1$$

$$\tilde{A}_{ij} = A_{ij} - \gamma\, u_i, \quad j = 3, \quad i = 2, 3$$
$$\tilde{A} = \begin{bmatrix} 0.8697 \\ 1.9348 \end{bmatrix} - \begin{bmatrix} 2.991 \\ 1.9348 \end{bmatrix} = \begin{bmatrix} -2.1212 \\ 0 \end{bmatrix} \tag{5.6.7}$$

so that the final result is

$$TA = \begin{bmatrix} -2.4495 & 1.2247 & -1.2247 \\ 0 & -2.1212 & -2.1212 \\ 0 & 0 & 0 \end{bmatrix} \tag{5.6.8}$$

which agrees with the Givens result except for the sign, which has no effect on $\hat{x}$ or P. Note that, as in the case for the Givens transformation, the observation residual is zero, implying that the observations were perfect.

5.6.4   A MORE ILLUSTRATIVE EXAMPLE

For this example we will include *a priori* information and observation errors. This will illustrate how to initialize the least squares procedure with *a priori* information using the Cholesky decomposition and orthogonal transformations and will also demonstrate the feature of computing the sum of squares of the observation residuals.

Assume that we are given observations

$$y = Hx + \epsilon$$

and that we have prewhitened the observation errors so that $\epsilon \sim (O, I)$. Values for y and H are

$$y = \begin{bmatrix} -1.1 \\ 1.2 \\ 1.8 \end{bmatrix}, \quad H = \begin{bmatrix} 1 & -2 \\ 2 & -1 \\ 1 & 1 \end{bmatrix}, \quad (5.6.9)$$

with assumed *a priori* information

$$\overline{P} = \begin{bmatrix} 100 & 0 \\ 0 & 100 \end{bmatrix}, \overline{x} = \begin{bmatrix} 2 \\ 2 \end{bmatrix}. \quad (5.6.10)$$

Our objective is to determine the best estimate, $\hat{x}$, of the constant vector x using the Cholesky decomposition, the Givens square root free, and Householder algorithms. We also wish to find the estimation error covariance matrix and the sum of squares of the observation residuals.

5.6.5 CHOLESKY DECOMPOSITION

If the observations are given unit weight, for example, $W = I$, the normal equations are

$$(H^T H + \overline{P}^{-1})x = H^T y + \overline{P}^{-1} \overline{x} \quad (5.6.11)$$

or

$$Mx = N \quad (5.6.12)$$

where

$$M = H^T H + \overline{P}^{-1} \quad (5.6.13)$$
$$N = H^T y + \overline{P}^{-1} \overline{x}. \quad (5.6.14)$$

Substituting H, $\overline{P}$, y, and $\overline{x}$ into Eqs. (5.6.13) and (5.6.14) yields

$$M = \begin{bmatrix} 6.01 & -3 \\ -3 & 6.01 \end{bmatrix}, \quad N = \begin{bmatrix} 3.12 \\ 2.82 \end{bmatrix}. \quad (5.6.15)$$

Using the Cholesky algorithm given by Eq. (5.2.6) to compute R, where $M = R^T R$, yields

$$R = \begin{bmatrix} 2.452 & -1.224 \\ 0 & 2.124 \end{bmatrix}. \tag{5.6.16}$$

From Eq. (5.2.4)

$$R^T \mathbf{z} = \mathbf{N},$$

from Eq. (5.2.7)

$$\mathbf{z} = \begin{bmatrix} 1.273 \\ 2.0609 \end{bmatrix}, \tag{5.6.17}$$

and from Eq. (5.2.8)

$$\hat{\mathbf{x}} = \begin{bmatrix} 1.003 \\ .970 \end{bmatrix}. \tag{5.6.18}$$

The covariance of the estimation error is given by Eq. (5.2.5),

$$P = E(\hat{\mathbf{x}} - \mathbf{x})(\hat{\mathbf{x}} - \mathbf{x})^T = R^{-1} R^{-T} = S S^T. \tag{5.6.19}$$

S is given by Eq. (5.2.9)

$$S = \begin{bmatrix} .408 & .235 \\ 0 & .471 \end{bmatrix}.$$

Hence,

$$P = \begin{bmatrix} .222 & .111 \\ .111 & .222 \end{bmatrix}. \tag{5.6.20}$$

The sum of squares of the observation errors, including errors in the *a priori* information, is given by

$$e^2 = (\hat{\mathbf{x}} - \bar{\mathbf{x}})^T \bar{P}^{-1} (\hat{\mathbf{x}} - \bar{\mathbf{x}}) + \sum_{i=1}^{3} (y_i - H_i \hat{\mathbf{x}})^2 \tag{5.6.21}$$

$$= .0205 + .0266 + .0269 + .0299$$

$$e^2 = .1039.$$

5.6.6   SQUARE ROOT FREE GIVENS TRANSFORMATION

The initialization procedure for the orthogonal transformation algorithms is described in Section 5.4.1. This involves computing the square root of the *a priori* information matrix, $\overline{R}$, and $\overline{\mathbf{b}} = \overline{R}\,\overline{\mathbf{x}}$. Recall that $\overline{P} = \overline{S}\,\overline{S}^T$ and $\overline{R} = \overline{S}^{-1}$. For this example

$$\overline{R} = \begin{bmatrix} 0.1 & 0 \\ 0 & 0.1 \end{bmatrix}, \quad \overline{\mathbf{b}} = \begin{bmatrix} 0.2 \\ 0.2 \end{bmatrix}. \tag{5.6.22}$$

We will use the square root free Givens rotation given by Eq. (5.4.70). For this we need $\overline{D}$ and $\overline{U}$, where $\overline{R} = \overline{D}^{\frac{1}{2}}\,\overline{U}$. Hence,

$$\overline{D} = \begin{bmatrix} .01 & 0 \\ 0 & .01 \end{bmatrix}, \quad \overline{U} = \begin{bmatrix} 1 & 0 \\ 0 & 1 \end{bmatrix}. \tag{5.6.23}$$

The *a priori* value of $\widetilde{\mathbf{b}}$ is given by

$$\begin{aligned} \overline{\widetilde{\mathbf{b}}} &= \overline{D}^{-\frac{1}{2}}\overline{\mathbf{b}} = \overline{D}^{-\frac{1}{2}}\overline{R}\,\overline{\mathbf{x}} \\ &= \overline{U}\,\overline{\mathbf{x}}, \end{aligned} \tag{5.6.24}$$

so

$$\overline{\widetilde{\mathbf{b}}} = \begin{bmatrix} 2 \\ 2 \end{bmatrix}.$$

The matrix we wish to compress is

$$\begin{bmatrix} \overline{U} & \overline{\widetilde{\mathbf{b}}} \\ H & \mathbf{y} \end{bmatrix} = \begin{bmatrix} 1 & 0 & 2 \\ 0 & 1 & 2 \\ 1 & -2 & -1.1 \\ 2 & -1 & 1.2 \\ 1 & 1 & 1.8 \end{bmatrix}. \tag{5.6.25}$$

Recall that Givens algorithm as defined by Eq. (5.4.70) operates on the matrix one row at a time. After processing the first observation we have

$$D = \begin{bmatrix} 1.0100 & 0 \\ 0 & 0.0496 \end{bmatrix}, \quad U = \begin{bmatrix} 1.00 & -1.9802 \\ 0 & 1.00 \end{bmatrix}$$

$$\widetilde{\mathbf{b}} = \begin{bmatrix} -1.0693 \\ 1.6407 \end{bmatrix}, \quad e = 0.0402.$$

Processing the second observation results in

$$D = \begin{bmatrix} 5.0100 & 0 \\ 0 & 1.8164 \end{bmatrix}, \quad U = \begin{bmatrix} 1.0 & -0.7984 \\ 0 & 1.0 \end{bmatrix}$$

$$\tilde{\mathbf{b}} = \begin{bmatrix} 0.2635 \\ 1.1418 \end{bmatrix}, \quad e = -0.1127.$$

After processing the third observation,

$$D = \begin{bmatrix} 6.010 & 0 \\ 0 & 4.5125 \end{bmatrix}, U - \begin{bmatrix} 1.0 & -0.4992 \\ 0 & 1.0 \end{bmatrix}$$

(5.6.26)

$$\tilde{\mathbf{b}} = \begin{bmatrix} 0.5191 \\ 0.9701 \end{bmatrix}, \quad e = -0.2994.$$

To compare with the Householder results we must multiply by $D^{1/2}$:

(5.6.27)

$$\begin{bmatrix} R & \mathbf{b} \\ 0 & e \end{bmatrix} = \begin{bmatrix} D^{\frac{1}{2}}U & D^{\frac{1}{2}}\tilde{\mathbf{b}} \\ 0 & e \end{bmatrix} \begin{matrix} \}2 \\ \}3 \end{matrix}$$

$$= \begin{bmatrix} 2.4515 & -1.2237 & 1.2727 \\ 0 & 2.1243 & 2.0607 \\ 0 & 0 & 0.0402 \\ 0 & 0 & -0.1127 \\ 0 & 0 & -0.2994 \end{bmatrix}.$$

In actual practice, to conserve computer storage, we would read and process the observations one at a time and store the observations residuals as

$$e^2 = \sum_{i=1}^{m} e_i^2.$$

$\hat{\mathbf{x}}$ is computed using Eq. (5.4.49),

$$U\hat{\mathbf{x}} = \tilde{\mathbf{b}}$$

or

$$\begin{bmatrix} 1.00 & -0.4992 \\ 0 & 1.00 \end{bmatrix} \begin{bmatrix} \hat{x}_1 \\ \hat{x}_2 \end{bmatrix} = \begin{bmatrix} 0.5191 \\ 0.9701 \end{bmatrix}$$

and

$$\begin{bmatrix} \hat{x}_1 \\ \hat{x}_2 \end{bmatrix} = \begin{bmatrix} 1.0034 \\ 0.9701 \end{bmatrix}. \tag{5.6.28}$$

The sum of squares of observation residuals is given by

$$e^2 = \sum_{i=1}^{3} e_i^2 = (0.0402)^2 + (-0.1127)^2 + (-0.2994)^2$$
$$= 0.1039.$$

The covariance of the estimation error is given by Eq. (5.4.80), $P = U^{-1} D^{-1} U^{-T}$. Equation (5.2.20) may be used to compute U^{-1},

$$U^{-1} = \begin{bmatrix} 1.0 & 0.4992 \\ 0 & 1.0 \end{bmatrix}$$

and

$$P = \begin{bmatrix} 1.0 & 0.4992 \\ 0 & 1.0 \end{bmatrix} \begin{bmatrix} .1664 & 0 \\ 0 & .2216 \end{bmatrix} \begin{bmatrix} 1.0 & 0 \\ 0.4992 & 1.0 \end{bmatrix}$$
$$= \begin{bmatrix} .2216 & .1106 \\ .1106 & .2216 \end{bmatrix}.$$

Notice that the results for $\hat{x}$, P, and e^2 agree with the Cholesky results.

5.6.7 THE HOUSEHOLDER TRANSFORMATION

The matrix we wish to transform is given by Eq. (5.6.25). In terms of $\overline{R}$ and $\overline{b}$ the matrix is given by

$$\begin{bmatrix} \overline{R} & \overline{b} \\ H & y \end{bmatrix} = \begin{bmatrix} 0.1 & 0 & 0.2 \\ 0 & 0.1 & 0.2 \\ 1 & -2 & -1.1 \\ 2 & -1 & 1.2 \\ 1 & 1 & 1.8 \end{bmatrix}. \tag{5.6.29}$$

The Householder transformation algorithm given by Eq. (5.5.29) nulls the elements of each column below the main diagonal. The first transformation yields

$$
\begin{bmatrix}
-2.4515 & 1.2237 & -1.2727 \\
0 & 0.1 & 0.2 \\
0 & -1.5204 & -1.6772 \\
0 & -0.0408 & 0.0457 \\
0 & 1.4796 & 1.2228
\end{bmatrix}.
$$

The second transformation results in

$$
\begin{bmatrix}
-2.4515 & 1.2237 & -1.2727 \\
0 & -2.1243 & -2.0607 \\
0 & 0 & -0.1319 \\
0 & 0 & 0.0871 \\
0 & 0 & -0.2810
\end{bmatrix}. \tag{5.6.30}
$$

Several points should be noted:

1. The Householder values of R and b are identical to the Givens results. Hence, the solution for $\hat{x}$ and P will be identical.

2. Although the individual values of e_i differ, both algorithms yield identical values of $e^2 = 0.1039$. This also agrees with the Cholesky result.

3. The Euclidean norm of each column is preserved by an orthogonal transformation.

4. The square root free Givens algorithm as derived here operates on the matrix row by row, whereas the Householder algorithm transforms column by column. The Givens algorithm can be modified to operate column by column (see Section 5.4.2).

5. The orthogonal transformations do not require the formation of $H^T H$. Hence, they will generally be more accurate than Cholesky decomposition.

5.7 SQUARE ROOT FILTER ALGORITHMS

Although the sequential estimation algorithms have had wide use in autonomous navigation and control applications, there has been a reluctance to adopt the sequential estimation algorithm for real-time orbit determination mission support.

The primary reason for this reluctance is due to the phenomenon of filter divergence, during which the estimate of the state can depart in an unbounded manner from the true value of the state. There are two fundamental reasons for filter divergence. The first of these is due to inaccuracies in the mathematical model used to describe the dynamic process or in the model used to relate the observations to the state.

A second factor that can cause filter divergence is associated with the errors that occur in the measurement update of the state error covariance matrix. In particular, this matrix can become nonpositive definite, a situation that is a theoretical impossibility due to the effects of using finite word length arithmetic to compute the update of the state error covariance matrix at the point where an observation is incorporated. Since this type of divergence is related to errors introduced during the computational procedure, it should be possible to reformulate the computational process to minimize the effects of such errors.

To accomplish this objective, several filter modifications, referred to as square root covariance filters, have been proposed in which the state error covariance matrix is replaced by its square root. The state error covariance matrix is obtained by multiplying the square root matrix by its transpose and will always be symmetric and positive semidefinite. Note that the algorithms presented in the remainder of this chapter are designed to accommodate scalar observations.

The motivation for considering the square root measurement update algorithms stems from the loss of significant digits that occurs in computing the measurement update of the state error covariance matrix at the observation epoch (Kaminski *et al.*, 1971) . When the eigenvalues have a wide spread, the error introduced in the computational process can destroy the symmetry and positive definite character of the covariance matrix and filter divergence may occur. The square root measurement update philosophy, which has been proposed to alleviate this condition, can be expressed as follows.

Define W, the state error covariance matrix square root such that

$$P = WW^T. \tag{5.7.1}$$

Note that P, if computed using Eq. (5.7.1), can never be nonpositive definite even in the presence of round-off or truncation errors. Furthermore, since P is symmetric and positive definite, there will exist an orthogonal matrix M such that

$$P^* = M^T P M, \tag{5.7.2}$$

where P^* is a diagonal matrix whose elements are the eigenvalues of P and M is the corresponding matrix of eigenvectors (Graybill, 1961). Define W^* as the matrix whose diagonal elements are equal to the square root of the diagonal elements of P^*:

$$W_{ii}^* = \sqrt{P_{ii}^*}\, i = 1, \ldots, n \tag{5.7.3}$$

where $P_{ii}^* > 0$. Then note that

$$W^* W^{*T} = P^* = M^T P M = M^T W W^T M.$$

Hence, $W^* = M^T W$ and, since M is an orthogonal matrix, it follows that

$$W = MW^*. \qquad (5.7.4)$$

If the matrix is symmetrical and positive definite, there are other methods of computing the square root matrix. For example, see the Cholesky decomposition discussed in Section 5.2.

The numerical conditioning of W is generally much better than that of P. The *conditioning number* $C(P)$ of P can be defined as (Lawson and Hanson, 1974)

$$C(P) = \gamma_{\max} / \gamma_{\min}, \qquad (5.7.5)$$

where γ_{max} is the maximum eigenvalue of P and γ_{min} is the minimum eigenvalue. In base 10 arithmetic with p significant digits, numerical difficulties with matrix inversion and the precision of $\hat{x}$ may be encountered as $C(P) \rightarrow 10^p$. However, for W,

$$C(W) = \sqrt{C(P)}.$$

Hence, numerical difficulties should not be encountered until

$$C(W) = 10^p$$

or

$$C(P) = 10^{2p}.$$

5.7.1 THE SQUARE ROOT MEASUREMENT UPDATE ALGORITHMS

Using these ideas, the covariance measurement update equation, Eq. (4.7.10), can be expressed in square root form as follows:

$$P = \overline{P} - \overline{P} H^T \left[H \overline{P} H^T + R \right]^{-1} H \overline{P}. \qquad (5.7.6)$$

Now, let $P = WW^T$ and make this substitution in Eq. (5.7.6) to obtain

$$WW^T = \overline{W}\, \overline{W}^T - \overline{W}\, \overline{W}^T H^T \left[H \overline{W}\, \overline{W}^T H^T + R \right]^{-1} H \overline{W}\, \overline{W}^T. \qquad (5.7.7)$$

Using the following definitions

$$\tilde{F} = \overline{W}^T H^T, \alpha = (\tilde{F}^T \tilde{F} + R)^{-1}, \qquad (5.7.8)$$

Eq. (5.7.7) can be expressed as

$$WW^T = \overline{W}[I - \tilde{F}\alpha\tilde{F}^T]\overline{W}^T. \tag{5.7.9}$$

If a matrix $\tilde{A}$ can be found such that

$$\tilde{A}\tilde{A}^T = I - \tilde{F}\alpha\tilde{F}^T, \tag{5.7.10}$$

then Eq. (5.7.9) can be expressed as

$$WW^T = \overline{W}\tilde{A}\tilde{A}^T\overline{W}^T. \tag{5.7.11}$$

Hence,

$$W = \overline{W}\tilde{A}. \tag{5.7.12}$$

The square root measurement update algorithm can be expressed as follows:

$$\tilde{F} = \overline{W}^T H^T$$
$$\alpha = (R + \tilde{F}^T\tilde{F})^{-1}$$
$$K = \overline{W}\tilde{F}\alpha$$
$$W = \overline{W}\tilde{A}$$
$$\hat{\mathbf{x}} = \overline{\mathbf{x}} + K(\mathbf{y} - H\overline{\mathbf{x}}),$$

where $\tilde{A} = [I - \tilde{F}\alpha\tilde{F}^T]^{1/2}$. The primary differences in the various algorithms for computing the measurement update in square root form lie in the manner in which the matrix $\tilde{A}$ is computed. The method first used in practice is that given by Potter (Battin, 1999).

The Potter Square Root Update

If attention is restricted to processing a single scalar observation, α will be a scalar. For this application, Potter considered the problem of finding the matrix $\tilde{A}$ such that

$$\tilde{A}\tilde{A}^T = [I - \alpha\tilde{F}\tilde{F}^T] = [I - \gamma\alpha\tilde{F}\tilde{F}^T][I - \gamma\alpha\tilde{F}\tilde{F}^T]^T, \tag{5.7.13}$$

where γ is an unspecified scalar parameter whose value is to be selected to satisfy Eq. (5.7.13). Expanding the right-hand side of Eq. (5.7.13) yields

$$I - \alpha\tilde{F}\tilde{F}^T = I - 2\alpha\gamma\tilde{F}\tilde{F}^T + \alpha^2\gamma^2(\tilde{F}\tilde{F}^T)(\tilde{F}\tilde{F}^T).$$

Canceling the identity and factoring $\tilde{F}\tilde{F}^T$ after noting that $\tilde{F}^T\tilde{F}$ is a scalar leads to

$$(1 - 2\gamma + \alpha\gamma^2\tilde{F}^T\tilde{F})\alpha\tilde{F}\tilde{F}^T = 0. \tag{5.7.14}$$

The solution, $\widetilde{F}\widetilde{F}^T = 0$, is trivial. Hence, the solution of Eq. (5.7.14) leads to the following condition:

$$1 - 2\gamma + \gamma^2 \alpha \widetilde{F}^T \widetilde{F} = 0. \qquad (5.7.15)$$

From Eq. (5.7.15), it follows that γ must satisfy the following relation

$$\gamma = \frac{2 \pm \sqrt{4 \quad 4\alpha \widetilde{F}^T \widetilde{F}}}{2\alpha \widetilde{F}^T \widetilde{F}}.$$

After some algebra, this can be simplified to

$$\gamma = \frac{1}{1 \pm \sqrt{R\alpha}}, \qquad (5.7.16)$$

where the $+$ sign is chosen to ensure that a singular value, $\gamma = \infty$, does not occur.

Using Eq. (5.7.16) the computational algorithm, based on the sequential estimation algorithm discussed in Chapter 4, can be expressed as follows:

Given $\overline{W}$, H, R, $\overline{\mathbf{x}}$ and $\mathbf{y}$

Compute:

$$\begin{aligned}
&1) && \widetilde{F} &= \overline{W}^T H^T \\
&2) && \alpha &= (\widetilde{F}^T \widetilde{F} + R)^{-1} \\
&3) && \gamma &= 1/(1 + \sqrt{R\alpha}) \\
&4) && K &= \alpha \overline{W} \widetilde{F} \\
&5) && \hat{\mathbf{x}} &= \overline{\mathbf{x}} + K(\mathbf{y} - H\overline{\mathbf{x}}) \\
&6) && W &= \overline{W} - \gamma K \widetilde{F}^T.
\end{aligned} \qquad (5.7.17)$$

Note that even if $\overline{W}$ is triangular, the computation involved in Eq. (5.7.17) will result in a nontriangular form for W. Also the computation of Eq. (5.7.17) involves two divisions and a square root that are not involved in the conventional Kalman algorithm. Consequently, Eq. (5.7.17) will be slower than the conventional algorithm; however, it will be more accurate. If $\overline{P}$ is given instead of $\overline{W}$, a Cholesky decomposition may be used to compute $\overline{W}$ in order to initialize the algorithm.

The time update for the state error covariance square root at time t_k for the Potter algorithm follows directly from the time update of the state error covariance. In the general case where there is process noise

$$\begin{aligned}
\overline{P}_k &= \Phi(t_k, t_{k-1})W_{k-1}W_{k-1}^T\Phi^T(t_k, t_{k-1}) \\
&\quad + \Gamma(t_k, t_{k-1})V_k V_k^T \Gamma^T(t_k, t_{k-1}) \\
&= \overline{W}_k \overline{W}_k^T
\end{aligned} \qquad (5.7.18)$$

where

$$Q_k = V_k V_k^T.$$

A Givens algorithm to compute an upper (or lower) triangular matrix, $\overline{W}_k$, is given in Section 5.8, Eqs. (5.8.6) through (5.8.13). If there is no process noise, set $V_k = 0$ in this algorithm. Methods for maintaining the measurement update in triangular form are discussed next.

Triangular Square Root Measurement Update

The algorithm for performing the measurement update on the state error co-variance matrix square root, $\overline{W}$, can be expressed as

$$W = \overline{W}\widetilde{A}. \qquad (5.7.19)$$

If $\overline{W}$ is lower triangular and if $\widetilde{A}$ is lower triangular, then W will be lower triangular, and the computational process expressed in Eq. (5.7.19) will be conducted in a more efficient manner. Choose $\widetilde{A}$ as the solution to

$$\widetilde{A}\widetilde{A}^T = [I - \alpha\widetilde{F}\widetilde{F}^T], \qquad (5.7.20)$$

and require that $\widetilde{A}$ be lower triangular. $\widetilde{A}$ will contain $(n^2 + n)/2$ unknowns. Since the right-hand side is symmetric, Eq. (5.7.20) contains $(n^2 + n)/2$ unique equations for the $(n^2 + n)/2$ unknowns. By analytically expanding the results we can determine the expressions for $\tilde{a}_{ij}$, the elements of $\widetilde{A}$. The procedure used is as follows.

If $\overline{W}$ is lower triangular, the equation

$$\widetilde{F} = \overline{W}^T H^T$$

can be expressed as

$$\begin{bmatrix} \widetilde{F}_1 \\ \vdots \\ \widetilde{F}_n \end{bmatrix} = \begin{bmatrix} \overline{W}_{11} & \cdots\cdots & \overline{W}_{n1} \\ & \overline{W}_{22} & \cdots\cdots & \overline{W}_{n2} \\ & & \ddots & \vdots \\ O & & & \overline{W}_{nn} \end{bmatrix} \begin{bmatrix} H_1 \\ \vdots \\ H_n \end{bmatrix}$$

$$= \begin{bmatrix} \displaystyle\sum_{j=1}^{n} \overline{W}_{j1} H_j \\ \vdots \\ \overline{W}_{nn} H_n \end{bmatrix} \tag{5.7.21}$$

or, for the general i^{th} element,

$$\widetilde{F}_i = \sum_{j=i}^{n} \overline{W}_{ji} H_j, \quad i = 1, \dots, n. \tag{5.7.22}$$

From Eq. (5.7.20), the matrix B can be defined as

$$B = \widetilde{A}\widetilde{A}^T = I - \alpha \widetilde{F}\widetilde{F}^T. \tag{5.7.23}$$

The general ij element of the product matrix in Eq. (5.7.23) can be expressed as

$$B_{ij} = \sum_{k=1}^{j} a_{ik} a_{jk} = \delta_{ij} - \alpha \widetilde{F}_i \widetilde{F}_j, i = 1, \dots, n; j = 1, \dots, i. \tag{5.7.24}$$

From Eq. (5.7.24), it follows that

$$
\begin{aligned}
B_{11} &= a_{11}^2 = 1 - (\widetilde{F}_1^2 \alpha) \\
B_{21} &= a_{21} a_{11} = -(\widetilde{F}_2 \widetilde{F}_1 \alpha) \\
B_{22} &= a_{21}^2 + a_{22}^2 = 1 - (\widetilde{F}_2^2 \alpha) \\
B_{31} &= a_{31} a_{11} = -(\widetilde{F}_3 \widetilde{F}_1 \alpha) \\
B_{32} &= a_{31} a_{21} + a_{32} a_{22} = -(\widetilde{F}_3 \widetilde{F}_2 \alpha) \\
B_{33} &= a_{31}^2 + a_{32}^2 + a_{33}^2 = 1 - (\widetilde{F}_3^2 \alpha) \\
B_{41} &= a_{41} a_{11} = -(F_4 F_1 \alpha).
\end{aligned}
\tag{5.7.25}
$$

$$\vdots$$

If the definition

$$\beta_i = \frac{1}{\alpha} - \sum_{j=1}^{i-1} \widetilde{F}_j^2 = R + \sum_{j=i}^{n} \widetilde{F}_j^2, \tag{5.7.26}$$

where

$$\frac{1}{\alpha} = R + \sum_{j=1}^{n} \widetilde{F}_j^2$$

is used, then

$$\beta_{i-1} = \beta_i + \tilde{F}_{i-1}^2 \, i = n+1, \dots, 2 \qquad (5.7.27)$$

where β_i satisfies the conditions

$$\beta_{n+1} = R, \beta_1 = \tfrac{1}{\alpha}. \qquad (5.7.28)$$

If Eq. (5.7.25) are solved recursively, starting with a_{11}, the solution can be expressed as

$$a_{ii} = \sqrt{\beta_{i+1}/\beta_i} \qquad (5.7.29)$$

$$a_{ij} = -\tilde{F}_i \tilde{F}_j / \sqrt{\beta_{j+1}\beta_j} \qquad \begin{array}{l} i = 1, \dots, n \\ j = 1, \dots, i-1. \end{array}$$

Once the elements of $\tilde{A}$ have been computed, the elements of Eq. (5.7.19) can be expressed as follows:

$$W_{ij} = \sum_{k=j}^{i} \overline{W}_{ik} a_{kj} \qquad \begin{array}{l} i = 1, \dots, n \\ j = 1, \dots, i. \end{array} \qquad (5.7.30)$$

The algorithm for the state estimate with a lower triangular square root covariance update is:

Given: $\overline{W}$, H, R, $\overline{\mathbf{x}}$, and y, where $\overline{W}$ is lower triangular
Compute:

1. $\tilde{F}_i = \displaystyle\sum_{j=1}^{n} \overline{W}_{ji} H_j \qquad i = 1, \ldots, n$

2. $\beta_{n+1} = R$

   $\beta_j = \beta_{j+1} + \tilde{F}_j^2 \qquad j = n, \ldots, 1$

   $\beta_1 = 1/\tilde{\alpha}$

3. $d_i = 1/\sqrt{\beta_{i+1}\beta_i} \qquad i = 1, \ldots, n$

   $a_{ii} = \beta_{i+1} d_i$ (5.7.31)

   $a_{ij} = -\tilde{F}_i \tilde{F}_j d_j \qquad j = 1, \ldots, i-1$

4. $W_{ij} = \displaystyle\sum_{k=j}^{i} \overline{W}_{ik} a_{kj} \qquad \begin{array}{l} i = 1, \ldots, n \\ j = 1, \ldots, i \end{array}$

5. $K_i = \alpha\left(\displaystyle\sum_{j=1}^{i} \overline{W}_{ij} \tilde{F}_j\right) \qquad i = 1, \ldots, n$

6. $\hat{\mathbf{x}}_i = \overline{\mathbf{x}} + K(y - H\overline{\mathbf{x}})$.

If $\mathbf{y}$ is a vector of observations, this procedure must be repeated for each element of $\mathbf{y}$.

The previous algorithm yields a lower triangular square root factorization of the measurement update error covariance matrix. An upper triangularization also can be used. The algorithm for this was developed by Carlson (1973). The version presented here is due to Bierman (1977) and is a slightly modified version of Carlson's algorithm to enhance accuracy.

Given: $\overline{W}$, H, R, $\overline{\mathbf{x}}$, and y, where $\overline{W}$ is upper triangular
Compute:

1. $\tilde{F} = \overline{W}^T H^T$

   $\beta_o = R, \quad K_2^T = \left[\overline{W}_{11}\tilde{F}_1, \overbrace{0, 0, \ldots 0}^{n-1}\right]$

   for $j = 1, \ldots n$ cycle through steps 2 through 7

   (for $j = 1$ do not evaluate steps 6 and 7)   (5.7.32)

2. $\beta_j = \beta_{j-1} + \widetilde{F}_j^2$

3. $d_j = (\beta_{j-1}/\beta_j)^{1/2}$

4. $\gamma_j = \widetilde{F}_j/(\beta_j d_j)$

5. $W_{jj} = \overline{W}_{jj} d_j$ (5.7.33)

6. $W_{ij} = \overline{W}_{ij} d_j - \gamma_j K_j(i)$ $i = 1, \ldots j - 1$

7. $K_{j+1}(i) = K_j(i) + \widetilde{F}_j \overline{W}_{ij}$ $i = 1, \ldots j$

8. $K = K_{n+1}/\beta_n$

9. $\hat{\mathbf{x}} = \overline{\mathbf{x}} + K(y - H\overline{\mathbf{x}})$.

Although at first glance it does not seem to matter whether one uses an upper or lower triangular form for W, Carlson (1973) points out that the time update for the upper triangular form is computationally more efficient. This can be seen by partitioning the state vector such that

$$\mathbf{x} = \begin{bmatrix} \mathbf{x}_1 \\ \mathbf{x}_2 \end{bmatrix},$$

where $\mathbf{x}_1$ contains the dynamical parameters such as position and velocity and $\mathbf{x}_2$ contains constants such as gravity coefficients or time-varying quantities such as gas leaks that affect the elements of $\mathbf{x}_1$ but are not affected by $\mathbf{x}_1$. Hence, the state transition matrix for $\mathbf{x}$ can be partitioned as

$$\Phi = \begin{bmatrix} \Phi_{11} & \Phi_{12} \\ O & \Phi_{22} \end{bmatrix},$$

where Φ_{11} is a dense matrix, Φ_{12} may contain some zeros, and Φ_{22} is typically diagonal. When performing the time update, if W is upper triangular, only the upper left partition of the product $\overline{W} = \Phi W$ becomes nontriangular and requires retriangularization.

Recall that these algorithms are designed to process the observations (acutally observation deviations) as scalars. Hence, if $\mathbf{y}$ is a vector the elements of $\mathbf{y}$ must be processed one at a time. We assume that the observations are uncorrelated so that R is a diagonal matrix.

If R is not a diagonal matrix, that is, the observation errors are correlated, we can perform a whitening and decorrelation transformation given by Bierman (1977), and described next.

Given a set of observations

$$\mathbf{y} - H\mathbf{x} = \epsilon \,, \tag{5.7.34}$$

where

$$E[\epsilon] = 0$$
$$E[\epsilon\epsilon^T] = R$$

and R is not diagonal but is positive definite. Compute the triangular square root of R

$$R = VV^T \,. \tag{5.7.35}$$

The Cholesky algorithm may be used to ensure that V is triangular. Next multiply Eq. (5.7.34) by V^{-1}

$$V^{-1}\mathbf{y} = V^{-1}H\mathbf{x} + V^{-1}\epsilon \tag{5.7.36}$$

let

$$\tilde{\mathbf{y}} = V^{-1}\mathbf{y}, \tilde{H} = V^{-1}H, \tilde{\epsilon} = V^{-1}\epsilon$$

then

$$\tilde{\mathbf{y}} = \tilde{H}x + \tilde{\epsilon} \tag{5.7.37}$$

It is easily shown that $\tilde{\epsilon}$ has zero mean and unit variance,

$$E[\tilde{\epsilon}] = V^{-1}E[\epsilon] = 0$$
$$E[\tilde{\epsilon}\,\tilde{\epsilon}^T] = V^{-1}E[\epsilon\,\epsilon^T]V^{-T} = V^{-1}RV^{-T} = I \tag{5.7.38}$$

Hence, we would process the observations, $\tilde{\mathbf{y}}$, instead of $\mathbf{y}$.

5.7.2   SQUARE ROOT FREE MEASUREMENT UPDATE ALGORITHMS

The square root operations in the Potter and Triangular algorithms shown earlier lead to increased computational load. Square root free equivalents of these algorithms can be used to obtain enhanced computational performance. A comprehensive discussion of these square root free algorithms is given by Thornton (1976) and Bierman (1977).

A square root free algorithm, known as the *U–D covariance factorization*, developed by Bierman and Thornton is described next. Let

$$P = UDU^T \tag{5.7.39}$$

where U is unit upper triangular and D is diagonal. Substituting Eq. (5.7.39) into the expression for the covariance matrix update

$$P = [I - KH]\overline{P} \tag{5.7.40}$$

and restricting attention to the scalar observation case leads to the following expression:

$$UDU^T = \overline{U}\,\overline{D}\,\overline{U}^T - \overline{U}\,\overline{D}\,\overline{U}^T H^T \alpha H \overline{U}\,\overline{D}\,\overline{U}^T \tag{5.7.41}$$

where

$$\alpha = (H\overline{P}H^T + R)^{-1} = \left(H\overline{U}\,\overline{D}\,\overline{U}^T H^T + R\right)^{-1}. \tag{5.7.42}$$

Now, by factoring, Eq. (5.7.41) can be expressed as follows:

$$UDU^T = \overline{U}\left[\overline{D} - \overline{D}\,\overline{U}^T H^T \alpha H \overline{U}\,\overline{D}\right]\overline{U}^T. \tag{5.7.43}$$

Let

$$F = \overline{U}^T H^T, V = \overline{D}F, \tag{5.7.44}$$

then Eq. (5.7.43) can be expressed as

$$P = UDU^T = \overline{U}\left[\overline{D} - V\alpha V^T\right]\overline{U}^T, \tag{5.7.45}$$

where

$$\overline{U} = \begin{bmatrix} 1\,\overline{U}_{12} & \cdots & \overline{U}_{1n} \\ & 1 & & \vdots \\ & & \ddots & \overline{U}_{n-1n} \\ & & & 1 \end{bmatrix}; \overline{D} = \begin{bmatrix} \overline{d}_1 & & 0 \\ & \overline{d}_2 & \\ & & \ddots \\ 0 & & & \overline{d}_n \end{bmatrix}. \tag{5.7.46}$$

$$F = \begin{bmatrix} F_1 \\ \vdots \\ F_n \end{bmatrix} = \overline{U}^T H^T; V = \overline{D}F = \begin{bmatrix} \overline{d}_1 F_1 \\ \overline{d}_2 F_2 \\ \vdots \\ \overline{d}_n F_n \end{bmatrix}. \tag{5.7.47}$$

Now let $\widetilde{U}$ and $\widetilde{D}$ be the factors of $[\overline{D} - V\alpha V^T]$. Then Eq. (5.7.45) can be written as

$$UDU^T = \overline{U}\widetilde{U}\,\widetilde{D}\widetilde{U}^T\overline{U}^T. \tag{5.7.48}$$

Since $\overline{U}$ and $\widetilde{U}$ are upper triangular, their product will be upper triangular, and can be expressed as

$$U = \overline{U}\widetilde{U} \tag{5.7.49}$$
$$D = \widetilde{D}\,.$$

Hence, the problem of factoring the covariance measurement update for P has been reduced to factoring the symmetric matrix $[\overline{D} - V\alpha V^T]$ into $\widetilde{U}$ and $\widetilde{D}$. This can be done using the square root free Cholesky decomposition or using the more computationally efficient algorithm presented in Thornton (1976) or Bierman (1977). This algorithm together with an example also appears in Maybeck (1979).

The computational algorithm for the measurement update of P can be summarized then as follows:

Given: $\overline{U}$, $\overline{D}$, H, R, $\overline{x}$, and y

Compute:

1. $F = \overline{U}^T H^T$

2. $V_i = \overline{d}_i F_i,$ $\qquad\qquad i = 1, \ldots, n$

3. $\alpha_1 = (R + V_1 F_1)$

   $d_1 = \dfrac{\overline{d}_1 R}{\alpha_1}$

   $b_1 = V_1$

4. $\alpha_j = \alpha_{j-1} + F_j V_j$

   $d_j = \overline{d}_j \alpha_{j-1} / \alpha_j$ $\qquad\qquad\left.\vphantom{\begin{matrix}1\\2\\3\\4\end{matrix}}\right\}$ $j = 2, \ldots, n$

   $b_j = V_j$

   $p_j = -F_j / \alpha_{j-1}$

5. $U_{ij} = \overline{U}_{ij} + b_i p_j$ $\qquad j = 2, \ldots, n$

   $b_i = b_i + \overline{U}_{ij} V_j$ $\qquad i = 1, \ldots, j-1$

6. $K = \mathbf{b} / \alpha_n$

7. $\hat{\mathbf{x}} = \overline{\mathbf{x}} + K(y - H\overline{\mathbf{x}}).$

Note that $\mathbf{b}$ is an n vector, $[b_1, b_2 \ldots b_n]^T$. If the estimation error covariance matrix is needed, it may be calculated by using Eq. (5.7.39).

Thus far, with the exception of the Potter Algorithm, we have not addressed the time update of the estimation error covariance matrix. This will be addressed in Sections 5.8 and 5.9.

5.8 TIME UPDATE OF THE ESTIMATION ERROR COVARIANCE MATRIX

Section 5.7 deals primarily with the measurement update of the estimation error covariance matrix, P. Various discrete algorithms have been developed for the time update. The most obvious approach for the time update would be to map WW^T or UDU^T using the conventional measurement update equation given in Chapter 4, for W_k and V_k,

$$
\begin{aligned}
\overline{P}_{k+1} &= \Phi(t_{k+1}, t_k) P_k \Phi^T(t_{k+1}, t_k) + \Gamma(t_{k+1}, t_k) Q_k \Gamma^T(t_{k+1}, t_k) \\
&= \Phi(t_{k+1}, t_k) W_k W_k^T \Phi^T(t_{k+1}, t_k) \\
&\quad + \Gamma(t_{k+1}, t_k) V_k V_k^T \Gamma^T(t_{k+1}, t_k)
\end{aligned}
\tag{5.8.1}
$$

where

$$
Q_k = V_k V_k^T .
$$

$\overline{P}_{k+1}$ may then be triangularized using a Cholesky transformation. After this, the measurement update is applied to process the next vector of observations.

Although this approach is computationally efficient, it is not as accurate as an orthogonal transformation approach that does not require the formation of the covariance matrix with the attendant numerical problems. A more accurate approach would be to find a propagation equation for the square root of $\overline{P}$. This can be accomplished by noting that the right-hand side of Eq. (5.8.1) may be written as

$$
\overline{P}_{k+1} = B_{k+1} B_{k+1}^T
\tag{5.8.2}
$$

where

$$
B_{k+1} = \left[\Phi(t_{k+1}, t_k) W_k \ \vdots \ \Gamma(t_{k+1}, t_k) V_k \right] .
\tag{5.8.3}
$$

Note, however, that B_{k+1} would not be a square matrix but would be of dimension $n \times (n+m)$, where n is the dimension of the state vector and m is the dimension of the process noise vector. Various methods for converting Eq. (5.8.3) into an $n \times n$ triangular matrix are described by Bierman (1977), Thornton (1976), Maybeck (1979), Dyer and McReynolds (1969), and Kaminski (1971). The key to the triangularization of Eq. (5.8.3) is to note that an $(n + m) \times (n + m)$ orthogonal transformation, T, can be applied to Eq. (5.8.2) so that (recall that $TT^T = I$)

$$
B_{k+1} T T^T B_{k+1}^T = \tilde{W}_{k+1} \tilde{W}_{k+1}^T
\tag{5.8.4}
$$

where

$$\tilde{W}_{k+1} = B_{k+1}T = \left[\phi : \overline{W}_{k+1}\right] \tag{5.8.5}$$

and where $\overline{W}_{k+1}$ is now an $n \times n$ upper triangular matrix and ϕ is an $n \times m$ null matrix. Various methods that may be used to accomplish the triangularization are the Gram-Schmidt orthogonalization (Bierman, 1977), the Householder transformation, or the Givens transformation (Thornton, 1976).

The following Givens square root factorization algorithm is taken from Thornton (1976), and yields an $n \times n$ upper triangularization for $\overline{W}$ and an $n \times m$ null matrix, ϕ.

Let B be a full rank $n \times (n+m)$ matrix with column vectors B_i, $i = 1 \ldots, n+m$. The following algorithm yields an $n \times n$ upper triangular factor $\overline{W}$ such that $\overline{W}\,\overline{W}^T = B\,B^T$. For $j = n, \ldots, 1$, cycle through Eqs. (5.8.6) through (5.8.12).

$$k = m + j. \tag{5.8.6}$$

For $i = k - 1, \ldots 1$ evaluate recursively Eqs. (5.8.7) through (5.8.12).

$$B'_k(j) = \sqrt{(B_i(j))^2 + (B_k(j))^2} \tag{5.8.7}$$

$$C = B_k(j)/B'_k(j) \tag{5.8.8}$$

$$S = B_i(j)/B'_k(j) \tag{5.8.9}$$

$$V = B_i \tag{5.8.10}$$

$$B_i = CB_i - SB_k \tag{5.8.11}$$

$$B_k = SV + CB_k. \tag{5.8.12}$$

When completed this algorithm yields

$$B = \left[\overset{\overbrace{m}}{\phi} : \overset{\overbrace{n}}{\overline{W}}\right]\Big\}n \tag{5.8.13}$$

where $\overline{W}$ is upper triangular.

As pointed out by Thornton (1976), a lower triangular factor $\overline{W}$ may be obtained from Eqs (5.8.6) through (5.8.12) if the indicies i and j are reordered so that $j = 1, \ldots, n$ and $i = j+1, \ldots, k$. At the conclusion of this algorithm B has the form

$$B = \left[\overset{\overbrace{n}}{\overline{W}} : \overset{\overbrace{m}}{\phi}\right]\Big\}n$$

where $\overline{W}$ is lower triangular.

The previous algorithms yield $\overline{W}$ as either upper or lower triangular. For the $U - D$ factorization we need $\overline{U}$ and $\overline{D}$ in order to perform the time update for P,

$$\overline{P} = \overline{U}\,\overline{D}\,\overline{U}^T$$

The following algorithm employs a generalized Gram-Schmidt orthogonalization to preserve numerical accuracy and was developed by Thornton and Bierman (1975), and it also appears in Maybeck (1979). The measurement update for P at the kth stage is given by

$$P_k = U_k\,D_k\,U_k^T. \tag{5.8.14}$$

Let

$$Y_{k+1} = [\Phi(t_{k+1}, t_k)U_k \vdots \Gamma(t_{k+1}, t_k)] \tag{5.8.15}$$

$$\tilde{D}_{k+1} = \begin{bmatrix} D_k & 0 \\ 0 & Q_k \end{bmatrix}. \tag{5.8.16}$$

Then it can be seen that $Y_{k+1}\tilde{D}_{k+1}Y_{k+1}^T$ satisfies Eq. (5.8.1). We may now apply the following algorithm to obtain $\overline{U}$ and $\overline{D}$, the time updated values of U and D. Let

$$Y_{k+1}^T = [\,\mathbf{a}_1 \vdots \mathbf{a}_2 \vdots \cdots \vdots \mathbf{a}_n\,] \tag{5.8.17}$$

Where each vector $\mathbf{a}_i$ is of dimension $n + m$. Again n is the dimension of the state vector and m is the dimension of the process noise vector.

For $\ell = n,\ n - 1, \ldots, 1$:

$$\begin{aligned}
\mathbf{C}_\ell &= \tilde{D}\mathbf{a}_\ell \quad (\text{i.e., } C_{\ell j} = \tilde{D}_{jj}a_{\ell j},\ \ j = 1, 2, \ldots, n + m) \\
\overline{D}_{\ell\ell} &= \mathbf{a}_\ell^T \mathbf{C}_\ell \\
\mathbf{d}_\ell &= \mathbf{C}_\ell / \overline{D}_{\ell\ell} \\
\overline{U}_{j\ell} &= \mathbf{a}_j^T \mathbf{d}_\ell \\
\mathbf{a}_j &\leftarrow \mathbf{a}_j - \overline{U}_{j\ell}\mathbf{a}_\ell
\end{aligned} \tag{5.8.18}$$

where $\leftarrow$ denotes replacement; that is, write over the old variable to reduce storage requirements. For the final iteration, $\ell = 1$ and only $\mathbf{C}_1$ and D_{11} are computed.

5.9   CONTINUOUS STATE ERROR COVARIANCE PROPAGATION

In this section a method that allows the integration of the continuous state error covariance differential equations in square root form is described. The derivation follows the approach used in Tapley and Choe (1976) and Tapley and Peters (1980), but the results are based on the $P \equiv UDU^T$ decomposition. This algorithm can be combined with a triangular measurement update algorithm to obtain a complete square root estimation algorithm for which square root operations are avoided. In addition, the effects of state process noise are included without approximation.

The differential equation for propagation of the state error covariance matrix can be expressed as

$$\dot{\overline{P}}(t) = A(t)\overline{P}(t) + \overline{P}(t)\, A^T(t) + Q(t) , \qquad (5.9.1)$$

where $\overline{P}(t)$ is the *a priori* state error covariance matrix, $A(t)$ is the $n \times n$ linearized dynamics matrix, and $Q(t)$ is the process noise covariance matrix. Each of the matrices in Eq. (5.9.1) is time dependent in the general case. However, for simplicity, the time dependence will not be noted specifically in the following discussion.

If the following definitions are used,

$$\overline{P} \equiv \overline{U}\,\overline{D}\,\overline{U}^T , \overline{Q} \equiv Q/2 , \qquad (5.9.2)$$

and if the first part of Eq. (5.9.2) is differentiated with respect to time and substituted into Eq. (5.9.1), the results can be rearranged to form

$$(\dot{\overline{U}}\,\overline{D} + \overline{U}\,\dot{\overline{D}}/2 - \overline{Q}\,\overline{U}^{-T} - A\overline{U}\,\overline{D})\overline{U}^T$$
$$+\overline{U}(\overline{D}\,\dot{\overline{U}}^T + \dot{\overline{D}}\,\overline{U}^T/2 - \overline{U}^{-1}\overline{Q}^T - \overline{D}\,\overline{U}^T A^T) = 0 . \qquad (5.9.3)$$

Noting that the first term of Eq. (5.9.3) is the transpose of the second term, and making the following definition

$$C(t) \equiv (\dot{\overline{U}}\,\overline{D} + \overline{U}\,\dot{\overline{D}}/2 - \overline{Q}\,\overline{U}^{-T} - A\overline{U}\,\overline{D})\overline{U}^T \qquad (5.9.4)$$

one obtains

$$C(t) + C^T(t) = 0 . \qquad (5.9.5)$$

Relation (5.9.5) requires that $C(t)$ be either the null matrix or, more generally, skew symmetric.

Equation (5.9.4) can be simplified by selectively carrying out the multiplication of the $-Q\overline{U}^{-T}$ term by $\overline{U}^T$ to yield, after terms are arranged

$$(\overline{U}\,\overline{D} + \overline{U}\,\dot{\overline{D}}/2 - A\overline{U}\,\overline{D})\overline{U}^T = \overline{Q} + C(t) \equiv \tilde{C}(t) \,. \tag{5.9.6}$$

Equation (5.9.6) defines the differential equations for $\overline{U}$ and $\overline{D}$ to the degree of uncertainty in $C(t)$. Since the unknown matrix $C(t)$ is skew symmetric, there exist $n(n-1)/2$ unknown scalar quantities in Eq. (5.9.6). The problem considered here is one of specifying the elements of $C(t)$ so that $\overline{U}$ is maintained in triangular form during the integration of Eq. (5.9.6). (The derivation pursued here assumes that $\overline{U}$ is lower triangular and $\overline{D}$ is diagonal, although an algorithm for an upper triangular $\overline{U}$ can be obtained as easily.) The following definitions are made to facilitate the solution to the problem posed:

$$T \equiv A\overline{U}\,\overline{D} \qquad M \equiv \dot{\overline{U}}\,\overline{D} + \overline{U}\,\dot{\overline{D}}/2 - T. \tag{5.9.7}$$

With these definitions, Eq. (5.9.6) is expressed as

$$M\overline{U}^T = \tilde{C}(t) = \overline{Q} + C(t) \,. \tag{5.9.8}$$

Since $\overline{U}$ and $\dot{\overline{U}}$ in Eq. (5.9.6) are lower triangular, and since from Eq. (5.9.5) $C(t)$ is skew symmetric, several observations can be made regarding Eq. (5.9.8). There are $n(n-1)/2$ unknown elements in $\tilde{C}$. The products $\dot{\overline{U}}\,\overline{D}$ and $\overline{U}\,\dot{\overline{D}}$ are lower triangular, creating $n(n+1)/2$ unknowns. Therefore, the $n \times n$ system of equations in Eq. (5.9.8) has $[n(n-1)/2 + n(n+1)/2] = n \times n$ unknowns that can be determined uniquely.

An expansion of Eq. (5.9.8) into matrix elements indicates the method of solution

$$\begin{bmatrix} M_{11} & -T_{12} & \cdots & -T_{1n} \\ M_{21} & M_{22} & \cdots & -T_{2n} \\ \vdots & \vdots & \ddots & \vdots \\ M_{n1} & M_{n2} & \cdots & M_{nn} \end{bmatrix} \begin{bmatrix} 1 & \overline{U}_{21} & \cdots & \overline{U}_{n1} \\ \vdots & 1 & \cdots & \overline{U}_{n2} \\ \vdots & \vdots & \ddots & \vdots \\ 0 & \vdots & \cdots & 1 \end{bmatrix}$$

$$= \begin{bmatrix} \overline{q}_{11} & -C_{21} & \cdots & -C_{n1} \\ C_{21} & \overline{q}_{22} & \cdots & -C_{n2} \\ \vdots & \vdots & \ddots & \vdots \\ C_{n1} & C_{n2} & \cdots & \overline{q}_{mn} \end{bmatrix} \,. \tag{5.9.9}$$

In Eq. (5.9.9), $\overline{Q}$ is assumed to be a diagonal matrix with elements $\overline{q}_{ii} = q_{ii}/2$

$(i = 1, \ldots, n)$. (This assumption can be generalized to allow other nonzero terms in the $\widetilde{Q}$ matrix with only a slight increase in algebraic complexity.) Each row of the upper triangular portion of the $\widetilde{C}$ matrix in Eq. (5.9.9) is determined as the product of the corresponding row of the M matrix with the appropriate column of the $\overline{U}^T$ matrix. After an upper triangular row of $\widetilde{C}$ is determined, the condition from Eq. (5.9.5) that $\widetilde{C}_{ij} = -\widetilde{C}_{ji}$ $(i = 1, \ldots, n; \ j = 1, \ldots, i-1)$ is invoked to evaluate the corresponding lower triangular column of $\widetilde{C}$. Then, a column of the lower triangular elements of M can be evaluated. Once the elements of the M matrix are determined, the new row of the upper triangular $\widetilde{C}$ elements can be computed along with a column of the $\dot{\overline{U}}$ and $\dot{\overline{D}}$ elements. This process is repeated until all $\dot{\overline{U}}$ and $\dot{\overline{D}}$ values are determined. The implementation of this approach proceeds as follows. From Eqs. (5.9.6) and (5.9.7) one can write

$$M + T = \dot{\overline{U}}\,\overline{D} + \overline{U}\,\dot{\overline{D}}/2. \tag{5.9.10}$$

The expansion of Eq. (5.9.10) in summation notation gives

$$M_{ij} + T_{ij} = \sum_{k=1}^{n} \dot{\overline{U}}_{ik}\overline{d}_{kj} + \sum_{k=1}^{n} \frac{\overline{U}_{ik}\dot{\overline{d}}_{kj}}{2}$$

$$(i = 1, \ldots, n; \ j = 1, \ldots, i). \tag{5.9.11}$$

But, since $\overline{D}$ is diagonal, Eq. (5.9.11) becomes

$$M_{ij} + T_{ij} = \dot{\overline{U}}_{ij}\overline{d}_{jj} + \overline{U}_{ij}\dot{\overline{d}}_{jj}/2$$

$$(i = 1, \ldots, n; j = 1, \ldots, i) \tag{5.9.12}$$

for $i = j$, $\overline{U}_{ij} \equiv 1$, and $\dot{\overline{U}}_{ij} \equiv 0$. Therefore, Eq. (5.9.12) becomes

$$\dot{\overline{d}}_{ii} = 2\,(M_{ii} + T_{ii}) \quad (i = 1, \ldots, n). \tag{5.9.13}$$

For $i > j$, Eq. (5.9.12) is rearranged to obtain the differential equation

$$\dot{\overline{U}}_{ij} = (M_{ij} + T_{ij} - \overline{U}_{ij}\dot{\overline{d}}_{jj}/2)/\overline{d}_{jj}$$

$$(i = 2, \ldots, n; \ j = 1, \ldots, i-1). \tag{5.9.14}$$

Equations (5.9.13) and (5.9.14) are the forms of the differential equations to be employed in the derivative routine of a numerical integrator. The elements of T_{ij} and M_{ij} are computed as defined in Eq. (5.9.7). The pertinent equations can be combined to obtain the following algorithm.

5.9.1 TRIANGULAR SQUARE ROOT ALGORITHM

Measurement Update

If the Extended Sequential Filter algorithm (for which the reference trajectory is updated at each observation point) is adopted, the measurement update algorithm for the UDU^T factorization has the following form. Using the observation $Y_{k+1} = G(\mathbf{X}_{k+1}, t_{k+1})$, calculate

$$H_{k+1} = [\partial G\left(\overline{\mathbf{X}}_{k+1}, t_{k+1}\right) / \partial \overline{\mathbf{X}}_{k+1}]. \tag{5.9.15}$$

For $i = 1, \ldots, n$

$$\overline{F}_i = H_i + \sum_{k=i+1}^{n} H_k \overline{U}_{ki} \tag{5.9.16}$$

$$V_i = \overline{d}_i \overline{F}_i. \tag{5.9.17}$$

Set $\beta_{n+1} = R_{k+1}$ (where R_{k+1} is the variance of the measurement noise at the $k + 1^{\text{st}}$ observation epoch) and calculate

$$\beta_i = \beta_{i+1} + V_i \overline{F}_i \quad (i = n, \ldots, 1). \tag{5.9.18}$$

Calculate diagonal covariance elements

$$d_i = \overline{d}_i \beta_{i+1} / \beta_i \quad (i = 1, \ldots, n) \tag{5.9.19}$$

$$\alpha = \beta_1. \tag{5.9.20}$$

For $i = 2, \ldots, n$ and $j = 1, \ldots, i - 1$, calculate

$$P_j = \overline{F}_j / \beta_{j+1} \tag{5.9.21}$$

$$B_{ij} = V_i + \sum_{k=j+1}^{i-1} \overline{U}_{ik} V_k \tag{5.9.22}$$

$$U_{ij} = \overline{U}_{ij} - B_{ij} P_j \quad (i = 2, \ldots n; \quad j = 1, \ldots, i - 1). \tag{5.9.23}$$

Compute the observation residual

$$y_{k+1} = Y_{k+1} - G\left(\overline{\mathbf{X}}_{k+1}, t_{k+1}\right). \tag{5.9.24}$$

Calculate gain, and update the state using

$$\overline{K}_i = V_i + \sum_{j=1}^{i-1} \overline{U}_{ij} V_j \quad (i = 1, \ldots, n) \tag{5.9.25}$$

$$\hat{\mathbf{X}}_i = \overline{\mathbf{X}}_i + \overline{K}_i \, y_{k+1} / \alpha \quad (i = 1, \ldots, n). \tag{5.9.26}$$

Propagation

Given the elements of the square root state error covariance in lower triangular $\overline{U}\,\overline{D}\,\overline{U}^T$ form, $\overline{Q} \equiv Q/2$ and $A(t)$, the differential equations $\dot{\overline{U}}_{ij}$ and $\dot{\overline{d}}_{ii}$ can be computed as follows:

$$T_{ij} = \sum_{k=j}^{n} A_{ik}\overline{U}_{kj}\overline{d}_{jj} \quad (i = 1, \ldots, n; \tag{5.9.27}$$

$$j = 1, \ldots, n)$$

$$\overline{C}_{ij} = \sum_{k=1}^{i} M_{ik}\overline{U}_{jk} - \sum_{k=i+1}^{j} T_{ik}\overline{U}_{jk} \quad (i = 1, \ldots, n; \tag{5.9.28}$$

$$j = i+1, \ldots, n)$$

$$M_{ii} = \overline{q}_{ii} - \sum_{k=1}^{i-1} M_{ik}\overline{U}_{ik} \quad (i = 1, \ldots, n) \tag{5.9.29}$$

$$M_{ij} = -\overline{C}_{ji} - \sum_{k=1}^{j-1} M_{ik}\overline{U}_{jk} \quad (i = 1, \ldots, n; \tag{5.9.30}$$

$$j = 1, \ldots, i-1)$$

$$\dot{\overline{d}}_{ii} = 2(M_{ii} + T_{ii}) \quad (i = 1, \ldots, n) \tag{5.9.31}$$

$$\dot{\overline{U}}_{ij} = (M_{ij} + T_{ij} - \overline{U}_{ij}\dot{\overline{d}}_{jj}/2)/\overline{d}_{jj} \quad (i = 1, \ldots, n; \tag{5.9.32}$$

$$j = 1, \ldots, i-1).$$

The algorithm summarized in Eqs. (5.9.15) through (5.9.32) defines a complete sequential estimation algorithm in which the covariance matrices P and $\overline{P}$ are replaced by the factors (U, D) and $(\overline{U}, \overline{D})$, respectively. The algorithm given here assumes that only a single scalar observation is processed at each observation epoch; however, the algorithm is applicable to the case of multiple observations at a given epoch if the observation errors are uncorrelated. To obtain

$$A(t) = \frac{F(\mathbf{X}, t)}{\partial \mathbf{X}},$$

the additional integration of the system of nonlinear differential equations, $\dot{\mathbf{X}} = F(\mathbf{X}, t)$, is required. At the initial point, the starting conditions $\overline{\mathbf{X}}_k = \hat{\mathbf{X}}_k$ are used.

5.10   THE SQUARE ROOT INFORMATION FILTER

In Sections 5.7, 5.8, and 5.9, attention has been directed to the problem of developing estimation algorithms based on the square root of the covariance matrix. Both measurement incorporation and time propagation algorithms were considered. In this section, we consider algorithms derived from the information equations, referred to as the normal equations in Chapter 4. Specifically, we consider algorithms that deal with factoring the information matrix. In general such an algorithm is referred to as a *square root information filter* or SRIF.

We will first consider the case where the state vector, $\mathbf{x}$, is independent of time; that is, a constant. Assume that *a priori* information $[\overline{\mathbf{x}}, \overline{\Lambda}]$ is given, where $\overline{\mathbf{x}}$ is the *a priori* value of $\mathbf{x}$ and $\overline{\Lambda}$ is the *a priori* information matrix, $\overline{\Lambda} = \overline{P}^{-1}$. The *a priori* information can be written in the form of a data equation by noting that

$$\overline{\mathbf{x}} = \mathbf{x} + \boldsymbol{\eta}, \qquad (5.10.1)$$

where $\boldsymbol{\eta}$ is the error in $\overline{\mathbf{x}}$ and is assumed to have the following characteristics

$$E[\boldsymbol{\eta}] = 0, \quad E[\boldsymbol{\eta}\boldsymbol{\eta}^T] = \overline{P} = \overline{\Lambda}^{-1}. \qquad (5.10.2)$$

Factoring the information matrix yields

$$\overline{\Lambda} = \overline{R}^T \overline{R}. \qquad (5.10.3)$$

Multiplying Eq. (5.10.1) by $\overline{R}$ yields

$$\overline{R}\overline{\mathbf{x}} = \overline{R}\mathbf{x} + \overline{R}\boldsymbol{\eta}. \qquad (5.10.4)$$

Define

$$\overline{\mathbf{b}} = \overline{R}\overline{\mathbf{x}} \quad \text{and} \quad \overline{\boldsymbol{\eta}} = \overline{R}\boldsymbol{\eta}. \qquad (5.10.5)$$

Then Eq. (5.10.4) assumes the standard form of the data equation

$$\overline{\mathbf{b}} = \overline{R}\mathbf{x} + \overline{\boldsymbol{\eta}}. \qquad (5.10.6)$$

Note that the error $\overline{\boldsymbol{\eta}}$ still has zero mean but now has unit variance,

$$\begin{aligned} E[\overline{\boldsymbol{\eta}}] &= \overline{R}E[\boldsymbol{\eta}] = 0 \\ E[\boldsymbol{\eta}\boldsymbol{\eta}^T] &= \overline{R}E[\boldsymbol{\eta}\boldsymbol{\eta}^T]\overline{R}^T = \overline{R}\,\overline{P}\,\overline{R}^T = I. \end{aligned} \qquad (5.10.7)$$

We now seek to determine the "best" estimate of $\mathbf{x}$ given the *a priori* information in the form of Eq. (5.10.6), and additional observation data

$$\mathbf{y} = H\mathbf{x} + \boldsymbol{\epsilon}, \qquad (5.10.8)$$

where $\epsilon \sim (O, I)$; the observations have been prewhitened as described in Section 5.7.1. This value of $\mathbf{x}$ will be defined as that which minimizes the least squares performance index,

$$
\begin{aligned}
J(\mathbf{x}) &= \|\epsilon\|^2 + \|\overline{\eta}\|^2 \\
&= \|H\mathbf{x} - \mathbf{y}\|^2 + \|\overline{R}\mathbf{x} - \overline{\mathbf{b}}\|^2 \\
&= \left\| \begin{bmatrix} \overline{R} \\ H \end{bmatrix} \mathbf{x} - \begin{bmatrix} \overline{\mathbf{b}} \\ \mathbf{y} \end{bmatrix} \right\|^2 .
\end{aligned}
\tag{5.10.9}
$$

Following the procedure of Section 5.4, we apply a series of orthogonal transformations to Eq. (5.10.9) such that

$$
T \begin{bmatrix} \overline{R} \; \overline{\mathbf{b}} \\ H \; \mathbf{y} \end{bmatrix} = \begin{bmatrix} R \, \mathbf{b} \\ O \, \mathbf{e} \end{bmatrix}
\tag{5.10.10}
$$

and the performance index becomes

$$
\begin{aligned}
J(\mathbf{x}) &= \left\| \begin{bmatrix} R \\ O \end{bmatrix} \mathbf{x} - \begin{bmatrix} \mathbf{b} \\ \mathbf{e} \end{bmatrix} \right\|^2 \\
&= \|R\mathbf{x} - \mathbf{b}\|^2 + \|\mathbf{e}\|^2 .
\end{aligned}
\tag{5.10.11}
$$

The value of the $\mathbf{x}$ that minimizes the performance index is

$$
\hat{\mathbf{x}} = R^{-1}\mathbf{b} .
\tag{5.10.12}
$$

Equation (5.10.12) is most easily solved by a backward substitution given by Eq. (5.2.8),

$$
\hat{x}_i = \left(b_i - \sum_{j=i+1}^{n} R_{ij} \, \hat{x}_j \right) / R_{ii}, \, i = n \ldots 1 .
\tag{5.10.13}
$$

The data processing algorithm based on the orthogonal transformation can be summarized as follows. Assume we are given the *a priori* information $\overline{R}$, $\overline{\mathbf{b}}$ and the measurements $y_j = H_j \mathbf{x} + \epsilon_j$, $j = 1, \ldots, \ell$, where each y_j is a scalar and where $\epsilon_j \sim (O, 1)$. Then the least squares estimate can be generated recursively as follows:

(a) Compute

$$
T_j \begin{bmatrix} \overline{R} \; \overline{\mathbf{b}} \\ H_j \; y_j \end{bmatrix} = \begin{bmatrix} R_j \, \mathbf{b}_j \\ 0 \; e_j \end{bmatrix} .
$$

(b) $R_j \hat{\mathbf{x}}_j = \mathbf{b}_j$.

Solve for $\hat{\mathbf{x}}_j$ by a backward substitution as given by Eq. (5.10.13) at each stage j. If the estimation error covariance matrix is desired at any stage, it is computed from

$$P_j = R_j^{-1} R_j^{-T}. \qquad (5.10.14)$$

(c) R_j and $\mathbf{b}_j$ become the *a priori* for the next stage.

(d) Repeat this process for $j = 1, \ldots, \ell$.

Note that the residual sum of squares of the *a priori* errors plus observation errors based on ℓ measurements is

$$J(\hat{\mathbf{x}}_\ell) = \sum_{i=1}^{\ell} (y_i - H_i \hat{\mathbf{x}}_\ell)^2 + (\hat{\mathbf{x}}_\ell - \overline{\mathbf{x}})^T \overline{R}^T \overline{R} (\hat{\mathbf{x}}_\ell - \overline{\mathbf{x}}) = \sum_{i=1}^{\ell} e_i^2. \qquad (5.10.15)$$

5.10.1 THE SQUARE ROOT INFORMATION FILTER WITH TIME-DEPENDENT EFFECTS

Consider the dynamic model of Eq. (4.9.46),

$$\mathbf{x}_k = \Phi(t_k, t_j) \mathbf{x}_j + \Gamma(t_k, t_j) \mathbf{u}_j \qquad (5.10.16)$$

where

$$E[\mathbf{u}_j] = \overline{\mathbf{u}}, \qquad E[(\mathbf{u}_j - \overline{\mathbf{u}})(\mathbf{u}_k - \overline{\mathbf{u}})^T] = Q \delta_{jk}.$$

Generally we assume that $\overline{\mathbf{u}} = 0$, but for this discussion we will assume that the random sequence $\mathbf{u}_j$ has a known mean, $\overline{\mathbf{u}}$. Q is assumed to be positive definite, and may vary with time, but we will assume that both Q and $\overline{\mathbf{u}}$ are constant here.

Assume that a sequence of scalar observation, y_i, is given such that

$$y_i = H_i \mathbf{x}_i + \epsilon_i, \qquad i = 1, \ldots, \ell \qquad (5.10.17)$$

and the observations have been prewhitened so that

$$E[\epsilon_i] = 0, \qquad E(\epsilon_i^2) = 1. \qquad (5.10.18)$$

Assume that at the initial time, say t_0, there is *a priori* information given in the form of the information array $[\overline{R}_0, \overline{\mathbf{b}}_0]$ from which $\overline{\mathbf{x}}_0$ and $\overline{P}_0$ are determined as (the *a priori* information may be $\overline{\mathbf{x}}_0$ and $\overline{P}_0$ from which $[\overline{R}_0, \overline{\mathbf{b}}_0]$ are computed)

$$\overline{\mathbf{x}}_0 = \overline{R}_0^{-1} \overline{\mathbf{b}}_0 \qquad \overline{P}_0 = \overline{R}_0^{-1} \overline{R}_0^{-T}. \qquad (5.10.19)$$

Because $E[\mathbf{u}(t_0)] = \overline{\mathbf{u}}$, the *a priori* quantities are mapped to t_1 by

$$\overline{\mathbf{x}}_1 = \Phi(t_1, t_0)\overline{\mathbf{x}}_0 + \Gamma(t_1, t_0)\overline{\mathbf{u}}$$
$$\mathbf{x}_1 = \Phi(t_1, t_0)\mathbf{x}_0 + \Gamma(t_1, t_0)\mathbf{u}_0$$
$$\overline{P}_1 = E[(\overline{\mathbf{x}}_1 - \mathbf{x}_1)(\overline{\mathbf{x}}_1 - \mathbf{x}_1)^T]$$
$$= \Phi(t_1, t_0)\overline{P}_0\Phi^T(t_1, t_0) + \Gamma(t_1, t_0)Q\Gamma^T(t_1, t_0). \quad (5.10.20)$$

For the case where there is no process noise, Q is assumed to be zero, as is $\mathbf{u}_j$, and the following solution is applicable.

Process Noise Absent

In this case both Q and $\overline{\mathbf{u}}$ are assumed to be zero and

$$\overline{P}_k = \Phi(t_k, t_j)P_j\Phi^T(t_k, t_j).$$

In terms of square root notation

$$\overline{P}_k = \overline{R}_k^{-1}\overline{R}_k^{-T} = \Phi(t_k, t_j)\,R_j^{-1}R_j^{-T}\Phi^T(t_k, t_j). \quad (5.10.21)$$

Hence,

$$\overline{R}_k^{-1} = \Phi(t_k, t_j)R_j^{-1} \quad (5.10.22)$$

and

$$\overline{R}_k = R_j\Phi^{-1}(t_k, t_j). \quad (5.10.23)$$

It follows then that the mapping relations are

$$\overline{\mathbf{x}}_k = \Phi(t_k, t_j)\hat{\mathbf{x}}_j \quad (5.10.24)$$
$$\overline{R}_k = R_j\Phi^{-1}(t_k, t_j). \quad (5.10.25)$$

Note also that the compressed observation can be obtained as

$$\overline{\mathbf{b}}_k = \overline{R}_k\overline{\mathbf{x}}_k = R_j\Phi^{-1}(t_k, t_j)\Phi(t_k, t_j)\hat{\mathbf{x}}_j$$
$$= R_j\hat{\mathbf{x}}_j = \mathbf{b}_j. \quad (5.10.26)$$

With these mapping equations we can set up the equations for the SRIF algorithm. Assume that at time, t_j , we have *a priori* information

$$\overline{\mathbf{b}}_j = \overline{R}_j\overline{\mathbf{x}}_j$$

where

$$\overline{\mathbf{x}}_j = \mathbf{x}_j + \boldsymbol{\eta}_j \quad (5.10.27)$$

and

$$E[\boldsymbol{\eta}_j] = 0$$
$$E[\boldsymbol{\eta}_j \boldsymbol{\eta}_j^T] = \overline{P}_j = \overline{R}_j^{-1} \, \overline{R}_j^{-T}.$$

By multiplying Eq. (5.10.27) by $\overline{R}_j$, we obtain

$$\overline{\mathbf{b}}_j = \overline{R}_j \mathbf{x}_j + \overline{\boldsymbol{\eta}}_j \tag{5.10.28}$$

where $\overline{\boldsymbol{\eta}}_j = \overline{R}_j \boldsymbol{\eta}_j$. Further, $\overline{\boldsymbol{\eta}}_j$ satisfies the condition $E(\overline{\boldsymbol{\eta}}_j) = 0$ and it is easily demonstrated that $E(\overline{\boldsymbol{\eta}}_j \overline{\boldsymbol{\eta}}_j^T) = I$. Now assume that we have the new observation, y_j, where

$$y_j = H_j \mathbf{x}_j + \epsilon_j$$

and ϵ_j satisfies the conditions given in Eq. (5.10.18). To obtain a best estimate of $\mathbf{x}_j$, the least squares performance index to be minimized is

$$J(\mathbf{x}_j) = \|\overline{\boldsymbol{\eta}}_j\|^2 + \epsilon_j^2, \tag{5.10.29}$$

or

$$J(\mathbf{x}_j) = \|\overline{R}_j \mathbf{x}_j - \overline{\mathbf{b}}_j\|^2 + (H_j \mathbf{x}_j - y_j)^2. \tag{5.10.30}$$

Following the procedure discussed in the previous section, $J(\mathbf{x}_j)$ can be written as

$$J(\mathbf{x}_j) = \left\| \begin{bmatrix} \overline{R}_j \\ H_j \end{bmatrix} \mathbf{x}_j - \begin{bmatrix} \overline{\mathbf{b}}_j \\ y_j \end{bmatrix} \right\|^2. \tag{5.10.31}$$

Multiplying by an orthogonal transformation, T, yields

$$J(\mathbf{x}_j) = \left[\begin{bmatrix} \overline{R}_j \\ H_j \end{bmatrix} \mathbf{x}_j - \begin{bmatrix} \overline{\mathbf{b}}_j \\ y_j \end{bmatrix} \right]^T T^T T \left[\begin{bmatrix} \overline{R}_j \\ H_j \end{bmatrix} \mathbf{x}_j - \begin{bmatrix} \overline{\mathbf{b}}_j \\ y_j \end{bmatrix} \right]. \tag{5.10.32}$$

We select T so that

$$T \begin{bmatrix} \overline{R}_j \\ H_j \end{bmatrix} = \begin{bmatrix} R_j \\ 0 \end{bmatrix}, T \begin{bmatrix} \overline{\mathbf{b}}_j \\ y_j \end{bmatrix} = \begin{bmatrix} \mathbf{b}_j \\ e_j \end{bmatrix} \tag{5.10.33}$$

where R_j is upper triangular. Eq. (5.10.30) can be expressed then as

$$J(\mathbf{x}_j) = \left\| \begin{bmatrix} R_j \\ 0 \end{bmatrix} \mathbf{x}_j - \begin{bmatrix} \mathbf{b}_j \\ e_j \end{bmatrix} \right\|^2 \tag{5.10.34}$$

or

$$J(\mathbf{x}_j) = (e_j)^2 + \|R_j\mathbf{x}_j - \mathbf{b}_j\|^2. \tag{5.10.35}$$

It follows that to minimize $J(\mathbf{x}_j)$,

$$\hat{\mathbf{x}}_j = R_j^{-1}\mathbf{b}_j, P_j = R_j^{-1}R_j^{-T}, J(\hat{\mathbf{x}}_j) = e_j^2. \tag{5.10.36}$$

where the elements of the vector $\hat{\mathbf{x}}_j$ are obtained by evaluating Eq. (5.10.13). The previous steps are equivalent to the measurement update for the Kalman filter. The time update is obtained by mapping $\hat{\mathbf{x}}_j$ and R_j forward to time t_k as follows

$$\begin{aligned}
\overline{\mathbf{x}}_k &= \Phi(t_k, t_j)\hat{\mathbf{x}}_j \\
\overline{R}_k &= R_j\Phi^{-1}(t_k, t_j) \\
\overline{\mathbf{b}}_k &= \overline{R}_k\overline{\mathbf{x}}_k = \mathbf{b}_j
\end{aligned} \tag{5.10.37}$$

and the measurement update process is repeated to obtain $\hat{\mathbf{x}}_k$. Note that the sum of squares of the estimation errors would be stored as (assuming we have processed k scalar observations)[1]

$$J(\mathbf{x}_k) = \sum_{i=1}^{k} e_i^2.$$

An alternate equation to Eq. (5.10.23) can be obtained by noting that for any t,

$$\overline{R}(t) = R_j\Phi^{-1}(t, t_j)$$

and

$$\dot{\overline{R}}(t) = R_j\dot{\Phi}^{-1}(t, t_j). \tag{5.10.38}$$

Substituting for $\dot{\Phi}^{-1}(t, t_j)$ leads to

$$\dot{\overline{R}}(t) = -R_j\Phi^{-1}(t, t_j)A(t) = -\overline{R}(t)A(t).$$

Hence, as an alternate to Eq. (5.10.23) one can integrate the equation

$$\dot{\overline{R}}(t) = -\overline{R}A(t).$$

with initial conditions

$$\overline{R}(t_j) = R_j. \tag{5.10.39}$$

---

[1]Again it is noted that $J(\mathbf{x}_k)$ includes the effects of *a priori* information (see Eq. 5.4.36). Also, if there are different observation types the values of $\sum e_i^2$ for each type should be stored in separate arrays.

The propagation can be accomplished either by using Eq. (5.10.37) or by integrating Eq. (5.10.38) while propagating the state by integrating

$$\dot{\overline{\mathbf{x}}} = A(t)\overline{\mathbf{x}}, \qquad (5.10.40)$$

with the initial conditions $\overline{\mathbf{x}}_j = \hat{\mathbf{x}}_j$. However, because $\overline{\mathbf{b}}_k = \mathbf{b}_j$ there is no need to map $\overline{\mathbf{x}}_j$. The *a priori* covariance $\overline{P}_k$ at t_k is obtained, in either case, as

$$\overline{P}_k = \overline{R}_k^{-1}\overline{R}_k^{-T}. \qquad (5.10.41)$$

If needed, a differential equation for $\overline{R}^{-1}$ is easily developed. From Eq. (5.10.37)

$$\overline{R}_k^{-1} = \Phi(t_k, t_j)R_j^{-1}$$

and

$$\dot{\overline{R}}^{-1}(t) = \dot{\Phi}(t, t_j)R_j^{-1}$$
$$= A(t)\Phi(t, t_j)R_j^{-1}$$
$$\dot{\overline{R}}^{-1}(t) = A(t)\overline{R}^{-1}(t), \quad \text{with I.C. } \overline{R}^{-1}(t_j) = R_j^{-1}. \qquad (5.10.42)$$

Finally, note that even though R_j is upper triangular, the propagated information matrix $\overline{R}_k$ will not be. An upper triangular time update for $\overline{R}_k$ can be developed by noting that

$$\overline{\Lambda}_k = \overline{P}_k^{-1} = \overline{R}_k^T\overline{R}_k.$$

From Eq. (5.10.37)

$$\overline{\Lambda}_k = \Phi^{-T}(t_k, t_j)R_j^T R_j \Phi^{-1}(t_k, t_j) \qquad (5.10.43)$$

multiplying by an orthogonal transformation yields

$$\overline{\Lambda}_k = \Phi^{-T}(t_k, t_j)R_j^T T^T T R_j \Phi^{-1}(t_k, t_j) = \overline{R}_k^T\overline{R}_k$$

where T is chosen so that

$$\overline{R}_k = T R_j \Phi^{-1}(t_k, t_j) \qquad (5.10.44)$$

is upper triangular. A Householder or Givens transformation may be used to accomplish this.

5.10.2 THE DYNAMIC CASE WITH PROCESS NOISE

The technique for including process noise in both the Potter and SRIF algorithms was developed by Dyer and McReynolds (1969). The state propagation equation for the case with process noise is given by Eq. (5.10.16),

$$\mathbf{x}_k = \Phi(t_k, t_{k-1})\mathbf{x}_{k-1} + \Gamma(t_k, t_{k-1})\mathbf{u}_{k-1}. \tag{5.10.45}$$

The SRIF for this case can be formulated as follows. Assume that at t_{k-1} a priori information in the form of an information array $[\overline{R}_{k-1}\, \overline{\mathbf{b}}_{k-1}]$ or equivalently $[\overline{P}_{k-1}\, \overline{\mathbf{x}}_{k-1}]$ is available,

$$\overline{\mathbf{b}}_{k-1} = \overline{R}_{k-1}\overline{\mathbf{x}}_{k-1} \tag{5.10.46}$$

but

$$\overline{\mathbf{x}}_{k-1} = \mathbf{x}_{k-1} + \boldsymbol{\eta}_{k-1} \tag{5.10.47}$$

where $\mathbf{x}_{k-1}$ is the true value. Also,

$$E[\boldsymbol{\eta}_{k-1}] = 0,\ E[\boldsymbol{\eta}_{k-1}\boldsymbol{\eta}_{k-1}^T] = \overline{P}_{k-1} = \overline{R}_{k-1}^{-1}\overline{R}_{k-1}^{-T}. \tag{5.10.48}$$

Substituting Eq. (5.10.47) into (5.10.46) yields the data equation for the a priori information

$$\overline{\mathbf{b}}_{k-1} = \overline{R}_{k-1}\mathbf{x}_{k-1} + \overline{\boldsymbol{\eta}}_{k-1} \tag{5.10.49}$$

where

$$E[\overline{\boldsymbol{\eta}}_{k-1}] = \overline{R}_{k-1}E[\boldsymbol{\eta}_{k-1}] = 0 \tag{5.10.50}$$

$$E[\overline{\boldsymbol{\eta}}_{k-1}\overline{\boldsymbol{\eta}}_{k-1}^T] = \overline{R}_{k-1}\overline{P}_{k-1}\overline{R}_{k-1}^T = I. \tag{5.10.51}$$

A scalar observation is given at t_{k-1}

$$y_{k-1} = H_{k-1}\mathbf{x}_{k-1} + \epsilon_{k-1} \tag{5.10.52}$$

where we assume that the observations have been prewhitened so that $\epsilon_{k-1} \sim [0, 1]$.

A priori information on $\mathbf{u}_{k-1}$ is given by covariance Q and by $\overline{\mathbf{u}}_{k-1}$, the mean value of $\mathbf{u}$. Generally it is assumed that $\mathbf{u}$ is a zero mean process so that the a priori value $\overline{\mathbf{u}} = 0$ at each stage. This information also may be written in the form of a data equation by noting that

$$\begin{aligned}\overline{\mathbf{u}} &= \overline{\mathbf{u}}_{k-1} \\ &= \mathbf{u}_{k-1} + \boldsymbol{\alpha}_{k-1}\end{aligned} \tag{5.10.53}$$

where $\bar{\mathbf{u}}_{k-1}$ is the *a priori* value and $\mathbf{u}_{k-1}$ is the true value. The error, $\boldsymbol{\alpha}_{k-1}$, has the properties

$$E[\boldsymbol{\alpha}_{k-1}] = 0,$$
$$E[\boldsymbol{\alpha}_{k-1}\boldsymbol{\alpha}_{k-1}^T] = Q. \tag{5.10.54}$$

We will assume that the process noise is uncorrelated in time; that is, $E[\boldsymbol{\alpha}_i \boldsymbol{\alpha}_j^T] = 0$ for $i \neq j$. Although it is not necessary, we will assume that both $\mathbf{u}$ and Q are constant in time. Factor Q such that

$$R_u^{-1} R_u^{-T} = Q. \tag{5.10.55}$$

After multiplying Eq. (5.10.53) by R_u, we may write the data equation for $\bar{\mathbf{u}}_{k-1}$ as

$$R_u \bar{\mathbf{u}}_{k-1} \equiv \bar{\mathbf{b}}_{u_{k-1}}$$
$$= R_u \mathbf{u}_{k-1} + \bar{\boldsymbol{\alpha}}_{k-1} \tag{5.10.56}$$

where

$$\bar{\boldsymbol{\alpha}}_{k-1} \sim [O, I].$$

Now define a performance index for the measurement update at t_{k-1}. We choose as our least squares performance index the sum of the squared errors given by Eqs (5.10.49), (5.10.52), and (5.10.56),

$$\hat{J}_{k-1} = \|\bar{\boldsymbol{\eta}}_{k-1}\|^2 + (\epsilon_{k-1})^2 + \|\bar{\boldsymbol{\alpha}}_{k-1}\|^2. \tag{5.10.57}$$

Note that including $\|\bar{\boldsymbol{\alpha}}\|^2$ in the performance index allows us to estimate $\mathbf{u}$ at each stage. The values of $\mathbf{x}_{k-1}$ that minimizes $\hat{J}_{k-1}$ is the filter value, $\hat{\mathbf{x}}_{k-1}$. At this point the value of $\mathbf{u}_{k-1}$ that minimizes $\hat{J}_{k-1}$ is just the *a priori* value, $\bar{\mathbf{u}}_{k-1}$. After we perform a time and measurement update at t_k we will have the necessary information to compute $\hat{\mathbf{u}}_{k-1}$. Use the defining equations for the errors to write $\hat{J}_{k-1}$ as

$$\hat{J}_{k-1} = \|\bar{R}_{k-1}\mathbf{x}_{k-1} - \bar{\mathbf{b}}_{k-1}\|^2 + (H_{k-1}\mathbf{x}_{k-1} - y_{k-1})^2$$
$$+ \|R_u \mathbf{u}_{k-1} - \bar{\mathbf{b}}_{u_{k-1}}\|^2. \tag{5.10.58}$$

Because $\mathbf{x}_{k-1}$ is independent of $\mathbf{u}_{k-1}$ it is convenient to write this as

$$\hat{J}_{k-1} = \left\| \begin{bmatrix} \bar{R}_{k-1} \\ H_{k-1} \end{bmatrix} \mathbf{x}_{k-1} - \begin{bmatrix} \bar{\mathbf{b}}_{k-1} \\ y_{k-1} \end{bmatrix} \right\|^2 + \|R_u \mathbf{u}_{k-1} - \bar{\mathbf{b}}_{u_{k-1}}\|^2. \tag{5.10.59}$$

Applying a series of orthogonal transformations to the first term of Eq. (5.10.59) results in

$$\hat{J}_{k-1} = \left\| \begin{bmatrix} \hat{R}_{k-1} \\ 0 \end{bmatrix} \mathbf{x}_{k-1} - \begin{bmatrix} \hat{\mathbf{b}}_{k-1} \\ e_{k-1} \end{bmatrix} \right\|^2 + \|R_u \mathbf{u}_{k-1} - \bar{\mathbf{b}}_{u_{k-1}}\|^2 \tag{5.10.60}$$

or

$$\hat{J}_{k-1} = (e_{k-1})^2 + \|\hat{R}_{k-1}\,\mathbf{x}_{k-1} - \hat{\mathbf{b}}_{k-1}\|^2 + \|R_u\mathbf{u}_{k-1} - \overline{\mathbf{b}}_{u_{k-1}}\|^2. \quad (5.10.61)$$

The minimum value of $\hat{J}_{k-1}$ is found by setting

$$\hat{R}_{k-1}\hat{\mathbf{x}}_{k-1} = \hat{\mathbf{b}}_{k-1} \qquad\qquad\qquad (5.10.62)$$

$$R_u\overline{\mathbf{u}}_{k-1} = \overline{\mathbf{b}}_{u_{k-1}}. \qquad\qquad\qquad (5.10.63)$$

As stated earlier, Eq. (5.10.63) returns the *a priori* value, $\overline{\mathbf{u}}_{k-1}$. The minimum value of $\hat{J}_{k-1}$ is given by

$$\hat{J}_{k-1} = (e_{k-1})^2. \qquad\qquad\qquad (5.10.64)$$

Also,

$$P_{k-1} = \hat{R}_{k-1}^{-1}\hat{R}_{k-1}^{-T}. \qquad\qquad\qquad (5.10.65)$$

Having completed the measurement update at t_{k-1}, we are ready to do the time update to t_k. In order to time update the performance index $\hat{J}_{k-1}$, we need to write Eq. (5.10.61) in terms of $\mathbf{x}_k$. Because $\mathbf{u}_{k-1}$ is not time dependent, the update of $\mathbf{u}$ to t_k will be handled in the measurement update. From Eq. (5.10.45) we may write $\mathbf{x}_{k-1}$ in terms of $\mathbf{x}_k$,

$$\mathbf{x}_{k-1} = \Phi^{-1}(t_k, t_{k-1})(\mathbf{x}_k - \Gamma(t_k, t_{k-1})\mathbf{u}_{k-1}). \qquad (5.10.66)$$

Substituting Eq. (5.10.66) into Eq. (5.10.61) yields the time update

$$\overline{J}_k = (e_{k-1})^2 + \|\hat{R}_{k-1}\Phi^{-1}(t_k, t_{k-1})(\mathbf{x}_k - \Gamma(t_k,t_{k-1})\mathbf{u}_{k-1}) - \hat{\mathbf{b}}_{k-1}\|^2$$
$$+\|R_u\mathbf{u}_{k-1} - \overline{\mathbf{b}}_{u_{k-1}}\|^2, \qquad\qquad (5.10.67)$$

which may be written as

$$\overline{J}_k = (e_{k-1})^2 + \qquad\qquad\qquad\qquad\qquad (5.10.68)$$

$$\left\| \begin{bmatrix} R_u & 0 \\ -\tilde{R}_k\Gamma(t_k,\,t_{k-1}) & \tilde{R}_k \end{bmatrix} \begin{bmatrix} \mathbf{u}_{k-1} \\ \mathbf{x}_k \end{bmatrix} - \begin{bmatrix} \overline{\mathbf{b}}_{u_{k-1}} \\ \hat{\mathbf{b}}_{k-1} \end{bmatrix} \right\|^2$$

where

$$\tilde{R}_k \equiv \hat{R}_{k-1}\Phi^{-1}(t_k, t_{k-1}). \qquad\qquad (5.10.69)$$

We now apply a series of q orthogonal transformations to the second term of Eq. (5.10.68), where q is the dimension of $\mathbf{u}_{k-1}$. This will partly upper triangularize Eq. (5.10.68) (for $q < n$) and will eliminate the explicit dependence of $\mathbf{x}_k$ on $\mathbf{u}_{k-1}$, i.e.

$$\overline{T}_k \begin{bmatrix} R_u & 0 & \overline{\mathbf{b}}_{u_{k-1}} \\ -\tilde{R}_k\Gamma(t_k, t_{k-1}) & \tilde{R}_k & \hat{\mathbf{b}}_{k-1} \end{bmatrix} = \begin{bmatrix} \overline{R}_{u_k} & \overline{R}_{ux_k} & \tilde{\mathbf{b}}_{u_k} \\ 0 & \overline{R}_k & \overline{\mathbf{b}}_k \end{bmatrix}. \qquad (5.10.70)$$

Then,

$$\overline{J}_k = (e_{k-1})^2 + \left\| \begin{bmatrix} \overline{R}_{u_k} & \overline{R}_{ux_k} \\ 0 & \overline{R}_k \end{bmatrix} \begin{bmatrix} \mathbf{u}_{k-1} \\ \mathbf{x}_k \end{bmatrix} - \begin{bmatrix} \widetilde{\mathbf{b}}_{u_k} \\ \overline{\mathbf{b}}_k \end{bmatrix} \right\|^2 . \tag{5.10.71}$$

The minimum value of J is obtained by setting

$$\overline{R}_{u_k} \mathbf{u}_{k-1} + \overline{R}_{ux_k} \mathbf{x}_k = \widetilde{\mathbf{b}}_{u_k} \tag{5.10.72}$$

$$\overline{R}_k \mathbf{x}_k = \overline{\mathbf{b}}_k . \tag{5.10.73}$$

Therefore,

$$\overline{\mathbf{x}}_k = \overline{R}_k^{-1} \overline{\mathbf{b}}_k . \tag{5.10.74}$$

Because $\overline{R}_{uk}$ is nonsingular we can find a value of $\mathbf{u}_{k-1}$ to satisfy Eq. (5.10.72) for any value of $\mathbf{x}_k$. We know that the value of $\overline{\mathbf{x}}_k$ that results from Eq. (5.10.74) is given by

$$\overline{\mathbf{x}}_k = \Phi(t_k, t_{k-1})\hat{\mathbf{x}}_{k-1} + \Gamma(t_k, t_{k-1})\overline{\mathbf{u}}_{k-1}$$

where $\overline{\mathbf{u}}_{k-1}$ is the mean or *a priori* value. Using this value of $\overline{\mathbf{x}}_k$ in Eq. (5.10.72) will yield $\overline{\mathbf{u}}_{k-1}$. This is to be expected since we get no information on $\overline{\mathbf{u}}_{k-1}$ until we process an observation at t_k.

Recall that the error covariance associated with $\overline{\mathbf{x}}_k$ is

$$\begin{aligned} \overline{P}_k &= E[(\overline{\mathbf{x}}_k - \mathbf{x}_k)(\overline{\mathbf{x}}_k - \mathbf{x}_k)^T] = \overline{R}_k^{-1} \overline{R}_k^{-T} \\ &= \Phi(t_k, t_{k-1})P_{k-1}\Phi^T(t_k, t_{k-1}) \\ &\quad + \Gamma(t_k, t_{k-1})Q\Gamma^T(t_k, t_{k-1}) . \end{aligned} \tag{5.10.75}$$

Equation (5.10.71) may be written as

$$\begin{aligned} \overline{J}_k &= (e_{k-1})^2 + \|\overline{R}_{u_k}\mathbf{u}_{k-1} + \overline{R}_{ux_k}\mathbf{x}_k - \widetilde{\mathbf{b}}_{u_k}\|^2 \\ &\quad + \|\overline{R}_k\mathbf{x}_k - \overline{\mathbf{b}}_k\|^2 . \end{aligned} \tag{5.10.76}$$

We may now do the measurement update at t_k. The least squares performance index for the measurement update is

$$\begin{aligned} \hat{J}_k &= \overline{J}_k + (\epsilon_k)^2 + \|\overline{\boldsymbol{\alpha}}_k\|^2 \\ &= \overline{J}_k + (H_k\mathbf{x}_k - y_k)^2 + \|R_u\mathbf{u}_k - \overline{\mathbf{b}}_{u_k}\|^2 . \end{aligned} \tag{5.10.77}$$

This may be written as

$$\begin{aligned} \hat{J}_k &= (e_{k-1})^2 + \|\overline{R}_{u_k}\mathbf{u}_{k-1} + \overline{R}_{ux_k}\mathbf{x}_k - \widetilde{\mathbf{b}}_{u_k}\|^2 \\ &\quad + \left\| \begin{bmatrix} \overline{R}_k \\ H_k \end{bmatrix} \mathbf{x}_k - \begin{bmatrix} \overline{\mathbf{b}}_k \\ y_k \end{bmatrix} \right\|^2 + \|R_u\mathbf{u}_k - \overline{\mathbf{b}}_{u_k}\|^2 . \end{aligned} \tag{5.10.78}$$

Applying orthogonal transformations to the third term yields

$$\hat{J}_k = (e_{k-1})^2 + \|\overline{R}_{u_k}\mathbf{u}_{k-1} + \overline{R}_{ux_k}\mathbf{x}_k - \tilde{\mathbf{b}}_{u_k}\|^2$$
$$+ \|\hat{R}_k\mathbf{x}_k - \hat{\mathbf{b}}_k\|^2 + (e_k)^2 + \|R_u\mathbf{u}_k - \overline{\mathbf{b}}_{u_k}\|^2. \qquad (5.10.79)$$

Once again $\mathbf{u}_{k-1}$, $\mathbf{u}_k$, and $\mathbf{x}_k$ may be chosen to null all but the $(e)^2$ terms. Hence, the minimum value of $\hat{J}_k$ is

$$\hat{J}_k = (e_{k-1})^2 + (e_k)^2. \qquad (5.10.80)$$

The time update to obtain $\overline{J}_{k+1}$ may now be obtained by substituting

$$\mathbf{x}_k = \Phi^{-1}(t_{k+1}, t_k)(\mathbf{x}_{k+1} - \Gamma(t_{k+1}, t_k)\mathbf{u}_k) \qquad (5.10.81)$$

for $\mathbf{x}_k$ in the term $\|\hat{R}_k\mathbf{x}_k - \hat{\mathbf{b}}_k\|^2$. Hence, the general expression for the time update at t_m after processing $m - 1$ observations is

$$\overline{J}_m = \sum_{i=1}^{m-1}(e_i)^2 + \sum_{i=1}^{m-1}\|\overline{R}_{u_i}\mathbf{u}_{i-1} + \overline{R}_{ux_i}\mathbf{x}_i - \tilde{\mathbf{b}}_{u_i}\|^2 \qquad (5.10.82)$$

$$+ \left\|\begin{bmatrix} R_u & 0 \\ -\tilde{R}_m\,\Gamma(t_m, t_{m-1}) & \tilde{R}_m \end{bmatrix}\begin{bmatrix} \mathbf{u}_{m-1} \\ \mathbf{x}_m \end{bmatrix} - \begin{bmatrix} \overline{\mathbf{b}}_{u_{m-1}} \\ \hat{\mathbf{b}}_{m-1} \end{bmatrix}\right\|^2$$

and the corresponding measurement update for processing m observations is obtained by upper triangularizing the third term of Eq. (5.10.82) and adding the data equations for the mth observation,

$$\hat{J}_m = \overline{J}_m + (\epsilon_m)^2 + \|\overline{\boldsymbol{\alpha}}_m\|^2$$
$$= \sum_{i=1}^{m-1}(e_i)^2 + \sum_{i=1}^{m}\|\overline{R}_{u_i}\mathbf{u}_{i-1} + \overline{R}_{ux_i}\mathbf{x}_i - \tilde{\mathbf{b}}_{u_i}\|^2 + \|R_u\mathbf{u}_m - \overline{\mathbf{b}}_{u_m}\|^2$$

$$+ \left\|\begin{bmatrix} \overline{R}_m \\ H_m \end{bmatrix}\mathbf{x}_m - \begin{bmatrix} \overline{\mathbf{b}}_m \\ y_m \end{bmatrix}\right\|^2. \qquad (5.10.83)$$

An orthogonal transformation is then applied to the last term of Eq. (5.10.83).
 Finally, continuing the time and measurement update through stage m yields

$$\hat{J}_m = \sum_{i=1}^{m}(e_i)^2 + \sum_{i=1}^{m}\|\overline{R}_{u_i}\mathbf{u}_{i-1} + \overline{R}_{ux_i}\mathbf{x}_i - \tilde{\mathbf{b}}_{u_i}\|^2$$
$$+ \|R_u\mathbf{u}_m - \overline{\mathbf{b}}_{u_m}\|^2 + \|\hat{R}_m\mathbf{x}_m - \hat{\mathbf{b}}_m\|^2, \qquad (5.10.84)$$

so that for the filtering problem $\hat{J}_m$ is minimized by choosing

$$\hat{\mathbf{x}}_m = \hat{R}_m^{-1}\,\hat{\mathbf{b}}_m \qquad (5.10.85)$$

and

$$\overline{R}_{u_i}\hat{\mathbf{u}}_{i-1} = \widetilde{\mathbf{b}}_{u_i} - \overline{R}_{ux_i}\hat{\mathbf{x}}_i;\, i = m, m-1\dots1\,. \qquad (5.10.86)$$

Notice that the third term in Eq. (5.10.84) is simply the addition of *a priori* information on $\mathbf{u}_m$ and does not affect the performance index until we perform a time and measurement update at t_{m+1}. Then it yields the estimate for $\hat{\mathbf{u}}_m$ given by Eq. (5.10.86) with $i = m + 1$.

If we were performing only a filtering operation and had a need for a filtered value of $\hat{\mathbf{u}}$, we would calculate it at each stage and not save the quantities $\overline{R}_u$, $\overline{\mathbf{b}}_u$, and $\overline{R}_{ux}$ needed to compute it. However, if we wish to perform smoothing, these quantities should be saved as described in Section 5.10.4.

5.10.3 SRIF COMPUTATIONAL ALGORITHM

The SRIF computational algorithm is summarized as follows. Assuming we have processed the observations at the $k - 1^{st}$ stage, the time update for t_k is obtained by applying a series of orthogonal transformations, $\overline{T}_k$, to

$$
\overline{T}_k \left[
\begin{array}{c|c|c}
\overbrace{R_u}^{q} & \overbrace{0}^{n} & \overbrace{\overline{\mathbf{b}}_{u_{k-1}}}^{1} \\
\hline
-\widetilde{R}_k\Gamma(t_k, t_{k-1}) & \widetilde{R}_k & \hat{\mathbf{b}}_{k-1}
\end{array}
\right]
\begin{array}{l} \}q \\ \}n \end{array}
$$

$$
= \left[
\begin{array}{ccc}
\overbrace{\overline{R}_{u_k}}^{q} & \overbrace{\overline{R}_{ux_k}}^{n} & \overbrace{\widetilde{\mathbf{b}}_{u_k}}^{1} \\
0 & \overline{R}_k & \overline{\mathbf{b}}_k
\end{array}
\right]
\begin{array}{l} \}q \\ \}n \end{array}
\qquad (5.10.87)
$$

where $\widetilde{R}_k$ is defined by Eq. (5.10.69) and q and n are the dimensions of the process noise and state vector, respectively. From Eqs. (5.10.70) and (5.10.71) we can write

$$
\begin{bmatrix} \overline{R}_{u_k} & \overline{R}_{ux_k} \\ 0 & \overline{R}_k \end{bmatrix}
\begin{bmatrix} \tilde{\mathbf{u}}_{k-1} \\ \overline{\mathbf{x}}_k \end{bmatrix}
=
\begin{bmatrix} \widetilde{\mathbf{b}}_{u_k} \\ \overline{\mathbf{b}}_k \end{bmatrix}
\qquad (5.10.88)
$$

from which $\tilde{\mathbf{u}}_{k-1}$, $\overline{\mathbf{x}}_k$ and $\overline{P}_k$ may be computed if desired. It is not necessary to compute these quantities. However, the quantities in the first row of Eq. (5.10.87) should be stored if smoothing is to take place.

The measurement update at t_k is obtained by applying a series of orthogonal transformations, $\hat{T}_k$, to

$$
\hat{T}_k
\begin{array}{c}
\overbrace{}^{n} \overbrace{}^{1} \\
\begin{bmatrix} \overline{R}_k & \overline{\mathbf{b}}_k \\ H_k & y_k \end{bmatrix}
\begin{array}{l} \}n \\ \}1 \end{array}
\end{array}
=
\begin{bmatrix} \hat{R}_k & \hat{\mathbf{b}}_k \\ 0 & e_k \end{bmatrix}
\tag{5.10.89}
$$

and

$$
\hat{R}_k \hat{\mathbf{x}}_k = \hat{\mathbf{b}}_k \tag{5.10.90}
$$

$$
P_k = \hat{R}_k^{-1} \hat{R}_k^{-T} \tag{5.10.91}
$$

$$
e = e + e_k^2 \tag{5.10.92}
$$

where e is the sum of squares of the observation residuals.

We can compute the filter value, $\hat{\mathbf{u}}_{k-1}$, by substituting $\hat{\mathbf{x}}_k$ into Eq. (5.10.86),

$$
\hat{\mathbf{u}}_{k-1} = \overline{R}_{u_k}^{-1} \left(\overline{\mathbf{b}}_{u_k} - \overline{R}_{ux_k} \hat{\mathbf{x}}_k \right). \tag{5.10.93}
$$

The time update at t_{k+1} may now be computed from Eq. (5.10.87) and the measurement update from Eq. (5.10.89) after changing the index from k to $k+1$. The procedure is continued until all observations have been processed.

5.10.4   SMOOTHING WITH THE SRIF

There are two approaches that may be taken to perform smoothing with the SRIF (Kaminski, 1971). The first of these uses the performance index given by Eq. (5.10.84). This performance index must be satisfied by the smooth states as well as the filter states. Hence, the equations needed for smoothing are Eqs. (5.10.85), (5.10.86), and (5.10.66). These are repeated using smoothing notation:

$$
\hat{R}_m \, \hat{\mathbf{x}}_m^m = \hat{\mathbf{b}}_m \tag{5.10.94}
$$

$$
\overline{R}_{u_i} \, \hat{\mathbf{u}}_{i-1}^m = \overline{\mathbf{b}}_{u_i} - \overline{R}_{ux_i} \, \hat{\mathbf{x}}_i^m \qquad i = m, m-1, \ldots, 1. \tag{5.10.95}
$$

$$
\hat{\mathbf{x}}_{i-1}^m = \Phi^{-1}(t_i, t_{i-1})(\hat{\mathbf{x}}_i^m - \Gamma(t_i, t_{i-1})\hat{\mathbf{u}}_{i-1}^m) \tag{5.10.96}
$$

where the notation $()_i^m$ means the smoothed value of the quantity at t_i based on m observations. Note that this is the same notation used for the conventional smoother in Section 4.15.

Starting with $\hat{\mathbf{x}}_m^m$ from Eq. (5.10.94), then $\hat{\mathbf{u}}_{m-1}^m$ is computed from Eq. (5.10.95) and $\hat{\mathbf{x}}_{m-1}^m$ is determined from Eq. (5.10.96). With this value of $\hat{\mathbf{x}}_{m-1}^m$, compute $\hat{\mathbf{u}}_{m-2}^m$ from Eq. (5.10.95) and $\hat{\mathbf{x}}_{m-2}^m$ from Eq. (5.10.96), and so on.

This backward sweep strategy may be shown to be equivalent to the smoother described in Section 4.15, and hence it is also equivalent to the Rauch, Tung, and Striebel smoother (1995).

This equivalence may be used to derive an expression for the smoothed covariance for $\hat{\mathbf{x}}_k^m$ using the filter time update equation

$$\overline{\mathbf{x}}_{k+1} = \Phi(t_{k+1}, t_k)\hat{\mathbf{x}}_k + \Gamma(t_{k+1}, t_k)\overline{\mathbf{u}}_k. \tag{5.10.97}$$

The smoothed solution must also satisfy this equation; hence

$$\hat{\mathbf{x}}_{k+1}^m = \Phi(t_{k+1}, t_k)\hat{\mathbf{x}}_k^m + \Gamma(t_{k+1}, t_k)\hat{\mathbf{u}}_k^m. \tag{5.10.98}$$

Using Eqs. (5.10.97) and (5.10.98) we may write

$$\hat{\mathbf{x}}_k^m - \hat{\mathbf{x}}_k = \Phi^{-1}(t_{k+1}, t_k)\left[\hat{\mathbf{x}}_{k+1}^m - \overline{\mathbf{x}}_{k+1}\right]$$
$$+\Phi^{-1}(t_{k+1}, t_k)\Gamma(t_{k+1}, t_k)\left[\overline{\mathbf{u}}_k - \hat{\mathbf{u}}_k^m\right]. \tag{5.10.99}$$

From Eq. (5.10.95)

$$\hat{\mathbf{u}}_k^m = \overline{R}_{u_{k+1}}^{-1}(\widetilde{\mathbf{b}}_{u_{k+1}} - \overline{R}_{ux_{k+1}}\hat{\mathbf{x}}_{k+1}^m). \tag{5.10.100}$$

This equation also yields the filter value of $\overline{\mathbf{u}}_k$,

$$\overline{\mathbf{u}}_k = \overline{R}_{u_{k+1}}^{-1}(\widetilde{\mathbf{b}}_{u_{k+1}} - \overline{R}_{ux_{k+1}}\overline{\mathbf{x}}_{k+1}). \tag{5.10.101}$$

From Eq. (5.10.100) and (5.10.101) we may write

$$\overline{\mathbf{u}}_k - \hat{\mathbf{u}}_k^m = \overline{R}_{u_{k+1}}^{-1}\overline{R}_{ux_{k+1}}(\hat{\mathbf{x}}_{k+1}^m - \overline{\mathbf{x}}_{k+1}). \tag{5.10.102}$$

Substituting Eq. (5.10.102) into (5.10.99) results in

$$\hat{\mathbf{x}}_k^m = \hat{\mathbf{x}}_k + \Phi^{-1}(t_{k+1}, t_k)\left[I + \Gamma(t_{k+1}, t_k)\overline{R}_{u_{k+1}}^{-1}\overline{R}_{ux_{k+1}}\right]$$
$$\times \left[\hat{\mathbf{x}}_{k+1}^m - \overline{\mathbf{x}}_{k+1}\right]. \tag{5.10.103}$$

Comparing Eq. (5.10.103) with Eqs. (4.15.9) and (4.15.10) yields the identity

$$S_k = P_k^k\Phi^T(t_{k+1}, t_k)\left(P_{k+1}^k\right)^{-1} \tag{5.10.104}$$
$$= \Phi^{-1}(t_{k+1}, t_k)\left[I + \Gamma(t_{k+1}, t_k)\overline{R}_{u_{k+1}}^{-1}\overline{R}_{ux_{k+1}}\right].$$

Recall that $\overline{\mathbf{x}}_{k+1} = \Phi(t_{k+1}, t_k)\hat{\mathbf{x}}_k^k$ in Eq. (4.15.9) since $\overline{\mathbf{u}}_k = 0$ by assumption. Hence, the smoothed covariance for the estimation error in $\hat{\mathbf{x}}_k^m$ may be obtained from Eq. (4.15.24),

$$P_k^m = P_k^k + S_k(P_{k+1}^m - P_{k+1}^k)S_k^T \qquad (5.10.105)$$

where P_k^k is the measurement update of the filter covariance at t_k and P_{k+1}^k is the time-updated value of P_k^k at t_{k+1}.

Note that this approach requires storage of the time and measurement update of the filter covariance. The time update is given by Eq. (5.10.75), where $\overline{P} \equiv P_{k+1}^k$. The measurement update is given by Eq. (5.10.91), where $P_k \equiv P_k^k$.

The smoothing operation also may be carried out, as described by Kaminski (1971), using an approach that yields the smoothed covariance directly and does not require storage of the filter covariance. Use Eq. (5.10.66) in Eq. (5.10.84) to eliminate $\mathbf{x}_m$. Then the performance index in terms of $\mathbf{x}_{m-1}^m$ and $\mathbf{u}_{m-1}^m$ becomes

$$\begin{aligned}
\hat{J}_{m-1}^m = e_m + \Sigma_{m-1} &+ \left\| \overline{R}_{u_m}^m \mathbf{u}_{m-1}^m + \overline{R}_{ux_m} \left[\Phi(t_m, t_{m-1})\mathbf{x}_{m-1}^m \right.\right. \\
&\left. + \Gamma(t_m, t_{m-1})\mathbf{u}_{m-1}^m \right] - \tilde{\mathbf{b}}_{u_m} \|^2 + \| \hat{R}_m \left[\Phi(t_m, t_{m-1})\mathbf{x}_{m-1}^m \right. \\
&\left.\left. + \Gamma(t_m, t_{m-1})\mathbf{u}_{m-1}^m \right] - \hat{\mathbf{b}}_m \right\|^2 .
\end{aligned} \qquad (5.10.106)$$

The term $\| R_u \mathbf{u}_m - \tilde{\mathbf{b}}_{u_m} \|^2$ has been dropped since its value is zero by definition; hence, it does not effect the value of $\hat{J}_{m-1}^m$. Also,

$$e_m \equiv \sum_{i=1}^{m}(e_i)^2 \qquad (5.10.107)$$

$$\Sigma_{m-1} \equiv \sum_{i-1}^{m-1} \left\| \overline{R}_{u_i}\mathbf{u}_{i-1} + \overline{R}_{ux_i}\mathbf{x}_i - \tilde{\mathbf{b}}_{u_i} \right\|^2 . \qquad (5.10.108)$$

Equation (5.10.106) may be written as

$$\hat{J}_{m-1}^m = \left\| \left[\begin{array}{c|c} \overline{R}_{u_m} + \overline{R}_{ux_m}\Gamma(t_m, t_{m-1}) & \overline{R}_{ux_m}\Phi(t_m, t_{m-1}) \\ \hline \hat{R}_m\Gamma(t_m, t_{m-1}) & \hat{R}_m\Phi(t_m, t_{m-1}) \end{array} \right] \right.$$
$$\left. \left[\begin{array}{c} \mathbf{u}_{m-1}^m \\ \mathbf{x}_{m-1}^m \end{array} \right] - \left[\begin{array}{c} \tilde{\mathbf{b}}_{u_m} \\ \hat{\mathbf{b}}_m \end{array} \right] \right\|^2 + e_m + \Sigma_{m-1}. \qquad (5.10.109)$$

A series of orthogonal transformation is used to upper triangularize the first term of Eq. (5.10.109),

$$T_{m-1}^* \left[\begin{array}{c|c} \overline{R}_{u_m} + \overline{R}_{ux_m}\Gamma(t_m, t_{m-1}) & \overline{R}_{ux_m}\Phi(t_m, t_{m-1}) \\ \hline \hat{R}_m\Gamma(t_m, t_{m-1}) & \hat{R}_m\Phi(t_m, t_{m-1}) \end{array} \right]$$

$$= \begin{bmatrix} R^*_{u_{m-1}} & R^*_{ux_{m-1}} & b^*_{u_{m-1}} \\ 0 & R^*_{m-1} & b^*_{m-1} \end{bmatrix}. \tag{5.10.110}$$

Hence,

$$\hat{J}^m_{m-1} = \left\| R^*_{u_{m-1}} u^m_{m-1} + R^*_{ux_{m-1}} x^m_{m-1} - b^*_{u_{m-1}} \right\|^2$$
$$+ \left\| R^*_{m-1} x^m_{m-1} - b^*_{m-1} \right\|^2 + e_m + \Sigma_{m-1}. \tag{5.10.111}$$

To minimize $\hat{J}^m_{m-1}$, choose

$$R^*_{u_{m-1}} \hat{u}^m_{m-1} = b^*_{u_{m-1}} - R^*_{ux_{m-1}} \hat{x}^m_{m-1} \tag{5.10.112}$$
$$R^*_{m-1} \hat{x}^m_{m-1} = b^*_{m-1}. \tag{5.10.113}$$

The elements of Σ_{m-1} can be nulled by the proper selection of $u_{m-2}....u_0$. Therefore, we do not need to consider this term. The covariance for the estimation error in $\hat{u}^m_{m-1}$ and $\hat{x}^m_{m-1}$ is given by

$$P^m_{m-1} \equiv \begin{bmatrix} P_u & P_{ux} \\ P_{xu} & P_x \end{bmatrix}^m_{m-1}$$
$$= \begin{bmatrix} R^*_{u_{m-1}} & R^*_{ux_{m-1}} \\ 0 & R^*_{m-1} \end{bmatrix}^{-1} \begin{bmatrix} R^*_{u_{m-1}} & R^*_{ux_{m-1}} \\ 0 & R^*_{m-1} \end{bmatrix}^{-T}. \tag{5.10.114}$$

Solving the equation

$$\begin{bmatrix} R^*_{u_{m-1}} & R^*_{ux_{m-1}} \\ 0 & R^*_{m-1} \end{bmatrix} \begin{bmatrix} A & B \\ 0 & C \end{bmatrix} = \begin{bmatrix} I & 0 \\ 0 & I \end{bmatrix}$$

for the matrices A, B, and C yields the inverse necessary to solve for P^m_{m-1}. Substituting the result into Eq. (5.10.114) results in

$$P^m_{u_{m-1}} = R^{*-1}_{u_{m-1}} R^{*-T}_{u_{m-1}}$$
$$+ R^{*-1}_{u_{m-1}} R^*_{ux_{m-1}} R^{*-1}_{m-1} R^{*-T}_{m-1} R^{*T}_{ux_{m-1}} R^{*-T}_{u_{m-1}} \tag{5.10.115}$$
$$P^m_{ux_{m-1}} = -R^{*-1}_{u_{m-1}} R^*_{ux_{m-1}} R^{*-1}_{m-1} R^{*-T}_{m-1} \tag{5.10.116}$$
$$P^m_{x_{m-1}} = R^{*-1}_{m-1} R^{*-T}_{m-1}. \tag{5.10.117}$$

We are now ready to write J^m_{m-1} in terms of x^m_{m-2} and u^m_{m-2}. Because u^m_{m-1} is independent of either of these quantities and we have already chosen its value to null the first term in Eq. (5.10.111), we may drop this term when writing J^m_{m-2}; hence,

$$\hat{J}^m_{m-2} = \|R^*_{m-1}\mathbf{x}_{m-1} - \mathbf{b}^*_{m-1}\|^2 + \|\overline{R}_{u_{m-1}}\mathbf{u}_{m-2}$$
$$+ \overline{R}_{ux_{m-1}}\mathbf{x}_{m-1} - \widetilde{\mathbf{b}}_{u_{m-1}}\|^2 + e_m + \Sigma_{m-2}. \quad (5.10.118)$$

Now use Eq. (5.10.45) to write $\mathbf{x}_{m-1}$ in Eq. (5.10.118) in terms of $\mathbf{x}_{m-2}$,

$$\hat{J}^m_{m-2} = \left\| R^*_{m-1}[\Phi(t_{m-2}, t_{m-1})\mathbf{x}_{m-2} + \Gamma(t_{m-2}, t_{m-1})\mathbf{u}_{m-2}] - \mathbf{b}^*_{m-1} \right\|^2$$
$$+ \left\| \overline{R}_{u_{m-1}}\mathbf{u}_{m-2} + \overline{R}_{ux_{m-1}}[\Phi(t_{m-2}, t_{m-1})\mathbf{x}_{m-2} \right.$$
$$\left. + \Gamma(t_{m-2}, t_{m-1})\mathbf{u}_{m-2}] - \widetilde{\mathbf{b}}_{u_{m-1}} \right\|^2 + e_m + \Sigma_{m-2}$$

$$= \left\| \begin{bmatrix} \overline{R}_{u_{m-1}} + \overline{R}_{ux_{m-1}}\Gamma(t_{m-2}, t_{m-1}) & \overline{R}_{ux_{m-1}}\Phi(t_{m-2}, t_{m-1}) \\ R^*_{m-1}\Gamma(t_{m-2}, t_{m-1}) & R^*_{m-1}\Phi(t_{m-2}, t_{m-1}) \end{bmatrix} \right.$$
$$\left. \begin{bmatrix} \mathbf{u}_{m-2} \\ \mathbf{x}_{m-2} \end{bmatrix} - \begin{bmatrix} \widetilde{\mathbf{b}}_{u_{m-1}} \\ \mathbf{b}^*_{m-1} \end{bmatrix} \right\|^2 + e_m + \Sigma_{m-2}. \quad (5.10.119)$$

Next a series of orthogonal tranformations is applied to Eq. (5.10.119). This yields the solution for $\hat{\mathbf{x}}^m_{m-2}$ and $\hat{\mathbf{u}}^m_{m-2}$ and the associated covariance matrices. This procedure is repeated until the initial stage is reached.

Hence, the procedure is continued recursively by applying orthogonal transformations so that

$$T^*_{k-1} \begin{bmatrix} \overline{R}_{u_k}\overline{R}_{ux_k}\Gamma(t_k, t_{k-1}) & \overline{R}_{ux_k}\Phi(t_k, t_{k-1}) & \widetilde{\mathbf{b}}_{u_k} \\ R^*_k\Gamma(t_k, t_{k-1}) & R^*_k\Phi(t_k, t_{k-1}) & \mathbf{b}^*_k \end{bmatrix}$$
$$= \begin{bmatrix} R^*_{u_{k-1}} & R^*_{ux_{k-1}} & \mathbf{b}^*_{u_{k-1}} \\ 0 & R^*_{k-1} & \mathbf{b}^*_{k-1} \end{bmatrix} \quad (5.10.120)$$

where $R^*_{u_{k-1}}$ and R^*_{k-1} are upper triangular. The smoothed solution is given by

$$R^*_{k-1}\hat{\mathbf{x}}^m_{k-1} = \mathbf{b}^*_{k-1} \quad (5.10.121)$$
$$R^*_{u_{k-1}}\hat{\mathbf{u}}^m_{k-1} = \mathbf{b}^*_{u_{k-1}} - R^*_{ux_{k-1}}\hat{\mathbf{x}}^m_{k-1}. \quad (5.10.122)$$

The smoothed covariance is given by

$$P^m_{x_{k-1}} = R^{*-1}_{k-1} R^{*-T}_{k-1} \quad (5.10.123)$$
$$P^m_{u_{k-1}} = R^{*-1}_{u_{k-1}} R^{*-T}_{u_{k-1}}$$
$$+ R^{*-1}_{u_{k-1}} R^*_{ux_{k-1}} R^{*-1}_{k-1} R^{*-T}_{k-1} R^{*T}_{ux_{k-1}} R^{*-T}_{u_{k-1}} \quad (5.10.124)$$

$$P^m_{ux_{k-1}} = -R^{*\,-1}_{u_{k-1}} R^*_{ux_{k-1}} R^{*\,-1}_{k-1} R^{*\,-T}_{k-1} . \tag{5.10.125}$$

The first row of the left-hand side of Eq. (5.10.120) is saved from the filtering solution and the second row is computed as part of the smoothing procedure.

5.11 PROCESS NOISE PARAMETER FILTERING/ SMOOTHING USING A SRIF

Along with the dynamic state parameters, it is often advantageous to include some other types of parameters in filtering satellite data to improve the solution. In this section we expand the results of the previous section to include in the state vector, bias parameters, and exponentially correlated process noise parameters. These additional parameters will be defined as

c : Bias parameters (constant acceleration parameters, ephemeris corrections, station coordinates, etc.)

p : Correlated process noise parameters; many random or unmodeled phenomena can be approximated quite well with first order exponentially correlated process noise, also referred to as a Gauss-Markov Process and sometimes as colored noise. Variables that commonly are modeled as a Gauss-Markov Process include

— Solar radiation pressure
— Mismodeled drag effects
— Leaky attitude control systems
— Moving station positions
— Polar motion parameters
— Clock errors
— Atmospheric path delays
— Earth rotation parameters

The recursive equation for mapping a discrete first order exponentially correlated process is (Tapley and Ingram, 1973; Bierman, 1977)

$$\mathbf{p}_{k+1} = M_{k+1}\,\mathbf{p}_k + \mathbf{w}_k . \tag{5.11.1}$$

M is the process noise parameter transition matrix and is assumed diagonal, with diagonals, m , given by

$$m = e^{-(t_{k+1}-t_k)/\tau} \tag{5.11.2}$$

where τ is the time constant of the process and represents how correlated a process noise parameter is from one time step to the next. The extremes for τ are related to m as follows:

$\tau \to 0 \qquad \to \qquad m \to 0$ White noise

(not correlated at all in time)

$\tau \to \infty \qquad \to \qquad m \to 1$ Random walk (no steady state,

strongly correlated in time).

$\mathbf{w}$ is called the process noise (not the same as $\mathbf{p}$) with

$$E\left[\mathbf{w}_j\right] = \overline{\mathbf{w}}_j . \tag{5.11.3}$$

In almost all applications, the *a priori* estimate of $\overline{\mathbf{w}}_j$ is zero, but this is not a necessary assumption for this development.

Next

$$E\left[(\mathbf{w}_j - \overline{\mathbf{w}}_j)(\mathbf{w}_k - \overline{\mathbf{w}}_k)^T\right] = Q\delta_{jk} \tag{5.11.4}$$

and

$$Q = R_w^{-1} R_w^{-T} \tag{5.11.5}$$

where Q is the process noise covariance and is diagonal with elements

$$q_i = (1 - m_i^2)\sigma_i^2 . \tag{5.11.6}$$

The variance corresponding to the particular process noise parameter p_i is σ_i^2 . The SRIF formulation of Section 5.10 can be adapted to handle bias parameters and first order exponentially correlated noise (Bierman, 1977). The state propagation equations are represented by

$$\begin{bmatrix} \mathbf{p} \\ \mathbf{x} \\ \mathbf{c} \end{bmatrix}_{k+1} = \begin{bmatrix} M & 0 & 0 \\ \Phi_p & \Phi_x & \Phi_c \\ 0 & 0 & I \end{bmatrix}_{k+1} \begin{bmatrix} \mathbf{p} \\ \mathbf{x} \\ \mathbf{c} \end{bmatrix}_k + \begin{bmatrix} \mathbf{w}_k \\ 0 \\ 0 \end{bmatrix} \tag{5.11.7}$$

where Φ_p, Φ_x, and Φ_c are state transition matrices that map perturbations in $\mathbf{p}$, $\mathbf{x}$, and $\mathbf{c}$ at t_k into perturbations in $\mathbf{x}$ at t_{k+1}. If we define

$$\mathbf{X}_{k+1} \equiv \begin{bmatrix} \mathbf{p} \\ \mathbf{x} \\ \mathbf{c} \end{bmatrix}_{k+1} \tag{5.11.8}$$

$$\Phi(t_{k+1}, t_k) \equiv \begin{bmatrix} M & 0 & 0 \\ \Phi_p & \Phi_x & \Phi_c \\ 0 & 0 & I \end{bmatrix}_{k+1} \qquad (5.11.9)$$

$$\Gamma(t_{k+1}, t_k) \equiv \begin{bmatrix} I \\ 0 \\ 0 \end{bmatrix}, \qquad (5.11.10)$$

Eq. (5.11.7) may be written as

$$\mathbf{X}_{k+1} = \Phi(t_{k+1}, t_k) \mathbf{X}_k + \Gamma(t_{k+1}, t_k) \mathbf{w}_k \qquad (5.11.11)$$

which is identical to Eq. (5.10.45). Hence the algorithms of Section 5.10.2 may be applied directly to this problem. In actual practice, however, it is customary to reduce computations by taking advantage of the fact that some parameters are constant and the process noise parameters are modeled as a Gauss-Markov Process (Bierman, 1977).

5.11.1 EXPONENTIALLY CORRELATED PROCESS NOISE SRIF

Recall that for a SRIF, R (the square root of the information matrix, Λ) is operated on for the measurement and time update rather than the covariance matrix, P, where

$$P = \Lambda^{-1} = R^{-1} R^{-T}. \qquad (5.11.12)$$

By keeping R upper triangular through orthogonal transformations, the state deviation estimate, $\hat{\mathbf{x}}$, in a standard SRIF is found by simple back substitution and is given by

$$\hat{\mathbf{x}} = R^{-1} \mathbf{b}.$$

We will derive the filter/smoother equations for the SRIF with bias parameters and process noise parameters shown explicitly. This will be accomplished by deriving and minimizing a least squares performance index.

As shown by Eq. (5.11.1), the recursive equation for a first-order exponentially correlated process is

$$\mathbf{p}_k = M_k \mathbf{p}_{k-1} + \mathbf{w}_{k-1}. \qquad (5.11.13)$$

Assume that *a priori* information for $\mathbf{w}_{k-1}$ is given by $\overline{\mathbf{w}}_{k-1}$. Generally it is assumed that $\overline{\mathbf{w}} = 0$ at each stage. Assume that the error in $\overline{\mathbf{w}}_{k-1}$ has zero mean (i.e., $\overline{\mathbf{w}}_{k-1}$ is the mean value of $\mathbf{w}_{k-1}$) and covariance Q. Thus,

$$\overline{\mathbf{w}}_{k-1} = \mathbf{w}_{k-1} + \gamma_{k-1} \qquad (5.11.14)$$

where $\mathbf{w}_{k-1}$ is the true value and γ_{k-1} is the error in $\overline{\mathbf{w}}_{k-1}$. Hence,

$$E[\gamma_{k-1}] = 0, \; E[\gamma_{k-1}\gamma_{k-1}^T] = Q. \tag{5.11.15}$$

We will assume that γ is uncorrelated in time so that

$$E[\gamma_k \gamma_j^T] = Q\delta_{kj}. \tag{5.11.16}$$

As stated in Eq. (5.11.5), define the square root of Q as

$$R_w^{-1} R_w^{-T} = Q. \tag{5.11.17}$$

Substituting $\mathbf{w}_{k-1}$ from Eq. (5.11.13) into Eq. (5.11.14) and multiplying by R_w yields a data equation for $\mathbf{w}_{k-1}$,

$$R_w \overline{\mathbf{w}}_{k-1} \equiv \overline{\mathbf{b}}_{w_{k-1}} = R_w(\mathbf{p}_k - M_k \mathbf{p}_{k-1}) + R_w \gamma_{k-1}$$

or

$$\overline{\mathbf{b}}_{w_{k-1}} = R_w(\mathbf{p}_k - M_k \, \mathbf{p}_{k-1}) + \overline{\gamma}_{k-1} \tag{5.11.18}$$

where

$$E\left[\overline{\gamma}_{k-1}\right] = R_w E\left[\gamma_{k-1}\right] = 0 \text{ and } E\left[\overline{\gamma}_{k-1}\overline{\gamma}_{k-1}^T\right] = I. \tag{5.11.19}$$

Assume further that at t_{k-1} we have *a priori* information arrays $[\overline{R}_p \; \overline{\mathbf{b}}_p]_{k-1}$, $[\overline{R}_x \; \overline{\mathbf{b}}_x]_{k-1}$, and $[\overline{R}_c \; \overline{\mathbf{b}}_c]_{k-1}$ for $\mathbf{p}$, $\mathbf{x}$, and $\mathbf{c}$, respectively. Assume that an observation is available at t_{k-1} given by

$$y_{k-1} = [H_p \; H_x \; H_c]_{k-1} \begin{bmatrix} \mathbf{p} \\ \mathbf{x} \\ \mathbf{c} \end{bmatrix}_{k-1} + \epsilon_{k-1} \tag{5.11.20}$$

where the observations have been prewhitened so that ϵ has zero means and unit variance.

Recall that the *a priori* information on $\mathbf{p}$, $\mathbf{x}$, and $\mathbf{c}$ may be written in the form of a data equation. For example, the *a priori* value, $\overline{\mathbf{p}}_{k-1}$, may be written in terms of the true value, $\mathbf{p}_{k-1}$, and the error, $\eta_{p_{k-1}}$, as

$$\overline{\mathbf{p}}_{k-1} = \mathbf{p}_{k-1} + \eta_{p_{k-1}} \tag{5.11.21}$$

where

$$E\left[\eta_{p_{k-1}}\right] = 0 \text{ and } E\left[\eta_p \eta_p^T\right]_{k-1} = \overline{P}_{p_{k-1}} \tag{5.11.22}$$

and

$$\overline{P}_{p_{k-1}} = (\overline{R}_p^{-1} \overline{R}_p^{-T})_{k-1}. \qquad (5.11.23)$$

Then the desired data equation is given by multiplying Eq. (5.11.21) by $\overline{R}_{p_{k-1}}$,

$$(\overline{R}_p \, \overline{\mathbf{p}})_{k-1} \equiv \overline{\mathbf{b}}_{p_{k-1}} = (\overline{R}_p \, \mathbf{p})_{k-1} + \overline{\eta}_{p_{k-1}} \qquad (5.11.24)$$

where $\overline{\eta}_{p_{k-1}}$ has zero mean and unit covariance. Similar data equations can be written to represent the *a priori* information for $\mathbf{x}$ and $\mathbf{c}$ at t_{k-1}.

Given the *a priori* information and the observation at t_{k-1} we wish to determine the corresponding filter values of $\mathbf{p}$, $\mathbf{x}$, and $\mathbf{c}$. The desired algorithm can be developed by minimizing the least squares performance index given by

$$\hat{J}_{k-1} \equiv \left\| \overline{\eta}_{p_{k-1}} \right\|^2 + \left\| \overline{\eta}_{x_{k-1}} \right\|^2 + \left\| \overline{\eta}_{c_{k-1}} \right\|^2 + \left\| \overline{\gamma}_{k-1} \right\|^2 + \left(\epsilon_{k-1} \right)^2. \quad (5.11.25)$$

Equation (5.11.25), which corresponds to the measurement update at t_{k-1}, may be written as (see Eqs. (5.11.18), (5.11.20), (5.11.24))

$$\hat{J}_{k-1} = \left\| \overline{R}_p \, \mathbf{p} - \overline{\mathbf{b}}_p \right\|_{k-1}^2 + \left\| \overline{R}_x \, \mathbf{x} - \overline{\mathbf{b}}_x \right\|_{k-1}^2 + \left\| \overline{R}_c \, \mathbf{c} - \overline{\mathbf{b}}_c \right\|_{k-1}^2$$

$$+ \left\| R_w (\mathbf{p}_k - M_k \mathbf{p}_{k-1}) - \overline{\mathbf{b}}_{w_{k-1}} \right\|^2$$

$$+ \left([H_p \; H_x \; H_c] \begin{bmatrix} \mathbf{p} \\ \mathbf{x} \\ \mathbf{c} \end{bmatrix} - y \right)^2_{k-1}. \qquad (5.11.26)$$

Because we may choose $\mathbf{p}_k$ to zero the next-to-last term in Eq. (5.11.26), we do not have to deal with it until we do the time update to t_k. We may write $\hat{J}_{k-1}$ as

$$\hat{J}_{k-1} = \left\| \begin{bmatrix} \overline{R}_p & 0 & 0 \\ 0 & \overline{R}_x & 0 \\ 0 & 0 & \overline{R}_c \\ H_p & H_x & H_c \end{bmatrix} \begin{bmatrix} \mathbf{p} \\ \mathbf{x} \\ \mathbf{c} \end{bmatrix} - \begin{bmatrix} \overline{\mathbf{b}}_p \\ \overline{\mathbf{b}}_x \\ \overline{\mathbf{b}}_c \\ y \end{bmatrix} \right\|_{k-1}^2$$

$$+ \left\| R_w (\mathbf{p}_k - M_k \, \mathbf{p}_{k-1}) - \overline{\mathbf{b}}_{w_{k-1}} \right\|^2. \qquad (5.11.27)$$

Applying a series of orthogonal transformations to the first term of Eq. (5.11.27) yields

$$T_{k-1} \begin{bmatrix} \overline{R}_p & 0 & 0 & \overline{\mathbf{b}}_p \\ 0 & \overline{R}_x & 0 & \overline{\mathbf{b}}_x \\ 0 & 0 & \overline{R}_c & \overline{\mathbf{b}}_c \\ H_p & H_x & H_c & y \end{bmatrix}_{k-1} = \begin{bmatrix} \hat{R}_p & \hat{R}_{px} & \hat{R}_{pc} & \hat{\mathbf{b}}_p \\ 0 & \hat{R}_x & \hat{R}_{xc} & \hat{\mathbf{b}}_x \\ 0 & 0 & \hat{R}_c & \hat{\mathbf{b}}_c \\ 0 & 0 & 0 & e \end{bmatrix}_{k-1} \qquad (5.11.28)$$

and Eq. (5.11.27) becomes

$$
\hat{J}_{k-1} = \left\| \begin{bmatrix} \hat{R}_p & \hat{R}_{px} & \hat{R}_{pc} \\ 0 & \hat{R}_x & \hat{R}_{xc} \\ 0 & 0 & \hat{R}_c \end{bmatrix} \begin{bmatrix} \mathbf{p} \\ \mathbf{x} \\ \mathbf{c} \end{bmatrix} - \begin{bmatrix} \hat{\mathbf{b}}_p \\ \hat{\mathbf{b}}_x \\ \hat{\mathbf{b}}_c \end{bmatrix} \right\|^2_{k-1}
$$

$$
+ \left\| R_w (\mathbf{p}_k - M_k \, \mathbf{p}_{k-1}) - \overline{\mathbf{b}}_{w_{k-1}} \right\|^2 + (e)^2_{k-1} .
$$

(5.11.29)

From Eq. (5.11.29) we could solve for $\hat{\mathbf{c}}$, $\hat{\mathbf{x}}$, and $\hat{\mathbf{p}}$ at the $k - 1^{\text{th}}$ stage.

We now perform the time update of $\hat{J}_{k-1}$ in order to obtain $\overline{J}_k$. To accomplish this we need to replace $\mathbf{x}_{k-1}$ with $\mathbf{x}_k$. Notice that our performance index already contains $\mathbf{p}_k$ and we will leave $\mathbf{p}_{k-1}$ in the performance index. This will conveniently yield the smoothed value of $\mathbf{p}_{k-1}$.

From Eq. (5.11.7)

$$
\mathbf{x}_k = \Phi_p(k)\mathbf{p}_{k-1} + \Phi_x(k)\mathbf{x}_{k-1} + \Phi_c(k)\mathbf{c}_{k-1}
$$

(5.11.30)

or

$$
\mathbf{x}_{k-1} = \Phi_x^{-1}(k) \left[\mathbf{x}_k - \Phi_p(k)\mathbf{p}_{k-1} - \Phi_c(k)\mathbf{c}_{k-1} \right]
$$

(5.11.31)

where

$$
\Phi(k) \equiv \Phi(t_k, t_{k-1}) .
$$

Also, $\mathbf{c}_{k-1} = \mathbf{c}_k$, since $\mathbf{c}$ is a constant. Substituting Eq. (5.11.31) into Eq. (5.11.29) yields the following equations

$$
\hat{R}_p \mathbf{p}_{k-1} + \overline{R}_{px} \left[\mathbf{x}_k - \Phi_p(k)\mathbf{p}_{k-1} - \Phi_c(k)\mathbf{c}_{k-1} \right] + \hat{R}_{pc}\mathbf{c}_{k-1} = \hat{\mathbf{b}}_{p_{k-1}}
$$

$$
\overline{R}_x \left[\mathbf{x}_k - \Phi_p(k)\mathbf{p}_{k-1} - \Phi_c(k)\mathbf{c}_{k-1} \right] + \hat{R}_{xc}\mathbf{c}_{k-1} = \hat{\mathbf{b}}_{x_{k-1}}
$$

(5.11.32)

$$
\hat{R}_c \mathbf{c}_{k-1} = \hat{\mathbf{b}}_{c_{k-1}} .
$$

Also,

$$
R_w (\mathbf{p}_k - M_k \, \mathbf{p}_{k-1}) = \overline{\mathbf{b}}_{w_{k-1}}
$$

where

$$
\overline{R}_{px} \equiv \hat{R}_{px} \Phi_x^{-1}(k)
$$

$$
\overline{R}_x \equiv \hat{R}_x \Phi_x^{-1}(k) .
$$

By regrouping and writing in matrix form (while noting that $c_{k-1} = c_k$), we may write Eq. (5.11.29) as

$$
\hat{J}_{k-1} = \left\| \left[\begin{array}{cccc} -R_w M_k & R_w & 0 & 0 \\ (\hat{R}_p - \overline{R}_{px} \Phi_p(k)) & 0 & \overline{R}_{px} & (\hat{R}_{pc} - \overline{R}_{px} \Phi_c(k)) \\ -\overline{R}_x \Phi_p(k) & 0 & \overline{R}_x & (\hat{R}_{xc} - \overline{R}_x \Phi_c(k)) \\ 0 & 0 & 0 & \hat{R}_c \end{array} \right]_{k-1} \right.
$$

$$
\left. \left[\begin{array}{c} \mathbf{p}_{k-1} \\ \mathbf{p}_k \\ \mathbf{x}_k \\ \mathbf{c}_k \end{array} \right] - \left[\begin{array}{c} \overline{\mathbf{b}}_w \\ \hat{\mathbf{b}}_p \\ \hat{\mathbf{b}}_x \\ \hat{\mathbf{b}}_c \end{array} \right]_{k-1} \right\|^2 + (e)_{k-1}^2 . \qquad (5.11.33)
$$

Applying a series of orthogonal transformations to upper triangularize the first term in Eq. (5.11.33) yields the time update

$$
\overline{J}_k = \left\| \left[\begin{array}{cccc} R_p^* & R_{pp}^* & R_{px}^* & R_{pc}^* \\ 0 & \overline{R}_p & \overline{R}_{px} & \overline{R}_{pc} \\ 0 & 0 & \overline{R}_x & \overline{R}_{xc} \\ 0 & 0 & 0 & \overline{R}_c \end{array} \right]_k \left[\begin{array}{c} \mathbf{p}_{k-1} \\ \mathbf{p}_k \\ \mathbf{x}_k \\ \mathbf{c}_k \end{array} \right] \right.
$$

$$
\left. - \left[\begin{array}{c} \mathbf{b}_p^* \\ \overline{\mathbf{b}}_p \\ \overline{\mathbf{b}}_x \\ \overline{\mathbf{b}}_c \end{array} \right]_k \right\|^2 + (e)_{k-1}^2 . \qquad (5.11.34)
$$

The $(\ )^*$ quantities are not used for filtering but are necessary if smoothing is to be done following the filtering. Hence, these quantities must be saved at each stage in the forward (filtering) sweep. Also, because $\mathbf{c}$ is a constant, it is unaffected by the mapping from t_{k-1} to t_k and need not be included in the time update procedure,

$$
\left[\hat{R}_c \ \hat{\mathbf{b}}_c \right]_{k-1} = \left[\overline{R}_c \ \overline{\mathbf{b}}_c \right]_k . \qquad (5.11.35)
$$

We may now perform the measurement update on $\overline{J}_k$ to obtain $\hat{J}_k$. This is accomplished by adding $\|\overline{\gamma}_k\|^2$ (see Eq. (5.11.18)) and $(\epsilon_k)^2$ to $\overline{J}_k$ given by Eq.

(5.11.34). Hence,

$$
\hat{J}_k = \left\|
\begin{bmatrix}
\bar{R}_p & \bar{R}_{px} & \bar{R}_{pc} \\
0 & \bar{R}_x & \bar{R}_{xc} \\
0 & 0 & \bar{R}_c \\
H_p & H_x & H_c
\end{bmatrix}_k
\begin{bmatrix}
\mathbf{p} \\
\mathbf{x} \\
\mathbf{c}
\end{bmatrix}_k
-
\begin{bmatrix}
\bar{\mathbf{b}}_p \\
\bar{\mathbf{b}}_x \\
\bar{\mathbf{b}}_c \\
y
\end{bmatrix}_k
\right\|^2
\tag{5.11.36}
$$

$$
+ \left\| R^*_{p_k} \mathbf{p}_{k-1} + R^*_{pp_k} \mathbf{p}_k + R^*_{px_k} \mathbf{x}_k + R^*_{pc_k} \mathbf{c}_k - \mathbf{b}^*_{p_k} \right\|^2
$$

$$
+ \left\| R_w (\mathbf{p}_{k+1} - M_{k+1} \mathbf{p}_k) - \bar{\mathbf{b}}_{w_k} \right\|^2 + (e)^2_{k-1}.
$$

Once again we can null the third term, $\|\bar{\gamma}_k\|$, by proper choice of $\mathbf{p}_{k+1}$; hence, we do not have to deal with this term until we perform the time update to t_{k+1}. Notice that we can choose $\mathbf{p}_{k-1}$ to null the second term. Later we will see that this is the smoothed value of $\mathbf{p}_{k-1}$.

A series of orthogonal transformations is now applied to Eq. (5.11.36) to yield

$$
\hat{J}_k = \left\|
\begin{bmatrix}
\hat{R}_p & \hat{R}_{px} & \hat{R}_{pc} \\
0 & \hat{R}_x & \hat{R}_{xc} \\
0 & 0 & \hat{R}_c
\end{bmatrix}
\begin{bmatrix}
\mathbf{p} \\
\mathbf{x} \\
\mathbf{c}
\end{bmatrix}
-
\begin{bmatrix}
\hat{\mathbf{b}}_p \\
\hat{\mathbf{b}}_x \\
\hat{\mathbf{b}}_c
\end{bmatrix}
\right\|^2_k
\tag{5.11.37}
$$

$$
+ \left\| R^*_{p_k} \mathbf{p}_{k-1} + R^*_{pp_k} \mathbf{p}_k + R^*_{px_k} \mathbf{x}_k + R^*_{pc_k} \mathbf{c}_k - \mathbf{b}^*_{p_k} \right\|^2
$$

$$
+ \left\| R_w (\mathbf{p}_{k+1} - M_{k+1} \mathbf{p}_k) - \bar{\mathbf{b}}_{w_k} \right\|^2 + (e)^2_{k-1} + (e)^2_k.
$$

This procedure of time and measurement updates is carried out until the desired number of observations has been processed.

After the Nth measurement update the performance index is given by

$$
\hat{J}_N = \Sigma_{N-1} + \left\| R^*_{p_N} \mathbf{p}_{N-1} + R^*_{pp_N} \mathbf{p}_N + R^*_{px_N} \mathbf{x}_N + R^*_{pc_N} \mathbf{c}_N - \mathbf{b}^*_{p_N} \right\|^2
$$

$$
+ \left\|
\begin{bmatrix}
\hat{R}_p & \hat{R}_{px} & \hat{R}_{pc} \\
0 & \hat{R}_x & \hat{R}_{xc} \\
0 & 0 & \hat{R}_c
\end{bmatrix}_N
\begin{bmatrix}
\mathbf{p}_N \\
\mathbf{x}_N \\
\mathbf{c}_N
\end{bmatrix}
-
\begin{bmatrix}
\hat{\mathbf{b}}_p \\
\hat{\mathbf{b}}_x \\
\hat{\mathbf{b}}_c
\end{bmatrix}_N
\right\|^2_N
$$

$$
+ \sum_{i=1}^{N} (e_i)^2
\tag{5.11.38}
$$

where

$$\Sigma_{N-1} \equiv \sum_{i=1}^{N-1} \| R_{p_i}^* \mathbf{p}_{i-1} + R_{pp_i}^* \mathbf{p}_i + R_{px_i}^* \mathbf{x}_i + R_{pc_i}^* \mathbf{c}_i - \mathbf{b}_{p_i}^* \|^2. \quad (5.11.39)$$

The best estimate of the state at t_N is obtained from

$$\hat{\mathbf{X}}_N = \hat{R}_N^{-1} \hat{\mathbf{b}}_N \quad (5.11.40)$$

where

$$\hat{\mathbf{X}}_N \equiv \begin{bmatrix} \hat{\mathbf{p}} \\ \hat{\mathbf{x}} \\ \hat{\mathbf{c}} \end{bmatrix}_N, \quad \hat{R}_N \equiv \begin{bmatrix} \hat{R}_p & \hat{R}_{px} & \hat{R}_{pc} \\ 0 & \hat{R}_x & \hat{R}_{xc} \\ 0 & 0 & \hat{R}_c \end{bmatrix}_N$$

$$\hat{\mathbf{b}}_N \equiv \begin{bmatrix} \hat{\mathbf{b}}_p \\ \hat{\mathbf{b}}_x \\ \hat{\mathbf{b}}_c \end{bmatrix}_N.$$

$$(5.11.41)$$

The filter covariance is given by

$$P_N = R_N^{-1} R_N^{-T}. \quad (5.11.42)$$

Since $\hat{R}_N$ is upper triangular, $\hat{\mathbf{X}}_N$ is obtained directly from a back substitution described by Eq. (5.2.8). If $\hat{P}_N$, $\hat{\mathbf{x}}_N$, and $\hat{\mathbf{c}}_N$ are used in the second term of Eq. (5.11.38), the smoothed value of P_{N-1} may be obtained. Note that it is not necessary to retain the first two terms of $\hat{J}_N$ if smoothing is not used.

5.11.2 SMOOTHING WITH A SRIF

Smoothing can now be done using Eq. (5.11.38) and the values of $\hat{\mathbf{X}}_N$ given by Eq. (5.11.40). From the second term of Eq. (5.11.38) we have the smoothed value of $\mathbf{p}_{N-1}$ based on N measurements

$$\hat{\mathbf{p}}_{N-1}^N = R_{p_N}^{*-1} \left[\mathbf{b}_p^* - R_{pp}^* \hat{\mathbf{p}}_N - R_{px}^* \hat{\mathbf{x}}_N - R_{pc}^* \hat{\mathbf{c}}_N \right]_N. \quad (5.11.43)$$

The smoothed value of $\mathbf{x}_{N-1}$ is obtained from Eq. (5.11.31)

$$\hat{\mathbf{x}}_{N-1}^N = \Phi_x^{-1}(N) \left[\hat{\mathbf{x}}_N - \Phi_p(N) \hat{\mathbf{p}}_{N-1}^N - \Phi_c(N) \hat{\mathbf{c}}_{N-1}^N \right] \quad (5.11.44)$$

and $\mathbf{c}$ smooths as a constant,

$$\hat{\mathbf{c}}_N = \hat{\mathbf{c}}_{N-1}^N = \hat{\mathbf{c}}_i^N \, i = 1, \ldots, N. \quad (5.11.45)$$

Hence, the general expression for smoothing is given by

$$\hat{\mathbf{p}}_i^N = R_{pi+1}^{*\,-1} \left[\mathbf{b}_p^* - R_{pp}^* \, \hat{\mathbf{p}}_{i+1}^N - R_{px}^* \, \hat{\mathbf{x}}_{i+1}^N - R_{pc}^* \, \hat{\mathbf{c}}_N \right]_{i+1}$$
$$i = N - 1, \dots, 1 \tag{5.11.46}$$

where the (   )* quantities have been saved at each value of t_i during the filtering process. Also,

$$\hat{\mathbf{x}}_i^N = \Phi_x^{-1}(i+1) \left[\hat{\mathbf{x}}_{i+1}^N - \Phi_p(i+1)\hat{\mathbf{p}}_i^N - \Phi_c(i+1)\hat{\mathbf{c}}_N \right]$$
$$\tag{5.11.47}$$
$$i = N - 1, \dots, 1 .$$

The state transition matrices also are saved during filtering.

Although the procedure just outlined yields the smoothed solutions, it does not yield a smoothed covariance. To obtain the covariance we use the procedure for the filter time update and substitute for $\mathbf{x}_N$ in terms of parameters at the $N - 1^{\text{st}}$ stage; from Eq. (5.11.30),

$$\mathbf{x}_N = \Phi_x(N)\mathbf{x}_{N-1} + \Phi_p(N)\mathbf{p}_{N-1} + \Phi_c(N)\mathbf{c}_{N-1} . \tag{5.11.48}$$

There is no need to substitute for $\mathbf{p}_N$ because we already have $\mathbf{p}_{N-1}$ available in Eq. (5.11.38) through the use of the data equation for the process noise parameters in the filter sweep.

Substituting Eq. (5.11.48) into Eq. (5.11.38) yields

$$\hat{J}_{N-1}^N = \sum_{i=1}^N (e_i)^2 + \Sigma_{N-1}$$

$$+ \left\| \begin{bmatrix} R_{pp}^* & R_p^* + R_{px}^*\Phi_p(N) & R_{px}^*\Phi_x(N) & R_{px}^*\Phi_c(N) + R_{pc}^* \\ \hat{R}_p & \hat{R}_{px}\Phi_p(N) & \hat{R}_{px}\Phi_x(N) & \hat{R}_{px}\Phi_c(N) + \hat{R}_{pc} \\ 0 & \hat{R}_x\Phi_p(N) & \hat{R}_x\Phi_x(N) & \hat{R}_x\Phi_c(N) + \hat{R}_{xc} \\ \\ 0 & 0 & 0 & \hat{R}_c \end{bmatrix}_N \begin{bmatrix} \mathbf{p}_N \\ \mathbf{p}_{N-1} \\ \mathbf{x}_{N-1} \\ \mathbf{c}_{N-1} \end{bmatrix} - \begin{bmatrix} \mathbf{b}_p^* \\ \hat{\mathbf{b}}_p \\ \hat{\mathbf{b}}_x \\ \hat{\mathbf{b}}_c \end{bmatrix}_N \right\|^2 .$$
$$\tag{5.11.49}$$

Because the smoothed value of c maps as a constant, we may drop the last row in the preceding matrix and use $\hat{R}_c$ and $\hat{c}_N$ when needed. Applying a series of orthogonal transformations to the third term yields

$$\hat{J}_{N-1}^N = \sum_{i=1}^{N}(e_i)^2 + \Sigma_{N-1} \tag{5.11.50}$$

$$+ \left\| \begin{bmatrix} R'_{pp} & R'_p & R'_{px} & R'_{pc} \\ 0 & \tilde{R}_p & \tilde{R}_{px} & \tilde{R}_{pc} \\ 0 & 0 & \tilde{R}_x & \tilde{R}_{xc} \end{bmatrix}_{N-1} \begin{bmatrix} \hat{\mathbf{p}}_N^N \\ \hat{\mathbf{p}}_{N-1}^N \\ \hat{\mathbf{x}}_{N-1}^N \\ \hat{\mathbf{c}}_N \end{bmatrix} - \begin{bmatrix} \mathbf{b}^{*\prime} \\ \mathbf{b}'_p \\ \tilde{\mathbf{b}}_p \\ \tilde{\mathbf{b}}_x \end{bmatrix}_{N-1} \right\|^2 .$$

Solving for the state vector that minimizes $\hat{J}_{N-1}^N$ yields the smoothes solution for the state. We may ignore the $(\ )'$ quantities because we already know the value of $\hat{\mathbf{p}}_N^N = \hat{\mathbf{p}}_N$. The desired solution is

$$\tilde{R}_{N-1}\hat{\mathbf{X}}_{N-1}^N = \tilde{\mathbf{b}}_{N-1} \tag{5.11.51}$$

where

$$\tilde{R}_{N-1} = \begin{bmatrix} \tilde{R}_p & \tilde{R}_{px} & \tilde{R}_{pc} \\ 0 & \tilde{R}_x & \tilde{R}_{xc} \\ 0 & 0 & \hat{R}_c \end{bmatrix} \tag{5.11.52}$$

$$\hat{\mathbf{X}}_{N-1}^N = \begin{bmatrix} \hat{\mathbf{p}}_{N-1}^N \\ \hat{\mathbf{x}}_{N-1}^N \\ \hat{\mathbf{c}}_N \end{bmatrix}, \qquad \tilde{\mathbf{b}}_{N-1} = \begin{bmatrix} \mathbf{b}'_p \\ \tilde{\mathbf{b}}_p \\ \tilde{\mathbf{b}}_x \end{bmatrix} .$$

The smoothed covariance at t_{N-1} is given by

$$P_{N-1}^N = \tilde{R}_{N-1}^{-1}\tilde{R}_{N-1}^{-T} . \tag{5.11.53}$$

Recall that the smoothed value of c never changes and is always the final filter value. Because the smoothed value of the state must also satisfy Eq. (5.11.48), we may substitute this into Eq. (5.11.50) in order to determine the smoothed state at t_{N-2}. Keep in mind that while we drop the $()'$ terms, we must now deal with the $N - 1^{\text{st}}$ term in Σ_{N-1} because it depends on the state at t_{N-1} and contains

$\mathbf{p}_{N-2}$. Hence, the performance index $\hat{J}_{N-2}^N$ becomes

$$\hat{J}_{N-2}^N = \Sigma_{N-2} + \left\| \begin{bmatrix} R_p^* & R_{pp}^* & R_{px}^* & R_{pc}^* \\ 0 & \tilde{R}_p & \tilde{R}_{px} & \tilde{R}_{pc} \\ 0 & 0 & \hat{R}_x & \hat{R}_{xc} \end{bmatrix}_{N-1} \begin{bmatrix} \mathbf{p}_{N-2} \\ \mathbf{p}_{N-1} \\ \mathbf{x}_{N-1} \\ \mathbf{c}_{N-1} \end{bmatrix} - \begin{bmatrix} \mathbf{b}_p^* \\ \hat{\mathbf{b}}_p \\ \tilde{\mathbf{b}}_x \end{bmatrix}_{N-1} \right\|^2 . \tag{5.11.54}$$

Substituting

$$\mathbf{x}_{N-1} = \Phi_x(N-1)\mathbf{x}_{N-2} + \Phi_p(N-1)\mathbf{p}_{N-2} + \Phi_c(N-1)\mathbf{c}_N \tag{5.11.55}$$

into Eq. (5.11.54) we obtain the $N - 2^{\text{nd}}$ stage of Eq. (5.11.49). Next we apply a series of orthogonal transformations to this equation to obtain the $N - 2^{\text{nd}}$ stage of Eq. (5.11.50). From this we obtain $\hat{\mathbf{X}}_{N-2}^N$ and the associated covariance matrix. This procedure is repeated until the initial stage is reached. Notice that we have dropped the term

$$\sum_{i=1}^N (e_i)^2$$

in Eq. (5.11.54). This is the sum of squares of residuals from the filter sweep and is not affected by smoothing. We do not obtain a smoothed sum of squares of residuals during the backward sweep.

5.12 REFERENCES

Battin, R., *An Introduction to the Mathematics and Methods of Astrodynamics*, American Institute of Aeronautics and Astronautics, Reston, VA, 1999.

Bierman, G. J., *Factorization Methods for Discrete Sequential Estimation*, Academic Press, New York, 1977.

Carlson, N. A., "Fast triangular formulation of the square root filter", *AIAA J.*, Vol. 11, No. 9, pp. 1239–1265, September 1973.

Dyer, P., and S. McReynolds, "Extension of square-root filtering to include process noise", *J. Optim. Theory Appl.*, Vol. 3, No. 6, pp. 444–458, 1969.

Gentleman, W. M., "Least squares computations by Givens transformations without square roots", *J. Inst. Math. Applic.*, Vol. 12, pp. 329–336, 1973.

Givens, W., "Computation of plane unitary rotations transforming a general matrix to triangular form," *J. Appl. Math.*, Vol. 6, pp. 26–50, 1958.

Golub, G. H., and C. F. Van Loan, *Matrix Computations*, Johns Hopkins University Press, 1996.

Graybill, F. A., *An Introduction to Linear Statistical Models, Volume I, McGraw-Hill Series in Probability and Statistics*, McGraw-Hill, New York, 1961.

Householder, A. S., "Unitary triangularization of a nonsymmetric matrix," *J. Assoc. Comput. Mach.*, Vol. 5, No. 4, pp. 339–342, October 1958.

Kaminski, P. G., A. E. Bryson, and S. F. Schmidt, "Discrete square root filtering: A survey of current techniques", *Trans. Auto. Cont.*, Vol. AC-16, No. 6, pp. 727–735, 1971.

Lawson, C. L., and R. J. Hanson, *Solving Least Squares Problems*, Prentice-Hall, Inc. Englewood Cliffs, NJ, 1974 (republished by SIAM, Philadelphia, PA, 1995).

Tapley, B. D., and C. Y. Choe, "An algorithm for propagating the square root covariance matrix in triangular form," *Trans. Auto. Cont.*, Vol. AC-21, pp. 122–123, 1976.

Tapley, B. D., and D. S. Ingram, "Orbit determination in the presence of unmodeled accelerations," *Trans. Auto. Cont.*, Vol. AC-18, No. 4, pp. 369–373, August, 1973.

Tapley, B. D., and J. G. Peters, "A sequential estimation algorithm using a continuous UDU^T covariance factorization," *J. Guid. Cont.*, Vol. 3, No. 4, pp. 326–331, July–August 1980.

Thornton, Thornton, C. L., *Triangular covariance factorizations for Kalman filtering*, Technical Memorandum, 33–798, Jet Propulsion Laboratory, Pasadena, CA, October 15, 1976.

5.13 EXERCISES

1. In Section 5.2 the algorithm for $\hat{x}$ is derived assuming the factorization

$$M = R^T R$$

(a) Rederive the algorithm assuming the factorization

$$M = RR^T$$

where R is $n \times n$ upper triangular and

$$M\hat{x} = N$$
$$RR^T\hat{x} = N$$
$$z \equiv R^T\hat{x}$$
$$Rz = N.$$

Answer:

$$z_i = \frac{(N_i - \sum_{j=i+1}^{n} R_{ij}z_j)}{R_{ii}} \qquad i = n,\, n-1 \ldots 1$$

$$\hat{x}_i = \frac{(z_i - \sum_{j=1}^{i-1} R_{ji}\hat{x}_j)}{R_{ii}} \qquad i = 1,\, 2 \ldots n.$$

2. Verify that the algorithm for R in Exercise 1 is given by

$$R_{jj} = \left(M_{jj} - \sum_{k=j+1}^{n} R_{jk}^2 \right)^{1/2} \qquad j = n,\, n-1, \ldots 1$$

$$R_{ij} = \frac{\left(M_{ij} - \sum_{k=j+1}^{n} R_{ik}R_{jk} \right)}{R_{jj}} \qquad i = j-1, \ldots 1.$$

3. Using Eq. (5.2.6), find R for

   (a) The matrix M given by

$$M = \begin{bmatrix} 1 & 2 & 3 \\ 2 & 8 & 2 \\ 3 & 2 & 14 \end{bmatrix}.$$

   (b) Compute R using the algorithm derived in Exercise 2. Note that these will be different matrices, thus illustrating the nonuniqueness of the matrix square root.

Answers:

Part (a)

$$R = \begin{bmatrix} 1 & 2 & 3 \\ 0 & 2 & -2 \\ 0 & 0 & 1 \end{bmatrix}$$

Part (b)

$$R = \begin{bmatrix} \dfrac{1}{3\sqrt{3}} & \dfrac{11}{3\sqrt{42}} & \dfrac{3}{\sqrt{14}} \\[2ex] 0 & 3\sqrt{\dfrac{6}{7}} & \dfrac{2}{\sqrt{14}} \\[2ex] 0 & 0 & \sqrt{14} \end{bmatrix}$$

4. Use the square root free Cholesky algorithm (Eqs. (5.2.11) and (5.2.12)) to determine the U and D factors of M for problem 3,

$$M = UDU^T.$$

Answer:

$$U = \begin{bmatrix} 1 & \dfrac{11}{54} & \dfrac{3}{14} \\[2ex] 0 & 1 & \dfrac{1}{7} \\[2ex] 0 & 0 & 1 \end{bmatrix}$$

$$D = \begin{bmatrix} \dfrac{1}{27} & 0 & 0 \\[2ex] 0 & \dfrac{54}{7} & 0 \\[2ex] 0 & 0 & 14 \end{bmatrix}$$

5. Given the function

$$f(t) = \sum_{i=0}^{4} a_i t^i + \sum_{i=1}^{4} A_i \cos(\omega_i t) + B_i \sin(\omega_i t)$$

where
$$\omega_1 = \frac{2\pi}{709}, \qquad \omega_2 = \frac{2\pi}{383}, \qquad \omega_3 = \frac{2\pi}{107}, \qquad \omega_4 = \frac{2\pi}{13}$$

and given measurements of f for $t = 0, 1, 2, \ldots, 1000$, do the following:

(a) Estimate the constants a_0, and a_i, A_i, B_i for $i = 1, 2, 3, 4$.

(b) Compare execution times and accuracy of the following four algorithms.

   i. Cholesky decomposition

   ii. Givens transformation

   iii. Givens square root free transformation

   iv. Householder transformation

   Generate your own perfect observations using the coefficients given:

   Exact Coefficients:

$$a_0 = -50$$
$$a_1 = 0.25 \qquad\qquad A_1 = -50 \qquad B_1 = 101$$
$$a_2 = -0.625 \times 10^{-3} \qquad A_2 = 1 \qquad B_2 = -0.5$$
$$a_3 = -0.4 \times 10^{-6} \qquad A_3 = -27 \qquad B_3 = -27$$
$$a_4 = 0.9 \times 10^{-9} \qquad A_4 = 4 \qquad B_4 = -3$$

(c) Redo (a) after adding Gaussian random noise with mean zero and variance $= 2$ to the data.

6. From Eq. (5.10.120), derive Eqs. (5.10.123), (5.10.124), and (5.10.125), which define the covariance for the smoothed solution including process noise. (Hint: see Eqs. (5.10.114) through Eq. (5.10.117)).

7. Generate one cycle of a sine wave with an amplitude of unity. Add white noise, $N(0, 0.05)$. Generate observation data by sampling the noisy sine wave 1000 times at equal intervals.

Using one process noise parameter, recover the sine wave using a SRIF both as a filter and a smoother. Try various combinations of τ and σ in Eqs. (5.11.2) and (5.11.6). For example try a large τ and small σ and estimate the amplitude as a constant (case 1). Try $\tau = 0$ and a very large σ to simulate the sine wave as a white noise process (case 2). Try a random walk process to describe the process noise parameter (i.e., $m = 1$, choose q (case 3)). Finally find a value of τ and σ that does a good job of replicating the sine wave (e.g., one for which the RMS of fit is near to the RMS of the noise on the data (case 4)). Assume *a priori* information, $\bar{p} = 0$ and $\overline{P}_0 = \infty$ and that the process noise, $\overline{w} = 0$, at each stage. Use the algorithms described in Sections 5.11.1 and 5.11.2 for filtering and smoothing, respectively.

Generate the following figures for these four cases:

(a) Plot the observations and the filter solutions. Compute the RMS difference.

(b) Plot the observations and the smoothed solutions. Compute the RMS difference.

(c) Compute the RMS differences between the true sine wave (without noise) and the filter and smoothed solutions.

(d) Plot the observations minus the smoothed solution.

(e) Plot w, the process noise.

For the optimal τ and σ, the plots in (d) and (e) should look similar. The optimal τ and σ result in the correlated signal in the observations being absorbed by p and the random noise by w.

* Solution hints for Exercise 7.

The measurement update at the ith stage is given by performing an orthogonal transformation (Householder or Givens),

$$T_i \begin{bmatrix} \overline{R}_{p_i} & \overline{b}_{p_i} \\ H & y_i \end{bmatrix}_{2\times 2} = \begin{bmatrix} \hat{R}_{p_i} & \hat{b}_{p_i} \\ 0 & e_i \end{bmatrix}_{2\times 2} \tag{5.11.28}$$

where the equation number refers to the corresponding equation in the text. At the epoch time $\overline{R}_{p_0} = 0$ and $\overline{b}_{p_0} = 0$ are given.

Next, a time update to the $i + 1^{\text{st}}$ stage results from a series of orthogonal transformations on

$$T_{i+1} \begin{bmatrix} -R_w m & R_w & \overline{b}_{w_i} \\ \hat{R}_{p_i} & 0 & \hat{b}_{p_i} \end{bmatrix}_{2\times 3} = \begin{bmatrix} R^*_{p_i} & R^*_{p_{i+1}} & b^*_{p_i} \\ 0 & \overline{R}_{p_{i+1}} & \overline{b}_{p_{i+1}} \end{bmatrix}_{2\times 3.} \tag{5.11.34}$$

The ( )* values must be saved for smoothing. Also,

$$R_w = \frac{1}{\sqrt{q}}$$

$$m = e^{-(t_{i+1}-t_i)/\tau} \tag{5.11.2}$$

$$q = (1 - m^2)\sigma^2 \tag{5.11.6}$$

$$H = 1 \quad (\text{the observation is } p)$$

$$\overline{P}_0 = \infty;$$

hence, $\overline{R}_{p_0} = 0$ and $\overline{b}_{p_0} = \overline{R}_{p_0}\overline{p}_0 = 0$. The filter value of p at each stage is computed from

$$\hat{p}_i = \frac{\hat{b}_{p_i}}{\hat{R}_{p_i}}.$$ (5.11.40)

After the final data point has been processed we may begin smoothing. The first smoothed value of p is given by (assuming N data points)

$$p^N_{N-1} = \frac{b^*_{p_N} - R^*_{p_N}\hat{p}_N}{R^*_{p_{N-1}}}.$$ (5.11.43)

The smoothed value of p at the ith stage is given by

$$p^N_i = \frac{b^*_{p_{i+1}} - R^*_{p_{i+1}}\hat{P}^N_{i+1}}{R^*_{p_i}}$$ (5.11.46)

$$i = N - 1, \ N - 2 \ldots 0.$$

A value of $\hat{w}^N_i$ may be determined from Eq. (5.11.13) by using the smoothed values of p,

$$\hat{w}^N_i = p^N_{i+1} - mp^N_i \quad i = N - 1, \ N - 2, \ldots 0.$$

8. Work Example 4.8.2, the spring mass problem, using the square root free Givens method to solve for $\hat{x}_0$ and the associated estimation error covariance matrix. You should get results almost identical to those given in Example 4.8.2.

Chapter 6

Consider Covariance Analysis

6.1 INTRODUCTION

The design and subsequent performance of the statistical estimation algorithms used for orbit determination, parameter identification, and navigation applications are dependent on the accuracy with which the dynamic system and the measurements used to observe the motion can be modeled. In particular, the design of orbit determination algorithms usually begins with the definition of the important error sources and a statistical description of these error sources. The effect of erroneous assumptions regarding (1) the mathematical description of the force model or the measurement model, (2) the statistical properties of the random errors, and (3) the accuracy of the numerical values assigned to the unestimated measurement and force model parameters, as well as the round-off and truncation characteristics that occur in the computation process, can lead to reduced estimation accuracy and, on occasion, to filter divergence. The general topic of covariance analysis treats the sensitivity of the estimation accuracy to these error sources. The first topic to be treated here is the effects of errors in the constant, but nonestimated, dynamic, and measurement model parameters. Errors of this nature lead to biased estimates. A second topic to be considered is the effect of errors in statistics such as the data noise covariance and the *a priori* state covariance.

On occasion it may be advantageous to ignore (i.e., not estimate) certain unknown or poorly known model parameters. *Consider covariance analysis* is a technique to assess the impact of neglecting to estimate these parameters on the accuracy of the state estimate.

The reasons for neglecting to estimate certain parameters are:

1. It is cost effective in computer cost, memory requirements, and execution time to use as small a parameter array as possible.

2. Large dimension parameter arrays may not be "totally" observable from an

observation set collected over a short time interval.

Consider covariance analysis is a "design tool" that can be used for sensitivity analysis to determine the optimal parameter array for a given estimation problem or to structure an estimation algorithm to achieve a more robust performance in the presence of erroneous force and/or measurement model parameters.

Covariance analysis is an outgrowth of the study of the effects of errors on an estimate of the state of a dynamical system. These errors manifest themselves as:

1. Large residuals in results obtained with a given estimation algorithm.

2. Divergence in the sequential estimation algorithms.

3. Incorrect navigation decisions based on an optimistic state error covariance.

As discussed in Chapter 4, the operational "fix" for the Kalman or sequential filter divergence problem caused by optimistic state error covariance estimates is the addition of process noise. In an application, the process noise model will cause the filter to place a higher weight on the most recent data. For the batch estimate, a short-arc solution, which reduces the time interval for collecting the batch of observations, can be used to achieve a similar result. Neither approach is very useful as a design tool.

The comments in the previous paragraphs can be summarized as follows. Consider covariance analysis, in the general case, attempts to quantify the effects of:

a. Nonestimated parameters, C, whose uncertainty is neglected in the estimation procedure.

b. Incorrect *a priori covariance* for the *a priori* estimate of X.

c. Incorrect *a priori covariance* for the measurement noise.

A *consider filter* will use actual data along with *a priori* information on certain consider parameters to improve the filter divergence characteristics due to errors in the dynamic and measurement models. The effects of bias in the unestimated model parameters is the most important of these effects and will be given primary emphasis in the following discussion.

6.2   BIAS IN LINEAR ESTIMATION PROBLEMS

Errors in the constant parameters that appear in the dynamic and/or measurement models may have a random distribution *a priori*. However, during any estimation procedure the values will be constant but unknown and, hence, must be treated as a bias. Bias errors in the estimation process are handled in one of three ways:

1. *Neglected.* The estimate of the state is determined, neglecting any errors in the nonestimated force model and measurement model parameters.

2. *Estimated.* The state vector is expanded to include dynamic and measurement model parameters that may be in error.

3. *Considered.* The state vector is estimated but the uncertainty in the nonestimated parameters is included in the estimation error covariance matrix. This assumes that the nonestimated parameters are constant and that their *a priori* estimate and associated covariance matrix is known.

In sequential filtering analysis, an alternate approach is to compensate for the effects of model errors through the addition of a process noise model, as discussed in Section 4.9 of Chapter 4.

6.3 FORMULATION OF THE CONSIDER COVARIANCE MATRIX

Consider the following partitioning of the generalized state vector $\mathbf{z}$, and observation-state mapping matrix H,

$$\mathbf{z} = \begin{bmatrix} \mathbf{x} \\ \mathbf{c} \end{bmatrix}, H = [H_x \vdots H_c] \qquad (6.3.1)$$

where $\mathbf{x}$ is an $n \times 1$ vector of state variables whose values are to be estimated and $\mathbf{c}$ is a $q \times 1$ vector of measurement and force model variables whose values are uncertain but whose values will not be estimated. Note that we are considering a linearized system so that $\mathbf{c}$ represents a vector of deviations between the true and nominal values of the *consider parameters*, $\mathbf{C}$,

$$\mathbf{c} = \mathbf{C} - \mathbf{C}^*. \qquad (6.3.2)$$

The measurement model for the ith observation,

$$\mathbf{y}_i = H_i \mathbf{z}_i + \boldsymbol{\epsilon}_i, i = 1, \dots, l$$

can be expressed as

$$\mathbf{y}_i = H_{x_i} \mathbf{x} + H_{c_i} \mathbf{c} + \boldsymbol{\epsilon}_i, i = 1, \dots, l. \qquad (6.3.3)$$

Assume that an *a priori* estimate of $\mathbf{x}$ and associated covariance (e.g., $(\overline{\mathbf{x}}, \overline{P}_x)$) is given along with $\overline{\mathbf{c}}$, an *a priori* estimate of $\mathbf{c}$. The filter equations can be derived by following the procedures used in Chapter 4. The relevant equations are

$$\mathbf{y} = H_x \mathbf{x} + H_c \mathbf{c} + \boldsymbol{\epsilon} \qquad (6.3.4)$$

where

$$\mathbf{y} = \begin{bmatrix} \mathbf{y}_1 \\ \mathbf{y}_2 \\ \vdots \\ \mathbf{y}_l \end{bmatrix}, H_x = \begin{bmatrix} H_{x_1} \\ H_{x_2} \\ \vdots \\ H_{x_l} \end{bmatrix},$$

$$H_c = \begin{bmatrix} H_{c_1} \\ H_{c_2} \\ \vdots \\ H_{c_l} \end{bmatrix}, \epsilon = \begin{bmatrix} \epsilon_1 \\ \epsilon_2 \\ \vdots \\ \epsilon_l \end{bmatrix} \tag{6.3.5}$$

and

$$
\begin{array}{rcl}
\mathbf{y}_i & = & p \times 1 \text{ vector of observations} \\
\mathbf{y} & = & lp \times 1 \text{ vector of observations} \\
H_{x_i} & = & p \times n \text{ matrix} \\
H_x & = & lp \times n \text{ matrix} \\
H_{c_i} & = & p \times q \text{ matrix} \\
H_c & = & lp \times q \text{ matrix} \\
\mathbf{x} & = & n \times 1 \text{ state vector} \\
\mathbf{c} & = & q \times 1 \text{ vector of consider parameters} \\
\epsilon & = & lp \times 1 \text{ vector of observation errors} \\
\epsilon_i & = & p \times 1 \text{ vector of observation errors.}
\end{array}
$$

Recall that unless the observation-state relationship and the state propagation equations are linear, $\mathbf{y}$, $\mathbf{x}$, and $\mathbf{c}$ represent observation, state, and consider parameter deviation vectors, respectively. Also,

$$E[\epsilon_i] = 0, \qquad E[\epsilon_i \epsilon_j^T] = R_i \delta_{ij},$$

$$E[\epsilon \epsilon^T] = R = \begin{bmatrix} R_1 & & & \\ & R_2 & & \\ & & \ddots & \\ & & & R_l \end{bmatrix}_{lp \times lp} \tag{6.3.6}$$

where δ_{ij} is the Kronecker delta.

A priori estimates for $\mathbf{x}$ and $\mathbf{c}$ are given by $\overline{\mathbf{x}}$ and $\overline{\mathbf{c}}$, where

$$\overline{\mathbf{x}} = \mathbf{x} + \boldsymbol{\eta}, \quad \overline{\mathbf{c}} = \mathbf{c} + \boldsymbol{\beta}. \tag{6.3.7}$$

The errors, $\boldsymbol{\eta}$ and $\boldsymbol{\beta}$, have the following statistical properties:

$$E[\boldsymbol{\eta}] = E[\boldsymbol{\beta}] = 0 \tag{6.3.8}$$
$$E[\boldsymbol{\eta}\boldsymbol{\eta}^T] = \overline{P}_x \tag{6.3.9}$$
$$E[\boldsymbol{\beta}\boldsymbol{\beta}^T] = \overline{P}_{cc} \tag{6.3.10}$$
$$E[\boldsymbol{\eta}\boldsymbol{\epsilon}^T] = E[\boldsymbol{\beta}\boldsymbol{\epsilon}^T] = 0 \tag{6.3.11}$$
$$E[\boldsymbol{\eta}\boldsymbol{\beta}^T] = \overline{P}_{xc}. \tag{6.3.12}$$

It is convenient to express this information in a more compact form, such as that of a data equation. From Eqs. (6.3.4) and (6.3.7)

$$\mathbf{y} = H_x\mathbf{x} + H_c\mathbf{c} + \boldsymbol{\epsilon}; \boldsymbol{\epsilon} \sim (0, R)$$
$$\overline{\mathbf{x}} = \mathbf{x} + \boldsymbol{\eta}; \qquad \boldsymbol{\eta} \sim (0, \overline{P}_x) \tag{6.3.13}$$
$$\overline{\mathbf{c}} = \mathbf{c} + \boldsymbol{\beta}; \qquad \boldsymbol{\beta} \sim (0, \overline{P}_{cc}).$$

Let

$$\mathbf{z} = \begin{bmatrix} \mathbf{x} \\ \mathbf{c} \end{bmatrix}; \quad \tilde{\mathbf{y}} = \begin{bmatrix} \mathbf{y} \\ \overline{\mathbf{x}} \\ \overline{\mathbf{c}} \end{bmatrix};$$

$$\tag{6.3.14}$$

$$H_z = \begin{bmatrix} H_x & H_c \\ I & 0 \\ 0 & I \end{bmatrix}; \quad \tilde{\boldsymbol{\epsilon}} = \begin{bmatrix} \boldsymbol{\epsilon} \\ \boldsymbol{\eta} \\ \boldsymbol{\beta} \end{bmatrix}.$$

It follows that the observation equations can be expressed as

$$\tilde{\mathbf{y}} = H_z\mathbf{z} + \tilde{\boldsymbol{\epsilon}}; \quad \tilde{\boldsymbol{\epsilon}} \sim (0, \tilde{R}) \tag{6.3.15}$$

where

$$\tilde{R} = \begin{bmatrix} R & 0 & 0 \\ 0 & \overline{P}_x & \overline{P}_{xc} \\ 0 & \overline{P}_{cx} & \overline{P}_{cc} \end{bmatrix}. \tag{6.3.16}$$

We wish to determine the weighted least squares estimate of z, obtained by choosing the value of z, which minimizes the performance index

$$J = 1/2\, \tilde{\epsilon}\tilde{R}^{-1}\tilde{\epsilon}^T. \tag{6.3.17}$$

Note that the weight has been specified as the *a priori* data noise covariance matrix. As illustrated in Section 4.3.3, the best estimate of z is given by

$$\hat{z} = \left(H_z^T\tilde{R}^{-1}H_z\right)^{-1}H_z^T\tilde{R}^{-1}\tilde{y} \tag{6.3.18}$$

with the associated estimation error covariance matrix

$$P_z = E\left[(\hat{z} - z)(\hat{z} - z)^T\right] = \left(H_z^T\tilde{R}^{-1}H_z\right)^{-1}. \tag{6.3.19}$$

Equations (6.3.18) and (6.3.19) may be written in partitioned form in order to isolate the quantities of interest. First define

$$\tilde{R}^{-1} = \begin{bmatrix} R^{-1} & 0 & 0 \\ 0 & \overline{M}_{xx} & \overline{M}_{xc} \\ 0 & \overline{M}_{cx} & \overline{M}_{cc} \end{bmatrix}. \tag{6.3.20}$$

From

$$\tilde{R}\tilde{R}^{-1} = I, \tag{6.3.21}$$

it can be shown that

$$
\begin{aligned}
\overline{M}_{xx} &= \overline{P}_x^{-1} + \overline{P}_x^{-1}\overline{P}_{xc}\overline{M}_{cc}\overline{P}_{cx}\overline{P}_x^{-1} \\
&= (\overline{P}_x - \overline{P}_{xc}\overline{P}_{cc}^{-1}\overline{P}_{cx})^{-1} \tag{6.3.22} \\
\overline{M}_{xc} &= -(\overline{P}_x - \overline{P}_{xc}\overline{P}_{cc}^{-1}\overline{P}_{cx})^{-1}\overline{P}_{xc}\overline{P}_{cc}^{-1} \\
&= -\overline{M}_{xx}\overline{P}_{xc}\overline{P}_{cc}^{-1} \tag{6.3.23} \\
\overline{M}_{cx} &= -(\overline{P}_{cc} - \overline{P}_{cx}\overline{P}_x^{-1}\overline{P}_{xc})^{-1}\overline{P}_{cx}\overline{P}_x^{-1} = \overline{M}_{xc}^T \tag{6.3.24} \\
\overline{M}_{cc} &= (\overline{P}_{cc} - \overline{P}_{cx}\overline{P}_x^{-1}\overline{P}_{xc})^{-1}. \tag{6.3.25}
\end{aligned}
$$

From Eq. (6.3.18) it follows that

$$\left(H_z^T\tilde{R}^{-1}H_z\right)\hat{z} = H_z^T\tilde{R}^{-1}\tilde{y} \tag{6.3.26}$$

or

$$\begin{bmatrix} (H_x^T R^{-1}H_x + \overline{M}_{xx}) & (H_x^T R^{-1}H_c + \overline{M}_{xc}) \\ (H_c^T R^{-1}H_x + \overline{M}_{cx}) & (H_c^T R^{-1}H_c + \overline{M}_{cc}) \end{bmatrix}\begin{bmatrix} \hat{x} \\ \hat{c} \end{bmatrix} \tag{6.3.27}$$

$$= \begin{bmatrix} H_x^T R^{-1} y + \overline{M}_{xx} \overline{x} + \overline{M}_{xc} \overline{c} \\ H_c^T R^{-1} y + \overline{M}_{cx} \overline{x} + \overline{M}_{cc} \overline{c} \end{bmatrix}.$$

Rewriting Eq. (6.3.27) as

$$\begin{bmatrix} M_{xx} & M_{xc} \\ M_{cx} & M_{cc} \end{bmatrix} \begin{bmatrix} \hat{\mathbf{x}} \\ \hat{\mathbf{c}} \end{bmatrix} = \begin{bmatrix} \mathbf{N}_x \\ \mathbf{N}_c \end{bmatrix} \qquad (6.3.28)$$

leads to the simultaneous equations

$$M_{xx}\hat{\mathbf{x}} + M_{xc}\hat{\mathbf{c}} = \mathbf{N}_x \qquad (6.3.29)$$
$$M_{cx}\hat{\mathbf{x}} + M_{cc}\hat{\mathbf{c}} = \mathbf{N}_c \qquad (6.3.30)$$

where

$$\begin{aligned} M_{xx} &= H_x^T R^{-1} H_x + \overline{M}_{xx} & (6.3.31) \\ M_{xc} &= H_x^T R^{-1} H_c + \overline{M}_{xc} \\ M_{cx} &= H_c^T R^{-1} H_x + \overline{M}_{cx} = M_{xc}^T \\ M_{cc} &= H_c^T R^{-1} H_c + \overline{M}_{cc} \\ \mathbf{N}_x &= H_x^T R^{-1} \mathbf{y} + \overline{M}_{xx} \overline{\mathbf{x}} + \overline{M}_{xc} \overline{\mathbf{c}} \\ \mathbf{N}_c &= H_c^T R^{-1} \mathbf{y} + \overline{M}_{cx} \overline{\mathbf{x}} + \overline{M}_{cc} \overline{\mathbf{c}}. \end{aligned}$$

From Eq. (6.3.29) it follows that

$$\hat{\mathbf{x}} = M_{xx}^{-1} \mathbf{N}_x - M_{xx}^{-1} M_{xc} \hat{\mathbf{c}}. \qquad (6.3.32)$$

If Eq. (6.3.32) is substituted into Eq. (6.3.30), the following result is obtained:

$$M_{cx}(M_{xx}^{-1} \mathbf{N}_x - M_{xx}^{-1} M_{xc} \hat{\mathbf{c}}) + M_{cc} \hat{\mathbf{c}} = \mathbf{N}_c$$

or

$$(M_{cc} - M_{cx} M_{xx}^{-1} M_{xc}) \hat{\mathbf{c}} = \mathbf{N}_c - M_{cx} M_{xx}^{-1} \mathbf{N}_x$$

and

$$\hat{\mathbf{c}} = (M_{cc} - M_{cx} M_{xx}^{-1} M_{xc})^{-1} (\mathbf{N}_c - M_{cx} M_{xx}^{-1} \mathbf{N}_x). \qquad (6.3.33)$$

The value for $\hat{\mathbf{c}}$ determined by Eq. (6.3.33) can be substituted into Eq. (6.3.32) to obtain $\hat{\mathbf{x}}$.

The covariance matrix for the errors in $\hat{\mathbf{x}}$ and $\hat{\mathbf{c}}$ given by Eq. (6.3.19) can be written in partitioned form as

$$
\begin{bmatrix} P_{xx} & P_{xc} \\ P_{cx} & P_{cc} \end{bmatrix} = \begin{bmatrix} M_{xx} & M_{xc} \\ M_{cx} & M_{cc} \end{bmatrix}^{-1}
\tag{6.3.34}
$$

where M_{xx}, M_{cx}, M_{xc}, and M_{cc} are defined by Eq. (6.3.31). From the equation

$$
\begin{bmatrix} M_{xx} & M_{xc} \\ M_{cx} & M_{cc} \end{bmatrix} \begin{bmatrix} P_{xx} & P_{xc} \\ P_{cx} & P_{cc} \end{bmatrix} = \begin{bmatrix} I & 0 \\ 0 & I \end{bmatrix}.
\tag{6.3.35}
$$

It can be shown that

$$
\begin{aligned}
P_{cx} &= -M_{cc}^{-1} M_{cx} P_{xx} & (6.3.36) \\
P_{xc} &= -M_{xx}^{-1} M_{xc} P_{cc} & (6.3.37) \\
P_{xx} &= (M_{xx} - M_{xc} M_{cc}^{-1} M_{cx})^{-1} & (6.3.38) \\
P_{cc} &= (M_{cc} - M_{cx} M_{xx}^{-1} M_{xc})^{-1}. & (6.3.39)
\end{aligned}
$$

Note that

$$
P_{xc} = P_{cx}^T.
\tag{6.3.40}
$$

The expression for P_{xx} can be written in an alternate form. Using the Schur identity (Theorem 4 of Appendix B) and letting

$$
\begin{aligned}
A &= M_{xx} \\
B &= -M_{xc} M_{cc}^{-1} \\
C &= M_{cx},
\end{aligned}
$$

it can be shown that

$$
P_{xx} = M_{xx}^{-1} + M_{xx}^{-1} M_{xc} (M_{cc} - M_{cx} M_{xx}^{-1} M_{xc})^{-1} M_{cx} M_{xx}^{-1}.
\tag{6.3.41}
$$

If we use the definition

$$
P_x = M_{xx}^{-1} = (H_x^T R^{-1} H_x + \overline{M}_{xx})^{-1},
\tag{6.3.42}
$$

and define S_{xc} as (in Section 6.4 we will show Eq. (6.3.43) to be the sensitivity matrix)

$$
S_{xc} \equiv -M_{xx}^{-1} M_{xc}
\tag{6.3.43}
$$

and using Eq. (6.3.39) to define P_{cc}, we can write Eq. (6.3.41) as

$$
P_{xx} = P_x + S_{xc} P_{cc} S_{xc}^T.
\tag{6.3.44}
$$

Also, from Eqs. (6.3.37) and (6.3.44) we have

$$P_{xc} = S_{xc} P_{cc} = P_{cx}^T. \tag{6.3.45}$$

Using Eqs. (6.3.39), (6.3.42), and (6.3.44) we may write Eqs. (6.3.32) and (6.3.33) as

$$\hat{x} = P_x \mathbf{N}_x + S_{xc} \hat{c} \tag{6.3.46}$$
$$\hat{c} = P_{cc}(\mathbf{N}_c + S_{xc}^T \mathbf{N}_x). \tag{6.3.47}$$

The consider estimate is obtained by choosing not to compute $\hat{c}$ and P_{cc} but to fix them at their *a priori* values $\overline{c}$ and $\overline{P}_{cc}$, respectively. In this case, the equations that require modification are

$$P_{cc} = \overline{P}_{cc} \tag{6.3.48}$$
$$P_{xx} = P_x + S_{xc} \overline{P}_{cc} S_{xc}^T \tag{6.3.49}$$
$$P_{xc} = S_{xc} \overline{P}_{cc} \tag{6.3.50}$$
$$\hat{x} = P_x \mathbf{N}_x + S_{xc} \overline{c}. \tag{6.3.51}$$

The *a priori* covariance of the state and consider parameters, $\overline{P}_{xc}$, generally is unknown and assumed to be the null matrix. For the case where $\overline{P}_{xc} = 0$, Eq. (6.3.31) reduces to

$$M_{xx} = H_x^T R^{-1} H_x + \overline{P}_x^{-1} \equiv P_x^{-1} \tag{6.3.52}$$
$$M_{xc} = H_x^T R^{-1} H_c. \tag{6.3.53}$$

Hence,

$$S_{xc} = -M_{xx}^{-1} M_{xc} \tag{6.3.54}$$

reduces to

$$S_{xc} = -P H_x^T R^{-1} H_c. \tag{6.3.55}$$

The computational algorithm for the batch consider filter is

Given: $\overline{x}, \overline{c}, \overline{P}_x, \overline{P}_{xc}, \overline{P}_{cc}, R_i, H_{x_i}, H_{c_i}$ and $y_i, i = 1 \cdots l$

$\overline{M}_{xx}, \overline{M}_{xc}$, and $\overline{M}_{cc}$ are given by Eqs. (6.3.22), (6.3.23), and (6.3.25), respectively.

Compute:

$$M_{xx} = H_x^T R^{-1} H_x + \overline{M}_{xx} \tag{6.3.56}$$

$$= \sum_{i=1}^{l} H_{x_i}^T R_i^{-1} H_{x_i} + \overline{M}_{xx} = P_x^{-1} \tag{6.3.57}$$

$$M_{xc} = H_x R^{-1} H_c + \overline{M}_{xc} \tag{6.3.58}$$

$$= \sum_{i=1}^{l} H_{x_i}^T R_i^{-1} H_{c_i} + \overline{M}_{xc} \tag{6.3.59}$$

$$M_{cx} = M_{xc}^T \tag{6.3.60}$$

$$M_{cc} = \sum_{i=1}^{l} H_{c_i}^T R_i^{-1} H_{c_i} + \overline{M}_{cc} \tag{6.3.61}$$

$$P_x = M_{xx}^{-1} \tag{6.3.62}$$

$$S_{xc} = -P_x M_{xc} \tag{6.3.63}$$

$$\mathbf{N}_x = \sum_{i=1}^{l} H_{x_i}^T R_i^{-1} y_i + \overline{P}_x^{-1} \overline{\mathbf{x}} \tag{6.3.64}$$

$$\hat{\mathbf{x}}_c = M_{xx}^{-1} \mathbf{N}_x - M_{xx}^{-1} M_{xc} \overline{\mathbf{c}} \tag{6.3.65}$$

$$P_{xx} = P_x + S_{xc} \overline{P}_{cc} S_{xc}^T \tag{6.3.66}$$

$$P_{xc} = S_{xc} \overline{P}_{cc} = P_{cx}^T. \tag{6.3.67}$$

Equation (6.3.66) may also be written as

$$P_{xx} = P_x + P_{xc} \overline{P}_{cc}^{-1} P_{xc}^T. \tag{6.3.68}$$

The complete *consider covariance matrix* may be written as

$$\mathbf{P_c} = E \left\{ \begin{bmatrix} \hat{\mathbf{x}}_c - \mathbf{x} \\ \overline{\mathbf{c}} - \mathbf{c} \end{bmatrix} \begin{bmatrix} (\hat{\mathbf{x}}_c - \mathbf{x})^T & (\overline{\mathbf{c}} - \mathbf{c})^T \end{bmatrix} \right\} = \begin{bmatrix} P_{xx} & P_{xc} \\ P_{cx} & \overline{P}_{cc} \end{bmatrix} \tag{6.3.69}$$

where P_{xx} is the consider covariance associated with the state vector of estimated parameters, $\mathbf{x}$.

Note also that the expression for $\hat{\mathbf{x}}_c$ can be written

$$\hat{\mathbf{x}}_c = P_x \mathbf{N}_x - P_x M_{xc} \overline{\mathbf{c}} \tag{6.3.70}$$

$$\hat{\mathbf{x}}_c = \hat{\mathbf{x}} - P_x M_{xc} \overline{\mathbf{c}} \tag{6.3.71}$$

or, from Eq. (6.3.53)

$$\hat{\mathbf{x}}_c = \hat{\mathbf{x}} - P_x H_x^T R^{-1} H_c \overline{\mathbf{c}} \tag{6.3.72}$$

where $\hat{\mathbf{x}}$ and P_x are the values of those parameters obtained from a batch processor, which assumes there are no errors in the consider parameters (i.e., $\mathbf{C}^* = \mathbf{C}$ and $P_{xc}, \overline{P}_{cc}, \mathbf{c}$, and $\overline{\mathbf{c}}$ are all zero).

6.4 THE SENSITIVITY AND PERTURBATION MATRICES

Several other matrices are often associated with the concept of consider analysis. The *sensitivity matrix* is defined to be

$$S_{xc} = \frac{\partial \hat{\mathbf{x}}_c}{\partial \hat{\mathbf{c}}} \tag{6.4.1}$$

which from Eq. (6.3.32) is

$$S_{xc} = -M_{xx}^{-1} M_{xc} . \tag{6.4.2}$$

From Eq. (6.3.52)

$$S_{xc} = -P M_{xc} , \tag{6.4.3}$$

and using Eq. (6.3.53) S_{xc} becomes

$$S_{xc} = -P H_x^T R^{-1} H_c . \tag{6.4.4}$$

Hence, Eq. (6.3.71) may be written as

$$\hat{\mathbf{x}}_c = \hat{\mathbf{x}} + S_{xc}\overline{\mathbf{c}} . \tag{6.4.5}$$

Recall from Eq. (6.3.66) that the consider covariance can be written in terms of the sensitivity matrix as

$$P_{xx} = P_x + S_{xc}\overline{P}_{cc}S_{xc}^T . \tag{6.4.6}$$

Also, using Eq. (6.3.67) the covariance, P_{xc}, can be written as

$$P_{xc} = S_{xc}\overline{P}_{cc} . \tag{6.4.7}$$

The sensitivity matrix describes how $\hat{\mathbf{x}}_c$ varies with respect to the consider parameters, **c**. Another commonly used matrix is the *perturbation matrix* defined by

$$\Gamma = S_{xc} \cdot [\text{diagonal} \, (\sigma_c)] \tag{6.4.8}$$

where the elements of the diagonal matrix are the standard deviations of the consider parameters. Each element, Γ_{ij}, gives the error in the estimate of $\mathbf{x}_i$ due to a one-sigma error in the consider parameter c_j.

Additional discussion of consider covariance analysis can be found in Bierman (1977).

6.4.1 EXAMPLE APPLICATION OF A SENSITIVITY AND PERTURBATION MATRIX

Assume we have a state vector comprised of coordinates x and y and a parameter α, and consider parameters γ and δ,

$$\mathbf{X} = \begin{bmatrix} x \\ y \\ \alpha \end{bmatrix}, \quad \mathbf{c} = \begin{bmatrix} \gamma \\ \delta \end{bmatrix}. \tag{6.4.9}$$

The perturbation matrix is defined by

$$\Gamma \equiv S_{xc} \cdot [\mathrm{diag}\ \sigma_c] = \begin{bmatrix} \dfrac{\partial \hat{x}}{\partial \gamma} & \dfrac{\partial \hat{x}}{\partial \delta} \\[2ex] \dfrac{\partial \hat{y}}{\partial \gamma} & \dfrac{\partial \hat{y}}{\partial \delta} \\[2ex] \dfrac{\partial \hat{\alpha}}{\partial \gamma} & \dfrac{\partial \hat{\alpha}}{\partial \delta} \end{bmatrix} \begin{bmatrix} \sigma_\gamma & 0 \\ 0 & \sigma_\delta \end{bmatrix}$$

or

$$\Gamma = \begin{bmatrix} \dfrac{\partial \hat{x}}{\partial \gamma}\sigma_\gamma & \dfrac{\partial \hat{x}}{\partial \delta}\sigma_\delta \\[2ex] \dfrac{\partial \hat{y}}{\partial \gamma}\sigma_\gamma & \dfrac{\partial \hat{y}}{\partial \delta}\sigma_\delta \\[2ex] \dfrac{\partial \hat{\alpha}}{\partial \gamma}\sigma_\gamma & \dfrac{\partial \hat{\alpha}}{\partial \delta}\sigma_\delta \end{bmatrix}. \tag{6.4.10}$$

Hence, the errors in the state estimate due to one-sigma errors in the consider parameters are given by

$$\begin{aligned} \Delta \hat{x} &= \frac{\partial \hat{x}}{\partial \gamma}\sigma_\gamma + \frac{\partial \hat{x}}{\partial \delta}\sigma_\delta \\[1ex] \Delta \hat{y} &= \frac{\partial \hat{y}}{\partial \gamma}\sigma_\gamma + \frac{\partial \hat{y}}{\partial \delta}\sigma_\delta \\[1ex] \Delta \hat{\alpha} &= \frac{\partial \hat{\alpha}}{\partial \gamma}\sigma_\gamma + \frac{\partial \hat{\alpha}}{\partial \delta}\sigma_\delta. \end{aligned} \tag{6.4.11}$$

Note that the information in Γ is meaningful only if $\overline{P}_{cc}$ (the covariance matrix of the errors in the consider parameters) is diagonal.

We may now form the consider covariance matrix. For simplicity, assume that

$\overline{P}_{cc}$ is diagonal. Using Eq. (6.4.6) we may write P_{xx} in terms of Γ,

$$P_{xx} = P_x + S_{xc}\overline{P}_{cc}S_{xc}^T = P_x + \Gamma\Gamma^T . \tag{6.4.12}$$

The contribution of the consider parameters is given by

$$S_{xc}\,\overline{P}_{cc}\,S_{xc}^T \;=\; P_{xc}\,S_{xc}^T \tag{6.4.13}$$

$$= \begin{bmatrix} \dfrac{\partial \hat{x}}{\partial \gamma}\sigma_\gamma^2 & \dfrac{\partial \hat{x}}{\partial \delta}\sigma_\delta^2 \\[2mm] \dfrac{\partial \hat{y}}{\partial \gamma}\sigma_\gamma^2 & \dfrac{\partial \hat{y}}{\partial \delta}\sigma_\delta^2 \\[2mm] \dfrac{\partial \hat{\alpha}}{\partial \gamma}\sigma_\gamma^2 & \dfrac{\partial \hat{\alpha}}{\partial \delta}\sigma_\delta^2 \end{bmatrix} \begin{bmatrix} \dfrac{\partial \hat{x}}{\partial \gamma} & \dfrac{\partial \hat{y}}{\partial \gamma} & \dfrac{\partial \hat{\alpha}}{\partial \gamma} \\[2mm] \dfrac{\partial \hat{x}}{\partial \delta} & \dfrac{\partial \hat{y}}{\partial \delta} & \dfrac{\partial \hat{\alpha}}{\partial \delta} \end{bmatrix} .$$

The diagonal terms of this matrix are

$$\tag{6.4.14}$$

$$= \begin{bmatrix} \left(\dfrac{\partial \hat{x}}{\partial \gamma}\right)^2\sigma_\gamma^2 + \left(\dfrac{\partial \hat{x}}{\partial \delta}\right)^2\sigma_\delta^2 & & \\[3mm] & \left(\dfrac{\partial \hat{y}}{\partial \gamma}\right)^2\sigma_\gamma^2 + \left(\dfrac{\partial \hat{y}}{\partial \delta}\right)^2\sigma_\delta^2 & \\[3mm] & & \left(\dfrac{\partial \hat{\alpha}}{\partial \gamma}\right)^2\sigma_\gamma^2 + \left(\dfrac{\partial \hat{\alpha}}{\partial \delta}\right)^2\sigma_\delta^2 \end{bmatrix}_{\text{consider}}$$

or in simplified notation, the upper triangular portion of this symmetric matrix is given by

$$S_{xc}\overline{P}_{cc}S_{xc}^T = \begin{bmatrix} \sigma_{\hat{x}}^2 & \sigma_{\hat{x}}\sigma_{\hat{y}}\rho_{xy} & \sigma_{\hat{x}}\sigma_{\hat{\alpha}}\rho_{x\alpha} \\[2mm] & \sigma_{\hat{y}}^2 & \sigma_{\hat{y}}\sigma_{\hat{\alpha}}\rho_{y\alpha} \\[2mm] & & \sigma_{\hat{\alpha}}^2 \end{bmatrix}_{\text{consider}} . \tag{6.4.15}$$

For example, the total consider variance for the state variable, x, is given by

$$\begin{aligned} \sigma_{\hat{x}}^2 &= [\sigma_{\hat{x}}^2]_{\text{data noise}} + \left(\dfrac{\partial \hat{x}}{\partial \gamma}\right)^2\sigma_\gamma^2 + \left(\dfrac{\partial \hat{x}}{\partial \delta}\right)^2\sigma_\delta^2 \\ &= [\sigma_{\hat{x}}^2]_{\text{data noise}} + [\sigma_{\hat{x}}^2]_{\text{consider}} . \end{aligned} \tag{6.4.16}$$

Comments:

1. If $\overline{P}_{cc}$ is diagonal, then the diagonal elements of P_{xx} consist of the variance due to data noise plus the sum of the squares of the perturbations due to each consider parameter.

2. Off diagonal terms of the consider portion of P_{xx} contain correlations between errors in the estimated parameters caused by the consider parameters.

3. If $\overline{P}_{cc}$ is a full matrix, $S_{xc}\,\overline{P}_{cc}\,S_{xc}^T$ becomes much more complex, and the diagonal terms become a function of the covariances (correlations) between consider parameters.

For example, consider the case where $\overline{P}_{cc}$ is a full matrix and $\mathbf{X} = [x\,y]^T$ and $\mathbf{c} = [c_1\,c_2]^T$ are 2×1 vectors. Then

$$
S_{xc}\,\overline{P}_{cc}\,S_{xc}^T =
\begin{bmatrix}
\dfrac{\partial \hat{x}}{\partial c_1} & \dfrac{\partial \hat{x}}{\partial c_2} \\[2ex]
\dfrac{\partial \hat{y}}{\partial c_1} & \dfrac{\partial \hat{y}}{\partial c_2}
\end{bmatrix}
\begin{bmatrix}
\sigma_{c_1}^2 & \mu_{12} \\[1ex]
\mu_{12} & \sigma_{c_2}^2
\end{bmatrix}
S_{xc}^T .
\tag{6.4.17}
$$

If this is expanded, the 1,1 element becomes

$$
(P_c)_{11} = \left(\frac{\partial \hat{x}}{\partial c_1}\right)^2 \sigma_{c_1}^2 + 2\frac{\partial \hat{x}}{\partial c_1}\frac{\partial \hat{x}}{\partial c_2}\mu_{12} + \left(\frac{\partial \hat{x}}{\partial c_2}\right)^2 \sigma_{c_2}^2 .
\tag{6.4.18}
$$

Depending on the sign of the components of the second term on the right-hand side, the consider uncertainty of x could be greater or less than that for the case where $\overline{P}_{cc}$ is diagonal. In general, inclusion of the correlations between the consider parameters (e.g., μ_{12}) results in a reduction in the consider variances.

6.5   INCLUSION OF TIME-DEPENDENT EFFECTS

In this section we will discuss consider covariance analysis under the assumption that the state vector is time dependent. However, in this chapter we will only "consider" the effect on the state vector of errors in constant measurement or model parameters.

The dynamical equations associated with the consider covariance model can be derived as follows. The differential equations for state propagation are given by (see Chapter 4):

$$
\dot{\mathbf{X}} = F(\mathbf{X}, \mathbf{C}, t) .
\tag{6.5.1}
$$

Expanding Eq. (6.5.1) in a Taylor series to first order about a nominal trajectory yields

$$
\dot{\mathbf{X}} = F(\mathbf{X}^*, \mathbf{C}^*, t) + \left[\frac{\partial F}{\partial \mathbf{X}}\right]^* (\mathbf{X} - \mathbf{X}^*) + \left[\frac{\partial F}{\partial \mathbf{C}}\right]^* (\mathbf{C} - \mathbf{C}^*) + \ldots
\tag{6.5.2}
$$

Define

$$\dot{\mathbf{x}} \equiv \dot{\mathbf{X}} - F(\mathbf{X}^*, \mathbf{C}^*, t) = \left[\frac{\partial F}{\partial \mathbf{X}}\right]^* (\mathbf{X} - \mathbf{X}^*) + \left[\frac{\partial F}{\partial \mathbf{C}}\right]^* (\mathbf{C} - \mathbf{C}^*). \quad (6.5.3)$$

This may be written as

$$\dot{\mathbf{x}} = A(t)\mathbf{x}(t) + B(t)\mathbf{c} \quad (6.5.4)$$

where

$$\begin{aligned}
\mathbf{x}(t) &= \mathbf{X}(t) - \mathbf{X}^*(t) & (6.5.5) \\
\mathbf{c} &= \mathbf{C} - \mathbf{C}^* & (6.5.6) \\
A(t) &= \left[\frac{\partial F}{\partial \mathbf{X}}\right]^* & \\
B(t) &= \left[\frac{\partial F}{\partial \mathbf{C}}\right]^*. &
\end{aligned}$$

For the conventional filter model $\mathbf{c} = 0$, and the solution for Eq. (6.5.4) is

$$\mathbf{x}(t) = \Phi(t, t_k)\mathbf{x}(t_k). \quad (6.5.7)$$

The general solution of Eq. (6.5.4) can be obtained by the method of variation of parameters to yield

$$\mathbf{x}(t) = \Phi(t, t_k)\mathbf{x}_k + \theta(t, t_k)\mathbf{c}. \quad (6.5.8)$$

In Eq. (6.5.8), the $n \times n$ mapping matrix, $\Phi(t, t_k)$, and the $n \times q$ mapping matrix, $\theta(t, t_k)$, are defined as

$$\Phi(t, t_k) = \frac{\partial \mathbf{X}(t)}{\partial \mathbf{X}(t_k)} \quad (6.5.9)$$

$$\theta(t, t_k) = \frac{\partial \mathbf{X}(t)}{\partial \mathbf{C}(t_k)}. \quad (6.5.10)$$

The corresponding differential equations for $\Phi(t, t_k)$ and $\theta(t, t_k)$ are obtained by differentiating Eq. (6.5.1) and recognizing that differentiation with respect to time and $\mathbf{X}(t_k)$ and $\mathbf{C}$ are interchangeable for functions whose derivatives are continuous. From Eq. (6.5.1)

$$\frac{\partial \dot{\mathbf{X}}}{\partial \mathbf{X}(t_k)} = \frac{\partial F}{\partial \mathbf{X}(t_k)}, \quad (6.5.11)$$

which may be written as

$$\frac{d}{dt}\frac{\partial \mathbf{X}(t)}{\partial \mathbf{X}(t_k)} = \left[\frac{\partial F}{\partial \mathbf{X}(t)}\right]^* \frac{\partial \mathbf{X}(t)}{\partial \mathbf{X}(t_k)} \quad (6.5.12)$$

or

$$\dot{\Phi}(t, t_k) = A(t)\Phi(t, t_k) \tag{6.5.13}$$

with initial conditions $\Phi(t_0, t_0) = I$. Likewise,

$$\frac{d}{dt}\frac{\partial \mathbf{X}(t)}{\partial \mathbf{C}(t_k)} = \left[\frac{\partial F}{\partial \mathbf{X}(t)}\right]^* \frac{\partial \mathbf{X}(t)}{\partial \mathbf{C}(t_k)} + \left[\frac{\partial F}{\partial \mathbf{C}}\right]^*$$

becomes

$$\dot{\theta}(t, t_k) = A(t)\theta(t, t_k) + B(t) \tag{6.5.14}$$

with initial conditions $\theta(t_0, t_0) = 0$. Note that if C is partitioned as

$$\mathbf{C}^T = [\mathbf{C}_d^T \vdots \mathbf{C}_m^T]$$

where $\mathbf{C}_d$ are dynamic model parameters and $\mathbf{C}_m$ are measurement model parameters, then θ can be written as

$$\theta = [\theta_d \vdots \theta_m]$$

where the solution for θ_m will be $\theta_m(t, t_k) = 0$, the null matrix. This follows since measurement model parameters do not appear in the dynamic equations.

The estimation errors for the conventional filter model (Eq. 6.5.7) and the consider model (Eq. 6.5.8), labeled here as Filter and Consider, are mapped according to

Filter: $\qquad\qquad \tilde{\mathbf{x}}(t) = \hat{\mathbf{x}}(t) - \mathbf{x}(t) = \Phi(t, t_k)\tilde{\mathbf{x}}_k \qquad\qquad (6.5.15)$

where $\tilde{\mathbf{x}}_k = \hat{\mathbf{x}}_k - \mathbf{x}_k$ and all quantities are computed assuming there are no consider parameters.

Consider: $\quad \begin{aligned} \tilde{\mathbf{x}}_c(t) &= \hat{\mathbf{x}}_c(t) - \mathbf{x}(t) \\ &= [\Phi(t, t_k)\hat{\mathbf{x}}_{c_k} + \theta(t, t_k)\bar{\mathbf{c}}] - [\Phi(t, t_k)\mathbf{x}_k + \theta(t, t_k)\mathbf{c}] \\ &= \Phi(t, t_k)(\hat{\mathbf{x}}_{c_k} - \mathbf{x}_k) + \theta(t, t_k)(\bar{\mathbf{c}} - \mathbf{c}) \qquad (6.5.16) \\ &= \Phi(t, t_k)\tilde{\mathbf{x}}_{c_k} + \theta(t, t_k)\beta \end{aligned}$

where $\tilde{\mathbf{x}}_{c_k} = \hat{\mathbf{x}}_{c_k} - \mathbf{x}_k$ and $\beta = \bar{\mathbf{c}} - \mathbf{c}$.

We have included the conventional filter results in this section to illustrate differences with the consider filter. Remember that the conventional filter results assume that there are no errors in the consider parameters ($\mathbf{C}^* = \mathbf{C}$). Hence, the true value, $\mathbf{x}_k$, for the filter results is different from the true value for the consider results.

The respective observation models are

Filter: $$\mathbf{y}_j = \tilde{H}_{x_j}\mathbf{x}_j + \boldsymbol{\epsilon}_j \quad j = 1, \ldots, l. \tag{6.5.17}$$

Consider: $$\mathbf{y}_j = \tilde{H}_{x_j}\mathbf{x}_j + \tilde{H}_{c_j}\mathbf{c} + \boldsymbol{\epsilon}_j \quad j = 1, \ldots, l. \tag{6.5.18}$$

The state vector, $\mathbf{x}_j$, is replaced by its value at the estimation epoch t_k, by using Eq. (6.5.7),

Filter: $$\mathbf{y}_j = \tilde{H}_{x_j}\Phi(t_j, t_k)\mathbf{x}_k + \boldsymbol{\epsilon}_j. \tag{6.5.19}$$

The consider expression for $\mathbf{y}_j$ is obtained by recognizing that

$$\mathbf{y}_j = \tilde{H}_{x_j}\left[\Phi(t_j, t_k)\mathbf{x}_k + \theta(t_j, t_k)\mathbf{c}\right] + \tilde{H}_{c_j}\mathbf{c} + \boldsymbol{\epsilon}_j. \tag{6.5.20}$$

This equation may be written as

$$\mathbf{y}_j = \left[\tilde{H}_{x_j}\ \tilde{H}_{c_j}\right] \begin{bmatrix} \Phi(t_j, t_k) & \theta(t_j, t_k) \\ 0 & I \end{bmatrix} \begin{bmatrix} \mathbf{x}_k \\ \mathbf{c} \end{bmatrix} + \boldsymbol{\epsilon}_j \tag{6.5.21}$$

or

$$\mathbf{y}_j = \left[\tilde{H}_{x_j}\Phi(t_j, t_k) \vdots \tilde{H}_{x_j}\theta(t_j, t_k) + \tilde{H}_{c_j}\right] \begin{bmatrix} \mathbf{x}_k \\ \mathbf{c} \end{bmatrix} + \boldsymbol{\epsilon}_j. \tag{6.5.22}$$

By using the definitions

$$H_{x_j} \equiv \tilde{H}_{x_j}\Phi(t_j, t_k) \quad j = 1, \ldots, l \tag{6.5.23}$$

$$H_{c_j} \equiv \tilde{H}_{x_j}\theta(t_j, t_k) + \tilde{H}_{c_j} \quad j = 1, \ldots, l \tag{6.5.24}$$

and

$$\mathbf{y} \equiv \begin{bmatrix} \mathbf{y}_1 \\ \vdots \\ \mathbf{y}_l \end{bmatrix}, \qquad H_x \equiv \begin{bmatrix} H_{x_1} \\ \vdots \\ H_{x_l} \end{bmatrix},$$

(6.5.25)

$$H_c \equiv \begin{bmatrix} H_{c_1} \\ \vdots \\ H_{c_l} \end{bmatrix}, \qquad \epsilon \equiv \begin{bmatrix} \epsilon_1 \\ \vdots \\ \epsilon_l \end{bmatrix}$$

the two cumulative observation models used in Section 6.3 can then be obtained,

Filter: $\mathbf{y} = H_x \mathbf{x}_k + \epsilon$ (6.5.26)

Consider: $\mathbf{y} = H_x \mathbf{x}_k + H_c \mathbf{c} + \epsilon$ (6.5.27)

where $E[\epsilon] = 0$ and $E[\epsilon \epsilon^T] = R$. The associated error covariance matrix for the filter model at the epoch, t_k, is

$$P_{xk} = E\big[(\hat{\mathbf{x}}_k - \mathbf{x}_k)(\hat{\mathbf{x}}_k - \mathbf{x}_k)^T\big] = \big[H_x^T R^{-1} H_x + \overline{P}_k^{-1}\big]^{-1}. \qquad (6.5.28)$$

To simplify, notation P_{xx} is being replaced by P_c, S_{xc} is being replaced by S, and P_x is being replaced by P for the remainder of this chapter.

The consider covariance is given by

$$P_{xx_k} \equiv P_{c_k} = E\big[(\hat{\mathbf{x}}_{c_k} - \mathbf{x}_k)(\hat{\mathbf{x}}_{c_k} - \mathbf{x}_k)^T\big].$$

From Eq. (6.4.5)

$$\hat{\mathbf{x}}_{c_k} = \hat{\mathbf{x}}_k + S_k \overline{\mathbf{c}}. \qquad (6.5.29)$$

However, the true value of $\mathbf{x}_k$ for the consider covariance analysis is given by

$$\mathbf{x}_k = \mathbf{x}_k^* + S_k \mathbf{c} \qquad (6.5.30)$$

where $\mathbf{x}_k^*$ is the true value of $\mathbf{x}_k$ for the conventional filter ($C^* = C$) and there are no errors in the consider parameters. $S_k \mathbf{c}$ is the contribution due to the true value of $\mathbf{c}$, the error in the consider parameters. Hence, $\Delta \hat{\mathbf{x}}_{c_k}$, the error in the consider estimate, $\hat{\mathbf{x}}_{c_k}$, is given by

$$\Delta \hat{\mathbf{x}}_{c_k} \equiv \hat{\mathbf{x}}_{c_k} - \mathbf{x}_k = \hat{\mathbf{x}}_k - \mathbf{x}_k^* + S_k(\overline{\mathbf{c}} - \mathbf{c}). \qquad (6.5.31)$$

Since $P_{c_k} = E[\Delta \hat{\mathbf{x}}_{c_k} \Delta \hat{\mathbf{x}}_{c_k}^T]$, it follows that

$$P_{c_k} = P_k + S_k \overline{P}_{cc} S_k^T \tag{6.5.32}$$

where

$$P_k = E\big[(\hat{\mathbf{x}}_k - \mathbf{x}_k^*)(\hat{\mathbf{x}}_k - \mathbf{x}_k^*)^T\big].$$

P_k is the conventional filter data noise covariance given by Eq. (6.5.28). Also,

$$E\big[(\hat{\mathbf{x}}_k - \mathbf{x}_k^*)(\bar{\mathbf{c}} - \mathbf{c})^T\big] = 0, \tag{6.5.33}$$

because errors in the conventional filter estimate are by definition independent of errors in the consider parameters.

6.6 PROPAGATION OF THE ERROR COVARIANCE

The models for propagating the covariance matrix from time t_j to t_k can be obtained by noting that

Filter: $$P_k = E\big[(\hat{\mathbf{x}}_k - \mathbf{x}_k)(\hat{\mathbf{x}}_k - \mathbf{x}_k)^T\big] \tag{6.6.1}$$

where $$\hat{\mathbf{x}}_k - \mathbf{x}_k = \Phi(t_k, t_j)[\hat{\mathbf{x}}_j - \mathbf{x}_j] \tag{6.6.2}$$

and

Consider: $$P_{c_k} = E\big[(\hat{\mathbf{x}}_{c_k} - \mathbf{x}_k)(\hat{\mathbf{x}}_{c_k} - \mathbf{x}_k)^T\big] \tag{6.6.3}$$

where

$$\hat{\mathbf{x}}_{c_k} - \mathbf{x}_k = \big[\Phi(t_k, t_j)\hat{\mathbf{x}}_{c_j} + \theta(t_k, t_j)\bar{\mathbf{c}}\big]$$
$$- \big[\Phi(t_k, t_j)\mathbf{x}_j + \theta(t_k, t_j)\mathbf{c}\big]. \tag{6.6.4}$$

Substituting Eqs. (6.6.2) and (6.6.4) into Eqs. (6.6.1) and (6.6.3), respectively, and performing the expected value operation leads to

Filter: $$P_k = \Phi(t_k, t_j) P_j \Phi^T(t_k, t_j) \tag{6.6.5}$$

and

Consider:

$$\begin{aligned}
P_{c_k} &= E\Big\{\big[\Phi(t_k, t_j)(\hat{\mathbf{x}}_{c_j} - \mathbf{x}_j) + \theta(t_k, t_j)(\bar{\mathbf{c}} - \mathbf{c})\big]\big[\ \ \big]^T\Big\} \\
&= \Phi(t_k, t_j) P_{c_j} \Phi^T(t_k, t_j) + \theta(t_k, t_j)\overline{P}_{cc}\theta^T(t_k, t_j) \\
&+ \Phi(t_k, t_j) S_j \overline{P}_{cc}\theta^T(t_k, t_j) + \theta(t_k, t_j)\overline{P}_{cc} S_j^T \Phi^T(t_k, t_j).
\end{aligned} \tag{6.6.6}$$

This result is obtained by noting from Eq. (6.4.7) that

$$P_{xc_j} = E\left[(\hat{\mathbf{x}}_{c_j} - \mathbf{x}_j)(\overline{\mathbf{c}} - \mathbf{c})^T\right] = S_j \overline{P}_{cc} \tag{6.6.7}$$

and

$$P_{cx_j} = \overline{P}_{cc} S_j^T \tag{6.6.8}$$

where we have used the notation

$$S_j \equiv S_{xc_j} .$$

By defining

$$S_k = \Phi(t_k, t_j) S_j + \theta(t_k, t_j) \tag{6.6.9}$$

and using

$$P_{cj} = P_j + S_j \overline{P}_{cc} S_j^T \tag{6.6.10}$$

from Eq. (6.5.32), we can write Eq. (6.6.6) for the consider covariance P_{ck} as

$$P_{c_k} = P_k + S_k \overline{P}_{cc} S_k^T . \tag{6.6.11}$$

The second term in the consider covariance will always be at least positive semidefinite and, as a consequence

$$\text{Trace}[P_{c_k}] > \text{Trace}[P_k] .$$

Hence, the effect of the uncertainty in the consider parameters will lead to a larger uncertainty in the value of the estimate at a given time.

P_{xc} is propagated as follows

$$\begin{aligned}
P_{xc_k} &= E\left[(\hat{\mathbf{x}}_{c_k} - \mathbf{x}_k)\boldsymbol{\beta}^T\right] \\
&= E\left[\Phi(t_k, t_j)(\hat{\mathbf{x}}_{c_j} - \mathbf{x}_j) + \theta(t_k, t_j)(\overline{\mathbf{c}} - \mathbf{c})\boldsymbol{\beta}^T\right] \\
&= \Phi(t_k, t_j) P_{xc_j} + \theta(t_k, t_j)\overline{P}_{cc}
\end{aligned} \tag{6.6.12}$$

or P_{xc_k} may be determined directly from S_k by using Eq. (6.6.7) and Eq. (6.6.9),

$$P_{xc_k} = S_k \overline{P}_{cc} . \tag{6.6.13}$$

Mapping of the complete consider covariance matrix may be written more compactly as follows. Define

$$\Psi(t_k, t_j) \equiv \begin{bmatrix} \Phi(t_k, t_j) & \theta(t_k, t_j) \\ 0 & I \end{bmatrix} \tag{6.6.14}$$

then

$$\mathbf{P}_{c_k} = \Psi(t_k, t_j)\mathbf{P}_{cj}\Psi^T(t_k, t_j).$$ (6.6.15)

Recall that

$$\mathbf{P}_{c_k} \equiv \begin{bmatrix} P_{c_k} & P_{xc_k} \\ P_{cx_k} & \overline{P}_{cc} \end{bmatrix}$$ (6.6.16)

where the expressions for P_{c_k} and P_{xc_k} are given by Eqs. (6.6.11) and (6.6.13) respectively.

6.7 SEQUENTIAL CONSIDER COVARIANCE ANALYSIS

The algorithms for consider covariance analysis developed in batch form also can be developed in sequential form. The batch algorithm for $\hat{\mathbf{x}}$ can be modified to function in sequential form as follows.

Write Eq. (6.3.72) at the kth stage

$$\hat{\mathbf{x}}_{c_k} = \hat{\mathbf{x}}_k - P_k H_{x_k}^T R_k^{-1} H_{c_k}\overline{\mathbf{c}}$$ (6.7.1)

where

$$P_k = \left(\overline{P}_k^{-1} + \sum_{i=1}^{k} H_{x_i}^T R_i^{-1} H_{x_i}\right)^{-1}$$ (6.7.2)

and

$$H_{x_k}^T R_k^{-1} H_{c_k} = \sum_{i=1}^{k} H_{x_i}^T R_i^{-1} H_{c_i}$$ (6.7.3)

that is, $\overline{P}_0$, H_{x_i}, and H_{c_i} all have been mapped to the appropriate time t_k (e.g., $\overline{P}_k = \Phi(t_k, t_0)\overline{P}_0, \Phi^T(t_k, t_0)$), and H_{x_i} and H_{c_i} are defined by Eq. (6.5.23) and (6.5.24), respectively,

$$H_{x_i} = \widetilde{H}_{x_i}\Phi(t_i, t_k).$$

The sensitivity matrix S_k is given by

$$S_k = \frac{\partial\hat{\mathbf{x}}_k}{\partial\overline{\mathbf{c}}} = -P_k H_{x_k}^T R_k^{-1} H_{c_k}.$$ (6.7.4)

Hence (see also Eq. (6.4.5)),

$$\hat{\mathbf{x}}_{c_k} = \hat{\mathbf{x}}_k + S_k\overline{\mathbf{c}}.$$ (6.7.5)

The *a priori* value of $\hat{\mathbf{x}}_{c_k}$ is given by

$$\overline{\mathbf{x}}_{c_k} = \Phi(t_k, t_{k-1})\hat{\mathbf{x}}_{c_{k-1}} + \theta(t_k, t_{k-1})\overline{\mathbf{c}}.$$ (6.7.6)

Substituting Eq. (6.7.5) at the $k - 1^{\text{st}}$ stage into Eq. (6.7.6) yields

$$\overline{\mathbf{x}}_{c_k} = \Phi(t_k, t_{k-1})\hat{\mathbf{x}}_{k-1} + \big(\Phi(t_k, t_{k-1})S_{k-1} + \theta(t_k, t_{k-1})\big)\overline{\mathbf{c}}. \qquad (6.7.7)$$

Define

$$\overline{S}_k = \Phi(t_k, t_{k-1})S_{k-1} + \theta(t_k, t_{k-1}), \qquad (6.7.8)$$

then

$$\overline{\mathbf{x}}_{c_k} = \overline{\mathbf{x}}_k + \overline{S}_k\overline{\mathbf{c}}. \qquad (6.7.9)$$

Recall that $\overline{\mathbf{x}}_k$ and $\hat{\mathbf{x}}_k$ are the values of these quantities assuming there are no errors in the consider parameters (i.e., $\mathbf{c} = \overline{\mathbf{c}} = 0$) and $\overline{P}_{cc}$ is the null matrix.

The deviation of $\overline{\mathbf{x}}_{ck}$ in Eq. (6.7.9) from the true value of $\mathbf{x}_c$ may be written in terms of the error in each component

$$\delta\overline{\mathbf{x}}_{c_k} = \delta\overline{\mathbf{x}}_k + \overline{S}_k\delta\overline{\mathbf{c}} \qquad (6.7.10)$$

Multiplying Eq. (6.7.10) by its transpose and taking the expected value yields

$$E\Big[\delta\overline{\mathbf{x}}_{c_k}\delta\overline{\mathbf{x}}_{c_k}^T\Big] = \overline{P}_{c_k} = E\Big[\delta\overline{\mathbf{x}}_k\delta\overline{\mathbf{x}}_k^T\Big] + \overline{S}_k E\Big[\boldsymbol{\beta}\boldsymbol{\beta}^T\Big]\overline{S}_k^T \qquad (6.7.11)$$

where $\delta\overline{\mathbf{c}} = \overline{\mathbf{c}} - \mathbf{c} = \boldsymbol{\beta}$, and all cross covariances are zero because $\delta\overline{\mathbf{x}}_k$ is by definition independent of the consider parameters. Hence,

$$\overline{P}_{c_k} = \overline{P}_k + \overline{S}_k\overline{P}_{cc}\overline{S}_k^T. \qquad (6.7.12)$$

Note that this equation is equivalent to Eq. (6.6.11) since both these equations simply map the consider covariance matrix forward in time.

The measurement update for $\overline{P}_{c_k}$ is defined by

$$P_{c_k} = E\big[(\hat{\mathbf{x}}_{c_k} - \mathbf{x}_k)(\hat{\mathbf{x}}_{c_k} - \mathbf{x}_k)^T\big]. \qquad (6.7.13)$$

Using Eq. (6.7.5), we may write the error in $\hat{\mathbf{x}}_{c_k}$ as

$$\delta\hat{\mathbf{x}}_{c_k} = \delta\hat{\mathbf{x}}_k + S_k\delta\overline{\mathbf{c}}. \qquad (6.7.14)$$

Hence Eq. (6.7.13) may be written as

$$P_{c_k} = E\Big[\delta\hat{\mathbf{x}}_{c_k}\delta\hat{\mathbf{x}}_{c_k}^T\Big] = E\Big[\delta\hat{\mathbf{x}}_k\delta\hat{\mathbf{x}}_k^T\Big] + S_k E\Big[\boldsymbol{\beta}\boldsymbol{\beta}^T\Big]S_k^T \qquad (6.7.15)$$

or

$$P_{c_k} = P_k + S_k\overline{P}_{cc}S_k^T. \qquad (6.7.16)$$

The cross covariances $\overline{P}_{xc_k}$ and P_{xc_k} are obtained from Eqs. (6.7.10) and (6.7.14), respectively,

$$
\begin{aligned}
\overline{P}_{xc_k} &= E\left[\delta\overline{\mathbf{x}}_{c_k}\beta^T\right] \\
&= \overline{S}_k\overline{P}_{cc} & (6.7.17) \\
P_{xc_k} &= E\left[\delta\hat{\mathbf{x}}_{c_k}\beta^T\right] \\
&= S_k\overline{P}_{cc}. & (6.7.18)
\end{aligned}
$$

To obtain an expression for the measurement update of $\overline{S}_k$, note that an analogy with the expression for $\hat{\mathbf{x}}_k$ of Eq. (4.7.16), $\hat{\mathbf{x}}_{c_k}$, may be written as

$$
\hat{\mathbf{x}}_{c_k} = \overline{\mathbf{x}}_{c_k} + K_k\left(\mathbf{y}_k - \widetilde{H}_{x_k}\overline{\mathbf{x}}_{c_k} - \widetilde{H}_{c_k}\overline{\mathbf{c}}\right). \qquad (6.7.19)
$$

Using Eq. (6.7.9) this may be written as

$$
\hat{\mathbf{x}}_{c_k} = \overline{\mathbf{x}}_k + \overline{S}_k\overline{\mathbf{c}} + K_k\left[\mathbf{y}_k - \widetilde{H}_{x_k}\left(\overline{\mathbf{x}}_k + \overline{S}_k\overline{\mathbf{c}}\right) - \widetilde{H}_{c_k}\overline{\mathbf{c}}\right] \qquad (6.7.20)
$$

and recognizing that $\hat{\mathbf{x}}_k = \overline{\mathbf{x}}_k + K_k(\mathbf{y}_k - \widetilde{H}_{x_k}\overline{\mathbf{x}}_k)$, Eq. (6.7.20) becomes

$$
\hat{\mathbf{x}}_{c_k} = \hat{\mathbf{x}}_k + (\overline{S}_k - K_k\widetilde{H}_{x_k}\overline{S}_k - K_k\widetilde{H}_{c_k})\overline{\mathbf{c}}. \qquad (6.7.21)
$$

Comparing this with Eq. 6.7.5), it is seen that

$$
S_k = (I - K_k\widetilde{H}_{x_k})\overline{S}_k - K_k\widetilde{H}_{c_k}. \qquad (6.7.22)
$$

We now have the equations needed to write the computational algorithm for the sequential consider covariance filter.

Given: $P_{k-1}, \hat{\mathbf{x}}_{k-1}, S_{k-1}, \widetilde{H}_{x_k}, \widetilde{H}_{c_k}, \mathbf{y}_k$

1. Compute the time updates

$$
\begin{aligned}
\overline{P}_k &= \Phi\left(t_k, t_{k-1}\right)P_{k-1}\Phi^T\left(t_k, t_{k-1}\right) & (6.7.23) \\
K_k &= \overline{P}_k\widetilde{H}_{x_k}^T\left(\widetilde{H}_{x_k}\overline{P}_k\widetilde{H}_{x_k}^T + R_k\right)^{-1} & (6.7.24) \\
\overline{S}_k &= \Phi(t_k, t_{k-1})S_{k-1} + \theta\left(t_k, t_{k-1}\right) & (6.7.25) \\
\overline{\mathbf{x}}_k &= \Phi(t_k, t_{k-1})\hat{\mathbf{x}}_{k-1} & (6.7.26) \\
\overline{\mathbf{x}}_{c_k} &= \overline{\mathbf{x}}_k + \overline{S}_k\overline{\mathbf{c}} & (6.7.27) \\
\overline{P}_{c_k} &= \overline{P}_k + \overline{S}_k\overline{P}_{cc}\overline{S}_k^T & (6.7.28) \\
\overline{P}_{xc_k} &= \Phi\left(t_k, t_{k-1}\right)P_{xc_{k-1}} + \theta(t_k, t_{k-1})\overline{P}_{cc} \\
&= \overline{S}_k\overline{P}_{cc}. & (6.7.29)
\end{aligned}
$$

2. Compute the measurement update

$$P_k = (I - K_k \tilde{H}_{x_k})\overline{P}_k \qquad (6.7.30)$$

$$S_k = (I - K_k \tilde{H}_{x_k})\overline{S}_k - K_k \tilde{H}_{c_k} \qquad (6.7.31)$$

$$\hat{x}_k = \overline{x}_k + K_k(y_k - \tilde{H}_{x_k}\overline{x}_k) \qquad (6.7.32)$$

$$\hat{x}_{c_k} = \hat{x}_k + S_k \overline{c} \qquad (6.7.33)$$

$$P_{c_k} = P_k + S_k \overline{P}_{cc} S_k^T \qquad (6.7.34)$$

$$P_{xc_k} = S_k \overline{P}_{cc}. \qquad (6.7.35)$$

Note that $\overline{x}_{c_0} = \overline{x}_0$ (from Eq. 6.7.9); the consider parameters do not affect the *a priori* value of $\overline{x}_{c_0}$, hence $\overline{S}_0 = 0$ and

$$S_0 = -K_0 \tilde{H}_{c_0}. \qquad (6.7.36)$$

Normally in a covariance analysis one would be interested only in computing the consider covariances; hence, the estimates for $\hat{x}$ and $\hat{x}_c$ need not be computed. Generally process noise is not included in a consider covariance analysis. However, if process noise has been included, Eq. (6.7.23) is replaced by

$$\overline{P}_k = \Phi(t_k, t_{k-1})P_{k-1}\Phi^T(t_k, t_{k-1}) + \Gamma(t_k, t_{k-1})Q_{k-1}\Gamma^T(t_k, t_{k-1}). \quad (6.7.37)$$

6.8 EXAMPLE: FREELY FALLING POINT MASS

A point mass is in free fall only under the influence of gravity. Observations of range, x, are made from a fixed referent point. Set up the consider analysis equations assuming that x is estimated and g is a consider parameter, and that $\overline{P}_0 = I$, $R = I$, $\overline{P}_{cc} = \Pi$, and $\overline{P}_{xc} = 0$.

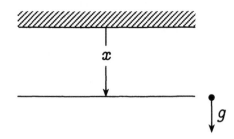

Equation of motion: $\ddot{x} = g$

State vector: $\mathbf{X} = \begin{bmatrix} x \\ v \end{bmatrix}$

System dynamics: $\dot{\mathbf{X}} = F(\mathbf{X}, t) = \begin{bmatrix} v \\ g \end{bmatrix}$

$$A = \frac{\partial F}{\partial \mathbf{X}} = \begin{bmatrix} \dfrac{\partial v}{\partial x} & \dfrac{\partial v}{\partial v} \\[2mm] \dfrac{\partial g}{\partial x} & \dfrac{\partial g}{\partial v} \end{bmatrix} = \begin{bmatrix} 0 & 1 \\ 0 & 0 \end{bmatrix}.$$

The consider parameter, C, is g. Hence,

$$B = \frac{\partial F}{\partial g} = \begin{bmatrix} \dfrac{\partial v}{\partial g} \\[2mm] \dfrac{\partial g}{\partial g} \end{bmatrix} = \begin{bmatrix} 0 \\ 1 \end{bmatrix}.$$

The observation equation is

$$\begin{aligned} Y(\mathbf{X}, t) &= G(\mathbf{X}, t) + \epsilon \\ &= x + \epsilon. \end{aligned}$$

Because, $G(\mathbf{X}, t) = x$

$$\begin{aligned} \tilde{H}_x &= \frac{\partial G}{\partial \mathbf{X}} = \begin{bmatrix} \dfrac{\partial x}{\partial x} & \dfrac{\partial x}{\partial v} \end{bmatrix} = [1 \ 0] \\ \tilde{H}_c &= \frac{\partial G}{\partial g} = 0. \end{aligned}$$

The differential equation for the state transition matrix is

$$\dot{\Phi} = A\Phi, \quad \Phi(t_o, t_0) = I$$

where

$$A = \begin{bmatrix} 0 & 1 \\ 0 & 0 \end{bmatrix}.$$

The solution for $\Phi(t, t_0)$ is

$$\Phi(t, t_0) = K e^{A(t - t_0)}$$

where K is a constant matrix. Since $\Phi(t_0, t_0) = I$, we have $K e^{A0} = I$ or $K = I$. Therefore,

$$\Phi(t, t_0) = e^{A(t - t_0)}$$

$$= I + A(t - t_0) + \frac{1}{2!}A^2(t - t_0)^2 + \dots$$

Now

$$A^2 = \begin{bmatrix} 0 & 1 \\ 0 & 0 \end{bmatrix} \begin{bmatrix} 0 & 1 \\ 0 & 0 \end{bmatrix} = \begin{bmatrix} 0 & 0 \\ 0 & 0 \end{bmatrix}.$$

Thus, A^2 and higher powers of A are all zero. Hence,

$$
\begin{aligned}
\Phi(t, t_0) &= I + A(t - t_0) \\
&= \begin{bmatrix} 1 & 0 \\ 0 & 1 \end{bmatrix} + \begin{bmatrix} 0 & 1 \\ 0 & 0 \end{bmatrix}(t - t_0) \\
&= \begin{bmatrix} 1 & (t - t_0) \\ 0 & 1 \end{bmatrix}.
\end{aligned}
$$

The equation for the consider parameter mapping matrix is

$$\dot{\theta} = A\theta + B, \quad \theta(t_0, t_0) = 0$$

or

$$\begin{bmatrix} \dot{\theta}_1 \\ \dot{\theta}_2 \end{bmatrix} = \begin{bmatrix} \theta_2 \\ 0 \end{bmatrix} + \begin{bmatrix} 0 \\ 1 \end{bmatrix}.$$

Hence,

$$
\begin{aligned}
\dot{\theta}_2 &= 1 \\
\theta_2(t, t_0) &= t + k_1. \quad \text{At } t = t_0, \quad \theta_2 = 0 \\
\theta_2(t, t_0) &= t - t_0.
\end{aligned}
$$

Also $\dot{\theta}_1 = \theta_2$; therefore,

$$\theta_1(t, t_0) = \frac{(t - t_0)^2}{2} + k_2.$$

At $t = t_0$, $\theta_1 = 0$; hence,

$$
\begin{aligned}
\theta_1 &= \frac{(t - t_0)^2}{2} \\
\theta(t, t_0) &= \begin{bmatrix} (t - t_0)^2/2 \\ (t - t_0) \end{bmatrix}.
\end{aligned}
$$

The transformation of observation-state partials to epoch, t_0, is obtained by using Eqs. (6.5.23) and (6.5.24):

$$
\begin{aligned}
H_x(t) &= \tilde{H}_x(t)\Phi(t, t_0) \\
&= \begin{bmatrix} 1 & 0 \end{bmatrix} \begin{bmatrix} 1 & (t - t_0) \\ 0 & 1 \end{bmatrix} \\
&= \begin{bmatrix} 1 & (t - t_0) \end{bmatrix} \\
H_c(t) &= \tilde{H}_x(t)\theta(t, t_0) + \tilde{H}_c(t) \\
&= \begin{bmatrix} 1 & 0 \end{bmatrix} \begin{bmatrix} (t - t_0)^2/2 \\ (t - t_0) \end{bmatrix} + 0 \\
&= (t - t_0)^2/2 .
\end{aligned}
$$

The (2×2) normal matrix of the state and the 2×1 normal matrix of the consider parameter at time t is given by

$$
\begin{aligned}
(H_x^T H_x)_t &= \begin{bmatrix} 1 \\ (t - t_0) \end{bmatrix} \begin{bmatrix} 1 & (t - t_0) \end{bmatrix} = \begin{bmatrix} 1 & (t - t_0) \\ (t - t_0) & (t - t_0)^2 \end{bmatrix} \\
(H_x^T H_c)_t &= \begin{bmatrix} 1 \\ (t - t_0) \end{bmatrix} \frac{(t - t_0)^2}{2} = \begin{bmatrix} (t - t_0)^2/2 \\ (t - t_0)^3/2 \end{bmatrix} .
\end{aligned}
$$

Given that $\overline{P}_0 = I$, $R = I$ and that three measurements are taken at $t = 0, 1,$ and 2, and let $t_0 = 0$. Then the accumulation of the normal matrices yields

$$
H_x^T H_x = \begin{bmatrix} 1 & 0 \\ 0 & 0 \end{bmatrix} + \begin{bmatrix} 1 & 1 \\ 1 & 1 \end{bmatrix} + \begin{bmatrix} 1 & 2 \\ 2 & 4 \end{bmatrix} = \begin{bmatrix} 3 & 3 \\ 3 & 5 \end{bmatrix}
$$

and

$$
H_x^T H_c = \begin{bmatrix} 0 \\ 0 \end{bmatrix} + \begin{bmatrix} 1/2 \\ 1/2 \end{bmatrix} + \begin{bmatrix} 2 \\ 4 \end{bmatrix} = \begin{bmatrix} 5/2 \\ 9/2 \end{bmatrix} .
$$

The computed covariance at epoch, t_0, is given by

$$
P_0 = \left(H_x^T H_x + \overline{P}_0^{-1} \right)^{-1} = \begin{bmatrix} 4 & 3 \\ 3 & 6 \end{bmatrix}^{-1} = \begin{bmatrix} 2/5 & -1/5 \\ -1/5 & 4/15 \end{bmatrix} .
$$

Assuming the variance of g to be Π, the consider covariance P_{c_o} at the epoch is given by Eq. (6.5.32) with $k = 0$,

$$P_{c_0} = P_0 + S_0 \Pi S_0^T.$$

The sensitivity matrix, S_0, is given by Eq. (6.4.4), with $R = I$,

$$
\begin{aligned}
S_0 &= -P_0(H_x^T H_c) \\
&= -\begin{bmatrix} 2/5 & -1/5 \\ -1/5 & 4/15 \end{bmatrix}\begin{bmatrix} 5/2 \\ 9/2 \end{bmatrix} = -\begin{bmatrix} 1/10 \\ 7/10 \end{bmatrix}.
\end{aligned}
$$

Therefore,

$$S_0 \Pi S_0^T = \frac{\Pi}{100}\begin{bmatrix} 1 & 7 \\ 7 & 49 \end{bmatrix}$$

and the value of P_{c_0} is given by

$$P_{c_0} = \begin{bmatrix} 2/5 & -1/5 \\ -1/5 & 4/15 \end{bmatrix} + \frac{\Pi}{100}\begin{bmatrix} 1 & 7 \\ 7 & 49 \end{bmatrix}.$$

Also,

$$P_{xc_0} = S_0\Pi = -\Pi\begin{bmatrix} 1/10 \\ 7/10 \end{bmatrix}.$$

Finally from Eq. (6.6.16),

$$
\begin{aligned}
\mathbf{P}_{c_0} &= \begin{bmatrix} P_{c_0} & P_{xc_0} \\ P_{cx_0} & \overline{P}_{cc} \end{bmatrix} \\
&= \begin{bmatrix} \dfrac{40+\Pi}{100} & \dfrac{-20+7\Pi}{100} & \dfrac{-\Pi}{10} \\[2mm] & \dfrac{80+147\Pi}{300} & \dfrac{-7\Pi}{10} \\[2mm] & & \Pi \end{bmatrix}.
\end{aligned}
$$

Thus, the consider standard deviations of the position and velocity,

$$\sqrt{\frac{(40+\Pi)}{100}}$$

and

$$\sqrt{\frac{(80 + 147\Pi)}{300}}$$

are greater than the respective data noise standard deviations; namely, $\sqrt{2/5}$ and $\sqrt{4/15}$.

6.8.1 PROPAGATION WITH TIME

The computed covariance is mapped from $t = 0$ to $t = 2$ by Eq. (6.6.10),

$$
\begin{aligned}
P_2 &= \Phi(t_2, t_0) P_0 \Phi^T(t_2, t_0) \\
&= \begin{bmatrix} 1 & 2 \\ 0 & 1 \end{bmatrix} \begin{bmatrix} 2/5 & -1/5 \\ -1/5 & 4/15 \end{bmatrix} \begin{bmatrix} 1 & 0 \\ 2 & 1 \end{bmatrix} = \begin{bmatrix} 2/3 & 1/3 \\ 1/3 & 4/15 \end{bmatrix}.
\end{aligned}
$$

The contribution to the propagated covariance from the consider parameter uncertainty is given by the last term of Eq. (6.6.11),

$$S_2 \Pi S_2^T$$

where, from Eq. (6.6.9)

$$
\begin{aligned}
S_2 &= \Phi(t_2, t_0) S_0 + \theta(t_2, t_0) \\
&= \begin{bmatrix} 1 & 2 \\ 0 & 1 \end{bmatrix} \begin{bmatrix} 1/10 \\ -7/10 \end{bmatrix} + \begin{bmatrix} 2 \\ 2 \end{bmatrix} \\
&= \begin{bmatrix} 1/2 \\ 13/10 \end{bmatrix}.
\end{aligned}
$$

Therefore,

$$
S_2 \Pi S_2^T = \begin{bmatrix} 1/2 \\ 13/10 \end{bmatrix} \Pi \begin{bmatrix} 1/2 & 13/10 \end{bmatrix} = \Pi \begin{bmatrix} 1/4 & 13/20 \\ 13/20 & 169/100 \end{bmatrix}.
$$

Thus, the consider covariance propagated to time $t = 2$ is

$$
P_{c_2} = P_2 + S_2 \Pi S_2^T = \begin{bmatrix} 2/3 & 1/3 \\ 1/3 & 4/15 \end{bmatrix} + \Pi \begin{bmatrix} 1/4 & 13/20 \\ 13/20 & 169/100 \end{bmatrix}.
$$

Also,

$$P_{xc_2} = S_2 \Pi = \frac{\Pi}{2} \begin{bmatrix} 1 \\ 13/5 \end{bmatrix}.$$

Hence,

$$\mathbf{P}_{c_2} = \begin{bmatrix} \dfrac{8 + 3\Pi}{12} & \dfrac{20 + 39\Pi}{60} & \dfrac{\Pi}{2} \\ & \dfrac{80 + 507\Pi}{300} & \dfrac{13\Pi}{10} \\ & & \Pi \end{bmatrix}$$

where again the consider standard deviation is greater than the computed or data noise standard deviation. Recall that in this development we have let $\sigma_g^2 = \Pi$. Also note that $P_{c_2} - P_2$ is positive definite as was the case at t_0.

6.8.2   THE SEQUENTIAL CONSIDER ALGORITHM

As a continuation of this example, compute $\mathbf{P}_{c_2}$ using the sequential consider filter algorithm. We will compute P_{c_0}, P_{c_1}, and P_{c_2} based on observations at $t = 0; t = 0, 1; t = 0, 1,$ and 2, respectively.

Beginning with $R = I, \overline{P}_0 = I, \overline{P}_{cc} = \Pi, P_{xc_0} = 0, \widetilde{H}_{x_0} = [1 \ 0], \widetilde{H}_{c_0} = 0,$ compute

$$K_0 = \overline{P}_0 \widetilde{H}_{x_0}^T (\widetilde{H}_{x_0} \overline{P}_0 \widetilde{H}_{x_0}^T + R)^{-1}$$

$$K_0 = \begin{bmatrix} 1/2 \\ 0 \end{bmatrix}$$

$$S_0 = -K_0 \widetilde{H}_{c_0}$$
$$= 0$$

$$P_{c_0} = P_0 + S_0 \Pi S_0^T$$
$$= P_0$$

where

$$P_0 = (I - K_0 \widetilde{H}_{x_0}) \overline{P}_0$$
$$= \begin{bmatrix} 1/2 & 0 \\ 0 & 1 \end{bmatrix}.$$

Hence, as expected, an error in g cannot affect the state at t_0 and the data noise and consider covariance for the state are identical. Therefore,

$$\mathbf{P}_{c_0} = \begin{bmatrix} 1/2 & 0 & 0 \\ 0 & 1 & 0 \\ 0 & 0 & \Pi \end{bmatrix}.$$

Now compute $\mathbf{P}_{c_1}$. The time update for P_0 is given by Eq. (6.7.23)

$$\begin{aligned}
\overline{P}_1 &= \Phi(t_1, t_0) P_0 \Phi^T(t_1, t_0) \\
&= \begin{bmatrix} 1 & 1 \\ 0 & 1 \end{bmatrix} \begin{bmatrix} 1/2 & 0 \\ 0 & 1 \end{bmatrix} \begin{bmatrix} 1 & 0 \\ 1 & 1 \end{bmatrix} \\
&= \begin{bmatrix} 3/2 & 1 \\ 1 & 1 \end{bmatrix}.
\end{aligned}$$

Also, from Eq. (6.7.24)

$$\begin{aligned}
K_1 &= \overline{P}_1 \tilde{H}_{x_1}^T (\tilde{H}_{x_1} \overline{P}_1 \tilde{H}_{x_1}^T + R)^{-1} \\
&= \begin{bmatrix} 3/5 \\ 2/5 \end{bmatrix}.
\end{aligned}$$

The measurement update for $\overline{P}_1$ is given by Eq. (6.7.30)

$$\begin{aligned}
P_1 &= (I - K_1 \tilde{H}_{x_1}) \overline{P}_1 \\
&= \left[I - \begin{bmatrix} 3/5 \\ 2/5 \end{bmatrix} \begin{bmatrix} 1 & 0 \end{bmatrix} \right] \begin{bmatrix} 3/2 & 1 \\ 1 & 1 \end{bmatrix} \\
&= \begin{bmatrix} 3/5 & 2/5 \\ 2/5 & 3/5 \end{bmatrix}.
\end{aligned}$$

From Eq. (6.7.25), the time update for S_0 is

$$\bar{S}_1 = \Phi(t_1, t_0) S_0 + \theta(t_1, t_0)$$

and the measurement update is given by Eq. (6.7.31)

$$S_1 = (I - K_1 \tilde{H}_{x_1}) \bar{S}_1 - K_1 \tilde{H}_{c_1}.$$

However, $S_0 = 0$, so

$$\bar{S}_1 = \begin{bmatrix} 1/2 \\ 1 \end{bmatrix}$$

and

$$S_1 = \begin{bmatrix} 1/5 \\ 4/5 \end{bmatrix}.$$

From Eq. (6.7.34)

$$\begin{aligned} P_{c_1} &= P_1 + S_1 \Pi S_1^T \\ &= \begin{bmatrix} 3/5 & 2/5 \\ 2/5 & 3/5 \end{bmatrix} + \Pi \begin{bmatrix} 1/25 & 4/25 \\ 4/25 & 16/25 \end{bmatrix} \end{aligned}$$

and from Eq. (6.7.35)

$$P_{xc_1} = S_1 \Pi = \begin{bmatrix} \dfrac{\Pi}{5} \\ \dfrac{4\Pi}{5} \end{bmatrix}.$$

Substituting these results into Eq. (6.6.16) yields the desired result

$$\mathbf{P}_{c_1} = \begin{bmatrix} \dfrac{15 + \Pi}{25} & \dfrac{10 + 4\Pi}{25} & \dfrac{\Pi}{5} \\[2mm] & \dfrac{15 + 16\Pi}{25} & \dfrac{4\Pi}{5} \\[2mm] & & \Pi \end{bmatrix}.$$

Finally, it can be shown that

$$P_2 = \begin{bmatrix} 2/3 & 1/3 \\ 1/3 & 4/15 \end{bmatrix}$$

$$\bar{S}_2 = \begin{bmatrix} 3/2 \\ 9/5 \end{bmatrix}$$

$$S_2 = \begin{bmatrix} 1/2 \\ 13/10 \end{bmatrix}$$

$$P_{c_2} = \begin{bmatrix} 2/3 & 1/3 \\ 1/3 & 4/15 \end{bmatrix} + \Pi \begin{bmatrix} 1/4 & 13/20 \\ 13/20 & 169/100 \end{bmatrix}$$

$$P_{xc_2} = \Pi \begin{bmatrix} 1/2 \\ 13/10 \end{bmatrix}$$

and

$$\mathbf{P}_{c_2} = \begin{bmatrix} \dfrac{8 + 3\Pi}{12} & \dfrac{20 + 39\Pi}{60} & \dfrac{\Pi}{2} \\[2ex] & \dfrac{80 + 507\Pi}{300} & \dfrac{13\Pi}{10} \\[2ex] & & \Pi \end{bmatrix}$$

which is in agreement with results from the batch processor.

6.8.3 PERTURBATION IN THE STATE ESTIMATE

Since the sensitivity matrix is by definition

$$S_0 = \frac{\partial \hat{\mathbf{X}}_0}{\partial g}.$$

It follows that

$$\Delta \hat{\mathbf{X}}_0 = \frac{\partial \hat{\mathbf{X}}_0}{\partial g} \sqrt{\Pi}.$$

In general, if σ_g is the 1σ uncertainty in the value of g, then

$$\Delta \hat{\mathbf{X}}_0 = \begin{bmatrix} \Delta \hat{x}_0 \\ \Delta \hat{\dot{x}}_0 \end{bmatrix} = S_0 \sigma_g = - \begin{bmatrix} 1/10 \\ 7/10 \end{bmatrix} \sigma_g$$

or

$$\Delta \hat{x}_0 = -\sigma_g / 10$$
$$\Delta \hat{\dot{x}}_0 = -7\sigma_g / 10$$

which illustrates how the estimate of the epoch state for the batch processor will be in error as a function of the uncertainty in the consider parameter g after processing the three observations of this example.

6.9 EXAMPLE: SPRING-MASS PROBLEM

A block of mass m is attached to two parallel vertical walls by two springs as shown in Fig. 4.8.2. k_1 and k_2 are the spring constants. h is the height of the position P on one of the walls, from which the distance, ρ, and the rate of change of distance of the block from P, $\dot{\rho}$ can be observed.

Let the horizontal distances be measured with respect to the point O where the line OP, the lengths of the springs, and the center of mass of the block are all assumed to be in the same vertical plane. Then, if $\bar{x}$ denotes the position of the block at static equilibrium, the equation of motion of the block is given by

$$\ddot{x} = -(k_1 + k_2)(x - \bar{x})/m. \qquad (6.9.1)$$

Let

$$\omega^2 = (k_1 + k_2)/m, \text{ and } \bar{x} = 0$$

so that Eq. (6.9.1) can be written as

$$\ddot{x} + \omega^2 x = 0. \qquad (6.9.2)$$

Consider the problem of estimating the position and the velocity of the block with respect to the reference point O, by using the range and range-rate measurements of the block from the point, P. To formulate this problem mathematically, the estimation state vector is taken as $\mathbf{X}^T = [x \quad v]$. Then the system dynamics are represented by

$$\dot{\mathbf{X}} = F(\mathbf{X}, t) = \begin{bmatrix} \dot{x} \\ \dot{v} \end{bmatrix} = \begin{bmatrix} v \\ -\omega^2 x \end{bmatrix}, \qquad (6.9.3)$$

or in state space form

$$\dot{\mathbf{X}} = \begin{bmatrix} \dot{x} \\ \dot{v} \end{bmatrix} = \begin{bmatrix} 0 & 1 \\ -\omega^2 & 0 \end{bmatrix} \begin{bmatrix} x \\ v \end{bmatrix}. \qquad (6.9.4)$$

The observation vector is

$$G(\mathbf{X}, t) = \begin{bmatrix} \rho \\ \dot{\rho} \end{bmatrix} = \begin{bmatrix} \sqrt{x^2 + h^2} \\ x\dot{x}/\rho \end{bmatrix}. \qquad (6.9.5)$$

Suppose that some parameters are not known exactly and that the effect of their errors on the state estimates are to be evaluated. For example, let m, k_2, and h be the set of "consider parameters." Then the consider vector $\mathbf{C}$ will be

$$\mathbf{C}^T = \left[m \vdots k_2 \vdots h \right]$$

and $\mathbf{c} = \mathbf{C} - \mathbf{C}^*$. The linearized state and observation equations, including the consider parameters, are given by

$$\delta \dot{\mathbf{X}} = A(t)\delta\mathbf{X} + B(t)\mathbf{c} \qquad (6.9.6)$$

$$\mathbf{y} = \tilde{H}_x \delta\mathbf{X} + \tilde{H}_c \mathbf{c} \qquad (6.9.7)$$

where

$$\delta\mathbf{X} = \begin{bmatrix} \delta x \\ \delta v \end{bmatrix}$$

$$A(t) = \begin{bmatrix} 0 & 1 \\ -\omega^2 & 0 \end{bmatrix}, \quad B(t) = \begin{bmatrix} 0 & 0 & 0 \\ \dfrac{\omega^2 x}{m} & \dfrac{-x}{m} & 0 \end{bmatrix} \qquad (6.9.8)$$

and

$$\tilde{H}_x = \begin{bmatrix} \dfrac{x}{\rho} & 0 \\ \dfrac{\dot{x}}{\rho} - \dfrac{x^2\dot{x}}{\rho^3} & \dfrac{x}{\rho} \end{bmatrix}, \quad \tilde{H}_c = \begin{bmatrix} 0 & 0 & \dfrac{h}{\rho} \\ 0 & 0 & -\dfrac{x\dot{x}h}{\rho^3} \end{bmatrix}.$$

The solution to Eq. (6.9.4) is given by

$$\begin{aligned} x(t) &= x_o \cos \omega t + \tfrac{v_o}{\omega} \sin \omega t \\ v(t) &= v_o \cos \omega t - x_o \omega \sin \omega t. \end{aligned} \qquad (6.9.9)$$

Note that the original differential equation for the state Eq. (6.9.2) is linear; hence the filter solution (i.e., $c = 0$) for

$$\delta\mathbf{X}(t) = \begin{bmatrix} \delta x(t) \\ \delta v(t) \end{bmatrix}$$

is also given by Eq. (6.9.9) when x_0 and v_0 are replaced by δx_0 and δv_0, respectively.

The differential equations for the filter state transition matrix are

$$\dot{\Phi}(t,0) = A(t)\Phi(t,0), \quad \Phi(0,0) = I$$

and the consider parameter transition matrix equations are

$$\dot{\theta}(t,0) = A(t)\theta(t,0) + B(t), \quad \theta(0,0) = 0.$$

Solutions to these are given by

$$\Phi(t,0) \quad = \quad \begin{bmatrix} \cos \omega t & \frac{1}{\omega} \sin \omega t \\ -\omega \sin \omega t & \cos \omega t \end{bmatrix} \qquad (6.9.10)$$

and

$$
\begin{aligned}
\theta_{11}(t,0) &= \frac{\beta_2}{\omega} \sin \omega t + t[-\beta_2 \cos \omega t + \beta_1 \omega \sin \omega t] \\
\theta_{12}(t,0) &= -\theta_{11}(t,0)\omega^2 \qquad\qquad\qquad (6.9.11) \\
\theta_{13}(t,0) &= 0 \\
\theta_{21}(t,0) &= \beta_1 \omega \sin \omega t + t[\beta_2 \omega \sin \omega t + \beta_1 \omega^2 \cos \omega t] \\
\theta_{22}(t,0) &= -\theta_{21}/\omega^2 \\
\theta_{23}(t,0) &= 0
\end{aligned}
$$

where

$$\beta_1 = \frac{x_0}{2m} \quad \text{and} \quad \beta_2 = \frac{\dot{x}_0}{2m}. \qquad (6.9.12)$$

Eleven perfect observations of ρ and $\dot{\rho}$ were simulated over a period of 10 seconds at one-second intervals and are given in Chapter 4 (Table 4.8.1), assuming the following values for the system parameters and the initial condition:

$$
\begin{aligned}
k_1 &= 2.5 \text{ N/m} \\
k_2 &= 3.7 \text{ N/m} \\
m &= 1.5 \text{ kg} \\
h &= 5.4 \text{ m} \qquad\qquad (6.9.13) \\
x_0 &= 3.0 \text{ m} \\
\dot{x}_0 &= 0.0 \text{ m/s}.
\end{aligned}
$$

The *a priori* values used are:

$$\bar{\mathbf{X}}_0 = \begin{bmatrix} 4.0 \\ 0.2 \end{bmatrix} \quad \text{and} \quad \bar{P}_o = \begin{bmatrix} 100.0 & 0 \\ 0 & 10.0 \end{bmatrix} . \tag{6.9.14}$$

When perfect values of the consider parameters, denoted by the vector $\mathbf{C}$, are used, the batch processor "recovers" the true estimate of the state, as shown in the example described in Section 4.8.2. However, when the consider parameters are in error, not only will the estimates of the state parameters deviate from their true values, but these errors will propagate with time. In the following cases the consider covariance is propagated with time under the influence of the indicated covariances on the consider parameters.

Twenty Percent Error in Consider Parameters

Assume that the values of m, k_2, and h actually used in the batch processor are

$$\mathbf{C}^* = \begin{bmatrix} m \\ k_2 \\ h \end{bmatrix} = \begin{bmatrix} 1.8 \\ 3.0 \\ 6.4 \end{bmatrix} ; \text{ hence, } \mathbf{c} = \begin{bmatrix} -.3 \\ .7 \\ -1 \end{bmatrix} . \tag{6.9.15}$$

$\mathbf{C}^*$ is approximately 20 percent in error relative to the true values given in Eq. (6.9.13). The corresponding consider covariance is assumed to be:

$$\bar{P}_{cc} = \begin{bmatrix} 0.09 & 0 & 0 \\ 0 & 0.49 & 0 \\ 0 & 0 & 1 \end{bmatrix} . \tag{6.9.16}$$

The solution for the epoch state using the observations of Table 4.8.1 and the values of the consider parameters given by Eq. (6.9.15) and *a priori* values given by Eq. (6.9.14) is

$$\hat{x}_0 = 2.309 \pm 0.623 \text{ m}$$
$$\hat{\dot{x}}_0 = -0.758 \pm 0.712 \text{ m/s}.$$

The error covariance matrices are propagated with time using Eqs. (6.6.10) and (6.6.11) with $\bar{P}_{cc}$ given by Eq. (6.9.16), and the results are shown in Fig. 6.9.1. The upper panel shows the error in the position estimate, and the positive values of the computed standard deviation and the consider standard deviation. The time span includes the measurement period (0–10 sec) and a prediction interval (10–20 sec). The center panel shows the previous three quantities for the

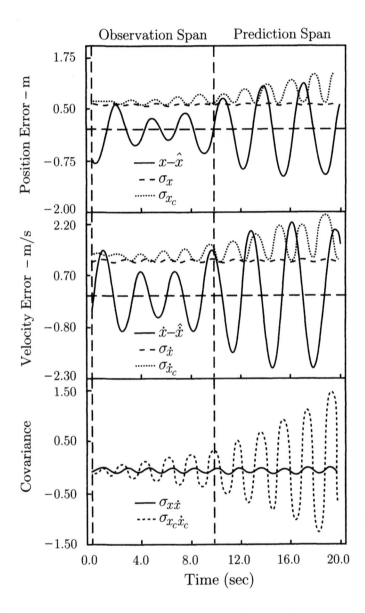

Figure 6.9.1: Consider analysis: Spring mass problem, 20% error in consider parameters.

velocity component. The computed and the consider covariance between the position and velocity estimates are shown in the lower panel. Note that the consider sigmas are always greater than the computed sigmas. The phase difference between the actual error and consider error is caused by the errors in m and k_2, which result in an error in frequency and phase. It is also interesting to note that the state errors are bounded above by the computed sigmas in the measurement interval but not in the prediction interval.

Five Percent Error in Consider Parameter

In this case, the only change is to $\mathbf{C}^*$ and $\overline{P}_{cc}$, with

$$\mathbf{C}^* = \begin{bmatrix} 1.6 \\ 3.5 \\ 5.7 \end{bmatrix} .$$

$\mathbf{C}^*$ is approximately 5 percent in error. The consider covariance is assumed to be:

$$\overline{P}_{cc} = \begin{bmatrix} 0.01 & 0 & 0 \\ 0 & 0.04 & 0 \\ 0 & 0 & 0.09 \end{bmatrix} .$$

The batch solution for the epoch state is:

$$\begin{aligned} \hat{x}_0 &= 2.873 \pm 0.505 \\ \hat{x}_0 &= -0.499 \pm 0.513. \end{aligned}$$

The propagation errors and the standard deviations corresponding to this case are shown in Fig. 6.9.2. The results are similar to the previous case, except that the magnitude of error is proportionally smaller.

6.10 ERRORS IN THE OBSERVATION NOISE AND *A PRIORI* STATE COVARIANCES

The effects of errors in the assumed values of the data noise and *a priori* state covariance matrices for the batch processor can be evaluated as follows. Assume that the values of these matrices used in the batch processor differ from the true values (denoted by $(\ \ )^*$) by the matrix δ,

$$\begin{aligned} R^* &= R + \delta R & (6.10.1) \\ \overline{P}_0^* &= \overline{P}_0 + \delta \overline{P}_0. & (6.10.2) \end{aligned}$$

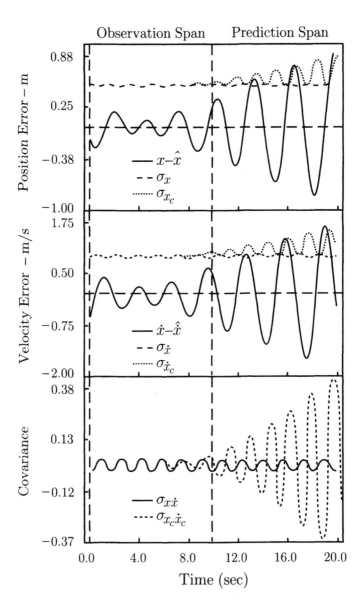

Figure 6.9.2: Consider analysis: spring mass problem, 5% error in consider parameters.

From Eq. (4.6.1),

$$\hat{\mathbf{x}} = \left(H^T R^{-1} H + \overline{P}_0^{-1}\right)^{-1} \left(H^T R^{-1} \mathbf{y} + \overline{P}_0^{-1} \overline{\mathbf{x}}\right) \tag{6.10.3}$$

where

$$\overline{\mathbf{x}} = \mathbf{x} + \eta \tag{6.10.4}$$
$$\mathbf{y} = H\mathbf{x} + \epsilon \tag{6.10.5}$$

and

$$E[\eta\,\eta^T] = \overline{P}_0^* \tag{6.10.6}$$
$$E[\epsilon\epsilon^T] = R^* .$$

Evaluating the estimation error covariance using the assumed values of R and $\overline{P}_0$ results in

$$\hat{\mathbf{x}} = \left[H^T R^{-1} H + \overline{P}_0^{-1}\right]^{-1} \left[H^T R^{-1}(H_x + \epsilon) + \overline{P}_0^{-1}(\mathbf{x} + \eta)\right] \tag{6.10.7}$$

or

$$\hat{\mathbf{x}} - \mathbf{x} = \left[H^T R^{-1} H + \overline{P}_0^{-1}\right]^{-1} \left[H^T R^{-1} \epsilon + \overline{P}_0^{-1} \eta\right] . \tag{6.10.8}$$

Assuming $E[\epsilon\eta^T] = 0$, we have

$$\begin{aligned}
P_c &= E\left[(\hat{\mathbf{x}} - \mathbf{x})(\hat{\mathbf{x}} - \mathbf{x})^T\right] \\
&= \left[H^T R^{-1} H + \overline{P}_0^{-1}\right]^{-1} \left[H^T R^{-1} E[\epsilon\epsilon^T] R^{-1} H\right] \\
&\quad + \left[\overline{P}_0^{-1} E[\eta\,\eta^T] \overline{P}_0^{-1}\right] \left[H^T R^{-1} H + \overline{P}_0^{-1}\right]^{-1} \\
&= P\left[H^T R^{-1} R^* R^{-1} H + \overline{P}_0^{-1} \overline{P}_0^* \overline{P}_0^{-1}\right] P . \tag{6.10.9}
\end{aligned}$$

Substituting Eqs. (6.10.1) and (6.10.2) for R^* and $\overline{P}_0^*$ yields

$$P_c = P + P\left[H^T R^{-1} \delta R R^{-1} H + \overline{P}_0^{-1} \delta \overline{P}_0 \overline{P}_0^{-1}\right] P . \tag{6.10.10}$$

The second term in Eq. (6.10.10) is the contribution to the consider covariance from errors in the data noise and *a priori* state covariance matrices.

6.11 ERRORS IN PROCESS NOISE, OBSERVATION NOISE, AND STATE COVARIANCE

The effects of errors in the process noise, observation noise, and *a priori* state covariance matrices can be evaluated using a sequential or Kalman filter algorithm. Following Heffes (1996), assume that the values of these quantities used in the filter are given by Q, R, and $\overline{P}_0$. Also,

$$
\begin{aligned}
Q^* &= Q + \delta Q \\
R^* &= R + \delta R \\
\overline{P}_0^* &= \overline{P}_0 + \delta \overline{P}_0
\end{aligned}
$$
(6.11.1)

where $(\ )^*$ represents the optimal or true value, and δ denotes the error in the assumed value.

We wish to find the covariance associated with the actual state error. The actual error in the estimated states at time t_k is given by

$$
\delta \mathbf{x}_k = \hat{\mathbf{x}}_k - \mathbf{x}_k
$$
(6.11.2)

where

$$
\begin{aligned}
\hat{\mathbf{x}}_k &= \overline{\mathbf{x}}_k + K_k(\mathbf{y}_k - \widetilde{H}_k \overline{\mathbf{x}}_k) & (6.11.3) \\
K_k &= \overline{P}_k \widetilde{H}_k^T (\widetilde{H}_k \overline{P}_k \widetilde{H}^T + R)^{-1} & (6.11.4)
\end{aligned}
$$

and

$$
\mathbf{y}_k = \widetilde{H}_k \mathbf{x}_k + \boldsymbol{\epsilon}_k .
$$
(6.11.5)

Here $\hat{\mathbf{x}}_k$ is the state estimate obtained using Q, R, and P_0 in Eq. (6.11.3), and $\mathbf{x}_k$ is the true value of the state at t_k.

Substituting Eqs. (6.11.3) and (6.11.5) into Eq. (6.11.2) yields

$$
\begin{aligned}
\delta \mathbf{x}_k &= \overline{\mathbf{x}}_k + K_k(\widetilde{H}_k \mathbf{x}_k + \boldsymbol{\epsilon}_k - \widetilde{H}_k \overline{\mathbf{x}}_k) - \mathbf{x}_k \\
&= (I - K_k \widetilde{H}_k)(\overline{\mathbf{x}}_k - \mathbf{x}_k) + K_k \boldsymbol{\epsilon}_k .
\end{aligned}
$$
(6.11.6)

The covariance of the actual estimation error is

$$
\begin{aligned}
P_k &\equiv E\left[\delta \mathbf{x}_k \delta \mathbf{x}_k^T\right] \\
&= (I - K_k \widetilde{H}_k)\overline{P}_k(I - K_k \widetilde{H}_k)^T + K_k R^* K_k^T
\end{aligned}
$$
(6.11.7)

where

$$
\overline{P}_k \equiv E\left[\delta \overline{\mathbf{x}}_k \delta \overline{\mathbf{x}}_k^T\right] = E\left[(\overline{\mathbf{x}}_k - \mathbf{x}_k)(\overline{\mathbf{x}}_k - \mathbf{x}_k)^T\right] .
$$
(6.11.8)

The equation relating $\overline{P}_k$ to P_{k-1} is found by using the expressions that relate the state at time t_{k-1} to the state at t_k,

$$
\begin{aligned}
\overline{\mathbf{x}}_k &= \Phi(t_k, t_{k-1})\hat{\mathbf{x}}_{k-1} & (6.11.9) \\
\mathbf{x}_k &= \Phi(t_k, t_{k-1})\mathbf{x}_{k-1} + \Gamma(t_k, t_{k-1})\mathbf{u}_{k-1} . & (6.11.10)
\end{aligned}
$$

Thus,

$$
\begin{aligned}
\delta\overline{\mathbf{x}}_k &= \overline{\mathbf{x}}_k - \mathbf{x}_k \\
&= \Phi(t_k, t_{k-1})\hat{\mathbf{x}}_{k-1} - \Phi(t_k, t_{k-1})\mathbf{x}_{k-1} - \Gamma(t_k, t_{k-1})\mathbf{u}_{k-1} \\
&= \Phi(t_k, t_{k-1})(\hat{\mathbf{x}}_{k-1} - \mathbf{x}_{k-1}) - \Gamma(t_k, t_{k-1})\mathbf{u}_{k-1} . \quad (6.11.11)
\end{aligned}
$$

Substituting this into Eq. (6.11.8) yields

$$
\begin{aligned}
\overline{P}_k &= \Phi(t_k, t_{k-1})P_{k-1}\Phi^T (t_k, t_{k-1}) \\
&+ \Gamma(t_k, t_{k-1})Q^*_{k-1}\Gamma^T(t_k, t_{k-1}) \quad (6.11.12)
\end{aligned}
$$

where $E[(\hat{\mathbf{x}}_{k-1} - \mathbf{x}_{k-1})\mathbf{u}^T_{k-1}\Gamma^T(t_k, t_{k-1})] = 0$ because $\mathbf{x}_{k-1}$ is dependent on $\mathbf{u}_{k-2}$ but not $\mathbf{u}_{k-1}$.

Hence, the covariance of the actual error given by Eqs. (6.11.12) and (6.11.7) is

$$
\begin{aligned}
\overline{P}_{k_a} &= \Phi(t_k, t_{k-1})P_{k-1_a}\Phi^T(t_k, t_{k-1}) \quad (6.11.13) \\
&+\Gamma(t_k, t_{k-1})Q^*_{k-1}\Gamma^T(t_k, t_{k-1}) \\
P_{k_a} &= (I - K_k\tilde{H}_k)\overline{P}_{k_a}(I - K_k\tilde{H}_k)^T + K_k R^* K_k^T . \quad (6.11.14)
\end{aligned}
$$

The recursion is initiated with $\overline{P}_{0_a} = \overline{P}^*_0$.

The gain matrix, K_k, at each stage is computed by using the suboptimal $\overline{P}$ and R, and is initiated with $\overline{P}_0$: K_k is given by

$$
K_k = \overline{P}_k\tilde{H}^T_k (\tilde{H}_k\overline{P}_k\tilde{H}^T_k + R)^{-1} . \quad (6.11.15)
$$

The suboptimal values of $\overline{P}_k$ and P_k are given by

$$
\begin{aligned}
\overline{P}_k &= \Phi(t_k, t_{k-1})P_{k-1}\Phi^T(t_k, t_{k-1}) \quad (6.11.16) \\
&+\Gamma(t_k, t_{k-1})Q_k\Gamma^T(t_k, t_{k-1}) \\
P_k &= (I - K_k\tilde{H}_k)\overline{P}_k . \quad (6.11.17)
\end{aligned}
$$

Hence, there will be three covariance matrices one can compute using Eqs. (6.11.13) and (6.11.14):

1. The optimal value of P based on $\overline{P}^*_0$, Q^*, R^*.

2. The actual value of P determined by using $\overline{P}^*_0$, Q^*, R^* and $\overline{P}_0$, Q, and R.

3. The suboptimal value of P determined from $\overline{P}_0$, Q, and R. Note that the suboptimal value of $\overline{P}_k$ must be computed in order to determine the actual value of P_k.

6.12   COVARIANCE ANALYSIS AND ORTHOGONAL TRANSFORMATIONS

The consider covariance formulation can also be developed in square-root form by using orthogonal transformations. Consider the dynamic and measurement models given by Eqs. (6.5.8) and (6.5.18):

$$\mathbf{x}(t) = \Phi(t, t_k)\mathbf{x}_k + \theta(t, t_k)\mathbf{c}_k \tag{6.12.1}$$

$$\mathbf{y}_j = \tilde{H}_{x_j}\mathbf{x}_j + \tilde{H}_{c_j}\mathbf{c} + \epsilon_j, \quad j = 1, \ldots, \ell. \tag{6.12.2}$$

By substituting Eq. (6.12.1) into Eq. (6.12.2) (see Eq. (6.5.22)), this set can be reduced to

$$\mathbf{y}_j = H_{x_j}\mathbf{x}_k + H_{c_j}\mathbf{c} + \epsilon_j, \quad j = 1, \ldots, \ell \tag{6.12.3}$$

where

$$H_{x_j} = \tilde{H}_{x_j}\Phi(t_j, t_k)$$
$$H_{c_j} = \tilde{H}_{x_j}\theta(t_j, t_k) + \tilde{H}_{c_j}, \quad j = 1, \ldots, \ell.$$

Then the set of ℓ observations can be combined as

$$\mathbf{y} = H_x\mathbf{x}_k + H_c\mathbf{c} + \epsilon. \tag{6.12.4}$$

where $\mathbf{y}$, H_x, H_c, and ϵ are defined in Eq. (6.5.25).

We next need to write the *a priori* information on the n-vector, $\mathbf{x}$, and the q-vector, $\mathbf{c}$, in terms of a data equation. Given *a priori* values $\bar{\mathbf{x}}$ and $\bar{\mathbf{c}}$, we may write

$$\begin{bmatrix} \bar{\mathbf{x}} \\ \bar{\mathbf{c}} \end{bmatrix} = \begin{bmatrix} \mathbf{x} \\ \mathbf{c} \end{bmatrix} + \begin{bmatrix} \eta \\ \beta \end{bmatrix} \tag{6.12.5}$$

where $\mathbf{x}$ and $\mathbf{c}$ are the true values. η and β have known mean and covariance,

$$E[\eta] = E[\beta] = 0$$

$$\bar{P} = E\left[\begin{bmatrix} \eta \\ \beta \end{bmatrix} \begin{bmatrix} \eta^T & \beta^T \end{bmatrix} \right] = E\begin{bmatrix} \eta\eta^T & \eta\beta^T \\ \beta^T\eta & \beta\beta^T \end{bmatrix}$$

$$= \begin{bmatrix} \bar{P}_x & \bar{P}_{xc} \\ \bar{P}_{cx} & \bar{P}_{cc} \end{bmatrix}. \tag{6.12.6}$$

We may now determine the upper triangular square root of the inverse of $\bar{P}$ with a Cholesky decomposition

$$\text{Chol} \begin{bmatrix} \overline{P}_x & \overline{P}_{xc} \\ \overline{P}_{cx} & \overline{P}_{cc} \end{bmatrix}^{-1} \equiv \begin{bmatrix} \overline{R}_x & \overline{R}_{xc} \\ 0 & \overline{R}_c \end{bmatrix}. \tag{6.12.7}$$

Multiplying Eq. (6.12.5) by Eq. (6.12.7) in order to write it as a data equation yields

$$\begin{bmatrix} \overline{R}_x & \overline{R}_{xc} \\ 0 & \overline{R}_c \end{bmatrix} \begin{bmatrix} \overline{\mathbf{x}} \\ \overline{\mathbf{c}} \end{bmatrix} = \begin{bmatrix} \overline{R}_x & \overline{R}_{xc} \\ 0 & \overline{R}_c \end{bmatrix} \begin{bmatrix} \mathbf{x} \\ \mathbf{c} \end{bmatrix}$$
$$+ \begin{bmatrix} \overline{R}_x & \overline{R}_{xc} \\ 0 & \overline{R}_c \end{bmatrix} \begin{bmatrix} \boldsymbol{\eta} \\ \boldsymbol{\beta} \end{bmatrix}. \tag{6.12.8}$$

Because we do not intend to estimate $\mathbf{c}$, we need to deal with only the first row of Eq. (6.12.8). Define

$$\begin{aligned} \overline{\mathbf{b}}_c &\equiv \overline{R}_c \overline{\mathbf{c}} \\ \overline{\mathbf{b}} &\equiv \overline{R}_x \overline{\mathbf{x}} + \overline{R}_{xc} \overline{\mathbf{c}} \\ \overline{\boldsymbol{\eta}} &\equiv \overline{R}_x \boldsymbol{\eta} + \overline{R}_{xc} \boldsymbol{\beta}. \end{aligned} \tag{6.12.9}$$

We will not use the first of Eq. (6.12.9) unless we estimate $\mathbf{c}$.

The first row of Eq. (6.12.8) becomes

$$\overline{\mathbf{b}} = \begin{bmatrix} \overline{R}_x & \overline{R}_{xc} \end{bmatrix} \begin{bmatrix} \mathbf{x} \\ \mathbf{c} \end{bmatrix} + \overline{\boldsymbol{\eta}}. \tag{6.12.10}$$

Using Eqs. (6.12.10) and (6.12.4) we may write the array to be upper triangular-ized,

$$\overbrace{}^{n} \; \overbrace{\phantom{R_{xc}}}^{q} \; \overbrace{}^{1}$$

$$\begin{bmatrix} \overline{R}_x & \overline{R}_{xc} & \overline{\mathbf{b}} \\ H_x & H_c & \mathbf{y} \end{bmatrix} \begin{matrix} \}n \\ \}\ell \end{matrix}. \tag{6.12.11}$$

We need only partially triangularize Eq. (6.12.11) by nulling the terms below the diagonal of the first n columns (n is the dimension of $\mathbf{x}$). Hence

$$
T \left[\begin{array}{ccc} \overline{R}_x & \overline{R}_{xc} & \mathbf{b} \\ H_x & H_c & \mathbf{y} \end{array} \right] \begin{array}{l} \}n \\ \}\ell \end{array} = \left[\begin{array}{ccc} \hat{R}_x & \hat{R}_{xc} & \hat{\mathbf{b}} \\ 0 & \tilde{R}_c & \tilde{\mathbf{b}}_c \end{array} \right] \begin{array}{l} \}n \\ \}q \end{array} .
$$

(6.12.12)

There are $\ell - q$ additional rows on the right-hand side of Eq. (6.12.12), which are not needed and are not shown here. Also, the terms $\tilde{R}_c$ and $\tilde{\mathbf{b}}_c$ are not needed to compute the consider covariance and are needed only if $\mathbf{c}$ is to be estimated. For consider covariance analysis $\overline{\mathbf{c}}$ is used wherever an estimate of $\mathbf{c}$ is needed. The first row of Eq. (6.12.12) yields

$$
\hat{R}_x \mathbf{x} + \hat{R}_{xc} \overline{\mathbf{c}} = \hat{\mathbf{b}} .
$$

(6.12.13)

In keeping with the notation of Section 6.3 and Eq. (6.3.70) we will refer to the estimate of $\mathbf{x}$ obtained from Eq. (6.12.13) as $\hat{\mathbf{x}}_c$; accordingly,

$$
\hat{\mathbf{x}}_c = \hat{R}_x^{-1} \hat{\mathbf{b}} - \hat{R}_x^{-1} \hat{R}_{xc} \overline{\mathbf{c}}.
$$

(6.12.14)

The estimate of $\mathbf{x}$ obtained by ignoring $\overline{\mathbf{c}}$ is called the computed estimate of $\mathbf{x}$ and is given by

$$
\hat{\mathbf{x}} = \hat{R}_x^{-1} \hat{\mathbf{b}}.
$$

(6.12.15)

Hence,

$$
\hat{\mathbf{x}}_c = \hat{\mathbf{x}} - \hat{R}_x^{-1} \hat{R}_{xc} \overline{\mathbf{c}}.
$$

(6.12.16)

Comparing Eq. (6.12.16) with Eq. (6.4.5) or by recalling that the sensitivity matrix is given by

$$
S = \frac{\partial \hat{\mathbf{x}}_c}{\partial \overline{\mathbf{c}}},
$$

we have

$$
S = -\hat{R}_x^{-1} \hat{R}_{xc}
$$

(6.12.17)

and Eq. (6.12.16) may be written as

$$
\hat{\mathbf{x}}_c = \hat{\mathbf{x}} + S \overline{\mathbf{c}}.
$$

(6.12.18)

From Eq. (6.12.15) the data noise covariance matrix is given by

$$
P = (\hat{R}_x^T \hat{R}_x)^{-1} = \hat{R}_x^{-1} \hat{R}_x^{-T}
$$

(6.12.19)

and the consider covariance for **x** is (see Eq. (6.3.64))

$$P_c = P + S\overline{P}_{cc}S^T .$$ (6.12.20)

The cross covariance is given by Eq. (6.3.65)

$$P_{xc} = S\overline{P}_{cc}$$ (6.12.21)

and the complete consider covariance matrix is given by Eq. (6.3.67)

$$\mathbf{P}_c = \begin{bmatrix} P_c & P_{xc} \\ P_{cx} & \overline{P}_{cc} \end{bmatrix} .$$ (6.12.22)

We could also determine the consider covariance directly by replacing $\tilde{R}_c$ with $\overline{R}_c$ in Eq. (6.12.12). Then,

$$
\begin{aligned}
\mathbf{P}_c &= \begin{bmatrix} \hat{R}_x & \hat{R}_{xc} \\ 0 & \overline{R}_c \end{bmatrix}^{-1} \begin{bmatrix} \hat{R}_x^T & 0 \\ \hat{R}_{xc}^T & \overline{R}_c^T \end{bmatrix}^{-1} \\[2mm]
&= \begin{bmatrix} \hat{R}_x^{-1} & -\hat{R}_x^{-1}\hat{R}_{xc}\overline{R}_c^{-1} \\ 0 & \overline{R}_c^{-1} \end{bmatrix} \begin{bmatrix} \hat{R}_x^{-T} & 0 \\ -\overline{R}_c^{-T}\hat{R}_{xc}^T\hat{R}_{xc}^{-T} & \overline{R}_c^{-T} \end{bmatrix} \\[2mm]
&= \begin{bmatrix} \hat{R}_x^{-1}\hat{R}_x^{-T} + S\overline{R}_c^{-1}\overline{R}_c^{-T}S^T & S\overline{R}_c^{-1}\overline{R}_c^{-T} \\ \overline{R}_c^{-1}\overline{R}_c^{-T}S^T & \overline{R}_c^{-1}\overline{R}_c^{-T} \end{bmatrix} \\[2mm]
&= \begin{bmatrix} P_c & P_{xc} \\ P_{cx} & \overline{P}_{cc} \end{bmatrix} .
\end{aligned}
$$ (6.12.23)

In component form

$$P_c = \hat{R}_x^{-1}\hat{R}_x^{-T} + S\overline{P}_{cc}S^T$$ (6.12.24)
$$P_{xc} = S\overline{P}_{cc}$$ (6.12.25)
$$\overline{P}_{cc} = \overline{R}_c^{-1}\overline{R}_c^{-T} .$$ (6.12.26)

If we wish to estimate **c** we would combine $\tilde{R}_c$ and $\tilde{\mathbf{b}}_c$ from Eq. (6.12.12) with the *a priori* information on **c** given by the first of Eq. (6.12.9). We then perform a series of orthogonal transformations on

$$T \begin{bmatrix} \overline{R}_c & \overline{\mathbf{b}}_c \\ \tilde{R}_c & \tilde{\mathbf{b}}_c \end{bmatrix} = \begin{bmatrix} \hat{R}_c & \hat{\mathbf{b}}_c \\ 0 & \mathbf{e} \end{bmatrix}$$ (6.12.27)

and

$$\hat{c} = \hat{R}_c^{-1}\hat{b}_c \tag{6.12.28}$$

the associated estimation error covariance for $\hat{c}$ is

$$P_{cc} = \hat{R}_c^{-1}\hat{R}_c^{-T}. \tag{6.12.29}$$

We could now replace $\bar{c}$ with $\hat{c}$ in Eq. (6.12.16) to obtain the proper estimate for x and replace $\overline{R}_c$ with $\hat{R}_c$ in the first of Eq. (6.12.23) to obtain the corresponding consider covariance matrices. This will result in $\overline{P}_{cc}$ being replaced by P_{cc} in Eqs. (6.12.24) and (6.12.25).

The computational algorithm for consider covariance analysis using orthogonal transformations is as follows.

Given *a priori* information $\overline{P}_x$, $\overline{P}_{xc}$, and $\overline{P}_{cc}$ at an epoch time, t_0, and observations $y_j = \tilde{H}_{x_j}x_j + \tilde{H}_{c_j}c + \epsilon_j$ $j = 1, \ldots, \ell$.

1. Use a Cholesky decompostion to compute the upper triangular matrix of Eq. (6.12.7).

2. Using the procedure indicated in Eqs. (6.12.8) through (6.12.10), form the matrix indicated by Eq. (6.12.11). The observations may be processed one at a time, in small batches, or simultaneously (see Chapter 5).

3. Partially upper triangularize the matrix of Eq. (6.12.11) by nulling the terms below the diagonal of the first n columns to yield the results of Eq. (6.12.12).

4. After processing all observations replace $\tilde{R}_c$ in Eq. (6.12.12) with the *a priori* value $\overline{R}_c$ and evaluate Eq. (6.12.23) to obtain $\mathbf{P}_c$.

Note that if only a consider covariance is needed, and no estimation is to be done, the last column of Eq. (6.12.12) containing $\overline{b}$ and y may be eliminated since the *a priori* state and the actual values of the measurements do not affect the estimation error covariance.

More extensive treatment of the square-root formulations of the covariance analysis problem including the effects of system model errors, errors in the *a priori* statistics, and errors in correlated process noise can be found in Curkendall (1972), Bierman (1977), Thornton (1976), and Jazwinski (1970).

6.13 REFERENCES

Bierman, G. J., *Factorization Methods for Discrete Sequential Estimation*, Academic Press, New York, 1977.

Curkendall, D. W., *Problems in estimation theory with applications to orbit determination*, UCLA-ENG-7275, UCLA School Engr. and Appl. Sci., Los Angeles, CA, September 1972.

Heffes, H., "The effects of erroneous models on the Kalman filter response", *Trans. Auto. Cont.*, Vol. AC-11, pp. 541–543, July 1966.

Jazwinski, A. H., *Stochastic Processes and Filtering Theory*, Academic Press, New York, 1970.

Thornton, C. L., *Triangular covariance factorizations for Kalman filtering*, Technical Memorandum, 33–798, Jet Propulsion Laboratory, Pasadena, CA, October 15, 1976.

6.14 EXERCISES

1. Map the consider covariance, $\mathbf{P}_{c_0}$, for the batch algorithm from t_0 to t_1 for Example 6.8. Now map $\mathbf{P}_{c_2}$, the result at t_2 from the sequential algorithm to t_1 and show that it agrees with the batch result.

   Answer:

$$
\mathbf{P}_{c_1} = \begin{bmatrix} \dfrac{80 + 27\Pi}{300} & \dfrac{20 - 27\Pi}{300} & -\dfrac{3\Pi}{10} \\[2mm] \dfrac{20 - 27\Pi}{300} & \dfrac{80 + 27\Pi}{300} & \dfrac{3\Pi}{10} \\[2mm] -\dfrac{3\Pi}{10} & \dfrac{3\Pi}{10} & \Pi \end{bmatrix}
$$

2. Compute the consider covariance for Example 6.8 at $t = 0$ using the square root formulation of Section 6.12. Compare your results to those given in Section 6.8.

3. Beginning with Eq. (6.3.38), derive the alternate equation for P_{xx} given by Eq. (6.3.41).

4. Beginning with Eqs. (6.6.3) and (6.6.4), derive Eqs. (6.6.6) and (6.6.11).

5. Given the pendulum problem shown here. Assume that θ_0 and $\dot{\theta}_0$ are to be estimated, that the observation is range magnitude, ρ, and x_0 and g are to be considered.

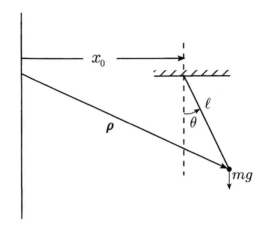

Assume: $\sigma(x_0) = \sigma(g) = \sigma(\rho) = 1$, and small oscillations so that the equation of motion is linear.

(a) Define: $\Phi(t, t_0)$, $\theta(t, t_0)$, H_x, H_c.

(b) Assuming $\sqrt{g/l} = 10$ and observations are taken at $t = 0, 1, 2$, compute the data noise covariance, the sensitivity, perturbation, and consider covariance matrices. Discuss the interpretation of these matrices. Assume $x_0 = l = 1$, $\overline{P} = R = I$, $\theta_0 = 0.01$ rad, and $\dot{\theta}_0 = 0$.

Answer:

$$
P_{c_0} = \begin{bmatrix}
0.539 & -0.0196 & -0.147 & 0.000154 \\
-0.0196 & 0.995 & -0.0122 & 0.000042 \\
-0.147 & -0.0122 & 1.0 & 0.0 \\
0.000154 & 0.000042 & 0.0 & 1.0
\end{bmatrix}
$$

6. Examine the problem of estimating the position and velocity of a spacecraft in low Earth orbit from ground-based range measurements. The dynamic model is chosen as a simple "flat Earth." Assume range measurements are taken from a single station. Derive the expression for the consider covariance matrix for the estimated and consider parameters shown.

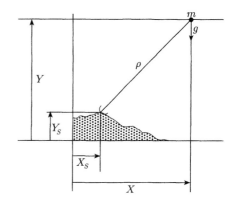

Estimate:
$X, Y, \dot{X}, \dot{Y}$

Consider:
g, X_s, Y_s

Observe: ρ

7. A vehicle travels in a straight line according to the following equation:

$$u(t) = u_0 + et + ft^2$$

where

$u = $ distance from origin

$t = $ time

$e, f = $ constants.

A tracking station location a distance s from the origin takes range measurements ρ.

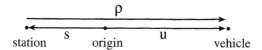

The true range at time t is

$$\rho(t) = u(t) + s.$$

The observed range, $y(t)$, has some error $\epsilon(t)$:

$$y(t) = \rho(t) + \epsilon(t)$$
$$= u(t) + s + \epsilon(t).$$

Assume that we wish to estimate $u(t_0) = u_0$ and e, and consider s and f. The following observations with measurement noise of zero mean and unit variance are given.

$$\text{Assume } \sigma_\epsilon = \sigma_s = \sigma_f = 1, \overline{P} = I \text{ and } \bar{c} = \begin{bmatrix} s \\ f \end{bmatrix} = \begin{bmatrix} 0 \\ 0 \end{bmatrix}$$

t	$y(t)$
0	1.5
1	10.8
2	24.8

Compute $\hat{u}_0$, $\hat{e}$, the computed and consider covariance matrices, and the sensitivity and perturbation matrix.

Answer:

$$\begin{bmatrix} \hat{u}_o \\ \hat{e} \end{bmatrix} = \begin{bmatrix} 2.76 \\ 8.69 \end{bmatrix},$$

$$\mathbf{P}_{c_0} = \begin{bmatrix} 0.80 & 0.20 & -0.60 & -0.20 \\ 0.20 & 2.27 & -0.20 & -1.40 \\ -0.60 & -0.20 & 1.0 & 0.0 \\ -0.20 & -1.40 & 0.0 & 1.0 \end{bmatrix}$$

8. Work the spring-mass problem of Example 6.9. Assume k_1 and k_2 are known exactly; estimate x_0, and v_0 and consider a 20 percent error in h.

9. Work Exercise 8 using the orthogonal transformation procedure of Section 6.12.

Appendix A

Probability and Statistics

A.1 INTRODUCTION

A *random variable* X is a real-valued function associated with the (chance) outcome of an experiment. That is, to each possible outcome of the experiment, we associate a real number x. The collection of numbers x_i is called the domain of the random variable X. A number x is referred to as a possible value or realization or sample value of the random variable X. Generally for this appendix, uppercase letters will represent random variables and lowercase letters will represent numerical values or realizations of the random variable.

Example 1:
A coin is flipped n times, let X denote the number of times heads appears. The domain of X is the set of numbers

$$\{x = 0, 1, \ldots, n\}.$$

Example 2:
A point on the surface of the Earth is chosen at random. Let X and Y denote its latitude and longitude. The domain of the random variable X and Y is the pair of numbers

$$\{X, Y, -\pi/2 \leq X \leq \pi/2, 0 \leq Y \leq 2\pi\}.$$

These two examples indicate the two types of random variables, discrete and continuous. The discrete random variable X has as its domain or possible outcomes the finite number of real numbers $x_1, x_2 \ldots x_n$ (see Example 1). For a continuous random variable, X has as its domain or outcome an interval on the real line, and to each of these intervals a probability may be assigned. The probability at a point is zero. In Example 2, there are an infinite number of real numbers between 0 and 2π; hence, the probability of obtaining any one of them is zero. Consequently, we must speak of the probability of X over an interval.

A.2 AXIOMS OF PROBABILITY

To define a probability associated with the possible values of a random variable, we introduce the fundamental axioms of probability. Let S represent the set of all possible outcomes of a trial or experiment. S is called the sample space of the experiment. Let A be any subset of points of the set S (i.e., a collection of one or more possible trials). For example, in one toss of a die the certain event S would be $s = 1, 2, 3, 4, 5, 6$, and a subset, A, may be $a = 3$.

Write p for a function that assigns to each event A a real number $p(A)$, called the probability of A; that is, $p(A)$ is the numerical probability of the event A occurring. For the previous example, $p(A) = 1/6$.

In modern probability theory, the following minimal set of axioms is generally met (Helstrom, 1984)

1. $p(A) \geq 0$

2. $p(S) = 1$

3. $p(A + B) = p(A) + p(B)$; A and B are any pair of mutually exclusive events.

Here, $p(A + B)$ indicates the probability of A or B or both occurring. For axiom 3, A and B cannot both occur since the events are mutually exclusive. If the events are not mutually exclusive, $p(A + B) = p(A) + p(B) - p(AB)$, where $p(AB)$ is the probability that both events A and B occur. Note that for mutually exclusive events $p(AB) = 0$ (by definition, both events cannot occur simultaneously). A *Venn diagram* (Freeman, 1963) can be used to demonstrate some of the concepts associated with probability. Consider an event A, then we define the complement of A as the event that A does not happen and denote it as $\overline{A}$. Thus A and $\overline{A}$ are said to be mutually exclusive—if A occurs, $\overline{A}$ cannot occur.

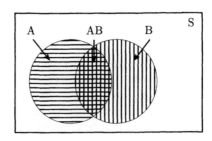

If A and B are any two events in a sample space S, we define AB to be the intersection of events A and B and assign to it all points in the space that belongs to both A and B; that is, the occurrence of AB implies that both events A and B have occurred. The intersection of A and B (meaning that both events have occurred) is sometimes denoted as $A \cap B = AB$. We define the union of A and B (written as $A \cup B = A + B$) as all points in the space for which A or B or both have occurred. For the *Venn diagram* shown,

$$
\begin{aligned}
S &: \quad \text{all points in the rectangle} \\
A &: \quad \text{points in horizontal hatch region} \\
B &: \quad \text{points in vertical hatch region} \\
AB &: \quad \text{points in cross hatch region} \\
A + B &: \quad \text{all points in hatched area}
\end{aligned}
$$

Example:
Throw 1 red and 1 green die. What is probability that at least one "1" will appear? Let

$$
\begin{aligned}
A &= \quad \text{1 on red die,} \\
\overline{A} &= \quad \text{2–6 on red die,} \\
B &= \quad \text{1 on green die,} \\
\overline{B} &= \quad \text{2–6 on green die.}
\end{aligned}
$$

The probability of at least one "1" occurring is the union of events A and B, $p(A + B)$. The probability of A or B or both occurring is seen from the Venn diagram to be

$$
p(A + B) = p(A) + p(B) - p(AB). \tag{A.2.1}
$$

Note that $p(A) + p(B)$ includes $2 \times p(AB)$. Now,

$$p\,(A) = 1/6$$
$$p\,(\overline{A}) = 5/6$$
$$p\,(B) = 1/6$$
$$p\,(\overline{B}) = 5/6$$
$$p\,(AB) = p\,(A)p\,(B) = 1/36\,.$$

Hence,

$$p(A + B) = 1/6 + 1/6 - 1/36 = 11/36\,.$$

Note that the events AB, $A\overline{B}$, $\overline{A}B$, $\overline{A}\overline{B}$ are exhaustive (i.e., one must occur), and they are exclusive (i.e., only one can occur). Hence,

$$p(AB,\ A\overline{B},\ \overline{A}B,\ \overline{A}\,\overline{B}) = p(AB) + p(A\overline{B}) + p(\overline{A}B) + p(\overline{A}\,\overline{B})$$
$$= 1/36 + 5/36 + 5/36 + 25/36 = 1\,.$$

Also, the probability of at least one "1" occurring can be expressed as

$$p(AB) + p(A\overline{B}) + p(\overline{A}B) = p(A)p(B) + p(A)p(\overline{B}) + p(\overline{A})p(B)$$
$$= (1/6)(1/6) + (1/6)(5/6) + (1/6)(5/6)$$
$$= 11/36\,.$$

Furthermore, the probability of two "1"s occurring is given by events A and B both occurring,

$$p\,(AB)^* \;=\; p\,(A)\,p\,(B) = 1/36\,.$$

Carrying on with this example, let

$A =$	1 on red die		$D =$	1 on green die
$B =$	2 on red die		$E =$	2 on green die
$C =$	3, 4, 5, 6 on red die		$F =$	3, 4, 5, 6 on green die

where $p\,(A) = p\,(B) = 1/6$ and $p\,(C) = 4/6$.

The probability that a "1" or a "2" occurs on either or both die (call this event G) is given by

$$p\,(G) = p(AD) + p(AE) + p(AF) + p(BD)$$
$$+ p(BE) + p(BF) + p(DC) + p(EC)$$
$$= 1/36 + 1/36 + 4/36 + 1/36 + 1/36$$
$$+ 4/36 + 4/36 + 4/36$$
$$= 20/36\,.$$

---

*Events A and B are independent (see definition in A.3).

Note that $p(G)$ also can be described by

$$p(G) = p(A) + p(B) + p(CD) + p(CE)$$
$$= 1/6 + 1/6 + 4/36 + 4/36 = 20/36.$$

If events A and B occur, it does not matter what happens on the green die, but if event C occurs, then either event D or E must occur. As stated earlier, if two events A and B are mutually exclusive, or disjoint, then the intersection, $A \cap B$, has no elements in common and $p(AB) = 0$; that is, the probability that both A and B occur is zero. Hence,

$$p(A + B) = p(A) + p(B).$$ (A.2.2)

For example, if we throw one die and let

$$A = \quad \text{a ``1'' appears}$$
$$B = \quad \text{a ``2'' appears.}$$

Then

$$p(A + B) = p(A) + p(B) = 2/6.$$

A.3 CONDITIONAL PROBABILITY

In wide classes of problems in the real world, some events of interest are those whose occurrence is conditional on the occurrence of other events. Hence, we introduce *conditional probability*, $p(A/B)$. By this we mean the probability that event A occurs, given that event B has occurred. It is given by

$$p(A/B) = \frac{p(AB)}{p(B)}.$$ (A.3.1)

We say that two events A and B are independent if

$$p(A/B) = p(A) \text{ and } p(B/A) = p(B);$$ (A.3.2)

that is,

$$p(AB) = p(A)p(B).$$ (A.3.3)

A.4 PROBABILITY DENSITY AND DISTRIBUTION FUNCTIONS

For a continuous random variable, we assume that all events of practical interest will be represented by intervals on the real line and to each of these intervals a probability may be assigned. Recall that the probability at a point is zero. We define a *probability density function*, $f(x)$, which represents the probability of X assuming a value somewhere in the interval $(x, \ x + dx)$. We define probability over the interval $x, x + dx$ in terms of area,

$$p(x \leq X \leq x + dx) = f(x)dx. \qquad (A.4.1)$$

For the continuous random variable, axioms 1 and 2 become

$$\begin{aligned} 1. \quad & f(x) \geq 0 & (A.4.2) \end{aligned}$$

$$2. \quad \int_{-\infty}^{\infty} f(x)\, dx = 1. \qquad (A.4.3)$$

The third axiom becomes

$$3. \quad p(a \leq X \leq c) = \int_a^c f(x)\, dx \qquad (A.4.4)$$

which for $a < b < c$

$$\begin{aligned} p(a \leq X \leq c) &= \int_a^b f(x)\, dx + \int_b^c f(x)\, dx \\ &= p(a \leq X \leq b) + p(b \leq X \leq c). \qquad (A.4.5) \end{aligned}$$

Note that for the continuous random variable we need not distinguish between $a \leq X \leq c, a \leq X < c, a < X \leq c$, and $a < X < c$, since the probability is the same for each (i.e., the probability at a point is zero).

Out of interest in the event $X \leq x$, we introduce $F(x)$, the *distribution function* of the continuous random variable X, and define it by

$$F(x) = p(X \leq x) = \int_{-\infty}^{x} f(t)\, dt. \qquad (A.4.6)$$

It follows that

$$F(-\infty) = 0 \text{ and } F(\infty) = 1.$$

From elementary calculus, at points of continuity of F

$$\frac{d\,F(x)}{dx} = f(x) \qquad (A.4.7)$$

which relates distribution and density functions for continuous random variables.

Consider the following sketch of the distribution and density function of a continuous random variable X.

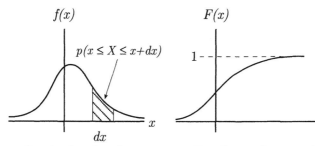

Density function of a
continuous random variable

Distribution function of a
continuous random variable

From the definition of the density and distribution functions we have

$$p\left(a \leq X \leq b\right) = \int_a^b f(x)\,dx = F\left(b\right) - F\left(a\right).$$

(A.4.8)

From axioms 1 and 2, we find

$$0 \leq F(x) \leq 1$$

(A.4.9)

and $F(x)$ is monotonically increasing.

A.5 EXPECTED VALUES

We will now discuss the characteristics of probability distributions. A probability distribution is a generic term used to describe the probability behavior of a random variable. We shall now define certain useful parameters associated with probability distributions for continuous random variables. The *expected value* or the *mean* of X is written $E\left(X\right)$, and is defined by

$$E(X) = \int_{-\infty}^{\infty} x f(x)\,dx.$$

(A.5.1)

Now consider a second random variable, where

$$Y = g\left(X\right)$$

then

$$E\left[g(X)\right] = \int_{-\infty}^{\infty} g(x)\,f(x)\,dx$$

(A.5.2)

which defines the expected value of a function of a random variable.

A.6　EXAMPLES AND DISCUSSION OF EXPECTATION

There are particular functions, $g(X)$, whose expectations describe important characteristics of the probability distribution of X. We will now consider several particular cases.

$E(X)$ is sometimes written as λ_1 or λ and is called the arithmetic mean of X. Geometrically, λ_1 is one of a number of possible devices for locating the "center" or centroid of the probability distribution with respect to the origin. The kth moment of X about the origin is

$$E[X^k] = \lambda_k = \int_{-\infty}^{\infty} x^k f(x)\,dx. \tag{A.6.1}$$

We also may speak of the kth moment of X about the mean λ_1. In this case, we define

$$\mu_k \equiv E(X - \lambda_1)^k = \int_{-\infty}^{\infty} (x - \lambda_1)^k f(x)dx. \tag{A.6.2}$$

Note that $\mu_1 = 0$. For $k = 2$

$$\mu_2 = E(X - \lambda_1)^2 = \int_{-\infty}^{\infty} (x - \lambda_1)^2 f(x)dx. \tag{A.6.3}$$

This is usually denoted as σ^2 or $\sigma^2(X)$ and called the *variance* of X, the variance of the probability distribution of X, or the second moment of X about the mean. Note that μ_2 is always greater than zero unless $p(X = \lambda_1) = 1$. In this case $\mu_2 = 0$. The positive square root of the variance, σ, is called the standard deviation of X. It is one measure of the dispersion of the distribution about its mean value.

We interpret the density function as a mass distribution, the first moment about the origin becomes the center of gravity of the distribution, and the variance is the moment of inertia about an axis through the mean. For example, consider the following density functions:

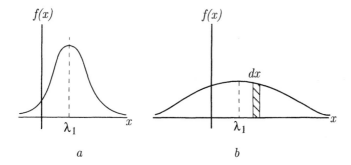

Both density functions a and b have their mean value indicated by λ_1. Note that the variance of b obviously will be much larger than that of a, since $(x - \lambda_1)^2$ generally will be larger than that of a for an increment of area dx.

A few useful results follow readily from the definition of the expected value (see Eq. (A.5.1)) and the fact that it is a linear operator

$$E(a + bX) = a + bE(X) \tag{A.6.4}$$

where a and b are constants. Also,

$$\mu_1 = \int_{-\infty}^{\infty} (x - \lambda_1)\, f(x)dx = \lambda_1 - \lambda_1 = 0 \tag{A.6.5}$$

$$\mu_2 = \int_{-\infty}^{\infty} (x - \lambda_1)^2 f(x)\, dx$$

$$= \int_{-\infty}^{\infty} (x^2 - 2x\lambda_1 + \lambda_1^2)f(x)dx$$

$$= \lambda_2 - 2\lambda_1^2 + \lambda_1^2$$

$$= \lambda_2 - \lambda_1^2. \tag{A.6.6}$$

The higher order moments are of theoretical importance in any distribution, but they do not have a simple geometric or physical interpretation as do λ_1 and μ_2.

All the information that can be known about the random variable X is contained in the probability density function. In guidance and estimation applications, we are most concerned with the first two moments, namely the mean and variance. Since the probability density function will change with time, the prediction of the future values for the state of the dynamic system can be obtained by propagating the joint density function (see Section A.9) forward in time and using it to calculate the mean and variance. The equations for propagating the mean and variance of a random vector X are discussed in Section 4.8 of the text.

A.7   MOMENT GENERATING FUNCTIONS

Consider the particular case of

$$E\left[g(X)\right] = \int_{-\infty}^{\infty} g(x)f(x)dx \tag{A.7.1}$$

for which

$$g(X) = e^{\theta x}$$

where θ is a dummy variable. Since

$$e^{\theta x} = 1 + \theta x + \frac{(\theta x)^2}{2!} + \cdots \frac{(\theta x)^n}{n!} + \cdots \tag{A.7.2}$$

substituting Eq. (A.7.2) into Eq. (A.7.1) results in

$$E\left(e^{\theta x}\right) = \lambda_0 + \theta\lambda_1 + \frac{\theta^2\lambda_2}{2!} + \cdots \frac{\theta^n\lambda_n}{n!} + \cdots \tag{A.7.3}$$

Thus $E(e^{\theta x})$ may be said to generate the moments $\lambda_0, \lambda_1 \ldots \lambda_n$ of the random variable X. It is called the *moment generating function* of X and is written $M_X(\theta)$. Note that

$$\frac{\partial^k M_X(\theta)}{\partial\theta^k}\bigg|_{\theta=0} = \lambda_k. \tag{A.7.4}$$

Accepting the fact that the moment generating function for the function $h(X)$ is given by

$$M_{h(X)}(\theta) = \int_{-\infty}^{\infty} e^{\theta h(X)} f(x)dx, \tag{A.7.5}$$

let $h(X) = X - \lambda_1$, then

$$M_{(X-\lambda_1)}(\theta) = e^{-\theta\lambda_1} M_X(\theta) \tag{A.7.6}$$

which relates moments about the origin to moments about the mean,

$$\mu_k = \frac{\partial^k M_{(X-\lambda_1)}(\theta)}{\partial\theta^k}\bigg|_{\theta=0}. \tag{A.7.7}$$

From Eqs. (A.7.3) and (A.7.6)

$$M_{(X-\lambda_1)}(\theta) = e^{-\theta\lambda_1}\left(\lambda_0 + \theta\lambda_1 + \frac{\theta^2\lambda_2}{2!} + \cdots \frac{\theta^n\lambda_n}{n!} \cdots\right)$$

and for example,

$$\mu_2 = \frac{\partial^2 M_{(X-\lambda_1)}(\theta)}{\partial\theta^2}\bigg|_{\theta=0}$$

$$= \lambda_1^2 e^{-\theta\lambda_1}(\lambda_0 + \theta\lambda_1) - \lambda_1 e^{-\theta\lambda_1}\lambda_1 - \lambda_1 e^{-\theta\lambda_1}\lambda_1 + e^{-\theta\lambda_1}\lambda_2\bigg|_{\theta=0}$$

$$= \lambda_2 - \lambda_1^2, \quad \text{recall that} \quad \lambda_0 = 1.$$

This is identical to the result in Eq. (A.6.6).

A.8 SOME IMPORTANT CONTINUOUS DISTRIBUTIONS

A.8.1 UNIFORM OR RECTANGULAR DISTRIBUTION

If X has equal probability over the range $a \leq X \leq b$, that is, every value of X in this range is equally likely to occur, we say that X is *uniformly distributed*. Its density function is

$$f(x) = \frac{1}{b - a}, \quad a \leq x \leq b \tag{A.8.1}$$
$$= 0 \quad \text{elsewhere}$$

$$F(x) = \int_{-\infty}^{a} f(x)dx + \int_{a}^{x} f(t)dt = \int_{a}^{x} \frac{dt}{b - a}$$

$$F(x) = 0 \quad x < a$$
$$= \frac{x - a}{b - a} \quad a \leq x \leq b$$
$$= 1 \quad x > b.$$

The first two moments (mean and variance) are

$$E(X) = \int_{a}^{b} xf(x)dx = \int_{a}^{b} \frac{xdx}{b - a}$$

$$= \left[\frac{x^2}{2(b - a)}\right]_{a}^{b} = \frac{a + b}{2} \tag{A.8.2}$$

$$\sigma^2(X) = \int_a^b [x - E(X)]^2 f(x) dx$$

$$= \int_a^b \left(x - \frac{a+b}{2}\right)^2 \frac{dx}{b-a} = \frac{(b-a)^2}{12}. \qquad (A.8.3)$$

Graphically the uniform density function is

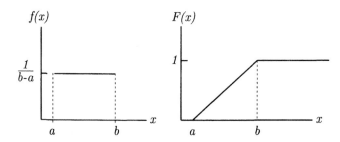

The uniform density function is hardly a likely description of the probability behavior of many physical systems. Its importance lies in its utility in statistical theory. Any continuous probability distribution can be converted first into a uniform distribution, then into a given continuous distribution. This often facilitates the study of the properties of the distribution, which themselves are somewhat intractable.

A.8.2   THE GAUSSIAN OR NORMAL DISTRIBUTION

One of the most important distributions in probability theory is the one for which

$$f(x) = \frac{1}{\sqrt{2\pi}b} \exp\left[-\frac{1}{2}\left(\frac{x-a}{b}\right)^2\right] \quad \begin{array}{c} -\infty < x < \infty \\ b > 0 \end{array}. \qquad (A.8.4)$$

The moment generating function for this distribution is

$$M_X(\theta) = E[e^{\theta x}] = \int_{-\infty}^{\infty} e^{\theta x} f(x)\, dx$$

$$M_X(\theta) = \int_{-\infty}^{\infty} \frac{e^{\theta x}}{\sqrt{2\pi}\, b} \exp\left[-\frac{1}{2}\left(\frac{x-a}{b}\right)^2\right] dx$$

$$= \exp\left[\frac{\theta^2 b^2}{2} + a\theta\right]. \qquad (A.8.5)$$

For details of the derivation of Eq. (A.8.5), see Freeman (1963).

From Eq. (A.8.5), we see that the mean of the normal distribution is

$$\frac{\partial M_X(\theta)}{\partial \theta}\bigg|_{\theta=0} = a$$

hence

$$\lambda_1 = a \, .$$

From Eq. (A.7.6), we have

$$M_{X-\lambda_1}(\theta) = e^{-a\theta} M_X(\theta)$$
$$= e^{\theta^2 b^2 / 2} \tag{A.8.6}$$

and from Eq. (A.8.6), the variance is

$$\frac{\partial^2 M_{X-\lambda_1}(\theta)}{\partial \theta^2}\bigg|_{\theta=0} = b^2 = \sigma^2$$

and $\sigma = b$. Hence, Eq. (A.8.4) may be written

$$f(x) = \frac{1}{\sqrt{2\pi}\sigma} \exp\left[-\frac{1}{2}\left(\frac{x-\lambda_1}{\sigma}\right)^2\right] \quad -\infty < x < \infty \, . \tag{A.8.7}$$

Using Eq. (A.8.6), it may be shown that

$$\mu_{2k+1} = 0,$$

that is, the odd moments of a normal random variable about its mean are zero. In fact the odd moments of any symmetric distribution about its mean are zero provided that they exist.

The normal distribution is depicted graphically here:

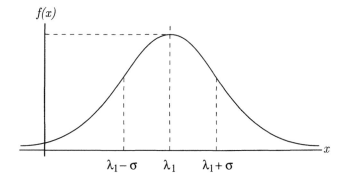

The density function has its maximum at $X = \lambda_1$, is symmetric about the line $X = \lambda_1$, and has two inflection points at $X = \lambda_1 \pm \sigma$ (i.e., points where $\frac{d^2 f(x)}{dx^2} = 0$), and the minimum occur at $\pm\infty$. Other interesting properties of the univariate normal distribution function are

$$p(\lambda_1 - \sigma \leq X \leq \lambda_1 + \sigma) = \int_{\lambda_1-\sigma}^{\lambda_1+\sigma} f(x)dx = .68268$$

$$p[(\lambda_1 - 2\sigma) \leq X \leq (\lambda_1 + 2\sigma)] = .95449$$

$$p[(\lambda_1 - 3\sigma) \leq X \leq (\lambda_1 + 3\sigma)] = .99730.$$

A.9   TWO RANDOM VARIABLES

The joint distribution function $F(x, y)$ of two random variables is defined as

$$F(x_0, y_0) = p\{X \leq x_0, Y \leq y_0\}. \tag{A.9.1}$$

It has the following properties

$$0 \leq F(x, y) \leq 1 \qquad \text{for all } x, y$$
$$F(-\infty, y) = F(x, -\infty) = 0, \; F[\infty, \infty] = 1 \tag{A.9.2}$$
$$F(\infty, y_0) = p\{Y \leq y_0\}, \qquad F(x_0, \infty) = p\{X \leq x_0\}$$

The concept of the density function now follows. Because $F(x, y)$ is continuous, a function $f(x, y)$ exists such that

$$F(x, y) = \int_{-\infty}^{y} \int_{-\infty}^{x} f(u, v) \, du \, dv \tag{A.9.3}$$

or the equivalent

$$f(x, y) = \frac{\partial F^2(x, y)}{\partial x \partial y}. \tag{A.9.4}$$

By definition,

$$p(a \leq X \leq b, c \leq Y \leq d) = \int_{c}^{d} \int_{a}^{b} f(x, y) \, dx \, dy \tag{A.9.5}$$

which is analogous to the relationship between the probability and the density function for the single random variable case. Eq. (A.9.5) may be written

$$p(x \leq X \leq x + dx, y \leq Y \leq y + dy) = f(x, y)dxdy. \tag{A.9.6}$$

In summary:

$$
\begin{aligned}
F\left(x,\,y\right) &\equiv & \text{joint distribution function of } X,\, Y.\\
f\left(x,\,y\right) &\equiv & \text{joint density function of } X,\, Y.\\
f\left(x,\,y\right)dx\,dy &\equiv & \text{joint probability element of } X,\, Y.
\end{aligned}
$$

A.10   MARGINAL DISTRIBUTIONS

We often want to determine the probability behavior of one random variable, given the joint probability behavior of two. This is interpreted to mean

$$
p\left(X \le x,\,\text{no condition on } Y\right) = F\left(x,\infty\right). \tag{A.10.1}
$$

For the continuous case

$$
F(x,\infty) = \int_{-\infty}^{x}\int_{-\infty}^{\infty} f(u,v)\,dv\,du = \int_{-\infty}^{x}\left[\int_{-\infty}^{\infty} f(u,v)\,dv\right] du
$$

$$
= \int_{-\infty}^{x} g\left(u\right)du. \tag{A.10.2}
$$

Hence,

$$
g\left(x\right) = \int_{-\infty}^{\infty} f(x,y)\,dy \tag{A.10.3}
$$

is called the marginal density function of X. Similarly,

$$
F\left(\infty,\,y\right) = \int_{-\infty}^{y}\left[\int_{-\infty}^{\infty} f\left(u,\,v\right)du\right] dv = \int_{-\infty}^{y} h\left(v\right)dv
$$

and

$$
h\left(y\right) = \int_{-\infty}^{\infty} f\left(x,\,y\right)dx \tag{A.10.4}
$$

is the marginal density function of Y. Hence, the marginal density function of a random variable is obtained from the joint density function by integrating over the unwanted variable.

A.11 INDEPENDENCE OF RANDOM VARIABLES

We have previously defined the independence of two events A and B by $p(A, B) = p(A) p(B)$. For the case of random variables X and Y, we say that they are independent if we can factor their joint density function into

$$f(x, y) = g(x)h(y) \tag{A.11.1}$$

where $g(x)$ and $h(y)$ are the marginal density functions of X and Y.

A.12 CONDITIONAL PROBABILITY

For the simple events A and B, we define $p(B/A)$ by

$$p(B/A) = \frac{p(AB)}{p(A)}. \tag{A.12.1}$$

In the spirit of Eq. (A.12.1), we wish to define a *conditional density function* for continuous random variables X and Y with density functions $f(x, y)$, $g(x)$, and $h(y)$. Accordingly,

$$g(x/y) = \frac{f(x, y)}{h(y)}, h(y/x) = \frac{f(x, y)}{g(x)}. \tag{A.12.2}$$

As an immediate consequence of Eq. (A.12.2), we have

$$p(a \le X \le b/Y = y) = \int_a^b g(x/y)\, dx = \frac{\int_a^b f(x, y)\, dx}{h(y)} \tag{A.12.3}$$

$$p(c \le Y \le d/X = x) = \int_c^d h(y/x)\, dy = \frac{\int_c^d f(x, y)\, dy}{g(x)}.$$

Note that in Eq. (A.12.3), we are talking about X and Y in the vicinity of the values x and y.

Also,

$$p(a \le X \le b/c \le Y \le d) = \frac{p(a \le X \le b, c \le Y \le d)}{p(c \le Y \le d)}$$

$$= \frac{\int_a^b \int_c^d f(x, y)\, dy\, dx}{\int_c^d h(y)\, dy}. \tag{A.12.4}$$

In the case of statistically independent variables, X and Y, the definition of statistical independence given by Eq. (A.11.1) leads to the following result from

Eq. (A.12.2)

$$g\left(x/y\right) = \frac{f\left(x, y\right)}{h\left(y\right)} = \frac{g\left(x\right)h\left(y\right)}{h\left(y\right)} = g\left(x\right)$$

(A.12.5)

$$h\left(y/x\right) = \frac{f\left(x, y\right)}{g\left(x\right)} = h\left(y\right).$$

A.13 EXPECTED VALUES OF BIVARIATE FUNCTIONS

Following the arguments for one random variable, we say that the mean or expected value $E\left[\phi\left(X, Y\right)\right]$ of an arbitrary function $\phi(X, Y)$ of two continuous random variables X and Y is given by

$$E\left[\phi\left(X, Y\right)\right] = \int\limits_{-\infty}^{\infty} \int\limits_{-\infty}^{\infty} \phi\left(x, y\right)f\left(x, y\right)dx\, dy. \tag{A.13.1}$$

As with one random variable, the expected value of certain functions is of great importance in identifying characteristics of joint probability distributions. Such expectations include the following. Setting

$$\phi\left(X, Y\right) = X^l Y^m$$

gives

$$E\left[X^l Y^m\right] = \int\limits_{-\infty}^{\infty} \int\limits_{-\infty}^{\infty} x^l y^m f\left(x, y\right)dx\, dy \equiv \lambda_{lm} \tag{A.13.2}$$

written λ_{lm}, the lm^{th} moment of X, Y about the origin. The lm^{th} moment about the mean is obtained by setting

$$\phi\left(X, Y\right) = [X - \lambda_{10}]^l [Y - \lambda_{01}]^m.$$

This results in

$$E\left\{[X - \lambda_{10}]^l [Y - \lambda_{01}]^m\right\}$$

$$= \int\limits_{-\infty}^{\infty} \int\limits_{-\infty}^{\infty} [x - \lambda_{10}]^l [y - \lambda_{01}]^m f(x, y)dx\, dy \equiv \mu_{lm} \tag{A.13.3}$$

written μ_{lm}, the lm^{th} moment of X, Y about the respective moments λ_{10} and λ_{01}. Particular cases of μ_{lm} and λ_{lm} often used are

l	m	
0	0	$\lambda_{00} = 1$
1	0	$\lambda_{10} = E(X)$, the mean of X
0	1	$\lambda_{01} = E(Y)$, the mean of Y
0	0	$\mu_{00} = 1$
1	1	$\mu_{11} = E\{[X - E(X)](Y - E(Y))\}$, the covariance of X and Y
2	0	$\mu_{20} = \sigma^2(X)$, the variance of X
0	2	$\mu_{02} = \sigma^2(Y)$, the variance of Y.

Consider as an example the computation of

$$
\begin{aligned}
\mu_{11} &= E(X - \lambda_{10})(Y - \lambda_{01}) \\
&= \int_{-\infty}^{\infty} \int_{-\infty}^{\infty} (x - \lambda_{10})(y - \lambda_{01}) f(x, y) dx\, dy \\
&= \int_{-\infty}^{\infty} \int_{-\infty}^{\infty} (xy - \lambda_{10}y - \lambda_{01}x + \lambda_{10}\lambda_{01}) f(x, y) dx\, dy \\
&= \lambda_{11} - 2\lambda_{10}\lambda_{01} + \lambda_{10}\lambda_{01} \\
&= \lambda_{11} - \lambda_{10}\lambda_{01} .
\end{aligned}
\tag{A.13.4}
$$

The result is analogous to Eq. (A.6.6).

A.14   THE VARIANCE-COVARIANCE MATRIX

The symmetric matrix

$$
\begin{aligned}
P &= E\left[\begin{bmatrix} X - E(X) \\ Y - E(Y) \end{bmatrix} \begin{bmatrix} X - E(X) & Y - E(Y) \end{bmatrix}\right] \\
&= E\begin{bmatrix} (X - E(X))^2 & (X - E(X))(Y - E(Y)) \\ (Y - E(Y))(X - E(X)) & (Y - E(Y))^2 \end{bmatrix} \\
P &= \begin{bmatrix} \sigma^2(X) & \mu_{11} \\ \mu_{11} & \sigma^2(Y) \end{bmatrix}
\end{aligned}
\tag{A.14.1}
$$

is called the *variance-covariance matrix* of the random variables X and Y. As seen from Eq. (A.14.1), the diagonals contain the variances of X and Y, and the off diagonal terms contain the covariances.

The covariance of the random variables X and Y often is written in terms of the *correlation coefficient* between X and Y, ρ_{XY}. The correlation coefficient is defined as

$$
\rho_{XY} \equiv \frac{E\{[X - E(X)][Y - E(Y)]\}}{\{E[X - E(X)]^2\}^{1/2}\{E[Y - E(Y)]^2\}^{1/2}}
$$
$$
= \frac{\mu_{11}}{\sigma(X)\sigma(Y)} . \tag{A.14.2}
$$

The variance-covariance matrix for an n-dimensional random vector, X, can be written as

$$
P = \begin{bmatrix} \sigma_1^2 & \rho_{12}\sigma_1\sigma_2 & \cdots & \rho_{1n}\sigma_1\sigma_n \\ \rho_{12}\sigma_1\sigma_2 & \sigma_2^2 & \cdots & \rho_{2n}\sigma_2\sigma_n \\ \vdots & & \ddots & \vdots \\ \rho_{1n}\sigma_1\sigma_n & \rho_{2n}\sigma_2\sigma_n & \cdots & \sigma_n^2 \end{bmatrix} \tag{A.14.3}
$$

where ρ_{ij} is a measure of the degree of linear correlation between X_i and X_j. It can also be written as

$$
\rho_{ij} \equiv \frac{P_{ij}}{\sigma_i\sigma_j}, \quad i \neq j . \tag{A.14.4}
$$

The correlation coefficient can be shown to be the covariance between the *standardized random variables*

$$
U \equiv \frac{X - E(X)}{\sigma(X)}, V \equiv \frac{Y - E(Y)}{\sigma(Y)} . \tag{A.14.5}
$$

That is,

$$
\rho_{XY} = \text{cov}(U, V) = E[(U - E(U))(V - E(V))] .
$$

This is easily demonstrated by noting that

$$
E(U) = E(V) = 0 . \tag{A.14.6}
$$

Thus

$$
E[(U - E(U))(V - E(V))] = E(UV) . \tag{A.14.7}
$$

So

$$
\text{cov}(UV) = E(UV) = E\left[\left(\frac{X - E(X)}{\sigma(X)}\right)\left(\frac{Y - E(Y)}{\sigma(Y)}\right)\right]
$$
$$
= \frac{\mu_{11}}{\sigma(X)\sigma(Y)} \equiv \rho_{XY} . \tag{A.14.8}
$$

Hence,

$$
\text{cov}(UV) = \rho_{XY} . \tag{A.14.9}
$$

Note also that

$$\sigma^2\,(U) = E[U - E(U)]^2 = E[U^2]$$
$$= \frac{E[X - E(X)]^2}{\sigma^2(X)} = \frac{\sigma^2(X)}{\sigma^2(X)} = 1\,. \qquad \text{(A.14.10)}$$

Likewise

$$\sigma^2(V) = 1\,.$$

So

$$\sigma^2(U) = \sigma^2(V) = 1\,. \qquad \text{(A.14.11)}$$

Note that the *standard deviation* is defined to be the positive square root of the variance. Consequently,

$$\rho_{UV} = \frac{\text{cov}\,(\text{UV})}{\sigma(U)\,\sigma(V)} = \text{cov}\,(\text{UV}) = \rho_{XY}\,. \qquad \text{(A.14.12)}$$

A.15   PROPERTIES OF THE CORRELATION COEFFICIENT

It is first convenient to prove two elementary relationships for a function of two random variables. For a and b constant

$$E(a\,X + b\,Y) = aE\,(X) + bE\,(Y)\,. \qquad \text{(A.15.1)}$$

This follows from the linear property of the expectation operator. Next,

$$\sigma^2(aX + bY) \equiv E[(aX + bY) - E(aX + bY)]^2$$
$$= E\,[a(X - E(X)) + b(Y - E(Y))]^2$$
$$= a^2 E[X - E(X)]^2 + 2ab\,E[(X - E(X))(Y - E(Y))] + b^2 E[Y - E(Y)]^2$$
$$= a^2\sigma^2(X) + 2ab\,\mu_{11} + b^2\sigma^2(Y)\,.$$

In terms of the correlation coefficient defined by Eq. (A.14.2), this result becomes

$$\sigma^2(aX + bY) = a^2\sigma^2(X) + 2ab\rho_{XY}\sigma(X)\sigma(Y) + b^2\sigma^2(Y)\,. \qquad \text{(A.15.2)}$$

Using Eqs. (A.15.1) and (A.15.2), we can demonstrate certain useful properties of ρ.

It will be convenient to use the standardized random variables defined by Eq. (A.14.5). From Eq. (A.15.2), we have that

$$\sigma^2(U + V) = \sigma^2(U) + 2\,\rho_{UV}\sigma(U)\sigma(V) + \sigma^2(V)$$
$$\sigma^2(U - V) = \sigma^2(U) - 2\,\rho_{UV}\sigma(U)\sigma(V) + \sigma^2(V)$$

but from Eq. (A.14.11)

$$\sigma^2(U) = \sigma^2(V) = 1$$

hence

$$\sigma^2(U \pm V) = 2(1 \pm \rho_{UV}).$$ (A.15.3)

Since by definition a variance is a nonnegative quantity

$$1 + \rho_{UV} \geq 0$$
$$1 - \rho_{UV} \geq 0.$$

Hence

$$-1 \leq \rho_{UV} \leq 1$$ (A.15.4)

and from Eq. (A.14.12)

$$-1 \leq \rho_{XY} \leq 1.$$ (A.15.5)

It can be shown that when ρ_{XY} assumes its extreme values $+1$ or -1, the relationship between X and Y is perfectly linear. That is, all values of the random variable pair X, Y lie on a straight line of positive or negative slope.

From Eq. (A.15.3), if $\rho_{UV} = +1$

$$\sigma^2(U - V) = 0.$$

Hence, $(U - V)$ is a constant with all probability concentrated at that constant. Or in terms of X and Y

$$\frac{X - E(X)}{\sigma(X)} - \frac{Y - E(Y)}{\sigma(Y)} = \text{const}.$$ (A.15.6)

This is an equation of the form

$$Y = a + bX$$

where $b = \sigma(Y)/\sigma(X)$, a positive constant.

A similar expression may be written for $\rho_{UV} = -1$. In this case, b is a negative constant $(-\sigma(Y)/\sigma(X))$. Also, the converse holds. Suppose that

$$Y = a \pm bX.$$

Then it can easily be shown that $\mu_{11} = \pm b\sigma^2(X)$ and that $\sigma^2(Y) = b^2 \sigma^2(X)$. Using the definition of ρ, we have $\rho_{XY} = \pm 1$.

Nonlinear functional relationships between random variables do not necessarily result in the correlation coefficient assuming a value of 1.

A.16   PROPERTIES OF COVARIANCE AND CORRELATION

From the definition of μ_{11}

$$\mu_{11} = E[(X - E(X))(Y - E(Y))]$$

we see that if large values of the random variable X are found paired generally with large values of Y in the function $f(x, y)$, and if small values of X are paired with small values of Y, μ_{11} and hence ρ_{XY} will be positive. Also, if large values of X are paired with small values of Y in $f(x, y)$, then μ_{11} and hence ρ_{XY} will be negative. Finally, if some large and small values of X and Y are paired then $\mu_{11} \simeq 0$. Graphically assume that we sample a value of X and Y and plot the results. Three cases are possible:

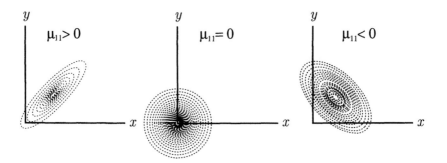

An example of positive correlation would be a sampling of human height and weight. Note that if $\rho_{XY} = 1$, the variance-covariance matrix will be singular.

A.17   BIVARIATE NORMAL DISTRIBUTION

The *bivariate normal density function* is given by

$$f(x, y) = \frac{1}{2\pi\,\sigma(x)\,\sigma(y)\,\sqrt{1-\rho^2}}\,\exp\left[-\frac{1}{2(1-\rho^2)}\left\{\left[\frac{x-\lambda_{10}}{\sigma(x)}\right]^2\right.\right.$$
$$\left.\left. -2\rho\frac{[x - \lambda_{10}][y - \lambda_{01}]}{\sigma(x)\sigma(y)} + \left[\frac{y - \lambda_{01}}{\sigma(y)}\right]^2\right\}\right] \qquad \begin{array}{l} -\infty < x < \infty \\ -\infty < y < \infty \end{array}$$

$$\text{(A.17.1)}$$

which has five parameters λ_{10}, λ_{01}, $\sigma(x)$, $\sigma(y)$, and ρ. From Eq. (A.17.1) note that if $\rho = 0$, $f(x, y)$ may be factored into $f(x, y) = g(x)h(y)$. Hence, $\rho = 0$ is a sufficient condition for statistical independence of bivariate normal variables. This is not true for most density functions.

A.18 MARGINAL DISTRIBUTIONS

It can be shown that each of the random variables in Eq. (A.17.1) is normally distributed. By carrying out the integral,

$$g(x) = \int_{-\infty}^{\infty} f(x, y) dy.$$ (A.18.1)

It can be shown that

$$g(x) = \frac{1}{\sqrt{2\pi}\sigma(x)} \exp\left[-\frac{1}{2}\left(\frac{x - \lambda_{10}}{\sigma(x)}\right)^2\right]$$ (A.18.2)

which is the normal density function of X. Similar results exist for the marginal distribution of Y. The converse is not true—if the marginal distributions $g(x)$ and $h(y)$ are normal, the joint density function $f(x, y)$ is not necessarily bivariate normal.

Now consider the conditional density function

$$h(y/x) = \frac{f(x, y)}{g(x)}$$ (A.18.3)

for the normal distribution. The numerator and denominator are given by Eqs. (A.17.1) and (A.18.2), respectively. Inserting these in Eq. (A.18.3) and simplifying, we obtain

$$h(y/x) = \frac{1}{\sigma(y)\sqrt{2\pi}\sqrt{1 - \rho^2}}$$

$$\times \exp{-\frac{1}{2}\left[\frac{y - \{\lambda_{01} + [\rho\sigma(y)/\sigma(x)][x - \lambda_{10}]\}}{\sigma(y)\sqrt{1 - \rho^2}}\right]^2}.$$ (A.18.4)

Hence, the conditional density function of Y is normal with conditional mean

$$E(Y/x) = \lambda_{01} + \rho\frac{\sigma(y)}{\sigma(x)}[x - \lambda_{10}]$$

and conditional standard deviation

$$\sigma(Y/x) = \sigma(y)\sqrt{1 - \rho^2}.$$

Thus the conditional as well as the marginal distribution of the bivariate normal distribution are normal. A graphic example of the conditional density function follows.

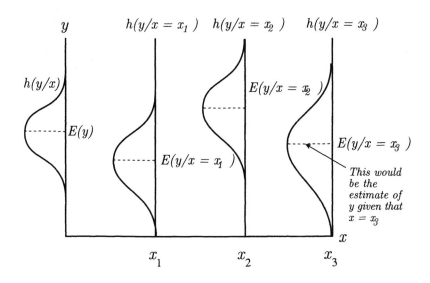

A.19   THE MULTIVARIATE NORMAL DISTRIBUTION

For the multivariate case, consider a vector of random variables; for example,

$$\mathbf{X} = \begin{bmatrix} x_1 \\ x_2 \\ \vdots \\ x_p \end{bmatrix}.$$

The *multivariate normal density function* is given by

$$f(x_1, x_2, \ldots, x_p) = \frac{1}{(2\pi)^{p/2}|V|^{1/2}} e^{-1/2\,(\mathbf{X} - \mathbf{\Lambda})^T V^{-1}\,(\mathbf{X} - \mathbf{\Lambda})}$$
$$-\infty < x_i < \infty \tag{A.19.1}$$

where

    V is the $p \times p$ variance-covariance matrix of the vector $\mathbf{X}$

    $|V|$ is the determinant of V

    $\mathbf{\Lambda}$ is a $p \times 1$ vector of mean values of X.

The matrix V is defined as

$$V = E\{[\mathbf{X} - \mathbf{\Lambda}][\mathbf{X} - \mathbf{\Lambda}]^T\} \tag{A.19.2}$$

in terms of the correlation coefficient

$$\rho_{ij} = \frac{\mu_{ij}}{\sigma(x_i)\sigma(x_j)} \tag{A.19.3}$$

$$V = \begin{bmatrix} \sigma^2(x_1) & \rho_{12}\sigma(x_1)\sigma(x_2) & \rho_{13}\sigma(x_1)\sigma(x_3) & \cdots & \rho_{1p}\sigma(x_1)\sigma(x_p) \\ \rho_{12}\sigma(x_1)\sigma(x_2) & \sigma^2(x_2) & \rho_{23}\sigma(x_2)\sigma(x_3) & \cdots & \rho_{2p}\sigma(x_2)\sigma(x_p) \\ \vdots & & & & \\ \rho_{1p}\sigma(x_1)\sigma(x_p) & \rho_{2p}\sigma(x_2)\sigma(x_p) & \cdots & \cdots & \sigma^2(x_p) \end{bmatrix}.$$

Equation (A.19.1) is called the multivariate normal density function of a vector **X** with mean **Λ** and variance-covariance V.

The following theorems illustrate useful properties of multivariate normal density functions.

Theorem 1: If the random variables $x_1, x_2 \ldots x_p$ are jointly normal, the joint marginal distribution of any subset of $s < p$ of the random variables is the s-variate normal. For a proof, see Chapter 3 of Graybill (1961).

Theorem 2: If the $p \times 1$ vector **X** has the multivariate normal distribution with mean **Λ** and covariance V, then the components, x_i, are jointly independent if and only if the covariance of x_i and x_j for all $i \neq j$ is zero, that is, if and only if the covariance matrix V is diagonal.

To prove this, we must show that the joint density function factors into the product of the marginal density functions, which by Theorem 1 are also normal. Hence, we must show that

$$f(x_1, x_2 \ldots x_p) = f(x_1)f(x_2) \ldots f(x_p) \tag{A.19.4}$$

where

$$f(x_1, x_2 \ldots x_p) = \kappa \, e^{-1/2(\mathbf{X}-\mathbf{\Lambda})^T V^{-1}(\mathbf{X}-\mathbf{\Lambda})}$$

and

$$f(x_i) = \frac{1}{\sqrt{2\pi V_{ii}}} e^{-1/2\frac{(x_i - \lambda_i)^2}{V_{ii}}}$$

$$\kappa = \frac{1}{(2\pi)^{p/2}|V|^{1/2}}$$

where V_{ii} indicates the diagonal elements of V.

If $V_{ij} = 0$ for $i \neq j$, we have

$$(\mathbf{X} - \mathbf{\Lambda})^T V^{-1}(\mathbf{X} - \mathbf{\Lambda}) = \sum_{i=1}^{p}(x_i - \lambda_i)^2 V_{ii}^{-1}.$$

Also,

$$|V|^{1/2} = \prod_{i=1}^{p}(V_{ii})^{1/2}$$

since V is a diagonal matrix. Consequently, the joint density function can be factored as indicated by Eq. (A.19.4) and the elements of X are independent.

A.19.1 THE CONDITIONAL DISTRIBUTION FOR MULTIVARIATE NORMAL VARIABLES

Theorem 3: If the $p \times 1$ vector $\mathbf{X}$ is normally distributed with mean Λ and covariance V and if the vector $\mathbf{X}$ is partitioned into two subvectors such that

$$\mathbf{X} = \begin{bmatrix} \mathbf{X}_1 \\ \mathbf{X}_2 \end{bmatrix}, \Lambda = \begin{bmatrix} \Lambda_1 \\ \Lambda_2 \end{bmatrix}, \text{ and } V = \begin{bmatrix} V_{11} & V_{12} \\ V_{21} & V_{22} \end{bmatrix}$$

then the conditional distribution of the $q \times 1$ vector $\mathbf{X}_1$, given that the vector $\mathbf{X}_2 = \mathbf{x}_2$, is the multivariate normal distribution with mean $\Lambda_1 + V_{12}V_{22}^{-1}(\mathbf{x}_2 - \Lambda_2)$ and the covariance matrix $(V_{11} - V_{12}V_{22}^{-1}V_{21})$ (Graybill, 1961); that is,

$$g\left(\mathbf{X}_1/\mathbf{X}_2 = \mathbf{x}_2\right) =$$
$$\frac{1}{\kappa^*}e^{-1/2[\mathbf{X}_1 - \Lambda_1 - V_{12}V_{22}^{-1}(\mathbf{x}_2 - \Lambda_2)]^T[V_{11} - V_{12}V_{22}^{-1}V_{21}]^{-1}[\mathbf{X}_1 - \Lambda_1 - V_{12}V_{22}^{-1}(\mathbf{x}_2 - \Lambda_2)]}$$

(A.19.5)

where

$$\kappa^* = (2\pi)^{q/2}|V_{11} - V_{12}V_{22}^{-1}V_{21}|^{1/2}.$$

Theorem 4: The covariance matrix of the conditional distribution of $\mathbf{X}_1$, given $\mathbf{X}_2 = \mathbf{x}_2$ does not depend on $\mathbf{x}_2$.

Proof: The proof of this is obvious from an examination of Eq. (A.19.5); that is, $V_{\mathbf{X}_1/\mathbf{X}_2 = \mathbf{x}_2} = V_{11} - V_{12}V_{22}^{-1}V_{21}$. From Theorem 3, we also have

$$E\left(\mathbf{X}_1/\mathbf{X}_2 = \mathbf{x}_2\right) = \Lambda_1 + V_{12}V_{22}^{-1}(\mathbf{x}_2 - \Lambda_2). \qquad (A.19.6)$$

If we were attempting to estimate $\mathbf{X}_1$, its mean value given by Eq. (A.19.6) would be a likely value to choose. Also, because the covariance of the conditional density function is independent of $\mathbf{x}_2$, we could generate the covariance without actually knowing the values of $\mathbf{X}_2$. This would allow us to perform an accuracy assessment of $\mathbf{X}_1$ without knowing the values of $\mathbf{X}_2$.

A.20 THE CENTRAL LIMIT THEOREM

If we have n independent random variables x_i, $i = 1 \ldots n$ that are identically distributed with common means $E[x_i] = \lambda$ and (finite) variance $\sigma^2 (x_i) = \sigma^2$, and we form the sum

$$W = x_1 + x_2 + \ldots x_n$$

whose mean and variance are given by

$$E[W] = n\lambda$$
$$E(W - E(W))^2 = \sigma^2(W) = n\sigma^2 .$$

The central limit theorem states that as $n \to \infty$ the standardized random variable of the sum

$$Z = \frac{W - E(W)}{\sigma(W)}$$

is normally distributed with mean 0 and variance 1 (Freeman, 1963). The important point is that W also is distributed normally. Hence, any random variable made up of the sum of enough independent random components from the same distribution will be distributed normally. Furthermore, if $n > 30$, Z (and W) will be distributed normally no matter what the shape of the distribution (Walpole *et al.*, 2002).

Another way of stating the theorem is that if sets of random samples are taken from any population, the means of these samples will tend to be distributed normally as the size of the samples becomes large (Davis, 1986).

The utility of the central limit theorem for orbit determination is that it gives some assurance that observation errors in tracking systems will tend to be distributed normally. This is because tracking system errors are usually the sum of a number of small random errors from a number of sources, including the hardware, the electronics, the mountings, and so on. This is a fundamental assumption in our development of statistical estimation algorithms.

A.21 BAYES THEOREM

One form of Bayes theorem is simply a statement of the conditional density functions given by Eq. (A.12.2),

$$g(x/y) = \frac{f(x, y)}{h(y)}$$

and

$$h(y/x) = \frac{f(x, y)}{g(x)} . \qquad (A.21.1)$$

Hence,

$$g\left(x/y\right) = \frac{h\left(y/x\right)g\left(x\right)}{h\left(y\right)}.$$

The last equations for $g\left(x/y\right)$ is the most elementary form of Bayes theorem. It is a useful starting point in the development of statistically based estimation criteria.

If we define X to be the state vector and Y to be the observation vector, then

$g\left(x/y\right) \equiv$ *a posteriori* density function

$h\left(y/x\right) \equiv$ *a priori* density function.

From a Bayesian viewpoint, we wish to develop a filter to propagate as a function of time the probability density function of the desired quantities conditioned on knowledge of the actual data coming from the measurement devices. Once such a conditional density function is propagated, the optimal estimate can be defined. Possible choices for the optimal estimate include:

1. The *mean* – the "center of the probability mass" distribution.

2. The *mode* – the value of x that has the highest probability, locating the peak of the density function.

3. The *median* – the value of x such that half the probability weight lies to the left and half to the right.

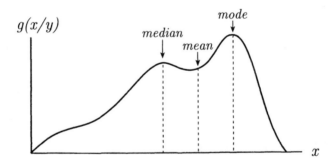

By generating the density function, some judgment can be made as to which criterion defines the most reasonable estimate for a given purpose.

It is useful to examine the difference between the mean and median of a density function. Recall that the mean is defined as

$$\lambda = \int\limits_{-\infty}^{\infty} x f\left(x\right)\, dx$$

and the median, M, is defined as

$$M = \int_{-\infty}^{M} f(x)dx = 1/2.$$ (A.21.2)

As an example, consider a bimodal distribution:

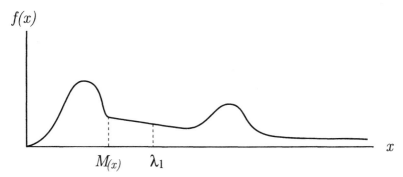

If we move the second mode further out without changing the density function between the origin and the median, the median remains fixed but the mean moves to the right.

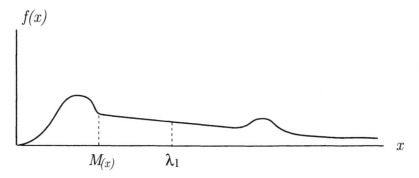

A.22 STOCHASTIC PROCESSES

Previously we have considered random variables that were not functions of time. Assume that we are given an experiment β specified by its outcomes η forming the space S and by the probabilities of these events. To every outcome η we now assign a time function $X(t, \eta)$. We have thus created a family of functions, one for each η. This family is called a *stochastic process*. A stochastic process is a function of two variables, t and η. The domain of η is S and the domain of t is a set of real numbers the time axis (Papoulis, 1991).

For a specific outcome η_i, the expression $X(t, \eta_i)$ signifies a single time function. For a specific time t_i, $X(t_i, \eta)$ is a quantity depending on η (i.e., a random variable). Finally, $X(t_i, \eta_i)$ is a mere number. In the future, $X(t, \eta)$ will be written $X(t)$.

As an example, consider a coin tossing experiment and define $X(t)$ so that

$$X(t) = \sin t, \quad \text{if } \eta = \text{heads}; X(t) = \cos t, \quad \text{if } \eta = \text{tails}.$$

Thus $X(t)$ consists of two regular curves; it is nevertheless a stochastic process.

From the preceding, we see that $X(t, \eta)$ represents four different things:

1. A family of time functions (t and η variable).

2. A single time function (t variable, η fixed).

3. A random variable (t fixed, η variable).

4. A single number (t fixed, η fixed).

We shall assume that $X(t)$ is a real process. For a specific t, $X(t)$ is a random variable. As in the case of random variables, we have a distribution function given by

$$F(x, t) = p(X(t) \leq x). \tag{A.22.1}$$

This is interpreted as: Given two real numbers x and t, the function $F(x, t)$ equals the probability of the event $\{X(t) \leq x\}$ consisting of all outcomes η such that at the specified time t, the functions $X(t)$ of our process do not exceed the given number x.

Associated with the distribution function is the corresponding density function

$$f(x, t) = \frac{\partial F(x, t)}{\partial x}. \tag{A.22.2}$$

These results hold at a given time and are known as the first order distribution and density function, respectively.

Now given two time instances t_1 and t_2, consider the random variable $X(t_1)$ and $X(t_2)$. Their joint distribution depends, in general, on t_1 and t_2 and will be denoted by $F(x_1, x_2, t_1, t_2)$

$$F(x_1, x_2, t_1, t_2) = p\{X(t_1) \leq x_1, X(t_2) \leq x_2\} \tag{A.22.3}$$

and will be called the second-order distribution function of the process $X(t)$. The corresponding density function is

$$f(x_1, x_2, t_1, t_2) = \frac{\partial^2 F}{\partial x_1 \, \partial x_2} (x_1, x_2, t_1, t_2). \tag{A.22.4}$$

The marginal distribution and density functions are given by

$$F(x_1, \infty, t_1, t_2) = F(x_1, t_1),$$

$$f(x_1, t_1) = \int\limits_{-\infty}^{\infty} f(x_1, x_2, t_1, t_2) \, dx_2. \qquad \text{(A.22.5)}$$

The conditional density is

$$f(X(t_1)/X(t_2) = (x_2)) = \frac{f(x_1, x_2, t_1, t_2)}{f(x_2, t_2)}. \qquad \text{(A.22.6)}$$

A.22.1 DEFINITIONS FOR STOCHASTIC PROCESSES

Given a stochastic process $X(t)$, its mean $\eta(t)$ is given by the expected value of the random variable $X(t)$,

$$\eta(t) = E\{X(t)\} = \int\limits_{-\infty}^{\infty} x f(x, t) dx. \qquad \text{(A.22.7)}$$

It is in general a function of time.

The *autocorrelation* $R(t_1, t_2)$ of a process $X(t)$ is the joint moment of the random variables $X(t_1)$ and $X(t_2)$

$$R(t_1, t_2) = E\{X(t_1)X(t_2)\}$$

$$= \int\limits_{-\infty}^{\infty} \int\limits_{-\infty}^{\infty} x_1 \, x_2 \, f(x_1, x_2, t_1, t_2) \, dx_1 \, dx_2 \qquad \text{(A.22.8)}$$

and is a function of t_1 and t_2. The autocorrelation is analogous to the moment about the origin.

The *autocovariance* of $X(t)$ is the covariance of the random variables $X(t_1)$ and $X(t_2)$,

$$C(t_1, t_2) = E\{[X(t_1) - \eta(t_1)][X(t_2) - \eta(t_2)]\}. \qquad \text{(A.22.9)}$$

If we have two stochastic processes $X(t)$ and $Y(t)$, these become the cross-correlation and cross-covariance, respectively. From Eq. (A.22.9), it is seen that

$$C(t_1, t_2) = R(t_1, t_2) - \eta(t_1)\eta(t_2). \qquad \text{(A.22.10)}$$

The variance of the random variable $X(t_1)$ is given by $(t_1 = t_2)$

$$C(t_1, t_1) = R(t_1, t_1) - \eta^2(t_1). \qquad \text{(A.22.11)}$$

Two random processes $X(t)$, $Y(t)$ are called uncorrelated if for any t_1 and t_2 we have

$$R_{XY}(t_1, t_2) = E[X(t_1) Y(t_2)] = \eta_x(t_1) \eta_y(t_2); \qquad (A.22.12)$$

that is,

$$C_{XY}(t_1, t_2) = 0. \qquad (A.22.13)$$

They are called *orthogonal* if

$$R_{XY}(t_1, t_2) = 0. \qquad (A.22.14)$$

A stochastic process $X(t)$ is *stationary* in the strict sense if its statistics are not affected by a shift in the time origin. This means that the two processes

$$X(t) \quad \text{and} \quad X(t + \epsilon)$$

have the same statistics for any ϵ.

As a result

$$f(x, t) = f(x, t + \epsilon) \qquad (A.22.15)$$

and since this is true for every ϵ, we must have the first-order density

$$f(x, t) = f(x) \qquad (A.22.16)$$

independent of time, and

$$E[X(t)] = \eta, \quad \text{a constant.}$$

The density of order two must be such that

$$f(x_1, x_2, t_1, t_2) = f(x_1, x_2, t_1 + \epsilon, t_2 + \epsilon). \qquad (A.22.17)$$

Because this must be true for any ϵ, it must be a function of only $t_1 - t_2$. This can be seen by noting that $t_1 + \epsilon - (t_2 + \epsilon) = t_1 - t_2$ is not dependent on ϵ, hence

$$f(x_1, x_2, t_1, t_2) = f(x_1, x_2, \tau) \qquad (A.22.18)$$

where $\tau = (t_1 - t_2)$. Thus, $f(x_1, x_2, \tau)$ is the joint density function of the random variables

$$X(t + \tau) \quad \text{and} \quad X(\tau). \qquad (A.22.19)$$

A.23 REFERENCES

Davis, J. C., *Statistics and Data Analysis in Geology*, John Wiley & Sons Inc., New York, 1986.

Freeman, H., *Introduction to Statistical Inference*, Addison-Wesley, 1963.

Graybill, F. A., *An Introduction to Linear Statistical Models*, McGraw-Hill, New York, 1961.

Helstrom, C. W., *Probability and Stochastic Processes for Engineers*, MacMillan, New York, 1984.

Papoulis, A., *Probability, Random Variables, and Stochastic Processes*, McGraw-Hill, New York, 1991.

Walpole, R. E., R. H. Myers, S. L. Myers, and Y. Keying, *Probability and Statistics for Engineers and Scientists*, Prentice Hall, Englewood Cliffs, NJ, 2002.

Appendix B

Review of Matrix Concepts

B.1 INTRODUCTION

The following matrix notation, definitions, and theorems are used extensively in this book. Much of this material is based on Graybill (1961).

- A matrix $\mathbf{A}$ will have elements denoted by a_{ij}, where i refers to the row and j to the column.

- $\mathbf{A}^T$ will denote the transpose of $\mathbf{A}$.

- $\mathbf{A}^{-1}$ will denote the inverse of $\mathbf{A}$.

- $|\mathbf{A}|$ will denote the determinant of $\mathbf{A}$.

- The dimension of a matrix is the number of its rows by the number of its columns.

- An $n \times m$ matrix $\mathbf{A}$ will have n rows and m columns.

- If $m = 1$, the matrix will be called an $n \times 1$ vector.

- Given the matrices $\mathbf{A} = (a_{ij})$ and $\mathbf{B} = (b_{ij})$, the product $\mathbf{AB} = \mathbf{C} = (c_{ij})$ is defined as the matrix $\mathbf{C}$ with the pqth element equal to

$$\sum_{s=1}^{m} a_{ps}\, b_{sq} \qquad (\text{B.1.1})$$

where m is the column dimension of $\mathbf{A}$ and the row dimension of $\mathbf{B}$.

- Given

$$\mathbf{A} = \begin{bmatrix} \mathbf{A}_{11} & \mathbf{A}_{12} \\ \mathbf{A}_{21} & \mathbf{A}_{22} \end{bmatrix}$$

and

$$\mathbf{B} = \begin{bmatrix} \mathbf{B}_{11} & \mathbf{B}_{12} \\ \mathbf{B}_{21} & \mathbf{B}_{22} \end{bmatrix}$$

then

$$\mathbf{AB} = \begin{bmatrix} \mathbf{A}_{11} & \mathbf{A}_{12} \\ \mathbf{A}_{21} & \mathbf{A}_{22} \end{bmatrix} \begin{bmatrix} \mathbf{B}_{11} & \mathbf{B}_{12} \\ \mathbf{B}_{21} & \mathbf{B}_{22} \end{bmatrix}$$

$$= \begin{bmatrix} \mathbf{A}_{11}\mathbf{B}_{11} + \mathbf{A}_{12}\mathbf{B}_{21} & \mathbf{A}_{11}\mathbf{B}_{12} + \mathbf{A}_{12}\mathbf{B}_{22} \\ \mathbf{A}_{21}\mathbf{B}_{11} + \mathbf{A}_{22}\mathbf{B}_{21} & \mathbf{A}_{21}\mathbf{B}_{12} + \mathbf{A}_{22}\mathbf{B}_{22} \end{bmatrix} \quad \text{(B.1.2)}$$

provided the elements of **A** and **B** are conformable.

- For **AB** to be defined, the number of columns in **A** must equal the number of rows in **B**.

- For **A** + **B** to be defined, **A** and **B** must have the same dimension.

- The transpose of $\mathbf{A}^T$ equals **A**; that is, $(\mathbf{A}^T)^T = \mathbf{A}$.

- The inverse of $\mathbf{A}^{-1}$ is **A**; that is, $(\mathbf{A}^{-1})^{-1} = \mathbf{A}$.

- The transpose and inverse symbols may be permuted; that is, $(\mathbf{A}^T)^{-1} = (\mathbf{A}^{-1})^T$.

- $(\mathbf{AB})^T = \mathbf{B}^T \mathbf{A}^T$.

- $(\mathbf{AB})^{-1} = \mathbf{B}^{-1} \mathbf{A}^{-1}$ if **A** and **B** are each nonsingular.

- A scalar commutes with every matrix; that is, $k\,\mathbf{A} = \mathbf{A}\,k$.

- For any matrix **A**, we have $\mathbf{IA} = \mathbf{AI} = \mathbf{A}$.

- All diagonal matrices of the same dimension are commutative.

- If $\mathbf{D}_1$ and $\mathbf{D}_2$ are diagonal matrices, then the product is diagonal.

- If **X** and **Y** are vectors and if **A** is a nonsingular matrix and if the equation $\mathbf{Y} = \mathbf{AX}$ holds, then $\mathbf{X} = \mathbf{A}^{-1}\mathbf{Y}$.

B.2 RANK

- The *rank* of a matrix is the dimension of its largest square nonsingular submatrix; that is, one whose determinant is nonzero.

- The rank of the product $\mathbf{AB}$ of the two matrices $\mathbf{A}$ and $\mathbf{B}$ is less than or equal to the rank of $\mathbf{A}$ and is less than or equal to the rank of $\mathbf{B}$.

- The rank of the sum of $\mathbf{A} + \mathbf{B}$ is less than or equal to the rank of $\mathbf{A}$ plus the rank of $\mathbf{B}$.

- If $\mathbf{A}$ is an $n \times n$ matrix and if $|\mathbf{A}| = 0$, then the rank of $\mathbf{A}$ is less than n.

- If the rank of $\mathbf{A}$ is less than n, then all the rows of $\mathbf{A}$ are not independent; likewise, all the columns of $\mathbf{A}$ are not independent ($\mathbf{A}$ is $n \times n$).

- If the rank of $\mathbf{A}$ is $m \leq n$, then the number of linearly independent rows is m; also, the number of linearly independent columns is m ($\mathbf{A}$ is $n \times n$).

- If $\mathbf{A}^T \mathbf{A} = 0$, then $\mathbf{A} = 0$.

- The rank of a matrix is unaltered by multiplication by a nonsingular matrix; that is, if $\mathbf{A}$, $\mathbf{B}$, and $\mathbf{C}$ are matrices such that $\mathbf{AB}$ and $\mathbf{BC}$ exist and if $\mathbf{A}$ and $\mathbf{C}$ are nonsingular, then $\rho(\mathbf{AB}) = \rho(\mathbf{BC}) = \rho(\mathbf{B})$. $\rho(\mathbf{B})$ = rank of $\mathbf{B}$.

- If the product $\mathbf{AB}$ of two square matrices is 0, then $\mathbf{A} = 0, \mathbf{B} = 0$, or $\mathbf{A}$ and $\mathbf{B}$ are both singular.

- If $\mathbf{A}$ and $\mathbf{B}$ are $n \times n$ matrices of rank r and s, respectively, then the rank of $\mathbf{AB}$ is greater than or equal to $r + s - n$.

- The rank of $\mathbf{AA}^T$ equals the rank of $\mathbf{A}^T \mathbf{A}$, equals the rank of $\mathbf{A}$, equals the rank of $\mathbf{A}^T$.

B.3 QUADRATIC FORMS

- The rank of the quadratic form $\mathbf{Y}^T \mathbf{AY}$ is defined as the rank of the matrix $\mathbf{A}$ where $\mathbf{Y}$ is a vector and $\mathbf{Y} \neq 0$.

- The quadratic form $\mathbf{Y}^T \mathbf{AY}$ is said to be *positive definite* if and only if $\mathbf{Y}^T \mathbf{AY} > 0$ for all vectors $\mathbf{Y}$ where $\mathbf{Y} \neq 0$.

- A quadratic form $\mathbf{Y}^T \mathbf{AY}$ is said to be *positive semidefinite* if and only if $\mathbf{Y}^T \mathbf{AY} \geq 0$ for all $\mathbf{Y}$, and $\mathbf{Y}^T \mathbf{AY} = 0$ for some vector $\mathbf{Y} \neq 0$.

- A quadratic form $\mathbf{Y}^T \mathbf{A} \mathbf{Y}$ that may be either positive definite or positive semidefinite is called *nonnegative definite*.

- The matrix $\mathbf{A}$ of a quadratic form $\mathbf{Y}^T \mathbf{A} \mathbf{Y}$ is said to be positive definite (semidefinite) when the quadratic form is positive definite (semidefinite).

- If $\mathbf{P}$ is a nonsingular matrix and if $\mathbf{A}$ is positive definite (semidefinite), then $\mathbf{P}^T \mathbf{A} \mathbf{P}$ is positive definite (semidefinite).

- A necessary and sufficient condition for the symmetric matrix $\mathbf{A}$ to be positive definite is that there exist a nonsingular matrix $\mathbf{P}$ such that $\mathbf{A} = \mathbf{P} \mathbf{P}^T$.

- A necessary and sufficient condition that the matrix $\mathbf{A}$ be positive definite, where

$$
\mathbf{A} = \begin{bmatrix}
a_{11} & a_{12} & \dots & a_{1n} \\
a_{21} & a_{22} & \dots & a_{2n} \\
\vdots & \vdots & \dots & \vdots \\
a_{n1} & a_{n2} & \dots & a_{nn}
\end{bmatrix}
$$

is that the following inequalities hold:

$$
a_{11} > 0, \quad \begin{vmatrix} a_{11} & a_{12} \\ a_{21} & a_{22} \end{vmatrix} > 0, \dots \quad \begin{vmatrix}
a_{11} & a_{12} & \dots & a_{1n} \\
a_{21} & a_{22} & \dots & a_{2n} \\
\vdots & \vdots & \dots & \vdots \\
a_{n1} & a_{n2} & \dots & a_{nn}
\end{vmatrix} > 0.
$$

- If $\mathbf{A}$ is an $m \times n$ matrix of rank $n < m$, then $\mathbf{A}^T \mathbf{A}$ is positive definite and $\mathbf{A} \mathbf{A}^T$ is positive semidefinite.

- If $\mathbf{A}$ is an $m \times n$ matrix of rank $k < n$ and $k < m$, then $\mathbf{A}^T \mathbf{A}$ and $\mathbf{A} \mathbf{A}^T$ are each positive semidefinite.

- A matrix that may be either positive definite or positive semidefinite is said to be nonnegative definite.

- If $\mathbf{A}$ and $\mathbf{B}$ are symmetric conformable matrices, $\mathbf{A}$ is said to be greater than $\mathbf{B}$ if $\mathbf{A} - \mathbf{B}$ is nonnegative definite.

B.4   DETERMINANTS

- For each square matrix $\mathbf{A}$, there is a uniquely defined scalar called the *determinant* of $\mathbf{A}$ and denoted by $|\mathbf{A}|$.

- The determinant of a diagonal matrix is equal to the product of the diagonal elements.

- If $\mathbf{A}$ and $\mathbf{B}$ are $n \times n$ matrices, then $|\mathbf{AB}| = |\mathbf{BA}| = |\mathbf{A}| |\mathbf{B}|$.

- If $\mathbf{A}$ is singular if and only if $|\mathbf{A}| = 0$.

- If $\mathbf{C}$ is an $n \times n$ matrix such that $\mathbf{C}^T \mathbf{C} = \mathbf{I}$, then $\mathbf{C}$ is said to be an *orthogonal* matrix, and $\mathbf{C}^T = \mathbf{C}^{-1}$.

- If $\mathbf{C}$ is an orthogonal matrix, then $|\mathbf{C}| = +1$ or $|\mathbf{C}| = -1$.

- If $\mathbf{C}$ is an orthogonal matrix, then $|\mathbf{C}^T \mathbf{AC}| = |\mathbf{A}|$.

- The determinant of a positive definite matrix is positive.

- The determinant of a triangular matrix is equal to the product of the diagonal elements.

- The determinant of a matrix is equal to the product of its eigenvalues.

- $|\mathbf{A}| = |\mathbf{A}^T|$

- $|\mathbf{A}^{-1}| = 1/|\mathbf{A}|$, if $|\mathbf{A}| \neq 0$.

- If $\mathbf{A}$ is a square matrix such that

$$\mathbf{A} = \begin{bmatrix} \mathbf{A}_{11} & \mathbf{A}_{12} \\ \mathbf{A}_{21} & \mathbf{A}_{22} \end{bmatrix}$$

where $\mathbf{A}_{11}$ and $\mathbf{A}_{22}$ are square matrices, and if $\mathbf{A}_{12} = 0$ or $\mathbf{A}_{21} = 0$, then $|\mathbf{A}| = |\mathbf{A}_{11}| |\mathbf{A}_{22}|$.

- If $\mathbf{A}_1$ and $\mathbf{A}_2$ are symmetric and $\mathbf{A}_2$ is positive definite and if $\mathbf{A}_1 - \mathbf{A}_2$ is positive semidefinite (or positive definite), then $|\mathbf{A}_1| \geq |\mathbf{A}_2|$.

B.5 MATRIX TRACE

- The *trace* of a matrix $\mathbf{A}$, which will be written $tr(\mathbf{A})$, is equal to the sum of the diagonal elements of $\mathbf{A}$; that is,

$$tr(\mathbf{A}) = \sum_{i=1}^{n} a_{ii}. \tag{B.5.1}$$

- tr(AB) = tr(BA).

- tr(ABC) = tr(CAB) = tr(BCA); that is, the trace of the product of matrices is invariant under any cyclic permutation of the matrices.

- Note that the trace is defined only for a square matrix.

- If $\mathbf{C}$ is an orthogonal matrix, $\text{tr}\,(\mathbf{C}^T\mathbf{A}\mathbf{C}) \;=\; \text{tr}\,(\mathbf{A})$.

B.6   EIGENVALUES AND EIGENVECTORS

- A *characteristic root (eigenvalue)* of a $p \times p$ matrix $\mathbf{A}$ is a scalar λ such that $\mathbf{A}\mathbf{X} \;=\; \lambda\mathbf{X}$ for some vector $\mathbf{X} \neq 0$.

- The vector $\mathbf{X}$ is called the *characteristic vector (eigenvector)* of the matrix $\mathbf{A}$.

- The characteristic root of a matrix $\mathbf{A}$ can be defined as a scalar λ such that $|\mathbf{A} - \lambda\mathbf{I}| \;=\; 0$.

- $|\mathbf{A} - \lambda\mathbf{I}|$ is a pth degree polynomial in λ.

- This polynomial is called the *characteristic polynomial*, and its roots are the characteristic roots of the matrix $\mathbf{A}$.

- The number of nonzero characteristic roots of a matrix $\mathbf{A}$ is equal to the rank of $\mathbf{A}$.

- The characteristic roots of $\mathbf{A}$ are identical with the characteristic roots of $\mathbf{C}\mathbf{A}\mathbf{C}^{-1}$. If $\mathbf{C}$ is an orthogonal matrix, it follows that $\mathbf{A}$ and $\mathbf{C}\mathbf{A}\mathbf{C}^T$ have identical characteristic roots; that is, $\mathbf{C}^T \;=\; \mathbf{C}^{-1}$.

- The characteristic roots of a symmetric matrix are real; that is, if $\mathbf{A} \;=\; \mathbf{A}^T$, the characteristic polynomial of $|\mathbf{A} \;=\; \lambda\mathbf{I}| \;=\; 0$ has all real roots.

- The characteristic roots of a positive definite matrix $\mathbf{A}$ are positive; the characteristic roots of a positive semidefinite matrix are nonnegative.

B.7   THE DERIVATIVES OF MATRICES AND VECTORS

- Let $\mathbf{X}$ be an $n \times 1$ vector and let Z be a scalar that is a function of $\mathbf{X}$. The derivative of Z with respect to the vector $\mathbf{X}$, which will be written $\partial Z\,/\,\partial\mathbf{X}$, will mean the $1 \times n$ row vector*

$$\mathbf{C} \equiv \left[\begin{array}{cccc} \frac{\partial Z}{\partial x_1} & \frac{\partial Z}{\partial x_2} & \cdots & \frac{\partial Z}{\partial x_n} \end{array}\right]. \tag{B.7.1}$$

---

*Generally this partial derivative would be defined as a column vector. However, it is defined as a row vector here because we have defined $\widetilde{H} = \frac{\partial G(\mathbf{X})}{\partial\mathbf{X}}$ as a row vector in the text.

- If $\mathbf{X}$, $\mathbf{C}$, and Z are as defined previously, then

$$\partial Z / \partial \mathbf{X} = \mathbf{C}. \tag{B.7.2}$$

- If $\mathbf{A}$ and $\mathbf{B}$ are $m \times 1$ vectors, which are a function of the $n \times 1$ vector $\mathbf{X}$, and we define

$$\frac{\partial (\mathbf{A}^T \mathbf{B})}{\partial \mathbf{X}}$$

to be a row vector as in Eq. (B.7.1), then

$$\partial (\mathbf{A}^T \mathbf{B}) / \partial \mathbf{X} = \mathbf{B}^T \frac{\partial \mathbf{A}}{\partial \mathbf{X}} + \mathbf{A}^T \frac{\partial \mathbf{B}}{\partial \mathbf{X}} \tag{B.7.3}$$

where

$$\frac{\partial \mathbf{A}}{\partial \mathbf{X}}$$

is an $m \times n$ matrix whose ij element is

$$\frac{\partial A_i}{\partial X_j}$$

and

$$\frac{\partial (\mathbf{A}^T \mathbf{B})}{\partial \mathbf{X}}$$

is a $1 \times n$ row vector.[†]

- If $\mathbf{A}$ is an $m \times 1$ vector that is a function of the $n \times 1$ vector $\mathbf{X}$, and W is an $m \times m$ symmetric matrix such that

$$Z = \mathbf{A}^T W \mathbf{A} = \mathbf{A}^T W^{1/2} W^{1/2} \mathbf{A}.$$

Let $\mathbf{B} \equiv W^{1/2} \mathbf{A}$, then

$$Z = \mathbf{B}^T \mathbf{B}.$$

From Eq. (B.7.3)

$$\frac{\partial Z}{\partial \mathbf{X}} = 2 \mathbf{B}^T \frac{\partial \mathbf{B}}{\partial \mathbf{X}} \tag{B.7.4}$$

where

$$\frac{\partial \mathbf{B}}{\partial \mathbf{X}} = W^{1/2} \frac{\partial \mathbf{A}}{\partial \mathbf{X}}.$$

[†]If $\frac{\partial Z}{\partial \mathbf{X}}$ is defined to be a column vector, $\frac{\partial (\mathbf{A}^T B)}{\partial \mathbf{X}}$ would be given by the transpose of Eq. (B.7.3).

- Let $\mathbf{A}$ be a $p \times 1$ vector, $\mathbf{B}$ be a $q \times 1$ vector, and C be a $p \times q$ matrix whose ij^{th} element equals c_{ij}. Let

$$Z = \mathbf{A}^T C \mathbf{B} = \sum_{m=1}^{q} \sum_{n=1}^{p} a_n c_{nm} b_m. \tag{B.7.5}$$

Then $\partial Z / \partial C = \mathbf{A}\mathbf{B}^T$.

Proof: $\partial Z / \partial C$ will be a $p \times q$ matrix whose ij^{th} element is $\partial Z / \partial c_{ij}$.

Assuming that C is not symmetric and that the elements of C are independent,

$$\frac{\partial Z}{\partial c_{ij}} = \frac{\partial \left(\sum\limits_{m=1}^{q} \sum\limits_{n=1}^{p} a_n c_{nm} b_m \right)}{\partial c_{ij}} = a_i \, b_j. \tag{B.7.6}$$

Thus the ij^{th} element of $\partial Z / \partial C$ is $a_i b_j$. Therefore, it follows that

$$\frac{\partial Z}{\partial C} = \mathbf{A}\mathbf{B}^T.$$

- The derivative of a matrix product with respect to a scalar is given by

$$\frac{d}{dt}\{\mathbf{A}(t)\mathbf{B}(t)\} = \frac{d\mathbf{A}(t)}{dt}\mathbf{B}(t) + \mathbf{A}(t)\frac{d\mathbf{B}(t)}{dt}. \tag{B.7.7}$$

See Graybill (1961) for additional discussion of the derivatives of matrices and vectors.

B.8   MAXIMA AND MINIMA

- If $y = f(x_1, x_2, ..., x_n)$ is a function of n variables and if all partial derivatives $\partial y / \partial x_i$ are continuous, then y attains its maxima and minima only at the points where

$$\frac{\partial y}{\partial x_1} = \frac{\partial y}{\partial x_2} = ... = \frac{\partial y}{\partial x_n} = 0. \tag{B.8.1}$$

- If $f(x_1, x_2, ..., x_n)$ is such that all the first and second partial derivatives are continuous, then at the point where

$$\frac{\partial f}{\partial x_1} = \frac{\partial f}{\partial x_2} = ... = \frac{\partial f}{\partial x_n} = 0 \tag{B.8.2}$$

the function has

- a minimum, if the matrix $\mathbf{K}$, where the ijth element of $\mathbf{K}$ is $\partial^2 f / \partial x_i \partial x_j$, is positive definite.
- a maximum, if the matrix $-\mathbf{K}$ is positive definite.

In these two theorems on maxima and minima, remember that the x_i are independent variables.

- If the x_i are not independent, that is, there are constraints relating them, we use the method of Lagrange multipliers. Suppose that we have a function $f(x_1, x_2, \ldots, x_n)$ we wish to maximize (or minimize) subject to the constraint that $h(x_1, x_2, \ldots, x_n) = 0$. The equation $h = 0$ describes a surface in space and the problem is one of maximizing $f(x_1, x_2, \ldots, x_n)$ as $x_1, x_2, \ldots, x_n$ vary on the curve of intersection of the two surfaces. At a maximum point the derivative of f must be zero along the intersection curve; that is, the directional derivative along the tangent must be zero. The directional derivative is the component of the vector ∇f along the tangent. Hence, ∇f must lie in a plane normal to the intersection curve at this point. This plane must also contain ∇h; that is, ∇f and ∇h are coplanar at this point. Hence, there must exist a scalar λ such that

$$\nabla f + \lambda \nabla h = 0 \qquad \text{(B.8.3)}$$

at the maximum point. If we define

$$F \equiv f + \lambda h$$

then Eq. (B.8.3) is equivalent to $\nabla F = 0$. Hence,

$$\frac{\partial F}{\partial x_1} = \frac{\partial F}{\partial x_2} = \cdots = \frac{\partial F}{\partial x_n} = 0.$$

These n equations together with $h = 0$ provide us with $n + 1$ equations and $n + 1$ unknowns $(x_1, x_2, \ldots, x_n, \lambda)$. We have assumed that all first partial derivatives are continuous and that $\partial h / \partial x_i \neq 0$ for all i at the point.

- If there are additional constraints we introduce additional Lagrange multipliers in Eq. (B.8.3); for example,

$$\nabla f + \lambda_1 \nabla h_1 + \lambda_2 \nabla h_2 + \cdots \lambda_k \nabla h_k = 0. \qquad \text{(B.8.4)}$$

B.9   USEFUL MATRIX INVERSION THEOREMS

Theorem 1: Let $\mathbf{A}$ and $\mathbf{B}$ be $n \times n$ positive definite (PD) matrices. If $\mathbf{A}^{-1} + \mathbf{B}^{-1}$ is PD, then $\mathbf{A} + \mathbf{B}$ is PD and

$$(\mathbf{A} + \mathbf{B})^{-1} = \mathbf{B}^{-1} (\mathbf{A}^{-1} + \mathbf{B}^{-1})^{-1} \mathbf{A}^{-1}$$

$$= \mathbf{A}^{-1} (\mathbf{A}^{-1} + \mathbf{B}^{-1})^{-1} \mathbf{B}^{-1}. \qquad (B.9.1)$$

Proof: From the identity

$$(\mathbf{A} + \mathbf{B})^{-1} = [\mathbf{A}(\mathbf{A}^{-1} + \mathbf{B}^{-1})\mathbf{B}]^{-1} = \mathbf{B}^{-1}(\mathbf{A}^{-1} + \mathbf{B}^{-1})^{-1}\mathbf{A}^{-1}$$

or

$$(\mathbf{A} + \mathbf{B})^{-1} = [\mathbf{B}(\mathbf{B}^{-1} + \mathbf{A}^{-1})\mathbf{A}]^{-1} = \mathbf{A}^{-1}(\mathbf{A}^{-1} + \mathbf{B}^{-1})^{-1}\mathbf{B}^{-1}.$$

Theorem 2: Let $\mathbf{A}$ and $\mathbf{B}$ be $n \times n$ PD matrices. If $\mathbf{A} + \mathbf{B}$ is PD, then $\mathbf{I} + \mathbf{AB}^{-1}$ and $\mathbf{I} + \mathbf{BA}^{-1}$ are PD and

$$\begin{aligned}(\mathbf{A} + \mathbf{B})^{-1} &= \mathbf{B}^{-1} - \mathbf{B}^{-1}(\mathbf{I} + \mathbf{AB}^{-1})^{-1}\mathbf{AB}^{-1} \\ &= \mathbf{A}^{-1} - \mathbf{A}^{-1}(\mathbf{I} + \mathbf{BA}^{-1})^{-1}\mathbf{BA}^{-1}. \qquad (B.9.2)\end{aligned}$$

Proof: From the identity

$$\mathbf{A}^{-1} = (\mathbf{A}^{-1} + \mathbf{B}^{-1}) - \mathbf{B}^{-1}$$

premultiply by $\mathbf{B}^{-1}(\mathbf{A}^{-1} + \mathbf{B}^{-1})^{-1}$ and use Theorem 1

$$\begin{aligned}\mathbf{B}^{-1}(\mathbf{A}^{-1} + \mathbf{B}^{-1})^{-1}\mathbf{A}^{-1} &= \mathbf{B}^{-1}(\mathbf{A}^{-1} + \mathbf{B}^{-1})^{-1}(\mathbf{A}^{-1} + \mathbf{B}^{-1}) \\ &\quad - \mathbf{B}^{-1}(\mathbf{A}^{-1} + \mathbf{B}^{-1})^{-1}\mathbf{B}^{-1} \\ &= \mathbf{B}^{-1} - \mathbf{B}^{-1}[\mathbf{A}^{-1}(\mathbf{I} + \mathbf{AB}^{-1})]^{-1}\mathbf{B}^{-1} \\ &= \mathbf{B}^{-1} - \mathbf{B}^{-1}(\mathbf{I} + \mathbf{AB}^{-1})^{-1}\mathbf{AB}^{-1}.\end{aligned}$$

The left-hand side of this equation is $(\mathbf{A} + \mathbf{B})^{-1}$ (from Theorem 1). Hence,

$$(\mathbf{A} + \mathbf{B})^{-1} = \mathbf{B}^{-1} - \mathbf{B}^{-1}(\mathbf{I} + \mathbf{AB}^{-1})^{-1}\mathbf{AB}^{-1}.$$

Theorem 3: If $\mathbf{A}$ and $\mathbf{B}$ are PD matrices of order n and m, respectively, and if $\mathbf{C}$ is of order $n \times m$, then

$$(\mathbf{C}^T\mathbf{A}^{-1}\mathbf{C} + \mathbf{B}^{-1})^{-1}\mathbf{C}^T\mathbf{A}^{-1} = \mathbf{BC}^T(\mathbf{A} + \mathbf{CBC}^T)^{-1} \qquad (B.9.3)$$

provided the inverse exists.

Proof: From the identity

$$\mathbf{C}^T(\mathbf{A}^{-1}\mathbf{CBC}^T + \mathbf{I})(\mathbf{I} + \mathbf{A}^{-1}\mathbf{CBC}^T)^{-1} \equiv \mathbf{C}^T$$

we have

$$(C^T A^{-1} CBC^T + C^T)(A^{-1}(A + CBC^T))^{-1} = C^T$$

or

$$(C^T A^{-1} C + B^{-1}) BC^T (A + CBC^T)^{-1} A = C^T.$$

Now premultiply by $(C^T A^{-1} C + B^{-1})^{-1}$ and postmultiply by A^{-1}, which yields

$$BC^T (A + CBC^T)^{-1} = (C^T A^{-1} C + B^{-1})^{-1} C^T A^{-1}.$$

Theorem 4: The Schur Identity or insideout rule. If A is a PD matrix of order n, and if B and C are any conformable matrices such that BC is order n, then

$$(A + BC)^{-1} = A^{-1} - A^{-1}B(I + CA^{-1}B)^{-1}CA^{-1}. \qquad (B.9.4)$$

Proof: Define

$$X = (A + BC)^{-1}.$$

Then

$$(A + BC) X = I \qquad (B.9.5)$$
$$AX + BCX = I.$$

Solve Eq. (B.9.5) for CX. First multiply by A^{-1} to yield

$$X + A^{-1}BCX = A^{-1}. \qquad (B.9.6)$$

Premultiply Eq. (B.9.6) by C

$$CX + CA^{-1}BCX = CA^{-1}.$$

Then

$$CX = (I + CA^{-1}B)^{-1}CA^{-1}. \qquad (B.9.7)$$

Substitute Eq. (B.9.7) into Eq. (B.9.6) to yield

$$X = (A + BC)^{-1} = A^{-1} - A^{-1}B(I + CA^{-1}B)^{-1}CA^{-1}.$$

B.10 REFERENCE

Graybill, F. A., *An Introduction to Linear Statistical Models*, McGraw-Hill, New York, 1961.

Appendix C

Equations of Motion

C.1 LAGRANGE PLANETARY EQUATIONS

If the perturbing force $\mathbf{f}$ is conservative, it follows that $\mathbf{f}$ is derivable from a *disturbing function*, D, such that $\mathbf{f} = \nabla D$. The force $\mathbf{f}$ will produce temporal changes in the orbit elements that can be expressed by Lagrange's Planetary Equations (e.g., Kaula, 1966):

$$\frac{da}{dt} = \frac{2}{na}\frac{\partial D}{\partial M}$$

$$\frac{de}{dt} = \frac{(1-e^2)^{1/2}}{na^2e}\left((1-e^2)^{1/2}\frac{\partial D}{\partial M} - \frac{\partial D}{\partial \omega}\right)$$

$$\frac{di}{dt} = \frac{1}{h\sin i}\left(\cos i \frac{\partial D}{\partial \omega} - \frac{\partial D}{\partial \Omega}\right)$$

$$\frac{d\Omega}{dt} = \frac{1}{h\sin i}\frac{\partial D}{\partial i}$$

$$\frac{d\omega}{dt} = -\frac{\cos i}{h\sin i}\frac{\partial D}{\partial i} + \frac{(1-e^2)^{1/2}}{na^2e}\frac{\partial D}{\partial e}$$

$$\frac{dM}{dt} = n - \frac{1-e^2}{na^2e}\frac{\partial D}{\partial e} - \frac{2}{na}\frac{\partial D}{\partial a}.$$

Note that $h = na^2[1-e^2]^{1/2}$.

C.2 GAUSSIAN FORM

If the perturbing force $\mathbf{f}$ is expressed as

$$\mathbf{f} = \hat{R}\,\overline{u}_r + \hat{T}\,\overline{u}_T + \hat{N}\,\overline{u}_n$$

485

where the unit vectors are defined by the RTN directions (radial, along-track, and cross-track) and $\hat{R}$, $\hat{T}$, $\hat{N}$ represent force components, the temporal changes in orbit elements can be expressed in the Gaussian form of Lagrange's Planetary Equations (e.g., Pollard, 1966) as:

$$\frac{da}{dt} = \frac{2a^2 e}{h} \sin f \hat{R} + \frac{2a^2 h}{\mu r} \hat{T}$$

$$\frac{de}{dt} = \frac{h}{\mu} \left[\sin f \hat{R} + \hat{T}(e + 2\cos f + e\cos^2 f)/(1 + e\cos f) \right]$$

$$\frac{di}{dt} = \frac{r}{h} \cos(\omega + f)\hat{N}$$

$$\frac{d\Omega}{dt} = \frac{r\sin(\omega + f)\hat{N}}{h\sin i}$$

$$\frac{d\omega}{dt} = -\frac{h}{\mu e} \cos f \hat{R} - \frac{r}{h} \cot i \sin(\omega + f) \hat{N}$$
$$+ \frac{(h^2 + r\mu)\sin f}{\mu e h} \hat{T}$$

$$\frac{dM}{dt} = n - \frac{1}{na} \left(\frac{2r}{a} - \frac{1 - e^2}{e} \cos f \right) \hat{R}$$
$$- \frac{1 - e^2}{nae} \left(1 + \frac{r}{p} \right) \sin f \hat{T}.$$

The Gaussian form applies to either conservative or nonconservative forces.

C.3 REFERENCES

Kaula, W. M., *Theory of Satellite Geodesy*, Blaisdell Publishing Co., Waltham, MA, 1966 (republished by Dover Publications, New York, 2000).

Pollard, H., *Mathematical Introduction to Celestial Mechanics*, Prentice-Hall, Inc., Englewood Cliffs, NJ, 1966.

Appendix D

Constants

D.1 PHYSICAL CONSTANTS*

The speed of light is an important fundamental constant since it effectively defines the length scale in a range measurement. As described in Chapter 3, the range measurement actually is based on a measurement of the time required for a signal to travel from the transmitter to the receiver. This time interval measurement is converted into a distance measurement (range) using the speed of light, c. The value of c was adopted by the International Astronomical Union (IAU) in 1976, to be a defining constant with the value

$$c = 299,792,458 \text{ m s}^{-1}.$$

The constant of gravitation, G, is experimentally determined. A recent value is (Mohr and Taylor, 2003)

$$G = 6.673 \times 10^{-11} \pm 1.0 \times 10^{-13} \text{m}^3 \text{ kg}^{-1} \text{ s}^{-2}.$$

The IAU (1976) System of Astronomical Constants can be found in The Astronomical Almanac for the Year 2000 on pages K6 and K7. Updated constants are given by Seidelmann (1992) and McCarthy (1996).

D.2 EARTH CONSTANTS

For an Earth-orbiting satellite, the normalized gravity coefficients ($\bar{C}_{lm}$ and $\bar{S}_{lm}$), GM, and a_e are required. Recent determinations include the WGS-84 (DMA, 1987), JGM-3 field (Tapley et al., 1996), EGM-96 (Lemoine et al., 1998),

---

*The physical constants given in this appendix use SI units (Le Systeme International d'Unites (SI), 1991).

487

and the GRIMS-C1 (Gruber *et al.*, 2000). The degree and order eight subset of JGM-3 is given in the Table D.1 with standard deviations. Conversion of normalized coefficients to conventional coefficients can be accomplished by:

$$C_{lm} = N_{lm}\bar{C}_{lm}$$
$$S_{lm} = N_{lm}\bar{S}_{lm}$$
$$N_{lm} = \sqrt{\frac{(l-m)!(2l+1)(2-\delta_{0m})}{(l+m)!}}$$

where δ_{0m} is the Kronecker delta function, which is zero when m is not zero, otherwise it is one. Furthermore, for zonal harmonics, the commonly used J_ℓ correspond to $m = 0$, and

$$J_l = -C_{l0}.$$

The ellipsoidal model of the Earth was described in Chapter 2. Current ellipsoid parameters adopted by the IERS (McCarthy, 1996) are:

$$a_e = 6378136.49 \pm 0.1 \text{ m}$$
$$1/f = 298.25645 \pm 0.00001.$$

D.3   LUNAR, SOLAR, AND PLANETARY MASSES

Additional parameters required for the description of satellite motion include the gravitational parameters for the Sun, Moon, and planets. The values that are used with the planetary ephemerides, such as DE-405 (Standish, *et al.*, 1997), are given in Table D.2. Additional information can be found in McCarthy (1996), Seidelmann (1992), and Standish *et al.* (1997). All mass parameters have been determined from observations; consult the references for uncertainties.

Table D.1

JGM-3 Earth Gravity Model

l m	$\overline{C}$	$\overline{S}$	$\sigma_{\overline{C}}$	$\sigma_{\overline{S}}$
2 0	$-0.48416954845647E-03$	$0.00000000000000E+00$	$0.4660E-10$	$0.0000E+00$
3 0	$0.95717059088800E-06$	$0.00000000000000E+00$	$0.3599E-10$	$0.0000E+00$
4 0	$0.53977706835730E-06$	$0.00000000000000E+00$	$0.1339E-09$	$0.0000E+00$
5 0	$0.68658987986543E-07$	$0.00000000000000E+00$	$0.8579E-10$	$0.0000E+00$
6 0	$-0.14967156178604E-06$	$0.00000000000000E+00$	$0.2428E-09$	$0.0000E+00$
7 0	$0.90722941643232E-07$	$0.00000000000000E+00$	$0.2604E-09$	$0.0000E+00$
8 0	$0.49118003174734E-07$	$0.00000000000000E+00$	$0.3996E-09$	$0.0000E+00$
2 1	$-0.18698764000000E-09$	$0.11952801000000E-08$	$0.0000E+00$	$0.0000E+00$
3 1	$0.20301372055530E-05$	$0.24813079825561E-06$	$0.1153E-09$	$0.1152E-09$
4 1	$-0.53624355429851E-06$	$-0.47377237061597E-06$	$0.8693E-10$	$0.8734E-10$
5 1	$-0.62727369697705E-07$	$-0.94194632134383E-07$	$0.2466E-09$	$0.2465E-09$
6 1	$-0.76103580407274E-07$	$0.26899818932629E-07$	$0.2702E-09$	$0.2752E-09$
7 1	$0.28028652203689E-06$	$0.94777317813313E-07$	$0.4361E-09$	$0.4344E-09$
8 1	$0.23333751687204E-07$	$0.58499274939368E-07$	$0.5070E-09$	$0.5137E-09$
2 2	$0.24392607486563E-05$	$-0.14002663975880E-05$	$0.3655E-10$	$0.3709E-10$
3 2	$0.90470634127291E-06$	$-0.61892284647849E-06$	$0.9378E-10$	$0.9375E-10$
4 2	$0.35067015645938E-06$	$0.66257134594268E-06$	$0.1559E-09$	$0.1560E-09$
5 2	$0.65245910276353E-06$	$-0.32333435244435E-06$	$0.2392E-09$	$0.2398E-09$
6 2	$0.48327472124892E-07$	$-0.37381591944355E-06$	$0.3145E-09$	$0.3160E-09$
7 2	$0.32976022742410E-06$	$0.93193696831045E-07$	$0.4635E-09$	$0.4587E-09$
8 2	$0.80070663931587E-07$	$0.65518559097464E-07$	$0.5185E-09$	$0.5323E-09$
3 3	$0.72114493982309E-06$	$0.14142039847354E-05$	$0.5755E-10$	$0.5720E-10$
4 3	$0.99086890577441E-06$	$-0.20098735484731E-06$	$0.7940E-10$	$0.7942E-10$
5 3	$-0.45183704808780E-06$	$-0.21495419346421E-06$	$0.1599E-09$	$0.1616E-09$
6 3	$0.57020965757974E-07$	$0.88894738008251E-08$	$0.2598E-09$	$0.2574E-09$
7 3	$0.25050152675038E-06$	$-0.21732010845254E-06$	$0.3656E-09$	$0.3736E-09$
8 3	$-0.19251764331400E-07$	$-0.86285836534248E-07$	$0.4947E-09$	$0.4918E-09$
4 4	$-0.18848136742527E-06$	$0.30884803690355E-06$	$0.7217E-10$	$0.7228E-10$
5 4	$-0.29512339302196E-06$	$0.49741427230934E-07$	$0.9264E-10$	$0.9288E-10$
6 4	$-0.86228032619800E-07$	$-0.47140511232148E-06$	$0.1656E-09$	$0.1663E-09$
7 4	$-0.27554096307403E-06$	$-0.12414151248516E-06$	$0.2665E-09$	$0.2656E-09$
8 4	$-0.24435806439297E-06$	$0.69857074850431E-07$	$0.4033E-09$	$0.4063E-09$
5 5	$0.17483157769990E-06$	$-0.66939293724911E-06$	$0.8139E-10$	$0.8131E-10$
6 5	$-0.26711227171966E-06$	$-0.53641016466390E-06$	$0.8465E-10$	$0.8510E-10$
7 5	$0.16440038146411E-08$	$0.18075335233506E-07$	$0.1832E-09$	$0.1835E-09$
8 5	$-0.25498410010257E-07$	$0.89090297494640E-07$	$0.2586E-09$	$0.2571E-09$
6 6	$0.95016518338557E-08$	$-0.23726147889522E-06$	$0.8021E-10$	$0.8081E-10$
7 6	$-0.35884263307918E-06$	$0.15177808443426E-06$	$0.5899E-10$	$0.5913E-10$
8 6	$-0.65859353864388E-07$	$0.30892064157956E-06$	$0.1566E-09$	$0.1569E-09$
7 7	$0.13795170564076E-08$	$0.24128594080773E-07$	$0.9709E-10$	$0.9747E-10$
8 7	$0.67262701848734E-07$	$0.74813196768710E-07$	$0.9308E-10$	$0.9378E-10$
8 8	$-0.12397061395498E-06$	$0.12044100668766E-06$	$0.1379E-09$	$0.1384E-09$

$GM_{\text{Earth}} = 3.986004415 \times 10^{14} \pm 8 \times 10^{5} \text{ m}^{3} \text{ s}^{-2}$
$a_e = 6378136.3 \text{ m}$

Table D.2

Lunar, Solar
and Planetary Masses

Planet	GM (m^3 s^{-2})
Mercury	2.203208×10^{13}
Venus	3.248586×10^{14}
Mars	4.282831×10^{13}
Jupiter	1.267128×10^{17}
Saturn	3.794063×10^{16}
Uranus	5.794549×10^{15}
Neptune	6.836534×10^{15}
Pluto	9.816009×10^{11}
Moon	4.902801×10^{12}
Sun	1.327124×10^{20}

The Earth-Moon mass ratio is 81.30056 and
1 Astronomical Unit $= 149, 597, 870, 691$ m,
as provided with DE-405.

Note that using the Earth-Moon mass ratio in Table D.2 yields $GM_{\text{Moon}} = 4.902804 \times 10^{12} m^3 s^{-2}$ based on the GM_{Earth} from JGM-3. The difference with GM_{Moon} in Table D.2 is caused by the GM_{Earth} adopted in DE-405.

D.4   REFERENCES

Defense Mapping Agency, *Department of Defense World Geodetic System 1984*, DMA Technical Report 8350.2, Washington, DC, September, 1987.

Gruber, T., A. Bode, C. Reigber, P. Schwintzer, G. Balmino, R. Biancale, J. Lemoine, "GRIM5-C1: Combination solution of the global gravity field to degree and order 120," *Geophys. Res. Ltrs.*, Vol. 27, No. 24, pp. 4005–4008, December 2000.

Lemoine, F., S. Kenyon, J. Factor, R. Trimmer, N. Pavlis, D. Chinn, C. Cox, S. Klosko, S. Luthcke, M. Torrence, Y. Wang, R. Williamson, E. Pavlis, R. Rapp, and T. Olson, *The development of the Joint NASA GSFC and the National Imagery and Mapping Agency (NIMA) Geopotential Model EGM96*, NASA/TP–1998–206861, Greenbelt, MD, July 1998.

Le Systeme International d'Unites (SI), Bureau International des Poids et Mesures, Sevres, France, 1991.

McCarthy, D. (ed.), *IERS Conventions (1996)*, IERS Technical Note 21, International Earth Rotation Service, Observatoire de Paris, July 1996.

Mohr, P. J., and B. Taylor, "The fundamental physical constants," *Physics Today*, Vol. 56, No. 8, Supplement, pp. BG6–BG13, August 2003.

Seidelmann, P. (ed.), *Explanatory Supplement to the Astronomical Almanac*, University Science Books, Mill Valley, CA, 1992.

Standish, E. M., X. Newhall, J. Williams, and W. Folkner, *JPL Planetary and Lunar Ephemerides* (CD-ROM), Willmann-Bell, Inc., Richmond, VA 1997.

Tapley, B. D., M. Watkins, J. Ries, G. Davis, R. Eanes, S. Poole, H. Rim, B. Schutz, C. Shum, R. Nerem, F. Lerch, J. Marshall, S. Klosko, N. Pavlis, and R. Williamson, "The Joint Gravity Model 3," *J. Geophys. Res.*, Vol. 101, No. B12, pp. 28,029–28,049, December 10, 1996.

Appendix E

Analytical Theory for Near-Circular Orbits

E.1 DESCRIPTION

An analytical theory for the variation of the orbit elements under the influence of the zonal harmonics J_2 through J_5 has been developed by Brouwer (1959). This appendix presents a simplification of his equations for the time variation of the orbit elements for a near-circular orbit under the influence of J_2. The classical elements e and ω used by Brouwer have been replaced by $h = e \sin \omega$ and $k = e \cos \omega$ since the argument of perigee is not well defined for a near-circular orbit. The analytical solutions are given by (Born *et al.*, 2001):

$$
\begin{aligned}
a(t) &= \bar{a} + k_1 \cos 2\bar{\beta} \\
h(t) &= \bar{h}(t) + k_2 \sin \bar{\beta} + k_3 \sin 3\bar{\beta} \\
k(t) &= \bar{k}(t) + k_4 \cos \bar{\beta} + k_3 \cos 3\bar{\beta} \\
i(t) &= \bar{\imath} + k_5 \cos 2\bar{\beta} \\
\Omega(t) &= \bar{\Omega} + k_6 \sin 2\bar{\beta} \\
\beta(t) &= \bar{\beta} + k_7 \sin 2\bar{\beta} + k_8 (t - t_0)^2 \\
\bar{h}(t) &= \bar{e} \sin \bar{\omega} \\
\bar{k}(t) &= \bar{e} \cos \bar{\omega} \\
\bar{\Omega} &= \bar{\Omega}(t_0) + k_9 (t - t_0) \\
\bar{\omega} &= \bar{\omega}(t_0) + k_{10} (t - t_0) \\
\bar{\beta} &= \bar{\beta}(t_0) + k_{11} (t - t_0)
\end{aligned}
\tag{E.1.1}
$$

where $\beta = \omega + M$ and $(\bar{\ })$ represents the mean value of the element. Mean values are given by

$$\bar{a} = a(t_0) - K_1 \cos 2\beta(t_0)$$
$$\bar{h}(t_0) = \bar{e} \sin \bar{\omega}(t_0) = h(t_0) - K_2 \sin \beta(t_0) - K_3 \sin 3\beta(t_0)$$
$$\bar{k}(t_0) = \bar{e} \cos \bar{\omega}(t_0) = k(t_0) - K_4 \cos \beta(t_0) - K_3 \cos 3\beta(t_0)$$
$$\bar{e} = \sqrt{\bar{h}^2(t_0) + \bar{k}^2(t_0)} \tag{E.1.2}$$
$$\bar{\omega}(t_0) = \text{atan} 2\ \left(\bar{h}(t_0), \bar{k}(t_0)\right)$$
$$\bar{i} = i(t_0) - K_5 \cos 2\beta(t_0)$$
$$\overline{\Omega}(t_0) = \Omega(t_0) - K_6 \sin 2\beta(t_0)$$
$$\bar{\beta}(t_0) = \beta(t_0) - K_7 \sin 2\beta(t_0).$$

Also,

$$K_1 = 3\bar{a}\gamma_2\delta^2$$
$$K_2 = \frac{\gamma_2}{2}\left(6 - \frac{21}{2}\delta^2\right)$$
$$K_3 = \frac{7}{4}\gamma_2\delta^2$$
$$K_4 = \frac{\gamma_2}{2}\left(6 - \frac{15}{2}\delta^2\right)$$
$$K_5 = \frac{3}{2}\gamma_2\theta\delta \tag{E.1.3}$$
$$K_6 = \frac{3}{2}\gamma_2\theta$$
$$K_7 = \frac{3}{4}\gamma_2(3 - 5\theta^2)$$
$$K_8 = \frac{3}{4}\frac{\rho v_{\text{rel}}^2}{\bar{a}}C_D\frac{A}{m}$$
$$K_9 = -3\bar{n}\theta\gamma_2$$
$$K_{10} = \frac{3}{2}\gamma_2\bar{n}(4 - 5\delta^2)$$
$$K_{11} = \bar{n}\left(1 + 3\gamma_2(3 - 4\delta^2)\right)$$

and

$$\gamma_2 = \frac{J_2}{2}\left(\frac{R}{\bar{a}}\right)^2$$
$$\theta = \cos\bar{i}$$
$$\delta = \sin\bar{i} \tag{E.1.4}$$

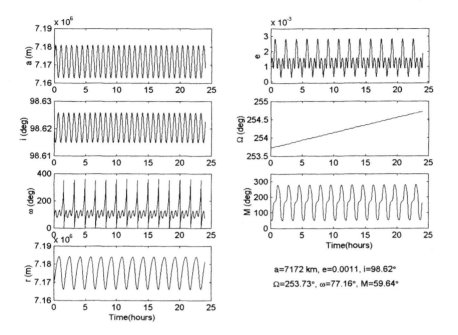

Figure E.1.1: Analytical solution results for the QUIKSCAT orbit using Eq. (E.1.1).

$$\bar{n} = \frac{\mu^{1/2}}{\bar{a}^{3/2}}, \qquad \mu = GM$$

ρ = average atmospheric density at satellite altitude

v_{rel} = average velocity of the satellite relative to the atmosphere.

This representation is based on the solution developed by Brouwer (1959), except for the last term in the $\beta(t)$ expression, $k_8(t - t_0)^2$. This quadratic term is introduced to compensate for drag.

Note that it is not necessary to iterate to determine mean values because this is a first-order theory in J_2. It is necessary to use $\bar{a}$ to compute $\bar{n}$ in order to avoid an error of $O(J_2)$ in computing K_{11}.

Other useful equations are

$$r(t) = \frac{a(t)\left[1 - \left(h(t)^2 + k^2(t)\right)\right]}{1 + k(t)\cos\beta(t) + h(t)\sin\beta(t)} \tag{E.1.5}$$

$$e(t) = \sqrt{h^2(t) + k^2(t)} \tag{E.1.6}$$

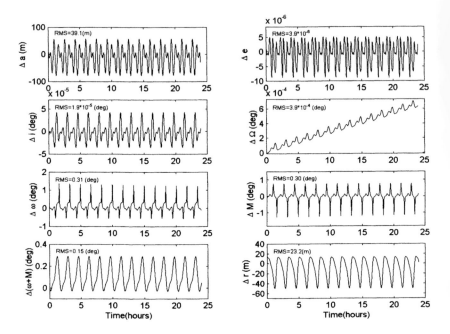

Figure E.1.2: Difference between QUIKSCAT analytical solutions and numerical integration.

$$\omega(t) = \text{atan} \, 2 \, (h(t), k(t)) \tag{E.1.7}$$
$$M(t) = \beta(t) - \omega(t).$$

The equation for $r(t)$ also can be written in terms of the mean elements as follows

$$r(t) = \bar{a}[1 - \bar{e}\cos\overline{M}(t)] + \frac{3}{4}\frac{R^2}{\bar{a}}J_2(3\sin^2\bar{\imath} - 2)$$
$$+\frac{1}{4}\frac{R^2}{a}J_2\sin^2\bar{\imath}\cos 2\bar{\beta} \tag{E.1.8}$$

where

$$\overline{M}(t) = \bar{\beta}(t) - \bar{\omega}(t). \tag{E.1.9}$$

E.2 EXAMPLE

Figure E.2.1 presents the classical orbit elements and radius time history for the QUIKSCAT satellite over one day computed by using Eq. (E.1.1). Initial conditions are given on the figure. Fig. E.2.2 presents the differences between the analytical solutions and numerical integration.

We have ignored terms of $O(eJ_2)$ and $O(J_2^2)$ in developing this theory. Here the values of e and J_2 are the same order of magnitude. However, errors of $O(eJ_2)$ are dominant since reducing the initial eccentricity to 10^{-4} reduces the periodic errors in the analytical solutions by a factor of two. Secular errors of $O(J_2^2)$ are apparent in Ω. Including secular rates of $O(J_2^2)$, given by Brouwer (1959), reduces the RMS error in Ω by an order of magnitude.

E.3 REFERENCES

Born, G. H., D. B. Goldstein and B. Thompson, An Analytical Theory for Orbit Determination, *J. Astronaut. Sci.*, Vol. 49, No. 2, pp. 345–361, April–June 2001.

Brouwer, D., "Solutions of the problem of artificial satellite theory without drag," *Astron. J.*, Vol. 64, No. 9, pp. 378–397, November 1959.

Appendix F

Example of State Noise and Dynamic Model Compensation

F.1 INTRODUCTION

Consider a target particle that moves in one dimension along the x axis in the positive direction.* Nominally, the particle's velocity is a constant 10 m/sec. This constant velocity is perturbed by a sinusoidal acceleration in the x direction, which is unknown and is described by:

$$\eta(t) = \frac{2\pi}{10} \cos\left(\frac{2\pi}{10}t\right) \text{ m/sec.}$$

The perturbing acceleration, perturbed velocity, and position perturbation (perturbed position—nominal position) are shown in Fig. F.1.1.

A measurement sensor is located at the known position $x = -10$ m. This sensor takes simultaneous range and range-rate measurements at a frequency of 10 Hz. The range measurements are corrupted by uncorrelated Gaussian noise having a mean of zero and a standard deviation of 1 m. Likewise, the range-rate measurements are corrupted by uncorrelated Gaussian noise having a mean of zero and a standard deviation of 0.1 m/sec. We want to estimate the state of the particle, with primary emphasis on position accuracy, using these observations given the condition that the sinusoidal perturbing acceleration is unknown. The following estimation results were generated using the extended sequential algorithm discussed in Section 4.7.3.

A simple estimator for this problem incorporates a two-parameter state vector

---

*We thank David R. Cruickshank of Lockheed-Martin Corp., Colorado Springs, Colorado, for providing this example.

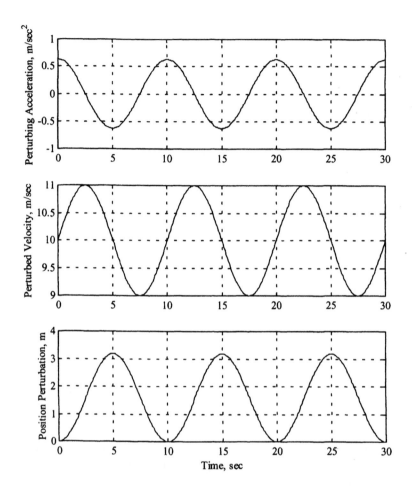

Figure F.1.1: Perturbed particle motion.

consisting of position and velocity:

$$\mathbf{X}(t) = \begin{bmatrix} x(t) \\ \dot{x}(t) \end{bmatrix} = \begin{bmatrix} x(t_0) + \dot{x}(t_0)(t - t_0) \\ \dot{x}(t_0) \end{bmatrix}.$$

The dynamic model assumes constant velocity for the particle motion and does not incorporate process noise. The state transition matrix for this estimator is:

$$\Phi(t, t_0) = \begin{bmatrix} 1 & t - t_0 \\ 0 & 1 \end{bmatrix}$$

and the observation/state mapping matrix is a two-by-two identity matrix:

$$\widetilde{H} = \begin{bmatrix} 1 & 0 \\ 0 & 1 \end{bmatrix}.$$

Since the sinusoidal acceleration is not included in the dynamic model and the filter has no means of compensation for the dynamic model deficiency, the filter quickly saturates and its estimation performance is poor. Fig. F.1.2 shows the actual position errors along with the one-sigma position standard deviation bounds estimated by the filter. The position error RMS is 0.9653 m, and only 18.60 percent of the actual position errors are within the estimated one-sigma standard deviation bounds. Fig. F.1.3 shows the corresponding velocity errors and one-sigma velocity standard deviation bounds. As with position, the velocity errors are generally well outside the standard deviation bounds.

F.2 STATE NOISE COMPENSATION

The *State Noise Compensation* (SNC) algorithm (see Section 4.9) provides a means to improve estimation performance through partial compensation for the unknown sinusoidal acceleration. SNC allows for the possibility that the state dynamics are influenced by a random acceleration. A simple SNC model uses a two-state filter but assumes that particle dynamics are perturbed by an acceleration that is characterized as simple white noise:

$$\ddot{x}(t) = \eta(t) = u(t)$$

where $u(t)$ is a stationary Gaussian process with a mean of zero and a variance of $\sigma_u^2 \delta(t - \tau)$, and $\delta(t - \tau)$ is the Dirac delta function. The Dirac delta function is not an ordinary function, and to be mathematically rigorous, white noise is a fictitious process. However, in linear stochastic models, it can be treated formally as an integrable function. Application of Eq. (4.9.1) to this case results in

$$\begin{bmatrix} \dot{x}(t) \\ \ddot{x}(t) \end{bmatrix} = \begin{bmatrix} 0 & 1 \\ 0 & 0 \end{bmatrix} \begin{bmatrix} x(t) \\ \dot{x}(t) \end{bmatrix} + \begin{bmatrix} 0 \\ 1 \end{bmatrix} u(t)$$

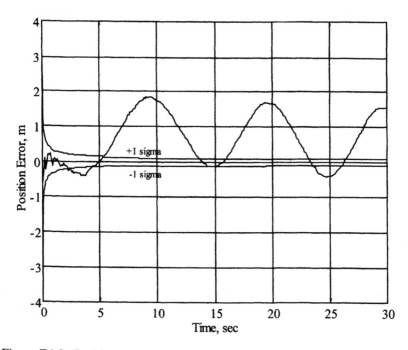

Figure F.1.2: Position errors and estimated standard deviation bounds from the two-state filter without process noise.

where the state propagation matrix A is identified as:

$$A = \begin{bmatrix} 0 & 1 \\ 0 & 0 \end{bmatrix}$$

and the process noise mapping matrix is:

$$B = \begin{bmatrix} 0 \\ 1 \end{bmatrix}.$$

The state transition matrix is the same:

$$\Phi(t, t_0) = \begin{bmatrix} 1 & t - t_0 \\ 0 & 1 \end{bmatrix}.$$

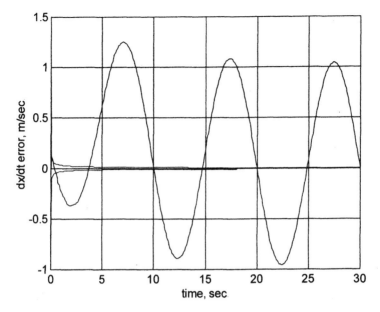

Figure F.1.3: Velocity errors and estimated standard deviation bounds from the two-state filter without process noise.

The process noise covariance integral (see Eq. (4.9.44)) needed for the time update of the estimation error covariance matrix at time t is expressed as:

$$Q_\eta(t) = \sigma_u^2 \int_{t_0}^t \Phi(t,\tau) B B^T \Phi^T(t,\tau) d\tau$$

where

$$\Phi(t,\tau) = \begin{bmatrix} 1 & (t-\tau) \\ 0 & 1 \end{bmatrix}.$$

Substituting for B and $\Phi(t,\tau)$ and evaluating results in the process noise covariance matrix:

$$Q_\eta(t) = \sigma_u^2 \begin{bmatrix} \frac{1}{3}(t-t_0)^3 & \frac{1}{2}(t-t_0)^2 \\ \frac{1}{2}(t-t_0)^2 & t-t_0 \end{bmatrix}$$

where $t - t_0$ is the measurement update interval; that is, $t_k - t_{k-1} = 0.1$ sec. The implication of this is that the original deterministic constant velocity model of particle motion is modified to include a random component that is a constant-diffusion Brownian motion process, σ_u^2 being known as the diffusion coefficient (Maybeck, 1979).

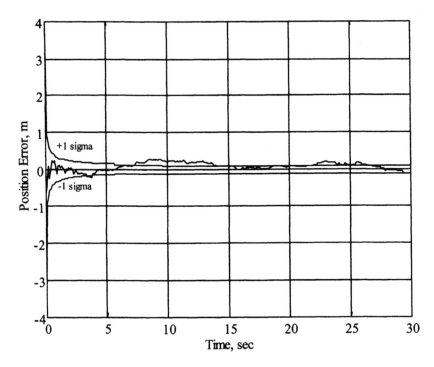

Figure F.2.1: Position errors and estimated standard deviation bounds from the two-state SNC filter.

The magnitude of the process noise covariance matrix and its effect on estimation performance are functions of this diffusion coefficient, hence σ_u is a tuning parameter whose value can be adjusted to optimize performance. Figures F.2.1 and F.2.2 show the result of the tuning process. The optimal value of σ_u is 0.42 m/sec$^{3/2}$ at which the position error RMS is 0.1378 m and 56.81 percent of the actual position errors fall within the estimated one-sigma standard deviation bounds. The large sinusoidal error signature in both position and velocity displayed by the uncompensated filter is eliminated by SNC.

Note that there is no noticable change in the position standard deviations; however, the velocity standard deviations show a significant increase when process noise is included. This increase in the velocity variances prevents the components of the Kalman gain matrix from going to zero with the attendant saturation of the filter.

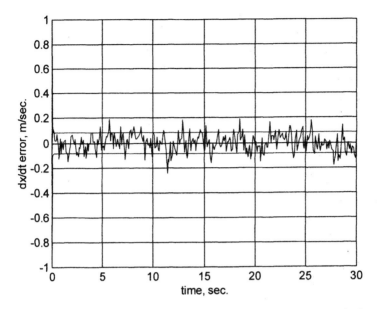

Figure F.2.2: Velocity errors and estimated standard deviation bounds from the two-state filter with SNC process noise.

F.3 DYNAMIC MODEL COMPENSATION

A more sophisticated process noise model is provided by the *Dynamic Model Compensation* (DMC) formulation. DMC assumes that the unknown acceleration can be characterized as a first-order linear stochastic differential equation, commonly known as a Langevin equation:

$$\dot{\eta}(t) + \beta\eta(t) = u(t),$$

where $u(t)$ is a white zero-mean Gaussian process as described earlier and β is the inverse of the correlation time:

$$\beta = \frac{1}{\tau}.$$

The solution to this Langevin equation is

$$\eta(t) = \eta_0 e^{-\beta(t-t_0)} + \int_{t_0}^{t} e^{-\beta(t-\tau)} u(\tau) d\tau.$$

This is the Gauss-Markov process (more precisely known as an Ornstein-Uhlenbeck process) described in Section 4.9. Note that, unlike SNC, the DMC process noise model yields a deterministic acceleration term as well as a purely random term. The deterministic acceleration can be added to the state vector and estimated along with the velocity and position; the augmented state vector becomes a three-state filter with

$$
\mathbf{X}(t) = \begin{bmatrix} x(t) \\ \dot{x}(t) \\ \eta_D(t) \end{bmatrix}
$$

where $\eta_D(t)$ is the deterministic part of $\eta(t)$; that is, $\eta_0 e^{-\beta(t-t_0)}$. Since the acceleration integrates into velocity and position increments, the dynamic model of the particle's motion becomes

$$
\mathbf{X}(t) = \begin{bmatrix} x_0 + \dot{x}_0(t - t_0) + \frac{\eta_0}{\beta}(t - t_0) + \frac{\eta_0}{\beta^2}\left(e^{-\beta(t-t_0)} - 1\right) \\ \dot{x}_0 + \frac{\eta_0}{\beta}\left(1 - e^{-\beta(t-t_0)}\right) \\ \eta_0 e^{-\beta(t-t_0)} \end{bmatrix}.
$$

The correlation time, τ, can also be added to the estimated state, resulting in a four-parameter state vector. However, a tuning strategy that works well for many DMC estimation problems is to set τ to a near-optimal value and hold it constant, or nearly constant, during the estimation span. For simplicity, τ is held constant in this case, allowing the use of the three-parameter state vector just noted. The observation/state mapping matrix is a simple extension of the two-state case:

$$
\tilde{H} = \begin{bmatrix} 1 & 0 & 0 \\ 0 & 1 & 0 \end{bmatrix}.
$$

Using the methods described in Example 4.2.3, the state transition matrix is found to be

$$
\Phi(t, t_0) = \frac{\partial X(t)}{\partial X(t_0)} = \begin{bmatrix} 1 & t - t_0 & \frac{1}{\beta}(t - t_0) + \frac{1}{\beta^2}\left(e^{-\beta(t-t_0)} - 1\right) \\ 0 & 1 & \frac{1}{\beta}\left(1 - e^{-\beta(t-t_0)}\right) \\ 0 & 0 & e^{-\beta(t-t_0)} \end{bmatrix}.
$$

The state propagation matrix for this case is

$$
A = \begin{bmatrix} 0 & 1 & 0 \\ 0 & 0 & 1 \\ 0 & 0 & -\beta \end{bmatrix}
$$

and the process noise mapping matrix is

$$B = \begin{bmatrix} 0 \\ 0 \\ 1 \end{bmatrix}.$$

With these, the process noise covariance integral becomes (in terms of the components of $\Phi(t, \tau)$)

$$Q_\eta(t) = \sigma_u^2 \times$$
$$\int_{t_0}^t \begin{bmatrix} \Phi_{1,3}^2(t,\tau) & \Phi_{1,3}(t,\tau)\Phi_{2,3}(t,\tau) & \Phi_{1,3}(t,\tau)\Phi_{3,3}(t,\tau) \\ \Phi_{2,3}(t,\tau)\Phi_{1,3}(t,\tau) & \Phi_{2,3}^2(t,\tau) & \Phi_{2,3}(t,\tau)\Phi_{3,3}(t,\tau) \\ \Phi_{3,3}(t,\tau)\Phi_{1,3}(t,\tau) & \Phi_{3,3}(t,\tau)\Phi_{2,3}(t,\tau) & \Phi_{3,3}^2(t,\tau) \end{bmatrix} d\tau .$$

Note that here τ designates the integration variable, not the correlation time. The expanded, integrated process noise covariance matrix terms follow:
position variance:

$$Q_{\eta 1,1}(t) = \sigma_u^2 \left(\frac{1}{3\beta^2}(t - t_0)^3 - \frac{1}{\beta^3}(t - t_0)^2 \right.$$
$$\left. + \frac{1}{\beta^4}(t - t_0)(1 - 2e^{-\beta(t-t_0)}) + \frac{1}{2\beta^5}(1 - e^{-2\beta(t-t_0)}) \right)$$

position/velocity covariance:

$$Q_{\eta 1,2}(t) = \sigma_u^2 \left(\frac{1}{2\beta^2}(t - t_0)^2 - \frac{1}{\beta^3}(t - t_0)(1 - e^{-\beta(t-t_0)}) \right.$$
$$\left. + \frac{1}{\beta^4}(1 - e^{-\beta(t-t_0)}) - \frac{1}{2\beta^4}(1 - e^{-2\beta(t-t_0)}) \right)$$

position/acceleration covariance:

$$Q_{\eta 1,3}(t) = \sigma_u^2 \left(\frac{1}{2\beta^3}(1 - e^{-2\beta(t-t_0)}) - \frac{1}{\beta^2}(t - t_0)e^{-\beta(t-t_0)} \right)$$

velocity variance:

$$Q_{\eta 2,2}(t) = \sigma_u^2 \left(\frac{1}{\beta^2}(t - t_0) - \frac{2}{\beta^3}\left(1 - e^{-\beta(t-t_0)}\right) + \frac{1}{2\beta^3}\left(1 - e^{-2\beta(t-t_0)}\right) \right)$$

velocity/acceleration covariance:

$$Q_{\eta 2,3}(t) = \sigma_u^2 \left(\frac{1}{2\beta^2}\left(1 + e^{-2\beta(t-t_0)}\right) - \frac{1}{\beta^2}e^{-\beta(t-t_0)} \right)$$

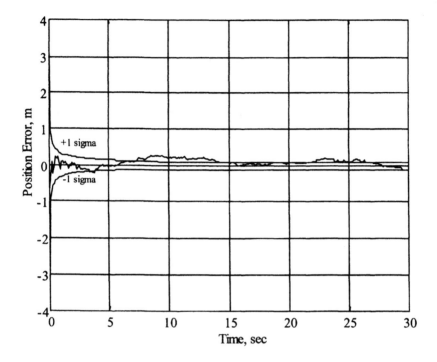

Figure F.3.1: Position errors and estimated standard deviation bounds from the three-state DMC filter.

acceleration variance:

$$Q_{\eta 3,3}(t) = \frac{\sigma_u^2}{2\beta} \left(1 - e^{-2\beta(t-t_0)} \right) .$$

With τ fixed at 200.0 sec (or $\beta = 0.005$ sec^{-1}), the optimal value of σ_u is 0.26 m/sec$^{5/2}$. This combination produces the position error results shown in Fig. F.3.1. The position error RMS is 0.1376 m and 56.81 percent of the actual position errors are within the estimated one-sigma standard deviation bounds. Although not shown here, the plot of velocity errors exhibits a behavior similar to the velocity errors for the SNC filter.

Figure F.3.2 is a plot of the RMS of the position error as a function of σ_u for both the two-state SNC filter and the three-state DMC filter. Although the optimally tuned SNC filter approaches the position error RMS performance of the DMC filter, it is much more sensitive to tuning. DMC achieves very good performance over a much broader, suboptimal range of σ_u values than does SNC.

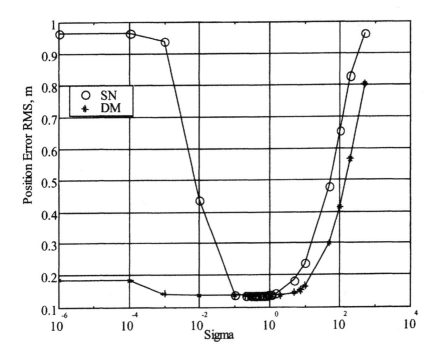

Figure F.3.2: RMS Position error as a function of σ_u for the two-state SNC and three-state DMC filters.

This is a significant advantage for DMC applications where the true state values are not available to guide the tuning effort, as they are in a simulation study, and the precisely optimal tuning parameter values are not known.

Aside from this, the DMC filter also provides a direct estimate of the unknown, unmodeled acceleration. This information could, in turn, be useful in improving the filter dynamic model. Figure F.3.3 is a plot of the estimated and true sinusoidal accelerations. This plot shows excellent agreement between the estimated and true accelerations; their correlation coefficient is 0.9306.

F.4 REFERENCE

Maybeck, P. S., *Stochastic Models, Estimation and Control*, Vol. 1, Academic Press, 1979.

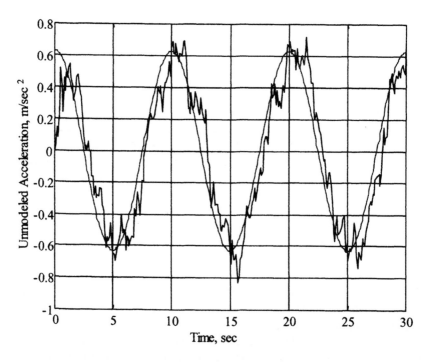

Figure F.3.3: Estimated and true sinusoidal accelerations.

Appendix G

Solution of the Linearized Equations of Motion

G.1 INTRODUCTION

The equations of motion for a satellite are given by:

$$\dot{\mathbf{X}} = F(\mathbf{X}, t),$$

where

$$\mathbf{X} = [X\ Y\ Z\ \dot{X}\ \dot{Y}\ \dot{Z}\ \beta]^T. \qquad\text{(G.1.1)}$$

$\mathbf{X}$ is the state vector containing six position and velocity elements and β, an m vector, represents all constant parameters such as gravity and drag coefficients that are to be solved for. Hence, $\mathbf{X}$ is a vector of dimension $n = m + 6$.

Equation (G.1.1) can be linearized by expanding about a reference state vector denoted by $\mathbf{X}^*$,

$$\dot{\mathbf{X}}(t) = \dot{\mathbf{X}}^*(t) + \left[\frac{\partial \dot{\mathbf{X}}(t)}{\partial \mathbf{X}(t)}\right]^* (\mathbf{X}(t) - \mathbf{X}^*(t)) + \text{h.o.t.} \qquad\text{(G.1.2)}$$

The $*$ indicates that the quantity is evaluated on the reference state. By ignoring higher-order terms (h.o.t.) and defining

$$\mathbf{x}(t) \equiv \mathbf{X}(t) - \mathbf{X}^*(t), \qquad\text{(G.1.3)}$$

we can write Eq. (G.1.2) as

$$\dot{\mathbf{x}}(t) = \left[\frac{\partial \dot{\mathbf{X}}(t)}{\partial \mathbf{X}(t)}\right]^* \mathbf{x}(t). \qquad\text{(G.1.4)}$$

Define

$$A(t) \equiv \left[\frac{\partial \dot{\mathbf{X}}(t)}{\partial \mathbf{X}(t)} \right]^{*},$$

then

$$\dot{\mathbf{x}}(t) = A(t)\mathbf{x}(t). \tag{G.1.5}$$

Equation (G.1.5) is a linear system of first-order differential equations with $A(t)$ being an $n \times n$ time varying matrix evaluated on the known reference state $\mathbf{X}^{*}(t)$. Note that $\dot{\beta} = 0$, so that

$$\frac{\partial \dot{\beta}}{\partial \mathbf{X}(t)} = 0.$$

Because Eq. (G.1.4) is linear[†] and of the form

$$\dot{\mathbf{x}}(t) = A(t)\mathbf{x}(t),$$

the solution can be written as

$$\mathbf{x}(t) = \frac{\partial \mathbf{x}(t)}{\partial \mathbf{x}_0} \mathbf{x}_0.$$

It is also true that

$$\mathbf{x}(t) = \frac{\partial \mathbf{X}(t)}{\partial \mathbf{X}_0} \mathbf{x}_0. \tag{G.1.6}$$

This follows from the fact that the reference state does not vary in this operation,

$$\frac{\partial \mathbf{x}(t)}{\partial \mathbf{x}_0} = \frac{\partial \left(\mathbf{X}(t) - \mathbf{X}^{*}(t) \right)}{\partial (\mathbf{X}_0 - \mathbf{X}_0^{*})}$$
$$= \frac{\partial \mathbf{X}(t)}{\partial \mathbf{X}_0}.$$

The conditions under which Eq. (G.1.6) satisfies Eq. (G.1.4) are demonstrated as follows. First define

$$\Phi(t, t_0) \equiv \frac{\partial \mathbf{X}(t)}{\partial \mathbf{X}_0}. \tag{G.1.7}$$

Then Eq. (G.1.6) can be written as

$$\mathbf{x}(t) = \Phi(t, t_0)\mathbf{x}_0. \tag{G.1.8}$$

[†] A differential equation of any order (order of highest-ordered derivative) and first degree (degree of its highest derivative) is said to be linear when it is linear in the dependent variable and its derivatives and their products; e.g., $\frac{d^2 x}{dt^2} = \frac{x}{t} + 3t^2$ is a linear second-order equation of first degree, and $x \frac{dx}{dt} = 3t + 4$ is first order and degree but nonlinear.

Differentiating Eq. (G.1.8) yields

$$\dot{\mathbf{x}}(t) = \dot{\Phi}(t, t_0)\mathbf{x}_0. \tag{G.1.9}$$

Equating Eq. (G.1.4) and Eq. (G.1.9) yields

$$\frac{\partial \dot{\mathbf{X}}(t)}{\partial \mathbf{X}(t)}\mathbf{x}(t) = \dot{\Phi}(t, t_0)\mathbf{x}_0. \tag{G.1.10}$$

Substituting Eq. (G.1.8) for $\mathbf{x}(t)$ into Eq. (G.1.10) results in

$$\left[\frac{\partial \dot{\mathbf{X}}(t)}{\partial \mathbf{X}(t)}\right]^{*} \Phi(t, t_0)\mathbf{x}_0 = \dot{\Phi}(t, t_0)\mathbf{x}_0.$$

Equating the coefficients of $\mathbf{x}_0$ in this equation yields the differential equation for $\dot{\Phi}(t, t_0)$,

$$\dot{\Phi}(t, t_0) = A(t)\Phi(t, t_0), \tag{G.1.11}$$

with initial conditions

$$\Phi(t_0, t_0) = I. \tag{G.1.12}$$

The matrix $\Phi(t, t_0)$ is referred to as the State Transition Matrix. Whenever Eqs. (G.1.11) and (G.1.12) are satisfied, the solution to $\dot{\mathbf{x}}(t) = A(t)\mathbf{x}(t)$ is given by Eq. (G.1.8).

G.2 THE STATE TRANSITION MATRIX

Insight into the $n \times n$ state transition matrix can be obtained as follows. Let

$$\Phi(t, t_0) \equiv \frac{\partial \mathbf{X}(t)}{\partial \mathbf{X}_0} \equiv \begin{bmatrix} \phi_1(t, t_0) \\ \phi_2(t, t_0) \\ \phi_3(t, t_0) \end{bmatrix} = \begin{bmatrix} \frac{\partial \mathbf{r}(t)}{\partial \mathbf{X}_0} \\ \frac{\partial \dot{\mathbf{r}}(t)}{\partial \mathbf{X}_0} \\ \frac{\partial \boldsymbol{\beta}(t)}{\partial \mathbf{X}_0} \end{bmatrix}. \tag{G.2.1}$$

Note that $\phi_3(t, t_0)$ is an $m \times n$ matrix of constants partitioned into an $m \times 6$ matrix of zeros on the left and an $m \times m$ identity matrix on the right, where m is the dimension of β and $\mathbf{X}$ is of dimension n. Because of this, it is only necessary to solve the upper $6 \times n$ portion of Eq. (G.1.11).

Equation (G.1.11) also can be written in terms of a second-order differential equation. This can be shown by differentiating Eq. (G.2.1):

$$\dot{\Phi}(t, t_0) = \frac{\partial \dot{\mathbf{X}}(t)}{\partial \mathbf{X}_0} = \begin{bmatrix} \dot{\phi}_1(t, t_0) \\ \dot{\phi}_2(t, t_0) \\ \dot{\phi}_3(t, t_0) \end{bmatrix} = \begin{bmatrix} \frac{\partial \dot{\mathbf{r}}(t)}{\partial \mathbf{X}_0} \\ \frac{\partial \ddot{\mathbf{r}}(t)}{\partial \mathbf{X}_0} \\ 0 \end{bmatrix}$$

$$= \begin{bmatrix} \frac{\partial \dot{\mathbf{r}}(t)}{\partial \mathbf{X}(t)} \\ \frac{\partial \ddot{\mathbf{r}}(t)}{\partial \mathbf{X}(t)} \\ 0 \end{bmatrix}_{n \times n} \begin{bmatrix} \frac{\partial \mathbf{X}(t)}{\partial \mathbf{X}_0} \end{bmatrix}_{n \times n}. \tag{G.2.2}$$

In these equations, 0 represents an $m \times n$ null matrix. Notice from the first of Eq. (G.2.2) that

$$\dot{\phi}_1 = \frac{\partial \dot{\mathbf{r}}}{\partial \mathbf{X}_0} = \dot{\phi}_2. \tag{G.2.3}$$

Hence, we could solve this second-order system of differential equations to obtain $\Phi(t, t_0)$,

$$\ddot{\phi}_1(t, t_0) = \frac{\partial \ddot{\mathbf{r}}(t)}{\partial \mathbf{X}_0} = \frac{\partial \ddot{\mathbf{r}}(t)}{\partial \mathbf{X}(t)} \frac{\partial \mathbf{X}(t)}{\partial \mathbf{X}_0} \tag{G.2.4}$$

$$= \begin{bmatrix} \frac{\partial \ddot{\mathbf{r}}(t)}{\partial \mathbf{r}(t)} & \frac{\partial \ddot{\mathbf{r}}(t)}{\partial \dot{\mathbf{r}}(t)} & \frac{\partial \ddot{\mathbf{r}}(t)}{\partial \boldsymbol{\beta}} \end{bmatrix}_{3 \times n} \begin{bmatrix} \frac{\partial \mathbf{r}(t)}{\partial \mathbf{X}_0} \\ \frac{\partial \dot{\mathbf{r}}(t)}{\partial \mathbf{X}_0} \\ \frac{\partial \boldsymbol{\beta}}{\partial \mathbf{X}_0} \end{bmatrix}_{n \times n}$$

or

$$\ddot{\phi}_1(t, t_0) = \frac{\partial \ddot{\mathbf{r}}(t)}{\partial \mathbf{r}(t)} \phi_1(t, t_0) + \frac{\partial \ddot{\mathbf{r}}(t)}{\partial \dot{\mathbf{r}}(t)} \dot{\phi}_1(t, t_0) + \frac{\partial \ddot{\mathbf{r}}(t)}{\partial \boldsymbol{\beta}} \phi_3(t, t_0). \tag{G.2.5}$$

With initial conditions

$$\phi_1(t_0, t_0) = \begin{bmatrix} [I]_{3 \times 3} [0]_{3 \times (n-3)} \end{bmatrix}$$

$$\dot{\phi}_1(t_0, t_0) = \phi_2(t_0, t_0) = \begin{bmatrix} [0]_{3 \times 3} [I]_{3 \times 3} [0]_{3 \times m} \end{bmatrix}.$$

We could solve Eq. (G.2.5), a $3 \times n$ system of second-order differential equations, instead of the $6 \times n$ first-order system given by Eq. (G.2.2). Recall that the partial derivatives are evaluated on the reference state and that the solution of the $m \times n$ system represented by $\dot{\phi}_3(t, t_0) = 0$ is trivial,

$$\phi_3(t, t_0) = \begin{bmatrix} [0]_{m \times 6} [I]_{m \times m} \end{bmatrix}.$$

In solving Eq. (G.2.5) we could write it as a system of $n \times n$ first-order equations,

$$\dot{\phi}_1(t, t_0) = \phi_2(t, t_0)$$

$$\dot{\phi}_2(t, t_0) = \frac{\partial \ddot{\mathbf{r}}(t)}{\partial \mathbf{r}(t)} \phi_1(t, t_0) + \frac{\partial \ddot{\mathbf{r}}(t)}{\partial \dot{\mathbf{r}}(t)} \phi_2(t, t_0) + \frac{\partial \ddot{\mathbf{r}}(t)}{\partial \boldsymbol{\beta}} \phi_3(t, t_0) \tag{G.2.6}$$

$$\dot{\phi}_3(t, t_0) = 0.$$

It is easily shown that Eq. (G.2.6) is identical to Eq. (G.1.11),

$$
\begin{bmatrix} \dot{\phi}_1(t, t_0) \\ \dot{\phi}_2(t, t_0) \\ \dot{\phi}_3(t, t_0) \end{bmatrix}_{n \times n} = \begin{bmatrix} [0]_{3\times3} & [I]_{3\times3} & [0]_{3\times m} \\ \left[\frac{\partial \ddot{\mathbf{r}}(t)}{\partial \mathbf{r}(t)}\right]_{3\times3} & \left[\frac{\partial \ddot{\mathbf{r}}(t)}{\partial \dot{\mathbf{r}}(t)}\right]_{3\times3} & \left[\frac{\partial \ddot{\mathbf{r}}(t)}{\partial \boldsymbol{\beta}}\right]_{3\times m} \\ [0]_{m\times3} & [0]_{m\times3} & [0]_{m\times m} \end{bmatrix}^{*}_{n \times n} \begin{bmatrix} \phi_1(t, t_0) \\ \phi_2(t, t_0) \\ \phi_3(t, t_0) \end{bmatrix}_{n \times n}
$$

or

$$
\dot{\Phi}(t, t_0) = A(t)\Phi(t, t_0) .
$$

Appendix H

Transformation between ECI and ECF Coordinates

H.1 INTRODUCTION

The transformation between ECI and ECF coordinates is defined by Eq. (2.4.12)

$$T^{xyz}_{XYZ} = T^{ECF}_{ECI} = WS'NP. \qquad (H.1.1)$$

The transformation matrices W, S', N, and P account for the following effects:

$W =$ the offset of the Earth's angular velocity vector with respect to the z axis of the ECF (see Section 2.4.2)

$S' =$ the rotation of the ECF about the angular velocity vector (see Sections 2.3.3 and 2.4.2)

$N =$ the nutation of the ECF with respect to the ECI (see Section 2.4.1)

$P =$ the precession of the ECF with respect to the ECI (see Section 2.4.1).

The ECI and ECF systems have fundamental characteristics described in Chapter 2. The realization of these reference systems in modern space applications is usually through the *International Celestial Reference Frame* (ICRF) for the ECI and the *International Terrestrial Reference Frame* (ITRF) for the ECF. These realizations infer certain characteristics for highest accuracy and consistency. For example, the ICRF is defined using the coordinates of extragalactic radio sources, known as quasars, where the coordinates are derived from observations obtained on the surface of the Earth with a set of large antennas known as the *Very Long Baseline Interferometry* (VLBI) network. The coordinates of the quasars are then

517

linked with visible sources to an optical star catalog and the coordinates of the celestial bodies in the solar system. In a similar manner, the ITRF is realized with ground networks of SLR, LLR, GPS, and DORIS (see Chapter 3), which establish the center of mass origin of the ITRF, as well as the reference meridian.

The following sections summarize the content of the matrices required for the transformation from an ECI system to an ECF system. This discussion follows the classical formulation and it will be assumed that the ECI is realized through J2000, which is not precisely the same as the actual ICRF adopted by the International Astronomical Union. If observed values of polar motion (x_p, y_p) and Δ(UT1) are used, the transformation matrix has an accuracy of better than one arc-sec. The satellite motion is especially sensitive to polar motion and the rate of change in UT1, which is related to length of day. This sensitivity enables the estimation of these parameters from satellite observations.

The *International Earth Rotation Service* (IERS) distributes *Earth Orientation Parameters* (EOP) derived from VLBI and satellite observations. The main EOP are (x_p, y_p), Δ(UT1) and corrections to the precession and nutation parameters. These parameters are empirical corrections derived from satellite and VLBI observations which account for changes resulting from tides and atmospheric effects, for example. The EOP are regularly distributed by the IERS as Bulletin B. Consult the IERS publications and web page for additional information. As the models adopted by the IERS are improved, they are documented in new releases of the IERS Standards (or "Conventions").

For the most part, the following discussion follows Seidelmann (1992) and McCarthy (1996). A comprehensive summary is given by Seeber (1993). A comparison of different implementations of the matrices, including a detailed review of the transformations, has been prepared by Webb (2002).

H.2   MATRIX P

For this discussion, P represents the transformation from J2000 to a mean-of-date system. If is a function of the precession of the equinoxes (see Section 2.4.1). The transformation matrix is dependent on three angles and is given by (where C represents cosine and S denotes sine):

$$P = \tag{H.2.1}$$

$$\begin{bmatrix} C\zeta_A C\theta_A Cz_A - S\zeta_A Sz_A & -S\zeta_A C\theta_A Cz_A - C\zeta_A Sz_A & -S\theta_A Cz_A \\ C\zeta_A C\theta_A Sz_A + S\zeta_A Cz_A & -S\zeta_A C\theta_A Sz_A + C\zeta_A Cz_A & -S\theta_A Sz_A \\ C\zeta_A S\theta_A & -S\zeta_A S\theta_A & C\theta_A \end{bmatrix},$$

and the precession angles are given by

$$\zeta_A = 2306\rlap{.}''2181t + 0\rlap{.}''30188t^2 + 0\rlap{.}''017998t^3$$
$$\theta_A = 2004\rlap{.}''3109t - 0\rlap{.}''42665t^2 - 0\rlap{.}''041833t^3 \qquad \text{(H.2.2)}$$
$$z_A = 2306\rlap{.}''2181t + 1\rlap{.}''09468t^2 + 0\rlap{.}''018203t^3$$

with t defined as

$$t = \frac{[TT - J2000.0](\text{days})}{36525}. \qquad \text{(H.2.3)}$$

In Eq. H.2.3, J2000.0 is January 1, 2000, 12:00 in TT, or JD 2451545.0. These equations are based on the work of Lieske, *et al.* (1977). Terrestrial Time (TT) is defined in Section 2.4.2.

H.3 MATRIX N

The transformation from a mean-of-date system to a true-of-date system is dependent on the nutations (see Section 2.4.1). Four angles are used: the mean obliquity of the ecliptic (ϵ_m), the true obliquity of the ecliptic (ϵ_t), the nutation in longitude ($\Delta\psi$), and the nutation in obliquity ($\Delta\epsilon$). The matrix N is given by

$$N = \qquad \qquad \qquad \qquad \qquad \qquad \qquad \qquad \qquad \qquad \qquad \text{(H.3.1)}$$

$$\begin{bmatrix} C\Delta\psi & -C\epsilon_m S\Delta\psi & -S\epsilon_m S\Delta\psi \\ C\epsilon_t S\Delta\psi & C\epsilon_m C\epsilon_t C\Delta\psi + S\epsilon_m S\epsilon_t & S\epsilon_m C\epsilon_t C\Delta\psi - C\epsilon_m S\epsilon_t \\ S\epsilon_t S\Delta\psi & C\epsilon_m S\epsilon_t C\Delta\psi - S\epsilon_m C\epsilon_t & S\epsilon_m S\epsilon_t C\Delta\psi + C\epsilon_m C\epsilon_t \end{bmatrix},$$

where the mean obliquity is

$$\epsilon_m = 84381\rlap{.}''448 - 46\rlap{.}''8150t - 0\rlap{.}''00059t^2 + 0\rlap{.}''001813t^3 \qquad \text{(H.3.2)}$$

and the true obliquity is

$$\epsilon_t = \epsilon_m + \Delta\epsilon. \qquad \text{(H.3.3)}$$

The expression for the nutation in longitude and in obliquity is given by McCarthy (1996), for example. The IAU 1980 theory of nutation is readily available in a convenient Chebychev polynomial approximation with the planetary ephemerides (Standish, *et al.*, 1997). Corrections to the adopted series, based on observations, are provided by the IERS in Bulletin B.

H.4 MATRIX S'

The true-of-date rotation to a pseudo-body-fixed system accounts for the 'spin' of the Earth. This diurnal rotation depends on a single angle α_G, which is referred to as the *Greenwich Mean Sidereal Time* (GMST). The matrix S' is identical to Eq. 2.3.20; namely,

$$S' = \begin{bmatrix} C\alpha_G & S\alpha_G & 0 \\ -S\alpha_G & C\alpha_G & 0 \\ 0 & 0 & 1 \end{bmatrix}. \tag{H.4.1}$$

In some cases, especially when a satellite velocity vector with respect to the rotating ECF is required, the time derivative of the time-dependent elements of each of the transformation matrices will be needed. However, the time derivative of S' is particularly significant since the angle α_G is the fastest changing variable in the matrices P, N, S' and W. The time derivative of S' is

$$\dot{S}' = \dot{\alpha}_G \begin{bmatrix} -S\alpha_G & C\alpha_G & 0 \\ -C\alpha_G & -S\alpha_G & 0 \\ 0 & 0 & 0 \end{bmatrix} \tag{H.4.2}$$

where $\dot{\alpha}_G$ is the rotation rate of the Earth. For some applications, this rate can be taken to be constant, but for high-accuracy applications the variability in this rate must be accounted for. According to the IERS, the average rate is $7.2921151467064 \times 10^{-5}$ rad/s, but this value should be adjusted using IERS Bulletin B to account for variations. The GMST for a given UT1 (UTC + Δ(UT1)) can be found from Kaplan, *et al.* (1981), modified to provide the result in radians:

$$\begin{aligned} \text{GMST(UT1)} = {} & 4.89496121282305875137570443 0 \tag{H.4.3} \\ & + \Delta T \big\{ 6.30038809898489 3552276513720 \\ & + \Delta T \big(5.0752099941135914780 53805523 \times 10^{-15} \\ & - 9.2530975681943356400671906 88 \times 10^{-24} \Delta T \big) \big\} \end{aligned}$$

where $\Delta T = \text{UT1} - \text{J2000.0}$, and where ΔT is in days, including the fractional part of a day.

An additional correction known as the equation of the equinoxes must be applied. For highest accuracy, additional terms amounting to a few milli-arc-sec amplitude should be applied (see McCarthy, 1996; or Simon, *et al.*, 1994) . With the equation of the equinoxes, the angle α_G becomes

$$\alpha_G = \text{GMST(UT1)} + \Delta \psi \cos \epsilon_m. \tag{H.4.4}$$

H.5 MATRIX W

The pseudo-body-fixed system is aligned with the instantaneous pole, equator and reference meridian of the Earth. As noted in Section 2.4.2, the Earth's spin axis is not fixed in the Earth and two angles (x_p, y_p) are used to describe the spin axis location with respect to an adopted ECF, such as an ITRF. Because of the small magnitude of the polar motion angles, the transformation matrix W can be represented by

$$T_{\text{PBF}}^{\text{BF}} = W = \begin{bmatrix} 1 & 0 & x_p \\ 0 & 1 & -y_p \\ -x_p & y_p & 1 \end{bmatrix}, \tag{H.5.1}$$

where the polar motion angles are expressed in radians. The angles are available in IERS Bulletin B, for example.

H.6 REFERENCES

Aoki, S., and H. Kinoshita, "Note on the relation between the equinox and Guinot's non-rotating origin," *Celest. Mech.*, Vol. 29, No. 4, pp. 335–360, 1983.

Kaplan, G. H. (ed.), *The IAU Resolutions on astronomical constants, time scales, and the fundamental reference frame,* USNO Circular No. 163, U.S. Naval Observatory, 1981.

Lambeck, K., *Geophysical Geodesy*, Clarendon Press, Oxford, 1988.

Lieske, J. H., T. Lederle, W. Fricke, and B. Morando, "Expressions for the precession quantities based upon the IAU (1976) System of Astronomical Constants," *Astronomy and Astrophysics*, Vol. 58, pp. 1–16, 1977.

McCarthy, D. (ed.), *IERS Conventions (1996)*, IERS Technical Note 21, International Earth Rotation Service, Observatoire de Paris, July 1996.

Seeber, G., *Satellite Geodesy: Foundations, Methods & Applications*, Walter de Gruyter, New York, 1993.

Seidelmann, P. K., "1980 IAU Nutation: The Final Report of the IAU Working Group on Nutation," *Celest. Mech.*, Vol. 27, No. 1, pp. 79–106, 1982.

Seidelmann, P. K., *Explanatory Supplement to the Astronautical Almanac*, University Science Books, Mill Valley, CA, 1992.

Simon, J. L., P. Bretagnon, J. Chapront, M. Chapront-Touzé, G. Francou, J. Laskar, "Numerical expressions for precession formulae and mean elements for the Moon and planets," *Astronomy and Astrophysics*, Vol. 282, pp. 663–683, 1994.

Standish, E. M., X. Newhall, J. Williams, and W. Folkner, *JPL Planetary and Lunar Ephemerides* (CD-ROM), Willmann-Bell, Inc. Richmond, 1997.

Webb, C.,*The ICRF-ITRF Transformation: A Comparison of Fundamental Earth Orientation Models found in MSODP and CALC,* The University of Texas at Austin, Center for Space Research Report CSR-02-01, Austin, TX, 2002.

Bibliography Abbreviations

Adv. Space Res.	*Advances in Space Research*
AIAA	American Institute of Aeronautics and Astronautics
AIAA J.	*AIAA Journal*
ASME	American Society of Mechanical Engineers
Astron. J.	*Astronomical Journal*
Celest. Mech.	*Celestial Mechanics and Dynamical Astronomy*
Geophys. J. Royal Astronom. Soc.	*Geophysical Journal of the Royal Astronomical Society*
Geophys. Res. Ltrs.	*Geophysical Research Letters*
IEEE	Institute for Electrical and Electronics Engineers
J. Appl. Math.	*Journal of Applied Mathematics*
J. Assoc. Comput. Mach.	*Journal of the Association of Computing Machinery*
J. Astronaut. Sci.	*Journal of the Astronautical Sciences*
J. Basic Engr.	*ASME Journal of Basic Engineering*
J. Geophys. Res.	*Journal of Geophysical Research*
J. Guid. Cont.	*Journal of Guidance and Control*
J. Guid. Cont. Dyn.	*Journal of Guidance, Control, and Dynamics*

J. Inst. Math. Applic.	*Journal of the Institute of Mathematics and Its Applications*
J. ION	*Journal of the Institute of Navigation*
J. Optim. Theory Appl.	*Journal of Optimization Theory and Applications*
J. Spacecr. Rockets	*Journal of Spacecraft and Rockets*
Mess. Math.	*Messenger of Mathematics*
SIAM	Society for Industrial and Applied Mathematics
SPIE	International Society for Optical Engineering
Trans. Auto. Cont.	*IEEE Transactions on Automatic Control*

Bibliography

[1] Aoki, S., and H. Kinoshita, "Note on the relation between the equinox and Guinot's non-rotating origin," *Celest. Mech.*, Vol. 29, No. 4, pp. 335–360, 1983.

[2] Ashby, N., "Relativity and the Global Positioning System," *Physics Today*, Vol. 55, No. 5, pp. 41–47, May 2002.

[3] Barlier, F., C. Berger, J. Falin, G. Kockarts, and G. Thuiller, "A thermospheric model based on satellite drag data," *Annales de Géophysique*, Vol. 34, No. 1, pp. 9–24, 1978.

[4] Battin, R., *An Introduction to the Mathematics and Methods of Astrodynamics*, American Institute of Aeronautics and Astronautics, Reston, VA, 1999.

[5] Bertiger, W., Y. Bar-Sever, E. Christensen, E. Davis, J. Guinn, B. Haines, R. Ibanez-Meier, J. Jee, S. Lichten, W. Melbourne, R. Muellerschoen, T. Munson, Y. Vigue, S. Wu, T. Yunck, B. Schutz, P. Abusali, H. Rim, W. Watkins, and P. Willis , "GPS precise tracking of TOPEX/Poseidon: Results and implications," *J. Geophys. Res.*, Vol. 99, No. C12, pp. 24,449–24,462, December 15, 1995.

[6] Beutler, G., E. Brockmann, W. Gurtner, U. Hugentobler, L. Mervart, M. Rothacher, and A. Verdun, "Extended orbit modeling techniques at the CODE Processing Center of the International GPS Service for Geodynamics (IGS): theory and initial results," *Manuscripta Geodaetica*, Vol. 19, No. 6, pp. 367–385, April 1994.

[7] Bierman, G. J., *Factorization Methods for Discrete Sequential Estimation*, Academic Press, New York, 1977.

[8] Björck, A., *Numerical Methods for Least Squares Problems*, SIAM, Philadelphia, PA, 1996.

[9] Bond, V., and M. Allman, *Modern Astrodynamics*, Princeton University Press, Princeton, NJ, 1996.

[10] Born, G. H., D. B. Goldstein, and B. Thompson, "An Analytical Theory for Orbit Determination", *J. Astronaut. Sci.*, Vol. 49, No. 2, pp. 345–361, April–June 2001.

[11] Born, G., J. Mitchell, and G. Hegler, "GEOSAT ERM – mission design," *J. Astronaut. Sci.*, Vol. 35, No. 2, pp. 119–134, April–June 1987.

[12] Born, G. H., B. D. Tapley, and M. L. Santee, "Orbit determination using dual crossing arc altimetry," *Acta Astronautica*, Vol. 13, No. 4, pp. 157–163, 1986.

[13] Boucher, C., Z. Altamini, and P. Sillard, *The International Terrestrial Reference Frame (ITRF97)*, IERS Technical Note 27, International Earth Rotation Service, Observatoire de Paris, May 1999.

[14] Brouwer, D., "Solutions of the problem of artificial satellite theory without drag," *Astron. J.*, Vol. 64, No. 9, pp. 378–397, November 1959.

[15] Brouwer, D., and G. Clemence, *Methods of Celestial Mechanics*, Academic Press, New York, 1961.

[16] Bryson, A. E., and Y. C. Ho, *Applied Optimal Control*, Hemisphere Publishing Corp., Washington, DC, 1975.

[17] Bucy, R., and P. Joseph, *Filtering for Stochastic Processes*, John Wiley & Sons, Inc., New York, 1968.

[18] Cajori, F., *A History of Mathematics*, MacMillan Co., New York, 1919.

[19] Carlson, N. A., "Fast triangular formulation of the square root filter", *AIAA J.*, Vol. 11, No. 9, pp. 1239-1265, September 1973.

[20] Chobotov, V. (ed.), *Orbital Mechanics*, American Institute of Aeronautics and Astronautics, Inc., Reston, VA, 1996.

[21] Christensen, E., B. Haines, K. C. McColl, and R. S. Nerem, "Observations of geographically correlated orbit errors for TOPEX/Poseidon using the Global Positioning System," *Geophys. Res. Ltrs.*, Vol. 21, No. 19, pp. 2175–2178, September 15, 1994.

[22] Cook, G. E., "Perturbations of near-circular orbits by the Earth's gravitational potential," *Planetary and Space Science*, Vol. 14, No. 5, pp. 433–444, May 1966.

[23] Cruickshank, D. R., *Genetic Model Compensation: Theory and Applications*, Ph. D. Dissertation, The University of Colorado at Boulder, 1998.

[24] Curkendall, D. W., *Problems in estimation theory with applications to orbit determination*, UCLA-ENG-7275, UCLA School Engr. and Appl. Sci., Los Angeles, CA, September 1972.

[25] Dahlquist, G., and A. Björck, *Numerical Methods*, Prentice-Hall, Englewood Cliffs, NJ, 1974 (translated to English: N. Anderson).

[26] Danby, J. M. A., *Fundamentals of Celestial Mechanics*, Willmann-Bell, Inc., Richmond, VA, 1988.

[27] Davis, J. C., *Statistics and Data Analysis in Geology*, John Wiley & Sons Inc., New York, 1986.

[28] Defense Mapping Agency, *Department of Defense World Geodetic System 1984*, DMA Technical Report 8350.2, Washington, DC, September, 1987.

[29] Degnan, J., and J. McGarry, "SLR2000: Eyesafe and autonomous satellite laser ranging at kilohertz rates," *SPIE Vol. 3218, Laser Radar Ranging and Atmospheric Lidar Techniques*, pp. 63–77, London, 1997.

[30] Desai, S. D., and B. J. Haines, "Near real-time GPS-based orbit determination and sea surface height observations from the Jason-1 mission," *Marine Geodesy*, Vol. 26, No. 3–4, pp. 187–199, 2003.

[31] Deutsch, R., *Estimation Theory*, Prentice-Hall, Inc., Englewood Cliffs, NJ, 1965.

[32] Dorroh, W. E., and T. H. Thornton, "Strategies and systems for navigation corrections," *Astronautics and Aeronautics*, Vol. 8, No. 5, pp. 50–55, May 1970.

[33] Dunn, C., W. Bertiger, Y. Bar-Sever, S. Desai, B. Haines, D. Kuang, G. Franklin, I. Harris, G. Kruizinga, T. Meehan, S. Nandi, D. Nguyen, T. Rogstad, J. Thomas, J. Tien, L. Romans, M. Watkins, S. C. Wu, S. Bettadpur, and J. Kim, "Instrument of GRACE," *GPS World*, Vol. 14, No. 2, pp. 16–28, February 2003.

[34] Dyer, P., and S. McReynolds, "Extension of square-root filtering to include process noise," *J. Optim. Theory Appl.*, Vol. 3, No. 6, pp. 444–458, 1969.

[35] Einstein, A., "Zur Elektrodynamik bewegter Körper," *Annalen der Physik*, Vol. 17, No. 10, pp. 891–921, 1905 (translated to English: Perrett, W., and G. Jeffery, *The Principle of Relativity*, Methuen and Co., 1923; republished by Dover, New York).

[36] El´Yasberg, P. E., *Introduction to the Theory of Flight of Artificial Earth Satellites*, translated from Russian, Israel Program for Scientific Translations, 1967.

[37] Fisher, R. A., "On an absolute criteria for fitting frequency curves," *Mess. Math.*, Vol. 41, pp. 155–160, 1912.

[38] Fliegel, H., T. Gallini, and E. Swift, "Global Positioning System radiation force models for geodetic applications", *J. Geophys. Res.*, Vol. 97, No. B1, pp. 559–568, January 10, 1992.

[39] Freeman, H., *Introduction to Statistical Inference*, Addison-Wesley, 1963.

[40] Gauss, K. F., *Theoria Motus Corporum Coelestium*, 1809 (Translated into English: Davis, C. H., *Theory of the Motion of the Heavenly Bodies Moving about the Sun in Conic Sections*, Dover, New York, 1963).

[41] Gelb, A. (ed.), *Applied Optimal Estimation*, Massachusetts Institute of Technology Press, Cambridge, MA, 1974.

[42] Gentleman, W. M., "Least squares computations by Givens transformations without square roots," *J. Inst. Math. Applic.*, Vol. 12, pp. 329–336, 1973.

[43] Givens, W., "Computation of plane unitary rotations transforming a general matrix to triangular form," *J. Appl. Math.*, Vol. 6, pp. 26–50, 1958.

[44] Goldstein, D. B., G. H. Born, and P. Axelrad, "Real-time, autonomous, precise orbit determination using GPS," *J. ION*, Vol. 48, No. 3, pp. 169–179, Fall 2001.

[45] Golub, G. H., and C. F. Van Loan, *Matrix Computations*, Johns Hopkins University Press, 1996.

[46] Goodyear, W. H., "Completely general closed form solution for coordinates and partial derivatives of the two-body problem," *Astron. J.*, Vol. 70, No. 3, pp. 189–192, April 1965.

[47] Graybill, F. A., *An Introduction to Linear Statistical Models*, McGraw-Hill, New York, 1961.

[48] Grewal, M. S., and A. P. Andrews, *Kalman Filtering: Theory and Practice*, Prentice Hall, 1993.

[49] Gruber, T., A. Bode, C. Reigber, P. Schwintzer, G. Balmino, R. Biancale, J. Lemoine, "GRIM5-C1: Combination solution of the global gravity field to degree and order 120," *Geophys. Res. Ltrs.*, Vol. 27, No. 24, pp. 4005–4008, December 2000.

[50] Heffes, H., "The effects of erroneous models on the Kalman filter response", *Trans. Auto. Cont.*, Vol. AC-11, pp. 541–543, July 1966.

[51] Heiskanen, W., and H. Moritz, *Physical Geodesy*, W. H. Freeman and Co., San Francisco, 1967.

[52] Helmert, F. R., "Zur Bestimmung kleiner Flächenstücke des Geoids aus Lothabweichungen mit Rücksicht auf Lothkrümmung", Sitzungsberichte Preuss. Akad. Wiss., Berlin, Germany, 1900.

[53] Helstrom, C. W., *Probability and Stochastic Processes for Engineers*, MacMillan, 1984.

[54] Herring, T., "Modeling atmospheric delays in the analysis of space geodetic data," in *Refraction of Transatmospheric Signals in Geodesy*, eds. J. C. DeMunck and T. A. Th. Spoelstra, Netherlands Geodetic Commission Publications in Geodesy, 36, pp. 157-164, 1992.

[55] Herring, T., B. Buffett, P. Mathews, and I. Shapiro, "Free nutations of the Earth: influence of inner core dynamics," *J. Geophys. Res.*, Vol. 96, No. B5, pp. 8259–8273, May 10, 1991.

[56] Hofmann-Wellenhof, B., H. Lichtenegger, and J. Collins, *Global Positioning System: Theory and Practice*, Springer-Verlag, Wien-New York, 1997.

[57] Householder, A. S., "Unitary triangularization of a nonsymmetric matrix," *J. Assoc. Comput. Mach.*, Vol. 5, No. 4, pp. 339–342, October 1958.

[58] Huang, C., J. C. Ries, B. Tapley, and M. Watkins, "Relativistic effects for near-Earth satellite orbit determination," *Celest. Mech.*, Vol. 48, No. 2, 167-185, 1990.

[59] Ingram, D. S., *Orbit determination in the presence of unmodeled accelerations*, Ph.D. Dissertation, The University of Texas at Austin, 1970.

[60] Ingram, D. S., and B. D. Tapley, "Lunar orbit determination in the presence of unmodeled accelerations," *Celest. Mech.*, Vol. 9, No. 2, pp. 191–211, 1974.

[61] Jacchia, L., *Thermospheric temperature, density and composition: new models*, Special Report 375, Smithsonian Astrophysical Observatory, Cambridge, MA, 1977.

[62] Jazwinski, A. H., *Stochastic Process and Filtering Theory*, Academic Press, New York, 1970.

[63] Kalman, R. E., "A New Approach to Linear Filtering and Prediction Theory," *J. Basic Eng.*, Vol. 82, Series E, No. 1, pp. 35–45, March, 1960.

[64] Kalman, R. E. and R. S. Bucy, "New Results in Linear Filtering and Prediction Theory," *J. Basic Eng.*, Vol. 83, Series D, No. 1, pp. 95–108, March, 1961.

[65] Kaminski, P. G., A. E. Bryson, and S. F. Schmidt, "Discrete square root filtering: A survey of current techniques," *Trans. Auto. Cont.*, Vol. AC-16, No. 6, pp. 727–736, December 1971.

[66] Kaplan, G. H. (ed.), *The IAU Resolutions on astronomical constants, time scales, and the fundamental reference frame*, USNO Circular No. 163, U.S. Naval Observatory, 1981.

[67] Kaula, W. M., *Theory of Satellite Geodesy*, Blaisdell Publishing Co., Waltham, 1966 (republished by Dover, New York, 2000).

[68] Kolmogorov, A. N., "Interpolation and Extrapolation of Stationary Random Sequences," *Bulletin of the Academy of Sciences of the USSR Math. Series*, Vol. 5, pp. 3–14, 1941.

[69] Kovalevsky, J., I. Mueller, and B. Kolaczek (eds.), *Reference Frames in Astronomy and Geophysics*, Kluwer Academic Publishers, Dordrecht, 1989.

[70] Kreyszig, E., *Advanced Engineering Mathematics*, John Wiley & Sons, Inc., New York, 1993.

[71] Lambeck, K., *The Earth's Variable Rotation*, Cambridge University Press, Cambridge, 1980.

[72] Lambeck, K., *Geophysical Geodesy*, Clarendon Press, Oxford, 1988.

[73] Laplace, P. S., *Théorie Analytique de Probabilités*, Paris, 1812 (The 1814 edition included an introduction, *Essai Philosophique sur les Probabilités*, which has been translated into English: Dale, A. I., *Philosophical Essay on Probabilities*, Springer-Verlag, New York, 1995).

[74] Lawson, C. L., and R. J. Hanson, *Solving Least Squares Problems*, Prentice-Hall, Inc. Englewood Cliffs, NJ, 1974 (republished by SIAM, Philadelphia, PA, 1995).

[75] Le Systeme International d'Unites (SI), Bureau International des Poids et Mesures, Sevres, France, 1991.

[76] Legendre, A. M., *Nouvelles méthodes pour la détermination des orbites des comètes*, Paris, 1806.

[77] Leick, A., *GPS Satellite Surveying*, J. Wiley & Sons, Inc., New York, 2003.

[78] Lemoine, F., S. Kenyon, J. Factor, R. Trimmer, N. Pavlis, D. Chinn, C. Cox, S. Klosko, S. Luthcke, M. Torrence, Y. Wang, R. Williamson, E. Pavlis, R. Rapp, and T. Olson, *The development of the Joint NASA GSFC and the National Imagery and Mapping Agency (NIMA) Geopotential Model EGM96*, NASA/TP–1998–206861, Greenbelt, MD, July 1998.

[79] Lemoine, F., D. Rowlands, S. Luthcke, N. Zelensky, D. Chinn, D. Pavlis, and G. Marr, "Precise orbit determination of GEOSAT follow-on using satellite laser ranging and intermission altimeter crossovers," NASA/CP-2001-209986, *Flight Mechanics Symposium*, John Lynch (ed.), NASA Goddard Space Flight Center, Greenbelt, MD, pp. 377–392, June 2001.

[80] Lichten, S. M., "Estimation and filtering for high precision GPS positioning applications," *Manuscripta Geodaetica*, Vol. 15, pp. 159–176, 1990.

[81] Liebelt, P. B., *An Introduction to Optimal Estimation*, Addison-Wesley, Reading, MA, 1967.

[82] Lieske, J. H., T. Lederle, W. Fricke, and B. Morando, "Expressions for the precession quantities based upon the IAU (1976) System of Astronomical Constants," *Astronomy and Astrophysics*, Vol. 58, pp. 1–16, 1977.

[83] Lundberg, J., and B. Schutz, "Recursion formulas of Legendre functions for use with nonsingular geopotential models," *J. Guid. Cont. Dyn.*, Vol. 11, No. 1, pp. 31–38, January–February 1988.

[84] Lutchke, S. B., N. P. Zelenski, D. D. Rowlands, F. G. Lemoine, and T. A. Williams, "The 1-centimeter orbit: Jason-1 precision orbit determination using GPS, SLR, DORIS and altimeter data," *Marine Geodesy*, Vol. 26, No. 3-4, pp. 399–421, 2003.

[85] Marini, J. W., and C. W. Murray, "Correction of laser range tracking data for atmospheric refraction at elevations above 10 degrees," *NASA GSFC X591-73-351*, Greenbelt, MD, 1973.

[86] Markov, A. A., "The law of large numbers and the method of Least Squares," (1898), *Izbr. Trudi., Izd. Akod. Nauk*, USSR, pp. 233-251, 1951.

[87] Marshall, J. A., F. J. Lerch, S. B. Luthcke, R. G. Williamson, and C. Chan, "An Assessment of TDRSS for Precision Orbit Determination," *J. Astonaut. Sci*, Vol. 44, No. 1, pp. 115–127, January–March, 1996.

[88] Marshall, J. A., N. P. Zelensky, S. M. Klosko, D. S. Chinn, S. B. Luthcke, K. E. Rachlin, and R. G. Williamson, "The temporal and spatial characteristics of TOPEX/Poseidon radial orbit error," *J. Geophys. Res.*, Vol. 99, No. C12, pp. 25,331–25,352, December 15, 1995.

[89] Maybeck, P. S., *Stochastic Models, Estimation and Control*, Vol. 1, Academic Press, 1979.

[90] McCarthy, D. (ed.), *IERS Conventions (1996)*, IERS Technical Note 21, International Earth Rotation Service, Observatoire de Paris, July 1996.

[91] Mikhail, E. M., *Observations and Least Squares*, University Press of America, Lanham, MD, 1976.

[92] Mohr, P. J., and B. Taylor, "The fundamental physical constants," *Physics Today*, Vol. 56, No. 8, Supplement, pp. BG6–BG13, August 2003.

[93] Moler, C., and C. Van Loan, "Nineteen dubious ways to compute the exponential of a matrix," *SIAM Review*, Vol. 20, No. 4, pp. 801–836, October 1978.

[94] Montenbruck, O., and E. Gill, *Satellite Orbits: Models, Methods, and Applications*, Springer-Verlag, Berlin, 2001.

[95] Moritz, H., and I. Mueller, *Earth Rotation: Theory and Observation*, Ungar Publishing Company, New York, 1987.

[96] Moulton, F. R., *An Introduction to Celestial Mechanics*, MacMillan Co., New York, 1914.

[97] Myers, K. A., *Filtering theory methods and applications to the orbit determination problem for near-Earth satellites*, Ph.D. Dissertation, The University of Texas at Austin, November 1973.

[98] Newton, I., *Philosophiae Naturalis Principia Mathematica*, 1687 (translated into English: A. Motte, 1729; revised by F. Cajori, University of California Press, 1962).

[99] Papoulis, A., *Probability, Random Variables, and Stochastic Processes*, McGraw-Hill, New York, 1991.

[100] Parkinson, B., J. Spilker, P. Axelrad, and P. Enge (eds.), *Global Positioning System: Theory and Applications*, Vols. 1–3, American Institute of Aeronautics and Astronautics, Inc., Washington, DC, 1966.

[101] Pines, S., "Uniform representation of the gravitational potential and its derivatives," *AIAA J.*, Vol. 11, No. 11, pp. 1508–1511, November 1973.

[102] Plummer, H. C., *An Introductory Treatise on Dynamical Astronomy*, Cambridge University Press, 1918 (republished by Dover Publications, New York, 1966).

[103] Pollard, H., *Mathematical Introduction to Celestial Mechanics*, Prentice-Hall, Inc., Englewood Cliffs, NJ, 1966.

[104] Press, W., B. Flannery, S. Teukolsky, and W. Vetterling, *Numerical Recipes*, Cambridge University Press, Cambridge, 1986.

[105] Prussing, J., and B. Conway, *Orbit Mechanics*, Oxford University Press, New York, 1993.

[106] Rausch, H. E., F. Tung, and C. T. Striebel, "Maximum likelihood estimates of linear dynamic systems," *AIAA J.*, Vol. 3, No. 7, pp. 1445–1450, August 1965.

[107] Reddy, J. N., and M. L. Rasmussen, *Advanced Engineering Analysis*, Robert E. Krieger Publishing Co., Malabar, FL 1990.

[108] Ries, J. C., C. Huang, M. M. Watkins, and B. D. Tapley, "Orbit determination in the relativistic geocentric reference frame," *J. Astronaut. Sci.*, Vol. 39, No. 2, pp. 173–181, April–June 1991.

[109] Ries, J. C., C. K. Shum, and B. Tapley, "Surface force modeling for precision orbit determination," *Geophysical Monograph Series, Vol. 73*, A. Jones (ed.), American Geophysical Union, Washington, DC, 1993.

[110] Rowlands, D. D., S. B. Luthcke, J. A. Marshall, C. M. Cox, R. G. Williamson, and S. C. Rowton, "Space Shuttle precision orbit determination in support of SLA-1 using TDRSS and GPS tracking data," *J. Astronaut. Sci.*, Vol. 45, No. 1, pp. 113–129, January–March 1997.

[111] Roy, A. E., *Orbital Motion*, John Wiley & Sons Inc., New York, 1988.

[112] Saastamoinen, J., "Atmospheric correction for the troposphere and stratosphere in radio ranging of satellites," *Geophysical Monograph Series, Vol. 15*, S. Henriksen, A. Mancini, and B. Chovitz (eds.), American Geophysical Union, Washington, DC, pp. 247–251, 1972.

[113] Schaub, H., and J. Junkins, *Analytical Mechanics of Space Systems*, American Institute of Aeronautics and Astronautics, Reston, VA, 2003.

[114] Schlee, F. H., C. J. Standish, and N. F. Toda, "Divergence in the Kalman filter," *AIAA J.*, Vol. 5, No. 6, pp. 1114–1120, June 1967.

[115] Schutz, B., B. D. Tapley, P. Abusali, H. Rim, "Dynamic orbit determination using GPS measurements from TOPEX/Poseidon," *Geophys. Res. Ltrs.*, Vol. 21, No. 19, pp. 2179–2182, 1994.

[116] Seeber, G., *Satellite Geodesy: Foundations, Methods & Applications*, Walter de Gruyter, New York, 1993.

[117] Seidelmann, P. K., "1980 IAU Nutation: The Final Report of the IAU Working Group on Nutation", *Celest. Mech.*, Vol. 27, No. 1, pp. 79–106, 1982.

[118] Seidelmann, P. K. (ed.), *Explanatory Supplement to the Astronomical Almanac*, University Science Books, Mill Valley, CA, 1992.

[119] Shampine, L., and M. Gordon, *Computer Solution of Ordinary Differential Equations, The Initial Value Problem*, W. H. Freeman and Co., San Francisco, 1975.

[120] Shum, C. K., B. Zhang, B. Schutz, and B. Tapley, "Altimeter crossover methods for precise orbit determination and the mapping of geophysical parameters," *J. Astronaut. Sci.*, Vol. 38, No. 3, pp. 355–368, July–September 1990.

[121] Simon, J. L., P. Bretagnon, J. Chapront, M. Chapront-Touzé, G. Francou, J. Laskar, "Numerical expressions for precession formulae and mean elements for the Moon and planets," *Astronomy and Astrophysics*, Vol. 282, pp. 663–683, 1994.

[122] Skolnik, M.I. (ed.), *Radar Handbook*, McGraw-Hill, New York, 1990.

[123] Smart, W. M., *Celestial Mechanics*, John Wiley & Sons Inc., New York, 1961.

[124] Sorenson, H. W., "Least squares estimation: from Gauss to Kalman," *IEEE Spectrum*, Vol. 7, No. 7, pp. 63–68, July, 1970.

[125] Sorenson, H. W. (ed.), *Kalman Filtering: Theory and Applications*, IEEE Press, 1985.

[126] Standish, E. M., X. Newhall, J. Williams, and W. Folkner, *JPL Planetary and Lunar Ephemerides* (CD-ROM), Willmann-Bell, Inc., Richmond, VA 1997.

[127] Swerling, P., "First order error propagation in a stagewise differential smoothing procedure for satellite observations," *J. Astronaut. Sci.*, Vol. 6, pp. 46–52, 1959.

[128] Szebehely, V., *Theory of Orbits*, Academic Press, New York, 1967.

[129] Szebehely, V., and H. Mark, *Adventures in Celestial Mechanics*, John Wiley & Sons, Inc., New York, 1998.

[130] Tapley, B. D., "Statistical orbit determination theory," in *Recent Advances in Dynamical Astronomy*, B. D. Tapley and V. Szebehely (eds.), D. Reidel, pp. 396–425, 1973.

[131] Tapley, B. D.," Fundamentals of orbit determination", in *Theory of Satellite Geodesy and Gravity Field Determination*, Vol. 25, pp. 235-260, Springer-Verlag, 1989.

[132] Tapley, B. D., and C. Y. Choe, "An algorithm for propagating the square root covariance matrix in triangular form," *Trans. Auto. Cont.*, Vol. AC-21, pp. 122–123, 1976.

[133] Tapley, B. D., and D. S. Ingram, "Orbit determination in the presence of unmodeled accelerations," *Trans. Auto. Cont.*, Vol. AC-18, No. 4, pp. 369–373, August 1973.

[134] Tapley, B. D., and J. G. Peters, "A sequential estimation algorithm using a continuous UDU^T covariance factorization," *J. Guid. Cont.*, Vol. 3, No. 4, pp. 326–331, July–August 1980.

[135] Tapley, B. D., J. Ries, G. Davis, R. Eanes, B. Schutz, C. Shum, M. Watkins, J. Marshall, R. Nerem, B. Putney, S. Klosko, S. Luthcke, D. Pavlis, R. Williamson, and N. P. Zelensky, "Precision orbit determination for TOPEX/Poseidon," *J. Geophys. Res.*, Vol. 99, No. C12, pp. 24,383–24,404, December 15, 1994.

[136] Tapley, B. D., M. Watkins, J. Ries, G. Davis, R. Eanes, S. Poole, H. Rim, B. Schutz, C. Shum, R. Nerem, F. Lerch, J. Marshall, S. Klosko, N. Pavlis, and R. Williamson, "The Joint Gravity Model 3," *J. Geophys. Res.*, Vol. 101, No. B12, pp. 28,029–28,049, December 10, 1996.

[137] Thompson, B., M. Meek, K. Gold, P. Axelrad, G. Born, and D. Kubitschek, "Orbit determination for the QUIKSCAT spacecraft," *J. Spacecr. Rockets*, Vol. 39, No. 6, pp. 852–858, November–December 2002.

[138] Thornton, C. L., *Triangular covariance factorizations for Kalman filtering*, Technical Memorandum, 33–798, Jet Propulsion Laboratory, Pasadena, CA, October 15, 1976.

[139] Torge, W., *Geodesy*, Walter de Gruyter, Berlin, 1980 (translated to English: Jekeli, C.).

[140] Vallado, D., *Fundamentals of Astrodynamics and Applications*, Space Technology Library, Microcosm Press, El Segundo, CA, 2001.

[141] Vigue, Y., B. Schutz, and P. Abusali, "Thermal force modeling for the Global Positioning System satellites using the finite element method," *J. Spacecr. Rockets*, Vol. 31, No. 5, pp. 866–859, 1994.

[142] Visser, P., and B. Ambrosius, "Orbit determination of TOPEX/Poseidon and TDRSS satellites using TDRSS and BRTS tracking," *Adv. Space Res.*, Vol. 19, pp. 1641–1644, 1997.

[143] Wahr, J. M., "The forced nutations of an elliptical, rotating, elastic, and oceanless Earth," *Geophys. J. of Royal Astronom. Soc.*, 64, pp. 705–727, 1981.

[144] Walpole, R. E., R. H. Myers, S. L. Myers, and Y. Keying, *Probability and Statistics for Engineers and Scientists*, Prentice Hall, Englewood Cliffs, NJ, 2002.

[145] Webb, C.,*The ICRF-ITRF Transformation: A Comparison of Fundamental Earth Orientation Models found in MSODP and CALC*, The University of Texas at Austin, Center for Space Research Report CSR-02-01, Austin, TX, 2002.

[146] Wells, D., *Guide GPS Positioning*, Canadian GPS Associates, Fredericton, 1987.

[147] Westfall, R., *Never at Rest: A Biography of Isaac Newton*, Cambridge University Press, Cambridge, 1980.

[148] Wiener, N., *The Extrapolation, Interpolation and Smoothing of Stationary Time Series*, John Wiley & Sons, Inc., New York, 1949.

[149] Wiesel, W. E., *Spaceflight Dynamics*, McGraw-Hill, 1997.

[150] Woolard, E., *Theory of the rotation of the Earth around its center of mass*, Astronomical Papers—American Ephemeris and Nautical Almanac, Vol. XV, Part I, U.S. Naval Observatory, Washington, DC, 1953.

Author Index

Index

Printed and bound by CPI Group (UK) Ltd, Croydon, CR0 4YY

08/05/2025

01864902-0001